THE PRINCETON ENCYCLOPEDIA OF DINOSAURS

SAUROPODS

THE PRINCETON ENCYCLOPEDIA OF DINOSAURS

Bryce Jones

Illustrated by **Egidio Viola**

SAUROPODS

PRINCETON UNIVERSITY PRESS
PRINCETON AND OXFORD

Published by Princeton University Press
41 William Street, Princeton, New Jersey 08540
99 Banbury Road, Oxford OX2 6JX
press.princeton.edu

GPSR Authorized Representative: Easy Access System Europe - Mustamäe tee 50, 10621 Tallinn, Estonia, gpsr.requests@easproject.com

All Rights Reserved

ISBN 9780691250236
ISBN (epub) 9780691288116
ISBN (PDF) 9780691250243
LCCN: 2025932468

British Library Cataloging-in-Publication Data is available

Editorial: Robert Kirk, Laura Lassen, and Megan Michel Mendonça
Production Editorial: Natalie Baan
Text Design: D & N Publishing, Wiltshire, UK, from original design by Bryce Jones
Cover Design: Wanda España
Production: Steven Sears
Publicity: Caitlyn Robson-Iszatt and Matthew Taylor
Copyeditor: Eva Silverfine
Cover Illustration: Egidio Viola

Cover: *Diplodocus carnegii*
Title page: *Camarasaurus supremus*

Malawisaurus dixeyi

This book has been composed in Minion Pro (main text and captions), and Veneer and Veneer Two (headings)
Printed in Italy
10 9 8 7 6 5 4 3 2 1

For all of the teachers
who inspired me to learn

CONTENTS

Preface 9

Introduction 10

How to Use This Book 11

Time Periods 12

Continental Movements 13

SAUROPODOMORPHA
19

MASSOPODA
43

SAUROPODA
75

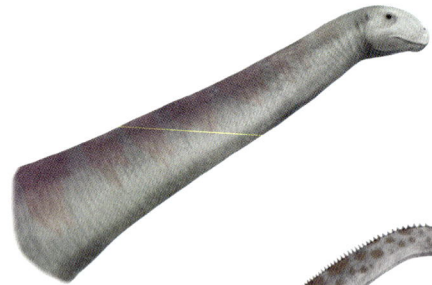

EUSAUROPODA
97

MAMENCHISAURIDAE
121

DIPLODOCOIDEA
141

REBBACHISAURIDAE
167

MACRONARIA
187

SOMPHOSPONDYLI
215

EUHELOPODIDAE
243

TITANOSAURIA
255

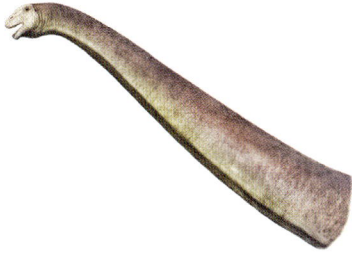

LITHOSTROTIA
285

RINCONSAURIA
317

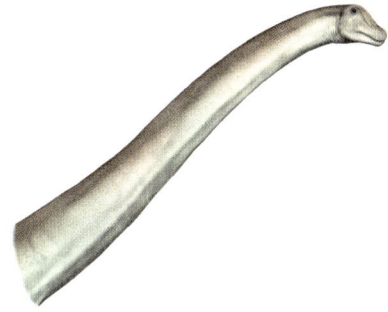

SALTASAUROIDEA
339

Nomina Dubia 361

Informally Named 367

Acknowledgments 371

Fossil Specimens 372

Subjective Synonyms 387

Selected Bibliography 388

Index 389

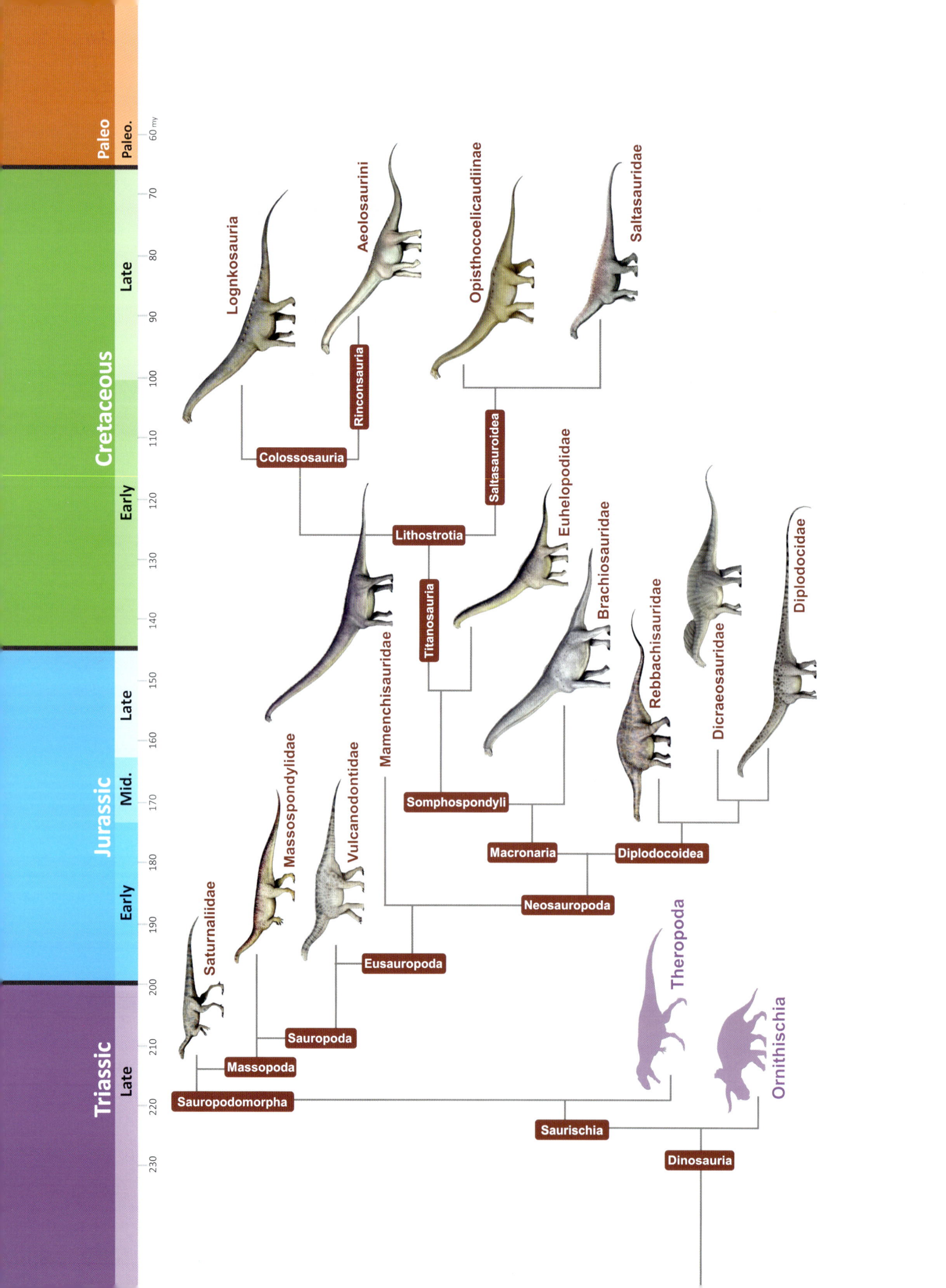

Paleo

Paleo.

60 my

70

Cretaceous

Late

80

90

100

110

Early

120

130

Late

140

150

160

Jurassic

Mid.

170

180

Early

190

200

Triassic

210

Late

220

230

Lognkosauria

Aeolosaurini

Opisthocoelicaudiinae

Saltasauridae

Rinconsauria

Colossosauria

Saltasauroidea

Euhelopodidae

Lithostrotia

Titanosauria

Brachiosauridae

Mamenchisauridae

Rebbachisauridae

Somphospondyli

Dicraeosauridae

Diplodocidae

Macronaria

Diplodocoidea

Massospondylidae

Vulcanodontidae

Neosauropoda

Saturnaliidae

Eusauropoda

Theropoda

Ornithischia

Sauropoda

Massopoda

Sauropodomorpha

Saurischia

Dinosauria

I have been a lifelong dinosaur enthusiast. At the age of four, I created what my mother has dubbed my "first research paper": a stack of illustrations of every different species I could find throughout the pages of my many dinosaur books.

I've collected dinosaur books throughout my life, and recently I embarked on a search for a very specific sort of tome: a truly comprehensive, informative, encyclopedic collection of all known dinosaurs. Needless to say, a wide variety of dinosaur-themed texts have been available throughout the years, but try as I might, I couldn't locate any volume that really fit the bill. I was able to locate a few texts that were incredible but didn't quite tackle the topic in the way I was looking for. Primarily, though, I found a great number of texts that were … well … subpar.

So I decided to make my own book, one that would provide the kind of information I'd be interested in learning. I wanted to know what made each species unique and noteworthy. I wanted to know how, exactly, dinosaur species were related. And I especially wanted to know how many fossils of each dinosaur had been uncovered.

Answering these questions meant perusing a great number of peer-reviewed scientific research papers … as well as, admittedly, many Wikipedia articles. Wikipedia is an amazing resource for finding which related professional papers exist out there, and for getting a good sense of which topics are worth investigating. However, both Wikipedia and published research papers present the same snag when it comes to

answering my first question: What makes each species unique and noteworthy?

The snag: mountains upon mountains of highly technical, ultradetailed anatomical descriptions. These details are, of course, crucial for professionals, but they are downright esoteric for the average dinosaur enthusiast. Therefore, I've done my best to trim away the expert-level jargon regarding, say, the positioning of foramena, or the shapes of spinal protuberances, and leave behind only the easily digestible, "big picture" view of what makes a particular species unique and interesting.

Answering my second big question—how each species is related to the others—proved to be difficult for a very fundamental reason: very few paleontologists actually *agree* on the exact relationships between specific dinosaurs. I've done my best to compile all of the various versions of family trees, as well as the different interpretations of phylogenetic data, and merge them into a single, easily absorbable, "most likely" version of reality. I've also attempted to document and justify each decision made during the process.

My third and final big query—exactly how many fossils of each dinosaur do we have—is (you guessed it) much more difficult to answer than you'd think it would be. Many scientific papers that describe new specimens (especially older papers) don't *actually* illustrate or describe each and every bone. *Some* papers are woefully barren in this regard. And, since there are only so many paleontologists in the world, and there are only so many hours in the day, these gaps in knowledge can potentially remain open for decades, even though the information is *technically* available … just sitting in a dark museum drawer somewhere.

However, all of this work and information would essentially be useless if it weren't accompanied by gorgeous illustrations that could enthrall the senses and capture the imagination. Many of the subpar dinosaur books I alluded to earlier either are devoid of illustrations entirely or feature only a distracting hodgepodge of inaccurate, amateurish drawings. I knew that I couldn't create the kind of dinosaur book I'd want to have without plenty of accurate, marvelous illustrations. Therefore, I'm extraordinarily grateful that Egidio Viola (aka CisioPurple) responded to my inquiries when I reached out to him, and that he has been such an extraordinary and flexible partner in bringing this book to life.

All told, I hope that this book, and the other volumes in this series, will be informative and interesting. Thanks for picking it up, and thanks for reading!

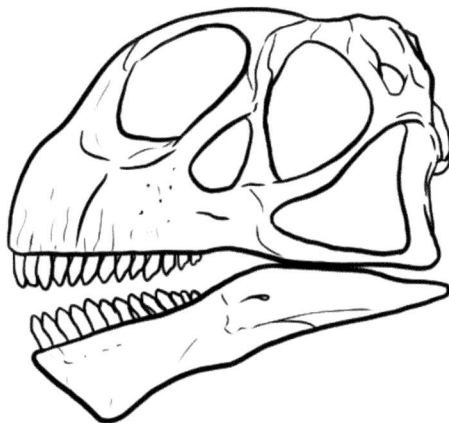

Camarasaurus supremus

What follows is a brief FAQ-style summary of general information that may be helpful to some readers.

How is a new species named?

In a nutshell, modern practices for naming new species are laid out by the International Commission on Zoological Nomenclature (ICZN), and names must be registered with the Official Register of Zoological Nomenclature (ZooBank). So, as long as paleontologists properly follow these guidelines, they can essentially name a new species whatever they like.

Is a named species permanent and official for all time?

Not at all. It all comes down to an informal consensus among scientists. If I assign a new name to a fossil specimen, "New Dino X," and follow the proper guidelines, the name is "official." However, if most paleontologists disagree with my conclusion and consistently refer to the fossil specimen as "Old Dino Y," then my newly created name would be, for all practical purposes, dead.

How is a phylogenetic analysis conducted to determine the relationship between species?

Rather than relying on subjective human interpretation, a modern analysis is based on numerical comparisons of objective physical properties. Datasets of fossil properties have been compiled by different teams for different comparisons. If I want to compare, say, a newly discovered femur to other similar bones, I would enter its precise values into the set, and, in a nutshell, a computer would tell me how similar or dissimilar the new bone is to all of the others.

How are the ages of fossils determined?

The subject of radiometric dating is too complex to be described here in detail, but in short, the relative levels of radioactive isotopes and their decay products are measured in certain materials, from which an absolute age can be derived. Typically, it is not the fossil itself that is dated, but the layer of rock in which it was found. Some geological formations have been precisely dated, but the ages of others remain stubbornly vague to this day. This is because not just any material can be used for accurate radiometric dating. Thus, the age of some fossils is ill defined; for example, a specimen might date "anywhere from 72 to 76 million years of age," bound by the known dates of surrounding rock layers.

Definition of terms:

Basal: In the direction toward the base of a cladogram, or "stemward."

Cladogram: A "family tree" diagram that shows the evolutionary relationship among species.

Crownward: In the direction away from the base of a cladogram.

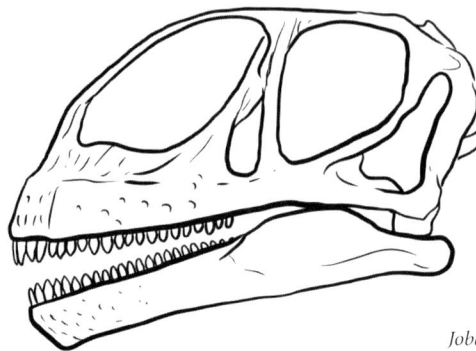

Jobaria tiguidensis

Derived: A trait that is found only among a particular lineage; a species that is placed in the direction away from the base of a cladogram.

Formation: A layer of similar rock types that were originally continuous and created by related events; characterized by its composition, how it looks, and how it is exposed over an area.

Genera: The plural form of "genus."

Holotype: The specific fossil specimen that serves to define the unique characteristics of a given species.

Ichnogenus: A classification used to identify footprints, burrows, or other indirect trace fossils.

Incertae sedis: A taxon of "uncertain placement," meaning its location on the family tree is uncertain.

Lectotype: A specimen chosen after the fact as the type of a species, if the author of the name fails to designate a type.

Monophyletic: A group of organisms descended from a single common evolutionary ancestor; this group contains any and all of that ancestor's descendants.

MYA: Million years ago.

Neotype: The specific fossil specimen that has been chosen to replace or supersede the holotype, typically when the holotype has been destroyed.

Nomen dubium: A species name that was once considered to be valid but is no longer, usually because there is a newer consensus that the fossil material lacks any important diagnostic features; a "doubtful name."

Nomen nudum: A name that was intended to be a valid species name but has not been validly published according to the guidelines of the ICZN; a "naked name."

Paraphyletic: A group of organisms descended from a single common evolutionary ancestor; this group does not contain all of that ancestor's descendants.

Paratype: Any fossil specimens used, along with the holotype, in the original designation of a new form; they must be part of the same series as the holotype (i.e., collected at the same immediate locality and at the same time).

Phylogenetics: The study of evolutionary relationships among biological organisms.

Syntype: Any of two or more specimens listed in the original description of a taxon when a holotype was not designated.

Name
The name of the dinosaur **genus**.

Scale
The top and side of each illustration section include a **scale bar**, measuring the animal's length and height.

Size Comparison
Gray silhouettes of modern creatures are included to provide a quickly grasped sense of the dinosaur's size.

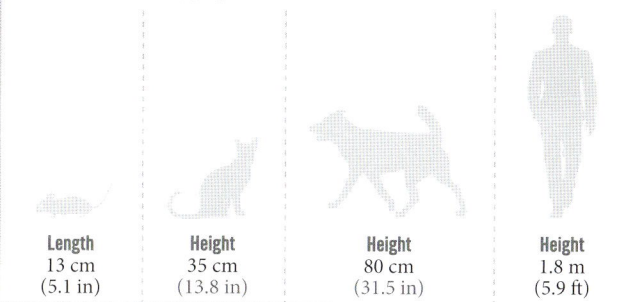

Length	Height	Height	Height
13 cm	35 cm	80 cm	1.8 m
(5.1 in)	(13.8 in)	(31.5 in)	(5.9 ft)

▶ BAGUALOSAURUS

B. agudoensis
(Pretto et al., 2019)
Length: 2.3 m (7.5 ft)
Height: 0.8 m (2.6 ft)
Hip height: 0.7 m (2.3 ft)
Body mass: 15 kg (33 lb)
Reconstruction:

TRIASSIC   JURASSIC   CRETACEOUS

Bagualosaurus agudoensis (meaning "stout lizard from Agudo") is among the largest known sauropodomorphs from the Carnian stage of the Triassic.

As *Bagualosaurus* shares some traits with sauropodomorphs that are more basal (or "primitive") and some that are more derived (or "advanced"), it is hypothesized to have had an intermediate or transitive position on the family tree. The morphology of its postcranial skeleton is very similar to that of the earliest ancestral sauropodomorphs, although it is significantly larger in size, being larger than most other animals in its environment. However, the animal's skull has some traits that are indicative of increased specialization in herbivory, some of which had previously only been observed in later sauropodomorphs.

This combination of features suggests that sauropodomorphs evolved mandibular traits related to herbivory before they evolved the adaptations that allowed them to grow to tremendous size. Although *Bagualosaurus*

was large for its time, it lacked the specializations (such as a robust foot) that would allow it to support a more massive body, like the sauropodomorphs of the latest Triassic.

Bagualosaurus is known from a single incomplete specimen (UFRGS-PV-1099-T) that was found in partial articulation. The specimen consists of a jaw, lower skull, vertebral, pelvic region, and hindlimb portions. It was discovered in 2007 but remained unprepared and unstudied until 2012. Some elements were already eroded by the time of their excavation. The bones showed that the animal's final resting position was unusual, as it was found lying on its back with its legs sticking forward.

The generic name *Bagualosaurus* comes from the Brazilian term "bagual," which is used to describe an animal that is stout or strong; this is combined with the Latin "saurus" (meaning "lizard"). The specific name *agudoensis* refers to the town of Agudo.

CLASSIFICATION
Dinosauria
Saurischia
Sauropodomorpha

LOCATION
Brazil

KNOWN REMAINS
Partial skull and skeleton

28
SAUROPODOMORPHA

B. agudoensis

(Pretto et al., 2019)

Length: 2.3 m (7.5 ft)
Height: 0.8 m (2.6 ft)
Hip height: 0.7 m (2.3 ft)
Body mass: 15 kg (33 lb)
Reconstruction:

The name of the **type species** of the particular genus.

The **author(s) and publication date** of the scientific paper that first described the species. Assume this citation applies to all text for the entry unless otherwise noted.

Estimated **physical measurements**, based on the most recent or credible sources. Bear in mind that, even when paleontologists have an entire skeleton to work with, estimates of an animal's size and mass can vary widely based on the methods and calculations used.

This **scale** is a rough estimate of how accurate any given reconstruction can possibly be, given the completeness of available fossil remains. A species known from a single bone will have a rating of 0.5 out of 4, while a species known from a complete skeleton will have a rating of 4 out of 4.

Timeline
This timeline of the Mesozoic Era highlights the period in which the animal lived and indicates the age or age range of the known fossil material.

Skeletal Reconstruction
This diagram indicates which elements of the animal's skeleton are currently known to science, combining material from every known specimen of the genus. (These diagrams are approximations and not anatomically exact.)

Classification
This lists the major phylogenetic clades that categorize the genus in question.

Maps
On the left is a world map of present-day Earth; on the right is a world map reflecting the appropriate period. A red dot indicates the region of origin of the fossil specimen(s).

A WORD ON PALEOART ILLUSTRATIONS

How accurate can an illustration of a dinosaur actually be? Let's briefly examine some common questions.

■ **COLOR** What color were dinosaurs? In almost all cases, we don't know. The only exceptions involve certain feathers that were fossilized in exquisite detail. So how do paleoartists decide how to color their illustrations? By examining the colorations of dinosaurs' closest living relatives: reptiles and birds. By doing so, artists know which colors, designs, and patterns would have been most probable and realistic.

■ **FEATHERS** Which dinosaurs had feathers? It seems that the very first protofeathers evolved prior to the dinosaurs, as we now know that certain pterosaurs (flying reptiles) possessed a fuzzy coat. Thus, any dinosaur could potentially have borne feathery structures. However, for the most part, direct evidence of substantial feather covering is known only from coelurosaurian theropods.

■ **FILLING IN THE BLANKS** If only a few bones of a species have been discovered, how do we know what the whole animal looks like? By using the animal's closest known relatives as a guide. By extrapolating, paleontologists can make an educated guess about what's missing—although they don't always predict correctly!

11

Broadly speaking, dinosaurs were the rulers of the Mesozoic Era, which spanned approximately 186 million years of Earth's history. The Mesozoic is divided into three periods: the Triassic, Jurassic, and Cretaceous, which, in turn, are further subdivided into smaller epochs and stages.

The Triassic Period (251.9–201.4 MYA) begins immediately following the worst mass extinction in Earth's history, the Permian–Triassic extinction, also known as the "Great Dying." This extinction would ultimately result in approximately 50% of animal genera vanishing from the fossil record, including at least 70% of vertebrate genera. The primary driver of the episode was likely a prolonged supervolcanic event, resulting in the formation of the igneous Siberian Traps.

The Triassic Period is divided into three epochs: the Early, Middle, and Late Triassic. Each of these is further subdivided into stages. For the most part, the boundaries between Mesozoic stages do not mark dramatic extinction or climatic events; rather, they are defined by the presence of certain index fossils—organisms that were ubiquitous only during a specific time span. Until the end of the Triassic, the fossilized teeth of conodonts (eel-like jawless fish) are a commonly used index fossil. Following this group's extinction, ammonites (spiral-shelled cephalopods) are used as the primary metric.

The end of the Triassic is marked by another mass extinction, one of the most severe on record. It was likely brought about by the Central Atlantic Magmatic Province, an episode of supervolcanic events spurred by the progressing breakup of the Pangea supercontinent. By the end of this event, dinosaurs had gone from being just one of many thriving groups of archosaurs to one of the few surviving groups. By some estimates, this extinction event wiped out just as many species as the one that would later end the reign of the dinosaurs.

The Jurassic Period (201.4–145 MYA) is also divided into three epochs (Early, Middle, and Late). The boundary that separates the Jurassic from the Cretaceous, though, is persistently puzzling. Originally, it was thought to be another sudden mass extinction like the one that began the Jurassic, but this does not appear to be the case. However, that is not to say that nothing of significance happened during this time. For instance, there was a large "turnover" of animal species, where certain families of organisms went extinct, only to be rapidly replaced by others. No fewer than three extraterrestrial-impact events occurred during this interval, along with various protracted supervolcanic events—whether these incidents played a significant role in shaping the Jurassic–Cretaceous turnover remains unclear (Tennant et al., 2017).

The Cretaceous Period (145–66 MYA) is divided into just two epochs: the Early and Late. Each of these epochs spans a considerable stretch of time. Two of the most notable "minor" extinction events of the Mesozoic—the Aptian extinction and the Cenomanian-Turonian boundary event—were also likely caused by various episodes of volcanism that led to periods of widespread oceanic anoxia. During such time spans, the world's oceans become depleted of dissolved oxygen, becoming increasingly toxic to many organisms. These events can be seen in the geologic record as layers of "black shale."

The Cretaceous ended abruptly, about 66 million years ago, due to the impact of a 10-kilometer-wide (6.2 mile) asteroid in the Yucatán Peninsula, resulting in the 180-kilometer-wide (110 mile) Chicxulub crater. The massive impact threw tremendous volumes of debris into the atmosphere, and the debris lingered for a prolonged period, blocking a significant portion of sunlight, altering global temperatures. Combined with a sudden increase in acid rain, which dramatically altered the pH of the global oceans, these occurrences collapsed food chains and devastated ecosystems, resulting in the most recent mass extinction in Earth's history, wiping out all of the nonavian dinosaurs.

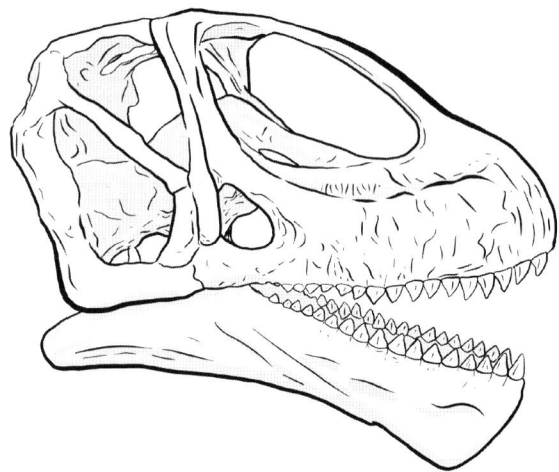

Turiasaurus riodevensis

During the 170-million-year rule of the dinosaurs, Earth did not remain static. The continents gradually drifted apart, sea levels fluctuated substantially, and biomes changed. Thus, the story of dinosaurian evolution is intermingled with the story of continental drift—as species became separated on isolated landmasses, evolution took them in different directions, and as those landmasses formed new connections, access to new territories led evolution in yet other directions.

The global supercontinent that dominated Earth during the Triassic Period is known as Pangea (or Pangaea). Pangea formed roughly 335 million years ago, during the Early Carboniferous. During the subsequent 100 million years, many terrestrial species achieved a near-global (or "cosmopolitan") distribution.

Beginning in the Triassic, though, Pangea began to fracture. By the end of the Mesozoic Era, the various modern continents were nearly as separate from one another as they are today. As a result, one may be led to ask sensible questions, such as "When, exactly, did Continent A separate from Continent B?" These kinds of questions, however, have complex, murky answers.

For starters, recall that tectonic plates move very, very slowly—on the order of just a few centimeters per year, on average. Thus, even neglecting all other factors, there would be no single moment when Continent A became completely inaccessible to every terrestrial species from Continent B. As seaways gradually expanded over millions of years, fewer and fewer sorts of organisms would be able to bridge the gap between landmasses.

But an even larger wrinkle must be considered: changing sea levels. The height of the ocean can change considerably, on both long time scales (10–20 million years) as well as shorter time scales (1–3 million years). These fluctuations can open and close various pathways for seaway crossings, resulting in the "connection" of the continents being highly variable.

Sea-level changes that take place over periods of tens of millions of years (called "second-order" cycles) can be explained through tectonic motions. Mantle plumes or mid-ocean ridges can raise the level of the seafloor, which in turn raises the surface sea level; variable rates of plate subduction transport various amounts of seawater down with them, into the mantle. During the Mesozoic specifically, average sea levels began at nearly modern levels during the Triassic and gradually rose throughout the era, culminating in a Late Cretaceous high of 250 meters above the modern level.

"Third-order" sea-level cycles, which take place on timescales between 0.5 and 3 million years, are much more difficult to explain, and there is currently no scientific consensus regarding their origins.

Fig.1 Triassic (Norian) 215 MYA

Fig.2 Jurassic (Sinemurian) 195 MYA

Fig.3 Jurassic (Callovian) 165 MYA

Fig.4 Jurassic (Tithonian) 145 MYA

While relatively brief in duration, third-order sea-level changes can be significant in scope. In only one million years, sea levels can plunge as much as 100 meters, although the average value is closer to 50 meters. In the Cretaceous alone, at least 58 third-order events have been recorded (Haq, 2014).

This means that a map like fig. 7 can be misleading. For example, one might assume that Africa and Europe were separated by an oceanic gulf that was far too wide for a large terrestrial animal to cross. However, we know that at least some abelisaurids (e.g., *Austrocheirus*) found their way to Europe from Africa around this time (Tortosa et al., 2014). How can this be? As fig. 9 shows, a variation in surface sea level of 100 meters can make all the difference, turning an impossible journey into a plausible one.

This still begs the question: What can cause 100-meter sea-level changes at fairly regular intervals over comparatively brief periods? For a planet with ice caps, the answer might be as simple as periodic warming/cooling phases, which lock up seawater at the poles. However, it is unlikely that Earth had ice caps in the Triassic or Jurassic, and the jury's still out on their presence during the Cretaceous (Haq, 2018a, 2018b). So what other options does that leave?

Periodic changes in Earth's rotation and orbit, known as Milankovitch Cycles, are known to alter Earth's climate slightly on timescales of tens of thousands to hundreds of thousands of years. However, there is evidence of even longer cycles (tens of millions of years), which, on a basic level, represent the same phenomenon. Another factor could be gradual changes in the distribution of Earth's water, going from the oceans into land-based groundwater and back again. The changing distribution of the planet's water mass would alter Earth's center of mass, which could also add perturbations to the Milankovitch Cycles (Boulila et al., 2021).

Regardless of sea-level concerns, however, Earth's modern continents did eventually separate geographically from their original Pangean form. North America, Europe, and Asia are often referred to as the "Laurasian" continents, after the northern sub-supercontinent **Laurasia**, which preceded Pangea. Although these landmasses were increasingly separated by seaways for most of the Mesozoic, faunal exchanges were still possible for most of this time, especially during portions of the Cretaceous when present-day Russia and Alaska were connected in the region of the modern Bering Strait.

The southern landmass of **Gondwana** took longer to fracture; the landmass that would become both South America and Africa separated from the landmass of Antarctica, Australia, India, and Madagascar only in the Cretaceous Period. It is possible that intermittent isthmuses still connected Antarctica with South America and Australia when the nonavian dinosaurs went extinct 66 million years ago (Ezcurra and Agnolín, 2012a, 2012b).

Fig. 5 Cretaceous (Hauterivian) 130 MYA

Fig. 6 Cretaceous (Aptian) 115 MYA

Fig. 7 Cretaceous (Santonian) 85 MYA

Fig. 8 Cretaceous (Maastrichtian) 75 MYA

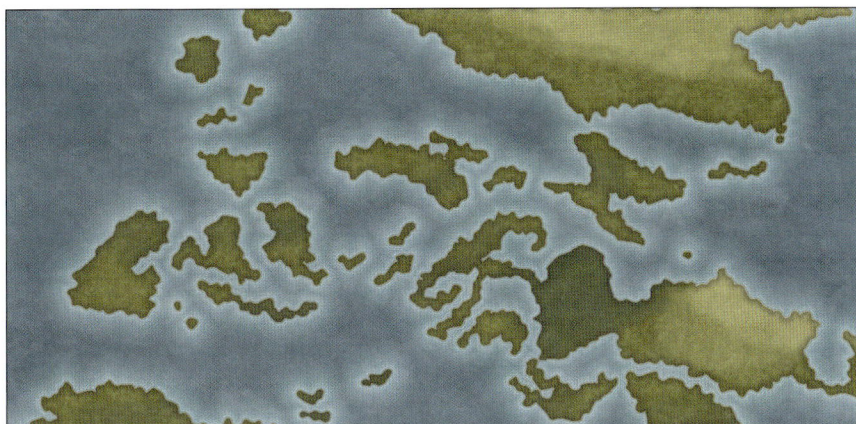

Fig. 9. The coastlines of what is now Europe, during a time of exceptionally high sea level (top) and exceptionally low sea level (bottom), during the Late Cretaceous.

As discussed above, sea levels are known to fluctuate by as much as 100 meters in as little as 0.5 to 3 million years. These changing landscapes could have enabled "faunal interchanges," where species that would otherwise have been isolated could explore and potentially thrive in new habitats.

(Coastlines based on and modified from Puértolas-Pascual et al. [2016])

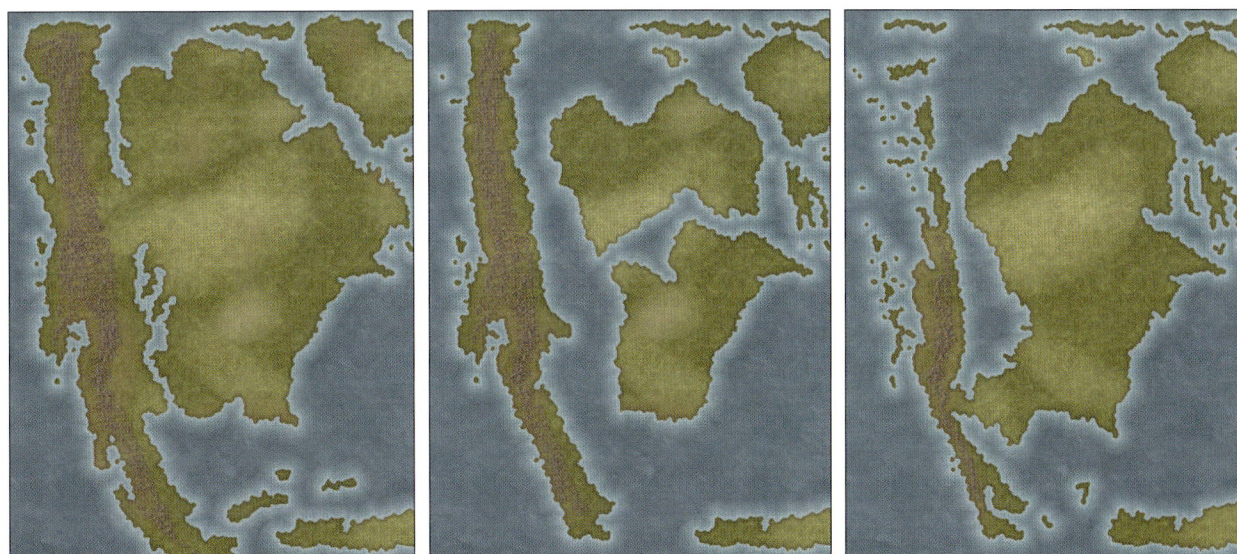

Fig. 10. The coastlines of what is now North America, highlighting the extent of the inland shallow sea known as the Western Interior Seaway, at 115 (left), 90 (middle), and 65 (right) million years ago. Throughout the Cretaceous Period, fluctuating sea levels caused the extent of the seaway to expand and contract multiple times. The seaway exists for the same fundamental reason the Rocky Mountains exist: As the Farallon and Kula tectonic plates were being subducted beneath the North American plate, they were not forced down into the mantle, as would normally be the case; rather, they continued eastward at a very shallow angle, essentially rubbing against the underside of the North American plate. This process warped the North American plate and produced drag, which both raised the foundations of the Rocky Mountains and lowered the terrain that would form the floor of the Western Interior Seaway. The western landmass produced when North America was divided in two is referred to as Laramidia; the eastern landmass is called Appalachia.

Information used to create the maps in this volume was compiled from Dalla Vecchia (2009), Novas et al. (2013), Upchurch et al. (2015), Wen et al. (2016), Puértolas-Pascual et al. (2016), Molina-Pérez et al. (2019), and Lovecchio et al. (2019).

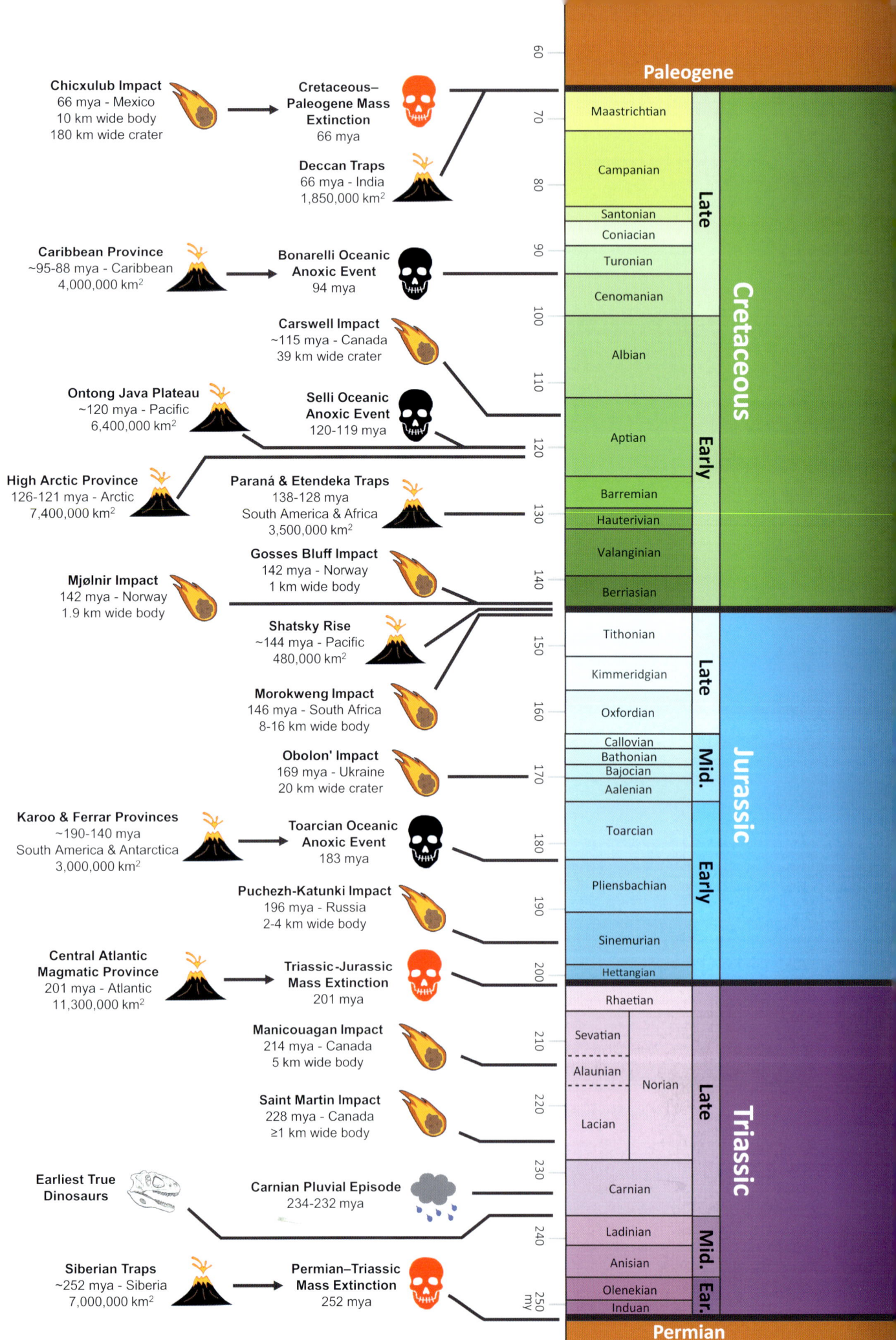

Mamenchisaurus

Sauropodomorpha

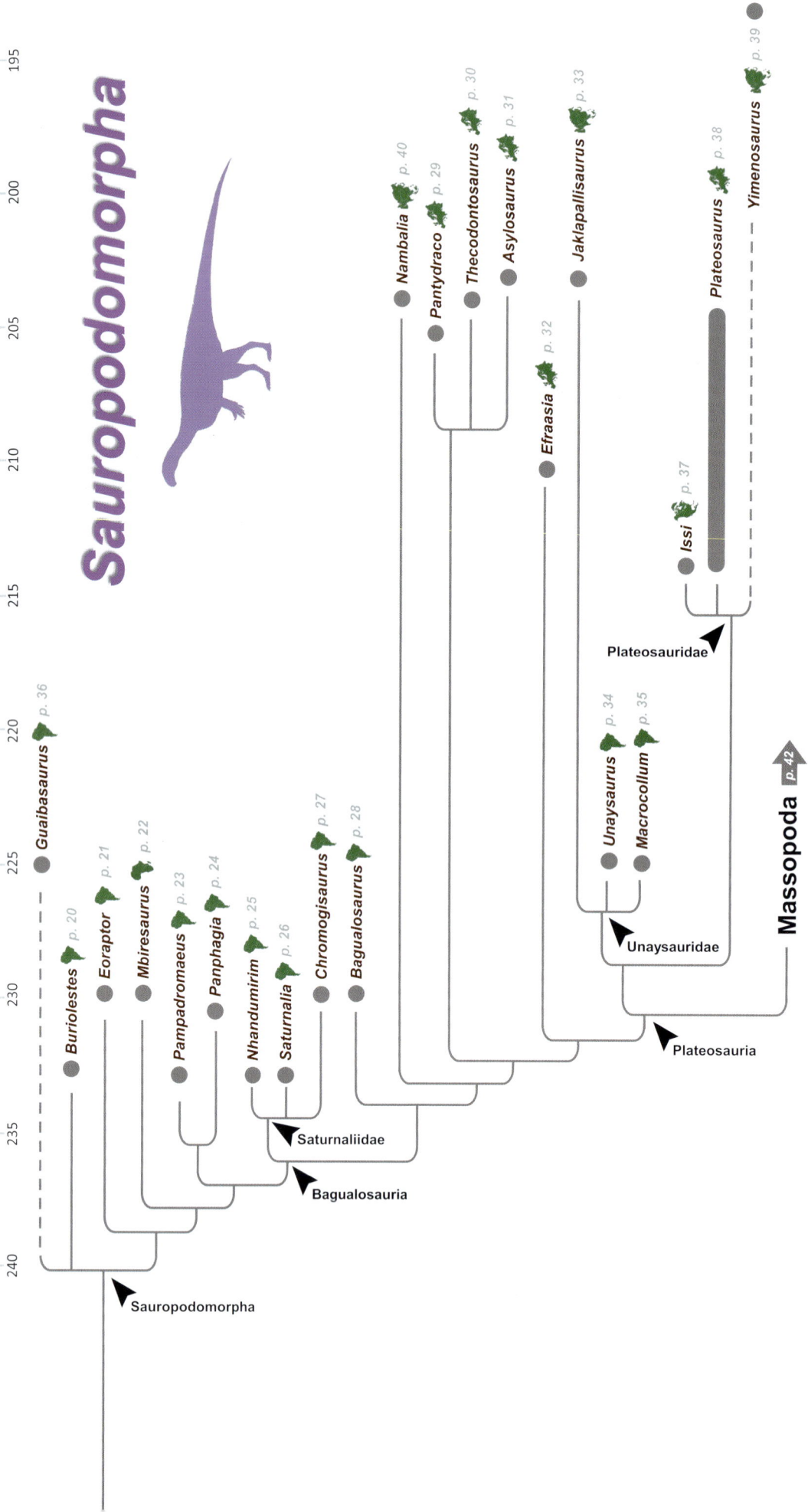

Jurassic

Early

Het. | Sinemurian

195 · 200

Triassic

Late

Rhaetian | Norian

205 · 210 · 215 · 220

Carnian

225 · 230 · 235

Middle

Ladinian

240

Anis.

Guaibasaurus p. 36

Buriolestes p. 20

Eoraptor p. 21

Mbiresaurus p. 22

Pampadromaeus p. 23

Panphagia p. 24

Nhandumirim p. 25

Saturnalia p. 26

Chromogisaurus p. 27

Bagualosaurus p. 28

Nambalia p. 40

Pantydraco p. 29

Thecodontosaurus p. 30

Asylosaurus p. 31

Efraasia p. 32

Jaklapallisaurus p. 33

Unaysaurus p. 34

Macrocollum p. 35

Issi p. 37

Plateosaurus p. 38

Yimenosaurus p. 39

Sauropodomorpha

Saturnaliidae

Bagualosauria

Unaysauridae

Plateosauria

Plateosauridae

Massopoda p. 42

The earliest ancestors of the sauropods were nimble carnivorous bipeds, difficult to distinguish from their theropod cousins. They swiftly differentiated themselves, rapidly increasing their size and vastly changing their diets in only a handful of millions of years, laying the groundwork for the giants yet to come.

Most analyses taking place after 2013 generally agree that *Eoraptor* is a sauropodomorph rather than a theropod. Whether or not *Eoraptor* or *Buriolestes* is the basalmost known sauropodomorph, though, is less certain (Müller et al., 2018b; Langer et al., 2019). Some studies even place the two genera, along with *Guaibasaurus*, in their own clade (Pretto et al., 2019; Griffin et al., 2022). The positioning of *Guaibasaurus* is much less certain; while the two aforementioned papers place it very basally, Müller and Garcia (2020) and Garcia et al. (2021) place it in a more derived clade alongside *Macrocollum*.

The initial description of *Mbiresaurus* places it as the next most derived sauropodomorph (Griffin et al., 2022). On the matter of *Pampadromaeus* and *Panphagia*, multiple studies have placed one or both of them as members of **Saturnaliidae** (Müller et al., 2018a; Langer et al., 2019), whereas some place them more basally as sister taxa (Bronzati et al., 2019; Pacheco et al., 2019; Müller and Garcia, 2020), and still others suggest other alternatives (Pretto et al., 2019; Beccari et al., 2021). Saturnaliidae is most commonly depicted as a valid taxon and is often shown to include *Saturnalia*, *Nhandumirim*, and *Chromogisaurus* (Pacheco et al., 2019; Müller and Garcia, 2020). *Bagualosaurus* has also been suggested as another member (Garcia et al., 2021) but is usually depicted in the next crownward position (Müller et al., 2018a; Langer et al., 2019; Pretto et al., 2019; Müller and Garcia, 2020; Griffin et al., 2022).

It is common to then see *Thecodontosaurus* and *Pantydraco* depicted either as sister taxa or subsequent taxa, followed immediately by *Efraasia* (Ezcurra, 2010; Müller et al., 2018b; Langer et al., 2019; McPhee et al., 2020), although some recent analyses have favored these genera in a noticeably more derived position (Müller and Garcia, 2020; Griffin et al., 2022). The often-excluded *Nambalia* has also been shown as a close relative of these (Novas et al., 2010; Müller et al., 2018a). *Asylosaurus* may or may not be synonymous with *Thecodontosaurus* (Ballell et al., 2020).

Macrocollum, *Jaklapallisaurus*, and *Unaysaurus* have been grouped into **Unaysauridae** by Müller et al. (2018a) and then Krupandan (2019), although the validity of this family has been questioned by some (McPhee et al., 2020), and many analyses have historically put *Unaysaurus* within Plateosauridae.

Unaysauridae has been placed in different positions across studies: basal to Plateosauridae (Müller et al., 2018a), as a sister group of it (Beccari et al., 2021), or as being more derived (Krupandan, 2019). The definition of **Plateosauridae** has changed as phylogenies have evolved; it presently includes *Plateosaurus* and *Issi* (Beccari et al., 2021) and potentially *Yimenosaurus*.

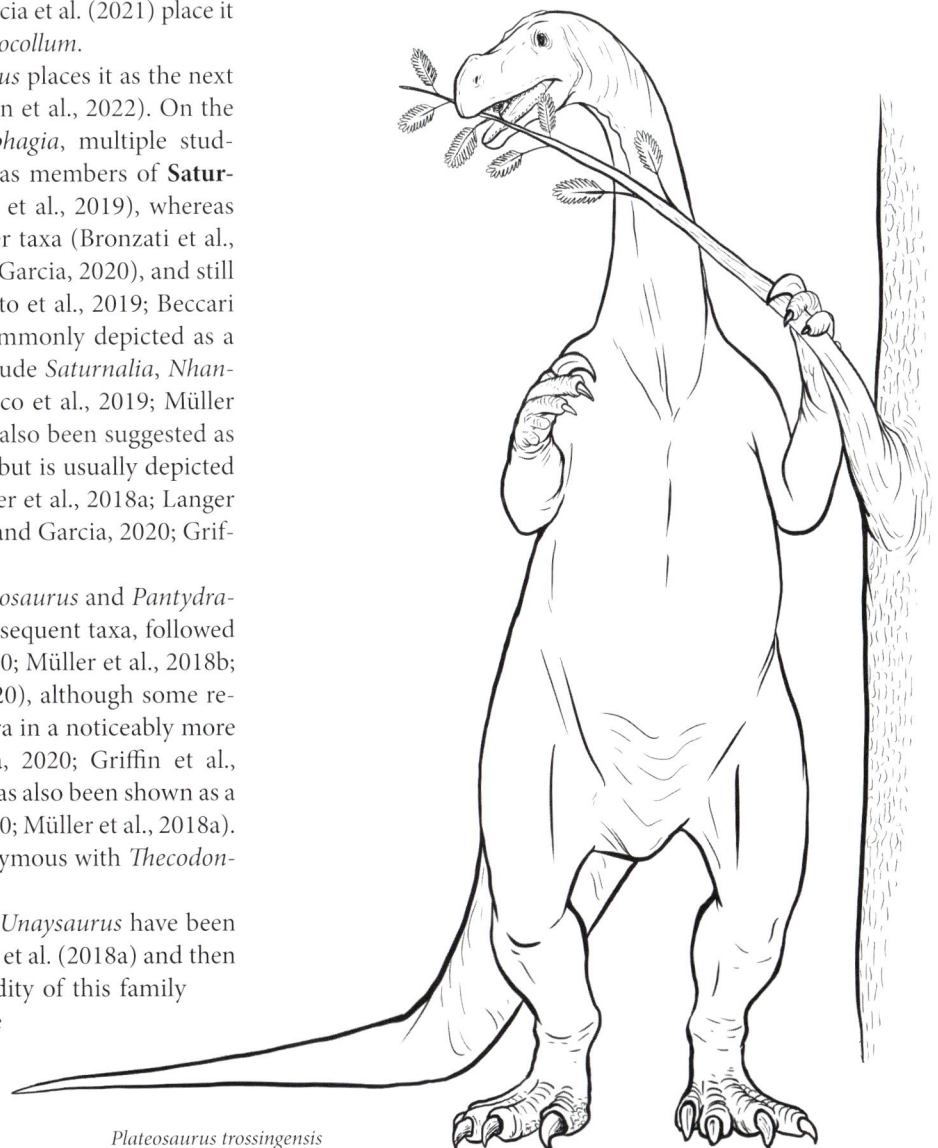

Plateosaurus trossingensis

50 cm   100 cm   150 cm

50 cm

B. schultzi

(Cabreira et al., 2016)
Length: 1.8 m (5.9 ft)
Height: 0.6 m (2 ft)
Hip height: 0.5 m (1.6 ft)
Body mass: 7.4 kg (16 lb)
Reconstruction: ▢▢▢▢▢

| 250 | 245 | 240 | **233** | 230 | 225 | 220 | 215 | 210 | 205 | 200 | 195 | 190 | 185 | 180 | 175 | 170 | 165 | 160 | 155 | 150 | 145 | 140 | 135 | 130 | 125 | 120 | 115 | 110 | 105 | 100 | 95 | 90 | 85 | 80 | 75 | 70 | 65 |

TRIASSIC   **JURASSIC**   **CRETACEOUS**

Buriolestes schultzi (meaning "Schultz's robber of Buriol") is, according to multiple analyses, the earliest ancestor of all sauropods.

With teeth that were serrated and curved, *Buriolestes* was undoubtedly a carnivorous animal. This makes *Buriolestes* the only known carnivorous sauropodomorph, as all other early sauropodomorphs that have been discovered were likely omnivorous.

Several skeletal features differentiate *Buriolestes* from the lineage of carnivorous theropods. For instance, the downturned tip of the animal's jaw is a sauropod trait; the shape of the muscle attachment sites seen on its humerus is also distinctively sauropodan in nature.

Still, the animal possesses some unique features that are unusual for theropods and sauropods alike: the upper portion of its tibia is angled backward and its pubis bone is exceptionally straight and unadorned.

The first *Buriolestes* specimen (ULBRA-PVT280) was found alongside a relative of dinosaurs, *Ixalerpeton*, which shows that the dinosaurs—as a group—did not immediately replace their most recent ancestors in their respective ecosystems. A second substantial *Buriolestes* specimen, along with several more fragmentary remains, was later uncovered from the same location (Müller et al., 2018a).

A detailed analysis of the animal's interior braincase showed that *Buriolestes* likely relied more upon its eyesight than its sense of smell and inferred that later sauropods might have lost some of their cognitive abilities as their bodies enlarged (Müller and Garcia, 2020).

The generic name *Buriolestes* refers to the Buriol family, who owned the land where the holotype was found; the name is combined with the Greek "lestes" (meaning "robber"). The specific name *schultzi* honors paleontologist Cezar Schultz.

CLASSIFICATION
Dinosauria
  Saurischia
   Sauropodomorpha

LOCATION
Brazil

KNOWN REMAINS
Nearly complete skeleton and skull

E. lunensis

(Sereno et al., 1993)
Length: 1.7 m (5.6 ft)
Height: 0.6 m (2 ft)
Hip height: 0.5 m (1.6 ft)
Body mass: 10 kg (22 lb)
Reconstruction: ☐☐☐☐☐

50 cm

TRIASSIC   JURASSIC   CRETACEOUS

Eoraptor lunensis (meaning "dawn thief inhabitant of the moon") is considered to be one of the earliest known sauropodomorphs. It had teeth that came in more than just one shape, suggesting that it might have been omnivorous.

Eoraptor was initially classified as the basalmost theropod in Paul Sereno's original 1993 paper. However, many differing viewpoints were expressed in the coming years, and in that time *Eoraptor* was also considered a sauropodomorph, or a basal saurischian (outside both Theropoda and Sauropodomorpha). In a 2013 comprehensive reanalysis of the holotype specimen (PVSJ 512), Sereno himself reclassified *Eoraptor* as a sauropodomorph (Sereno et al., 2013).

Several factors played into this reassessment. In addition to advancements made in technology and assessment techniques, new clarity had been reached due to the examination of various referred specimens; the bones now recognized as *Eodromaeus* were reassigned out of the genus,

which had previously muddied the waters. In addition, new specimens of *Eoraptor* have yielded additional important anatomical clues.

Most importantly, back in 1993, the holotype specimen had been only partially exposed. Details that had been obscured were later able to be uncovered. Of particular note, *Eoraptor* was found to bear features of the digits and front-lower jaw that are only known in **sauropodomorphs**.

The generic name *Eoraptor* is a combination of the Greek word "eós" (meaning "dawn," in reference to the dinosaur's primitive nature) and the Latin word "raptor" (meaning "thief" or "plunderer"). The specific name *lunensis* is derived from the Latin words "luna" ("moon") and the suffix "ensis" ("inhabitant"), referring to the holotype's place of discovery, the Valle de la Luna ("Valley of the Moon"). The location received its name due to its otherworldly, arid appearance, which can be evocative of a desolate lunar landscape.

CLASSIFICATION
Dinosauria
 Saurischia
  Sauropodomorpha

LOCATION
Argentina

KNOWN REMAINS
Nearly complete skeleton and skull

21

50 cm · 100 cm · 150 cm

50 cm

M. raathi

(Griffin et al., 2022)
Length: 1.7 m (5.6 ft)
Height: 0.5 m (1.6 ft)
Hip height: 0.4 m (1.3 ft)
Body mass: 6 kg (13 lb)
Reconstruction: ▢▢▢▢

250 245 240 235 230 225 220 215 210 205 200 195 190 185 180 175 170 165 160 155 150 145 140 135 130 125 120 115 110 105 100 95 90 85 80 75 70 65

TRIASSIC | **JURASSIC** | **CRETACEOUS**

Mbiresaurus raathi (meaning "Raath's lizard from Mbire") is among the very earliest dinosaurs ever discovered from Africa. It is quite comparable to similarly aged creatures found in South America and India, which helps to confirm that these locations—which shared a similar latitude during the Late Triassic—all had similar ecologies.

The discoverers of *Mbiresaurus* expressly set out to find early dinosaurs that inhabited this latitude-governed ecosystem. They chose their excavation area based on previous work done by paleontologist Michael Raath, who had unearthed reptilian remains and a possible shard of dinosaur bone from the area in the past.

Two specimens of *Mbiresaurus* were excavated between 2017 and 2019. The holotype (NHMZ 2222) is a mostly complete adult specimen; a second, more fragmentary

individual was approximately 15% larger than the holotype. This could indicate that *Mbiresaurus* continued to increase in size after the animal reached maturity, or simply that not every adult reached its maximum growth potential.

Based on its teeth, *Mbiresaurus* was already at least partially herbivorous. The shape of the indentations on the teeth is more similar to that seen in later sauropodomorphs, such as *Plateosaurus*, than it is to contemporaries such as *Eoraptor*.

The phylogenetic analysis conducted by the describing authors unequivocally placed *Mbiresaurus* as being among the earliest **sauropodomorphs**.

The generic name *Mbiresaurus* refers to the Mbire District of Zimbabwe combined with the Latin "sauros" (meaning "reptile"). The specific name *raathi* honors Michael Raath, one of the fossil's discoverers.

CLASSIFICATION
Dinosauria
 Saurischia
  Sauropodomorpha

LOCATION
Zimbabwe

KNOWN REMAINS
Nearly complete skeleton and skull

50 cm   100 cm   150 cm

50 cm

P. barberenai

(Cabreira et al., 2011)
Length: 1.4 m (4.6 ft)
Height: 0.4 m (1.3 ft)
Hip height: 0.3 m (1 ft)
Body mass: 3 kg (7 lb)
Reconstruction: ☐ ☐ ☐ ☐

250 245 240 233 230 225 220 215 210 205 200 195 190 185 180 175 170 165 160 155 150 145 140 135 130 125 120 115 110 105 100 95 90 85 80 75 70 65

TRIASSIC   **JURASSIC**   **CRETACEOUS**

Pampadromaeus barberenai (meaning "Barberena's plains runner") is a very early sauropodomorph that nonetheless possesses many theropod-like traits, showing the close relationship between these two groups that would later give rise to vastly different animals.

Like a theropod, *Pampadromaeus* was likely a relatively swift animal; the elongation of its lower legs in comparison to its upper legs suggests a highly cursorial lifestyle. Also, like other early theropods, the tip of its snout is downturned and features a distinctive notch.

Yet, *Pampadromaeus* has consistently been classified as a sauropodomorph, not as a theropod. Although its teeth are sharp and serrated, their bases overlap one another in a decidedly non-theropod manner. The teeth are also designed to be retained without shedding for a greater period of time, which is a trait most commonly seen in herbivores. Other skeletal traits, such as the size and shape of its upper arm bones and tail vertebrae, also link *Pampadromaeus* firmly to the sauropodomorphs.

The nearly complete holotype (ULBRA-PVT016) remains the only substantial *Pampadromaeus* specimen known, although two separate fossilized femora have also been assigned to the genus.

The exact phylogenetic placement of *Pampadromaeus* among the early **sauropodomorphs** is somewhat uncertain. Some analyses favor it being a member of the Saturnaliidae family (Langer et al., 2019), while others do not (Griffin et al., 2022).

The generic name *Pampadromaeus* combines the indigenous Quechua word "pampa" (meaning "plains" or "grasslands") with the Greek "dromeus" (meaning "runner"). The specific name honors paleontologist Mário Barberena.

CLASSIFICATION
Dinosauria
  Saurischia
    Sauropodomorpha

LOCATION
Brazil

KNOWN REMAINS
Mostly complete skeleton and skull

50 cm ⎯ 100 cm ⎯ 150 cm ⎯ 200 cm

50 cm

P. protos

(Martínez and Alcober, 2009)
Length: 2 m (6.6 ft)
Height: 0.6 m (2 ft)
Hip height: 0.5 m (1.6 ft)
Body mass: 7.8 kg (17 lb)
Reconstruction: ☐☐☐☐☐

250 245 240 235 **231** 225 220 215 210 205 200 195 190 185 180 175 170 165 160 155 150 145 140 135 130 125 120 115 110 105 100 95 90 85 80 75 70 65

TRIASSIC	JURASSIC	CRETACEOUS

Panphagia protos (meaning "first all-eating one") was, at the time of its discovery, the basalmost unambiguous sauropodomorph.

Of particular note for its describers, *Panphagia* appeared to share a number of traits in common with both *Saturnalia* (a very basal but not-quite-as-basal-as *Panphagia* sauropodomorph) and *Eoraptor* (an animal that confusingly blurs the line between the theropods and the sauropods). These dual similarities helped to further the understanding of how the earliest sauropodomorphs evolved from very small, nimble forms into larger forms.

The holotype and only known specimen of *Panphagia* (PVSJ 874) contains a mixture of various disarticulated skeletal elements. This individual is thought to have been immature at the time of its death, making the full size of an adult specimen difficult to gauge. Although the recovered portions of the skull were lacking teeth, the nearly complete jaw still held numerous teeth. Those present at the front of the mouth were longer and more pointed, while those at the back of the mouth were smaller and more "leaf-shaped." This tooth arrangement led the describing authors to interpret *Panphagia* as an omnivorous animal, possibly representing a transitional form between its predatory theropod-like ancestors and its herbivorous sauropod descendants (Martínez et al., 2012b).

The exact phylogenetic placement of *Panphagia* among the early **sauropodomorphs** is somewhat uncertain. Some analyses favor it being a member of the Saturnaliidae family (Langer et al., 2019), while others do not (Griffin et al., 2022).

The generic name *Panphagia* combines the Greek "pan" (meaning "all"), and "phagein" (meaning "to eat"), in reference to the animal's perceived omnivorous diet. The specific name *protos* is Greek for "first," alluding to the animal's early position on the sauropodomorph family tree.

CLASSIFICATION
Dinosauria
　Saurischia
　　Sauropodomorpha

LOCATION
Argentina

KNOWN REMAINS
Partial skull and skeleton

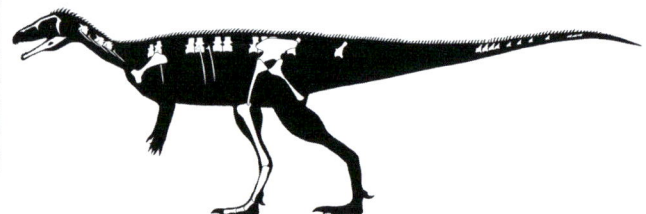

| 25 cm | 50 cm | 75 cm | 100 cm |

25 cm

N. waldsangae

(Marzola et al., 2018)
Length: 1 m (3.3 ft)
Height: 0.3 m (11 in)
Hip height: 0.2 m (7 in)
Body mass: 1.5 kg (3 lb)
Reconstruction: ☐☐☐☐

| 250 | 245 | 240 | **233** | 230 | 225 | 220 | 215 | 210 | 205 | 200 | 195 | 190 | 185 | 180 | 175 | 170 | 165 | 160 | 155 | 150 | 145 | 140 | 135 | 130 | 125 | 120 | 115 | 110 | 105 | 100 | 95 | 90 | 85 | 80 | 75 | 70 | 65 |

| TRIASSIC | JURASSIC | CRETACEOUS |

Nhandumirim waldsangae (meaning "small rhea from Waldsanga") is a small-bodied example of an early sauropodomorph. With relatively long, gracile legs, the animal is assumed to have been swift and nimble.

The sole known remains of *Nhandumirim* (LPRP/USP 0651), unearthed in 2012, consist of fragmentary portions of the pelvis, leg, foot, and scattered vertebrae. The remains were only partially articulated, but all were located within close proximity of one another. It is believed that the individual was not yet mature at the time of its death. Researchers ruled out the possibility that the specimen simply represented a juvenile of an already known genus, such as *Saturnalia*, by noting numerous differentiating skeletal details that set it apart.

The describers performed multiple versions of various phylogenetic analyses and favored the conclusion that *Nhandumirim* was a very basal theropod. However, numerous studies from subsequent papers have consistently concluded that *Nhandumirim* is a sauropodomorph in the family **Saturnaliidae** (Pacheco et al., 2019; Garcia et al., 2021; Novas et al., 2021).

This sort of phylogenetic confusion is not unreasonable. It is currently unclear how, exactly, the lineages of the theropods and the sauropods diverged from one another. Numerous saurischian species known from this time period share traits of both groups, and untangling the knot that this creates in the family tree is an ongoing area of research. Although it is tempting to assign all such species to either one category or the other, it remains a possibility that some species represent members of more basal lineages that belong to neither branch.

The generic name *Nhandumirim* combines the indigenous Tupi-Guarani words "nhandu" (referring to the common rhea, an ostrich-like flightless bird) and "mirim" (meaning "small"). The specific name *waldsangae* refers to the site of the fossil's discovery, known as Waldsanga.

CLASSIFICATION
Dinosauria
 Saurischia
  Sauropodomorpha
  Saturnaliidae

LOCATION
Brazil

KNOWN REMAINS
Fragments

25

50 cm   100 cm   150 cm

50 cm

S. tupiniquim

(Langer et al., 1999)
Length: 1.7 m (5.6 ft)
Height: 0.55 m (1.8 ft)
Hip height: 0.4 m (1.3 ft)
Body mass: 6 kg (13 lb)
Reconstruction: ☐☐☐☐

250 | 245 | 240 | 233 | 230 | 225 | 220 | 215 | 210 | 205 | 200 | 195 | 190 | 185 | 180 | 175 | 170 | 165 | 160 | 155 | 150 | 145 | 140 | 135 | 130 | 125 | 120 | 115 | 110 | 105 | 100 | 95 | 90 | 85 | 80 | 75 | 70 | 65

TRIASSIC   **JURASSIC**   **CRETACEOUS**

Saturnalia tupiniquim (meaning "carnival native") is another example of a creature that blurs the lines separating the sauropods form the theropods. Various early studies disagreed on the animal's exact phylogenetic placement, although recent works agree on a firm placement within Sauropodomorpha (Griffin et al., 2022).

Three partial specimens from the same locality were first described in 1999. Thorough analysis of the animal's skull would have to wait another 20 years: the disarticulated skull fragments discovered from a single specimen were too delicate to remove from their rocky housing, and thus many details were hidden until detailed CT scans could be used to generate three-dimensional models of the partial bones. This revealed that the skull of *Saturnalia* was quite small in relation to its body—a trait that would later become common throughout the sauropod lineage. However, the structure of its teeth suggests that *Saturnalia* was a mostly carnivorous animal, like the early theropods (Bronzati et al., 2019).

One proposal suggests that the small skull of *Saturnalia* was an adaptation to allow for swift movements of the head and neck, in order to better catch small prey animals. Having a small skull then became foundational for the evolution of longer and longer necks in *Saturnalia's* descendants, which in turn enabled highly efficient herbivorous characteristics. Thus, it could well be that a trait that originated in relation to carnivory was crucial for the later development of herbivorous specialization (Bronzati et al., 2019).

The generic name *Saturnalia* is Latin for "carnival," as the paratype fossils were discovered during a traditional feasting celebration. The specific name *tupiniquim* means "native" in the indigenous Guarani language (which has Portuguese lingual influences).

CLASSIFICATION
Dinosauria
  Saurischia
   Sauropodomorpha
   Saturnaliidae

LOCATION
Brazil

KNOWN REMAINS
Mostly complete skeleton and skull

50 cm   100 cm   150 cm

50 cm

C. novasi

(Ezcurra, 2010)
Length: 1.9 m (6.2 ft)
Height: 0.6 m (2 ft)
Hip height: 0.5 m (1.6 ft)
Body mass: 12 kg (26 lb)
Reconstruction: ☐☐☐☐☐

250 245 240 235 **230** 225 220 215 210 205 200 195 190 185 180 175 170 165 160 155 150 145 140 135 130 125 120 115 110 105 100 95 90 85 80 75 70 65

TRIASSIC   **JURASSIC**   **CRETACEOUS**

Chromogisaurus novasi (meaning "Novas's colored-earth lizard") is among the earliest known dinosaurs, and more specifically, one of the earliest known sauropodomorphs. Its discovery helped to show that, although dinosaurs were only a small part of the ecosystem during the earliest Late Triassic, their diversity was already considerable, as all of the primary dinosaurian lineages already seem to have been present at this time. It shared its environment with other early sauropodomorphs such as *Panphagia* and *Eoraptor*.

The holotype and only known specimen of *Chromogisaurus* (PVSJ 845) is fragmentary and not in pristine condition. It consists of the animal's pelvic region, along with portions of the legs and feet, several tail vertebrae, and fragments of the arm. (A later study suggested that the fragment initially interpreted as a partial ulna was instead merely a reptilian jaw fragment [Martínez et al., 2012a].) The fossils were found in 1988, from the Ischigualasto geological formation; during the Carnian stage of the Triassic, the region featured numerous lakes and rivers and had a temperate, seasonal environment.

The describers found *Chromogisaurus* to be the sister taxon of *Saturnalia*, as part of the family **Saturnaliidae**, a result that has also been upheld by subsequent studies (Müller and Garcia, 2020). The describers placed Saturnaliidae within a larger group, Guaibasauridae, but this arrangement is not always supported by more recent studies.

The generic name *Chromogisaurus* combines the Greek words "chroma" (meaning "color"), "gi" (meaning "ground" or "land"), and "saurus" (meaning "reptile"); the name refers to the Valle Pintado, or Painted Valley, where the specimen was discovered. The specific name *novasi* honors paleontologist Fernando Novas.

CLASSIFICATION
Dinosauria
  Saurischia
    Sauropodomorpha
    Saturnaliidae

LOCATION
Argentina

KNOWN REMAINS
Fragments

SAUROPODOMORPHA

50 cm | 100 cm | 150 cm | 200 cm | 250 cm

50 cm

B. agudoensis

(Pretto et al., 2019)
Length: 2.3 m (7.5 ft)
Height: 0.8 m (2.6 ft)
Hip height: 0.7 m (2.3 ft)
Body mass: 15 kg (33 lb)
Reconstruction: ▢▢▢▢▢

| 250 | 245 | 240 | 235 | **230** | 225 | 220 | 215 | 210 | 205 | 200 | 195 | 190 | 185 | 180 | 175 | 170 | 165 | 160 | 155 | 150 | 145 | 140 | 135 | 130 | 125 | 120 | 115 | 110 | 105 | 100 | 95 | 90 | 85 | 80 | 75 | 70 | 65 |

TRIASSIC | **JURASSIC** | **CRETACEOUS**

Bagualosaurus agudoensis (meaning "stout lizard from Agudo") is among the largest known sauropodomorphs from the Carnian stage of the Triassic.

As *Bagualosaurus* shares some traits with sauropodomorphs that are more basal (or "primitive") and some that are more derived (or "advanced"), it is hypothesized to have had an intermediate or transitive position on the family tree. The morphology of its postcranial skeleton is very similar to that of the earliest ancestral sauropodomorphs, although it is significantly larger in size, being larger than most other animals in its environment. However, the animal's skull has some traits that are indicative of increased specialization in herbivory, some of which had previously only been observed in later sauropodomorphs.

This combination of features suggests that sauropodomorphs evolved mandibular traits related to herbivory before they evolved the adaptations that allowed them to grow to tremendous size. Although *Bagualosaurus*

was large for its time, it lacked the specializations (such as a robust foot) that would allow it to support a more massive body, like the sauropodomorphs of the latest Triassic.

Bagualosaurus is known from a single incomplete specimen (UFRGS-PV-1099-T) that was found in partial articulation. The specimen consists of a jaw, lower skull, vertebral, pelvic region, and hindlimb portions. It was discovered in 2007 but remained unprepared and unstudied until 2012. Some elements were already eroded by the time of their excavation. The bones showed that the animal's final resting position was unusual, as it was found lying on its back with its legs sticking forward.

The generic name *Bagualosaurus* comes from the Brazilian term "bagual," which is used to describe an animal that is stout or strong; this is combined with the Latin "saurus" (meaning "lizard"). The specific name *agudoensis* refers to the town of Agudo.

CLASSIFICATION
Dinosauria
  Saurischia
    Sauropodomorpha

LOCATION
Brazil

KNOWN REMAINS
Partial skull and skeleton

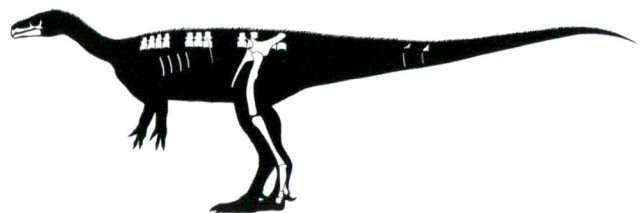

50 cm   100 cm   150 cm

50 cm

P. caducus

(Galton et al., 2007b)

Length: 1.5 m (4.9 ft)
Height: 0.5 m (1.6 ft)
Hip height: 0.3 m (1 ft)
Body mass: 6 kg (13 lb)
Reconstruction: ▢▢▢▢

250 245 240 235 230 225 220 215 210 **205** 200 195 190 185 180 175 170 165 160 155 150 145 140 135 130 125 120 115 110 105 100 95 90 85 80 75 70 65

| TRIASSIC | JURASSIC | CRETACEOUS |

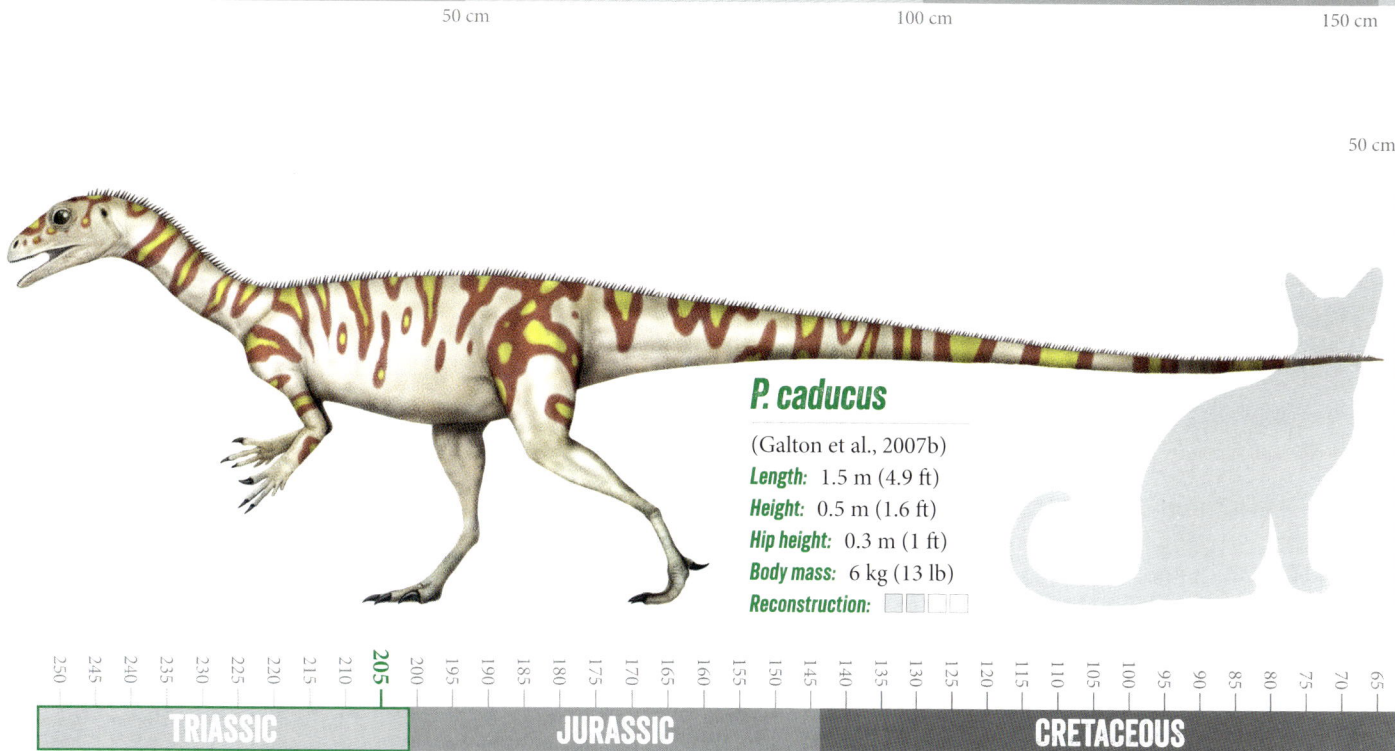

Pantydraco caducus (meaning "fallen dragon from Pant-y-ffynnon") is a basal sauropodomorph that inhabited an island ecosystem during the latest Triassic. Based on the structure of its teeth, which lacked any specializations for processing tough plant material, *Pantydraco* was likely omnivorous. The claws of its first manual digit were slightly enlarged.

The holotype remains were first unearthed in 1952 and were described in 1983 as belonging to the species *Thecodontosaurus antiquus*. In 2003, paleontologist Adam Yates used the remains as the basis for a new species within the genus, *T. caducus*. A few years later, the remains were once again reclassified as belonging to a distinct new genus, *Pantydraco*.

The holotype specimen (BMNH P24) would have been approximately 1.5 meters in length. According to the describing authors, the specimen was a juvenile, based partially on the relative proportions of the animal's skull; an adult specimen would have hypothetically been at least twice as long (Galton and Kermack, 2010).

However, a later analysis has suggested that the animal was actually an adult and that its relatively small size was due to its island habitat. As small islands cannot support a population of large animals, many island species are subject to "island dwarfism," whereby species shrink in size compared with their mainland counterparts (Keeble et al., 2018).

One study in which numerous specimens of *Thecodontosaurus* were examined suggested that *Pantydraco* may be an invalid genus and that the fossil remains attributed to it might actually represent juvenile specimens of *T. antiquus*, just as was originally hypothesized in 1983 (Ballell et al., 2020).

The generic name *Pantydraco* abbreviates the location of the specimen's discovery, the Pant-y-ffynnon (itself meaning "valley of the spring" in Welsh) quarry, which is then combined with the Latin "draco" (meaning "dragon"). The specific name *caducus* is Latin for "fallen" or "perished," as the animal had come to rest at the bottom of a "fissure" in the rock.

CLASSIFICATION
Dinosauria
  Saurischia
    Sauropodomorpha

LOCATION
Wales, UK

KNOWN REMAINS
Partial skull and skeleton

1 m   2 m   3 m

1 m

T. antiquus

(Riley and Stutchbury, 1836)
Length: 3 m (9.8 ft)
Height: 0.8 m (2.6 ft)
Hip height: 0.7 m (2.3 ft)
Body mass: 34 kg (75 lb)
Reconstruction: ☐☐☐☐

TRIASSIC	JURASSIC	CRETACEOUS

Thecodontosaurus antiquus (meaning "ancient socket-toothed lizard") was discovered in 1834, prior to the coining of the term "Dinosauria" by Sir Richard Owen in 1842. The first bones were unearthed by workers in a limestone quarry and quickly garnered the attention of early paleontologists. *Thecodontosaurus* was originally interpreted as being some sort of crocodile-like lizard and was only recognized as a dinosaur in 1870 by Thomas Huxley.

A great many fossils, the majority of which are fragmentary, have been assigned to *Thecodontosaurus* over the decades. Numerous species have been named, although today only *T. antiquus* is typically considered valid. The animal's average length has been reported as being anywhere from 1.2 to 2.5 meters, with the largest known individual being approximately 3 meters in length. Some individuals seem to have been more robust than others, which has been interpreted as potentially indicating sexual dimorphism (Benton et al., 2000).

Thecodontosaurus was a swift and agile animal that had not yet developed the limb traits that would allow its descendants to reach gargantuan sizes: its femur was S-shaped rather than straight (Ballell et al., 2020) and it lacked rigorous musculature (Ballell et al., 2022). Its arms were well suited for grasping and manipulating vegetation; aspects of the animal's skull and teeth suggest that it was primarily herbivorous, although occasional faunivorous habits have not been ruled out.

The generic name *Thecodontosaurus* combines the Greek terms "theke" (meaning "sheath" or "socket"), "odous" (meaning "tooth"), and "sauros" (meaning "lizard"). This name refers to the deep sockets that anchor the teeth, quite unlike those of modern reptiles. The specific name *antiquus* is from the Latin "antiqua" (meaning "ancient").

CLASSIFICATION
Dinosauria
  Saurischia
    Sauropodomorpha

LOCATION
England, UK

KNOWN REMAINS
Nearly complete

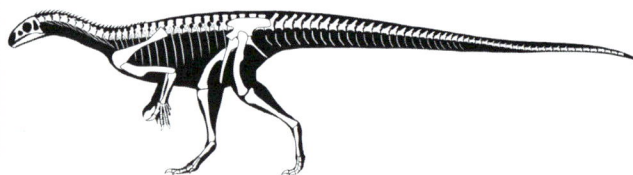

SAUROPODOMORPHA

50 cm 100 cm 150 cm 200 cm

50 cm

A. yalensis

(Galton, 2007a)

Length: 2.2 m (7.4 ft)
Height: 0.7 m (2.3 ft)
Hip height: 0.5 m (1.7 ft)
Body mass: 14 kg (31 lb)
Reconstruction: ☐☐☐☐

250 245 240 235 230 225 220 215 210 **202** 200 195 190 185 180 175 170 165 160 155 150 145 140 135 130 125 120 115 110 105 100 95 90 85 80 75 70 65

TRIASSIC **JURASSIC** **CRETACEOUS**

Asylosaurus yalensis (meaning "unharmed lizard from Yale") is an ambiguous basal sauropodomorph.

The original specimens (YPM 2195) were discovered in 1834 inside an ancient fissure-fill and were first described in 1836 as belonging to *Thecodontosaurus*. It would not be until 2007 that they were interpreted as being a distinct new genus, *Asylosaurus*. At this time, a smattering of other bones were also assigned to *Asylosaurus*. It was assumed that *Asylosaurus* would have avoided competition with similar animals, such as *Pantydraco* and *Thecodontosaurus*, by occupying a slightly different ecological niche.

According to a more recent analysis, though, the traits that were used to establish *Asylosaurus* as a distinct taxon may not actually be valid. Defining characteristics of the humerus were questioned based on the incompleteness of the bone samples and the possibility of distortion during fossilization; characteristics of the proportions of the hand were cast into doubt; and traits of the hip were rejected because the hip bones that were originally described as belonging to *Asylosaurus* were not actually part of the main skeletal specimen and were referred to the genus without sufficient justification. Thus, the argument has been made that *Asylosaurus* should be considered *nomen dubium* since the bones cannot be confidently distinguished from those of *Thecodontosaurus* (Ballell et al., 2020).

The generic name *Asylosaurus* combines the Greek "asylos" (meaning "unharmed" or "safe") with "sauros" (meaning "lizard"). The specific name *yalensis* is meant to mean "of Yale College." The binomial name refers to how paleontologist O. C. Marsh brought the specimen from Europe to Yale, thus saving it from destruction by World War II air raids.

CLASSIFICATION

Dinosauria
  Saurischia
    Sauropodomorpha

LOCATION

England, UK

KNOWN REMAINS

Partial skeleton

2 m   4 m   6 m

2 m

E. minor

(Galton, 1973)
Length: 6.5 m (21.3 ft)
Height: 1.6 m (5.2 ft)
Hip height: 1.4 m (4.6 ft)
Body mass: 365 kg (805 lb)
Reconstruction: ▢▢▢▢

| 250 | 245 | 240 | 235 | 230 | 225 | 220 | 215 | 210 | 205 | 200 | 195 | 190 | 185 | 180 | 175 | 170 | 165 | 160 | 155 | 150 | 145 | 140 | 135 | 130 | 125 | 120 | 115 | 110 | 105 | 100 | 95 | 90 | 85 | 80 | 75 | 70 | 65 |

| TRIASSIC | JURASSIC | CRETACEOUS |

Efraasia minor (meaning "Eberhard Fraas's smaller one") is among the earliest discovered multimeter sauropodomorphs.

As is the case with many dinosaur species that were first described in the early 1900s, the naming history of various *Efraasia* specimens has been complex and fraught with changes. Specimens collected between 1907 and 1909 were once assigned to either *Teratosaurus*, *Sellosaurus*, *Thecodontosaurus*, or *Paleosaurus*. The name *Efraasia* was coined in 1973 (Galton, 1973), although for a time this name was considered to be a junior synonym for *Sellosaurus* (Galton and Bakker, 1985). In 2003, a reexamination of various early sauropodomorph specimens from Germany concluded that a number of these specimens were most properly assigned to *Efraasia minor* (Yates, 2003). Taken together, these remains represent the majority of the animal's skeleton, with only some vertebral sections and skull portions being unaccounted for (Bronzati and Rauhut, 2018).

Older sources state that the length of *Efraasia* would have been somewhere between 2 and 3 meters in total. However, since the majority of known specimens are actually of juveniles, these measurements are likely substantially smaller than the animal's maximum size. Based on one fossilized femur, the adult size of *Efraasia* is now estimated to be approximately 6.5 meters in length (Yates, 2003).

There is debate over just how bipedal or quadrupedal *Efraasia* would have been. Some research has shown that the animal's wrists would have been able to support its weight (Palmer, 1999), while other research has shown that the animal's forearms would not have been able to rotate sufficiently enough for its hands to have rested on the ground. Its fingers were fairly long and thin, and it is hypothesized that *Efraasia* could easily use its hands to grab and manipulate foliage.

The generic name *Efraasia* refers to paleontologist Eberhard Fraas, who discovered the original sets of remains. The specific name *minor* is a holdover from when the holotype was originally known as *Teratosaurus minor*, so named because of its smaller size in comparison with the already described *Teratosaurus suevicus*.

CLASSIFICATION

Dinosauria
 Saurischia
  Sauropodomorpha

LOCATION

Germany

KNOWN REMAINS

Mostly complete

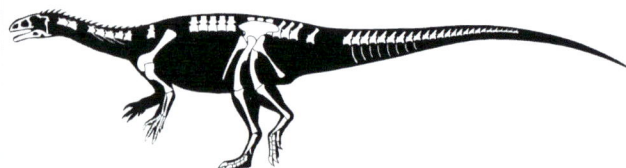

1 m 2 m 3 m 4 m

2 m

1 m

J. asymmetrica

(Novas et al., 2010)
Length: 4.5 m (14.8 ft)
Height: 2 m (6.6 ft)
Hip height: 1.2 m (3.9 ft)
Body mass: 240 kg (529 lb)
Reconstruction:

| 250 | 245 | 240 | 235 | 230 | 225 | 220 | 215 | 210 | 203 | 200 | 195 | 190 | 185 | 180 | 175 | 170 | 165 | 160 | 155 | 150 | 145 | 140 | 135 | 130 | 125 | 120 | 115 | 110 | 105 | 100 | 95 | 90 | 85 | 80 | 75 | 70 | 65 |

| TRIASSIC | JURASSIC | CRETACEOUS |

Jaklapallisaurus asymmetrica (meaning "asymmetrical lizard from Jaklapalli") is a little-known, midsized early sauropodomorph.

Jaklapallisaurus is known from only very sparse remains. The holotype (ISI R274) consists of only two partial vertebrae and portions of a leg and foot. The bones of the animal's foot were noted as being relatively robust in relation to most other early sauropodomorphs, which is in line with its increased size.

The lack of skeletal remains means that few conclusions can be drawn about the appearance or habits of *Jaklapallisaurus*. Based on the characteristics of the closely related *Unaysaurus* and *Plateosaurus*, *Jaklapallisaurus* was likely herbivorous and bipedal, with a moderately elongated neck.

The tentative phylogenetic conclusion reached by the original analysis of *Jaklapallisaurus* was that the animal was the most basal member of Plateosauridae, along with *Unaysaurus*. A newer analysis, though, has proposed a slightly different phylogenetic position for *Jaklapallisaurus*, within a new group dubbed **Unaysauridae**. This new distinction was made possible by the more recent description of *Macrocollum* (Müller et al., 2018a).

A subsequent study has questioned the validity of the criteria that unite the members of Unaysauridae (McPhee et al., 2020). Otherwise, the sparse condition of the skeletal remains has precluded *Jaklapallisaurus* from being a part of any other phylogenetic study.

The generic name *Jaklapallisaurus* refers to the town of Jaklapalli, India, near which the fossils were discovered. The specific name *asymmetrica* refers to the animal's highly asymmetrical astragalus (ankle bone) when viewed distally.

CLASSIFICATION

Dinosauria
  Saurischia
   Sauropodomorpha
    Plateosauria
     Unaysauridae

LOCATION

India

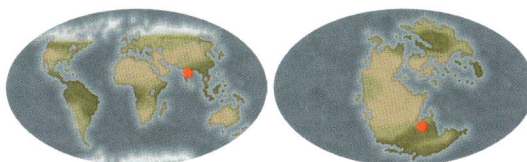

KNOWN REMAINS

Fragments

1 m

U. tolentinoi

(Leal et al., 2004)
Length: 2.9 m (9.5 ft)
Height: 1 m (3.3 ft)
Hip height: 0.8 m (2.6 ft)
Body mass: 65 kg (143 lb)
Reconstruction: ☐☐☐☐

| 250 | 245 | 240 | 235 | 230 | 225 | 220 | 215 | 210 | 205 | 200 | 195 | 190 | 185 | 180 | 175 | 170 | 165 | 160 | 155 | 150 | 145 | 140 | 135 | 130 | 125 | 120 | 115 | 110 | 105 | 100 | 95 | 90 | 85 | 80 | 75 | 70 | 65 |

TRIASSIC   **JURASSIC**   **CRETACEOUS**

Unaysaurus tolentinoi (meaning "Tolentino's black-water lizard") is an early sauropodomorph known from a single specimen (UFSM 11069). This set of remains includes a nearly complete skull. At the time of the animal's description, this was the most complete dinosaur skull ever found in Brazil. Still, the skull was disarticulated, and many of the fragments were splintered and fragile. The skull held leaf-shaped, overlapping teeth, which indicates that *Unaysaurus* was primarily herbivorous. The teeth show that herbivory, as a trait, was already well developed among the very early sauropodomorphs, indicating that this evolutionary development was wildly successful.

Upon its original description, *Unaysaurus* was thought to be most closely related to the famous *Plateosaurus*, which lived 15 million years later, on the other side of Pangaea. There do indeed exist many similarities between *Unaysaurus* and *Plateosaurus*, showing that early sauropodomorphs

quickly spread far and wide across Pangaea (McPhee et al., 2020).

A newer analysis, though, has proposed a slightly more basal phylogenetic position for *Unaysaurus*, within a new group dubbed **Unaysauridae**. This new distinction was made possible by the more recent descriptions of *Macrocollum* and *Jaklapallisaurus* (Müller et al., 2018a). A subsequent study, though, questioned the validity of the criteria that were used to unite the members of Unaysauridae (McPhee et al., 2020).

The generic name *Unaysaurus* is derived from "unay" (u-na-hee), meaning "black water" in the indigenous Tupi language; black water refers to the region where the fossil was discovered, which is known as "Agua Negra." The specific name *tolentinoi* honors the fossil's discoverer, Tolentino Marafiga, who found the specimen during the process of constructing a new road in 1998.

CLASSIFICATION
Dinosauria
 Saurischia
  Sauropodomorpha
   Plateosauria
    Unaysauridae

LOCATION
Brazil

KNOWN REMAINS
Partial skull and skeleton

1 m   2 m   3 m

1 m

M. itaquii

(Müller et al., 2018a)
Length: 3.4 m (11.2 ft)
Height: 1.1 m (3.6 ft)
Hip height: 0.8 m (2.6 ft)
Body mass: 95 kg (209 lb)
Reconstruction: ☐ ☐ ☐ ☐

| 250 | 245 | 240 | 235 | 230 | 225 | 220 | 215 | 210 | 205 | 200 | 195 | 190 | 185 | 180 | 175 | 170 | 165 | 160 | 155 | 150 | 145 | 140 | 135 | 130 | 125 | 120 | 115 | 110 | 105 | 100 | 95 | 90 | 85 | 80 | 75 | 70 | 65 |

TRIASSIC   **JURASSIC**   **CRETACEOUS**

Macrocollum itaquii (meaning "Itaqui's long-necked one") is the oldest known example of a long-necked sauropodomorph. Known from three relatively complete specimens that were found in close association, this grouping seems to suggest that *Macrocollum* spent at least some of its time in groups. This is the earliest such evidence of gregarious dinosaurian behavior.

Macrocollum lived only eight million years after the earliest known sauropodomorphs, such as *Buriolestes*, and yet it had already evolved significant traits that would later become typical of sauropods in general. *Macrocollum* was much larger than *Buriolestes*, with a femur that was 230% longer; moreover, the length of its neck had increased significantly, being proportionally twice as long as that of *Buriolestes*. Some of its neck vertebrae were four times as long as they were tall. This elongation of the neck would have been made possible in part due to the reduced size of the animal's skull, which is less than half the length of its femur—a much smaller proportion than its earlier relatives.

Unlike its recent carnivorous ancestors, *Macrocollum* was primarily herbivorous. Its overlapping teeth and coarse tooth serrations would have been well adapted for processing vegetation. With these traits, combined with its long neck, it seems clear that high-grazing herbivory was rapidly becoming a winning strategy for sauropodomorphs.

Macrocollum was also developing hindlimb traits in support of its increased size and slower lifestyle, having a femur that was both elongated and straight. However, its feet were still rather long and gracile.

The generic name *Macrocollum* combines the Greek "macro" (meaning "large" or "long") and the Latin "collum" (meaning "neck"). The specific name *itaquii* honors José Jerundino Machado Itaqui, a founder of the paleontological research center at UFSM (Universidade Federal de Santa Maria).

CLASSIFICATION
Dinosauria
 Saurischia
  Sauropodomorpha
   Plateosauria
    Unaysauridae

LOCATION
Brazil

KNOWN REMAINS
Complete

1 m   2 m   3 m

1 m

G. candelariensis

(Bonaparte et al., 1999)
Length: 3 m (9.8 ft)
Height: 1.1 m (3.6 ft)
Hip height: 0.7 m (2.3 ft)
Body mass: 35 kg (77 lb)
Reconstruction:

| 250 | 245 | 240 | 235 | 230 | **225** | 220 | 215 | 210 | 205 | 200 | 195 | 190 | 185 | 180 | 175 | 170 | 165 | 160 | 155 | 150 | 145 | 140 | 135 | 130 | 125 | 120 | 115 | 110 | 105 | 100 | 95 | 90 | 85 | 80 | 75 | 70 | 65 |

| TRIASSIC | JURASSIC | CRETACEOUS |

Guaibasaurus candelariensis (meaning "lizard from Candelária, Guaíba") is another example of an early saurischian that seems to blur the lines that seek to separate the theropods from the sauropods.

The phylogenetic placement of *Guaibasaurus* has been a contentious topic, and clear conclusions have not yet been reached. In the original analysis, *Guaibasaurus* was suggested to be a basal theropod, and this notion has again been brought up by some subsequent studies (Langer et al., 2019). Other analyses, though, have instead found *Guaibasaurus* to be a basal sauropodomorph (Ezcurra, 2010; Griffin et al., 2022). Still others have recovered it as a stem saurischian (Cabreira et al., 2016) or as a sauropodomorph in a more derived position (Müller and Garcia, 2020).

Guaibasaurus is known from several sets of remains that, when taken together, preserve nearly the entire animal, with the exception of the head and neck. Thus, the overall length of the animal is quite uncertain. The illustration seen above interprets the animal as a slightly derived sauropodomorph; other reconstructions have shown a shorter-necked, more theropod-like animal.

The sets of fossil remains were unearthed from various sites within the Caturrita geological formation of Brazil. These strata have been found to contain numerous reptilian and early mammalian remains. Another sauropodomorph, *Unaysaurus*, was also discovered here. During the early Late Triassic, the region was a humid floodplain that was dominated by braided river systems.

The generic name *Guaibasaurus* refers to the Rio Guaíba area in which the holotype was recovered. Similarly, the specific name *candelariensis* refers to the nearby city of Candelária.

CLASSIFICATION
Dinosauria
  Saurischia
    Sauropodomorpha (?)

LOCATION
Brazil

KNOWN REMAINS
Nearly complete skeleton

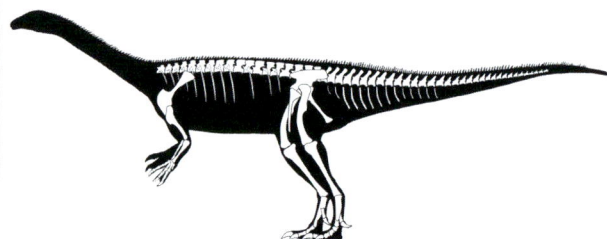

I. saaneq

(Beccari et al., 2021)
Length: 4.7 m (15.4 ft)
Height: 1.6 m (5.2 ft)
Hip height: 1.2 m (3.9 ft)
Body mass: 260 kg (573 lb)
Reconstruction: ☐☐☐☐

| 250 | 245 | 240 | 235 | 230 | 225 | 220 | **214** | 210 | 205 | 200 | 195 | 190 | 185 | 180 | 175 | 170 | 165 | 160 | 155 | 150 | 145 | 140 | 135 | 130 | 125 | 120 | 115 | 110 | 105 | 100 | 95 | 90 | 85 | 80 | 75 | 70 | 65 |

TRIASSIC | **JURASSIC** | **CRETACEOUS**

Issi saaneq (meaning "cold bones") is the northernmost known Late Triassic sauropodomorph, and the first new dinosaur genus to be described from Greenland.

Issi is known from two preserved skulls, NHMD 164741 (the holotype), along with NHMD 164758. The former, belonging to a subadult, was first discovered in 1991 and was identified in 1994 as belonging to *Plateosaurus* (Novas et al., 1994). The latter specimen, belonging to a juvenile, was discovered in 1995 from the same locality but would remain undescribed until 2021. Both specimens were reasonably complete and mostly articulated, although they had been compressed. (Several skeletons, likely belonging to the same species, were recovered in 2012 but thus far have remained unprepared or undescribed.)

Doubts regarding the first specimen's *Plateosaurus* designation were first raised in 2018 (Marzola et al., 2018). When CAT scans were utilized by researchers to digitally reconstruct both of the specimens, it became clear that the animal in question was distinct and separate from *Plateosaurus* and that these differences were not due to intraspecies variation or from deformations of the fossils. Indeed, some features of the bones more closely resembled the Brazilian sauropodomorphs *Unaysaurus* and *Macrocollum*, as opposed to the European *Plateosaurus*.

Phylogenetically, *Issi* was found to be the sister genus of *Plateosaurus*, although the authors noted that the relationships among **Plateosauridae** (and early sauropodomorphs in general) are poorly understood because of numerous unresolved issues and that their classifications are subject to change.

The generic name *Issi* means "cold" in the Inuit language Kalaallisut (also known as West Greenlandic), and the specific name *saaneq* means "bone"; the excavators experienced cold and unpleasant conditions while working at the fossil site.

CLASSIFICATION

Dinosauria
  Saurischia
    Sauropodomorpha
      Plateosauria
        Plateosauridae

LOCATION

Greenland

KNOWN REMAINS

Skull and jaw

P. trossingensis

(von Meyer, 1837)
Length: 9.4 m (30.8 ft)
Height: 3.5 m (11.5 ft)
Hip height: 2 m (6.6 ft)
Body mass: 2,600 kg (2.9 t)
Reconstruction: ☐☐☐☐

TRIASSIC	JURASSIC	CRETACEOUS

Plateosaurus (meaning "broad lizard") is among the most prolific of all known dinosaur genera. Known from hundreds of specimens discovered at numerous sites across central Europe, *Plateosaurus* seems to have been a very successful and adaptable creature.

The first recognized *Plateosaurus* remains were unearthed in 1834 and were named by Hermann von Meyer in 1837. Numerous fossil specimens were lost or destroyed during World War II.

The size of adult *Plateosaurus* specimens varies widely. Fully grown individuals range anywhere from 4.8 meters in length to nearly 10 meters. This degree of variability is quite unusual for dinosaur species.

Plateosaurus has often been depicted as a partially quadrupedal animal, one which could spend time moving on either two or four limbs. However, biomechanical studies of the animal's forelimbs have shown that *Plateosaurus* was physically incapable of placing its hands flat on the ground, thus showing that sauropodomorphs of this time were obligate bipeds (Bonnan and Senter, 2007).

As is often the case with early described dinosaurs, the history of the classification and naming of *Plateosaurus* is a daunting saga. There exist many synonymous genus names, such as *Sellosaurus* and *Pachysaurus*. For a long while, the type species was considered *P. engelhardti*, until the species' type specimen was determined to be undiagnostic (Galton, 2012). Currently, three species are most often considered valid: *P. trossingensis* (type), *P. gracilis*, and *P. longiceps*.

The proper interpretation of the name *Plateosaurus* is ambiguous, as the animal's original description did not clarify its meaning; one prominent interpretation is that it combines the Greek terms "plata" (meaning "broad") and "sauros" (meaning "lizard").

CLASSIFICATION

Dinosauria
  Saurischia
    Sauropodomorpha
      Plateosauria
        Plateosauridae

LOCATION

Germany; Switzerland; France

KNOWN REMAINS

Complete

2 m   4 m   6 m   8 m   4 m

2 m

Y. youngi

(Bai et al., 1990)
Length: 8.9 m (29 ft)
Height: 2.6 m (8.5 ft)
Hip height: 2.2 m (7.2 ft)
Body mass: 1,800 kg (2 t)
Reconstruction: ☐☐☐☐

TRIASSIC   **JURASSIC**   **CRETACEOUS**

Yimenosaurus youngi (meaning "Young's lizard from Yimen") is an understudied Jurassic sauropodomorph that has been compared to *Lufengosaurus* and *Plateosaurus*. Like in these animals, the teeth of *Yimenosaurus* were long and spoon shaped.

The remains of *Yimenosaurus* were unearthed from the Fengjiahe Formation of China, a locality that has yielded several impressive dinosaurian specimens. In 1987, numerous disarticulated specimens were located embedded within silty mudstone that was exposed to the surface. Many elements had been exposed to intense weathering and erosion, damaging the remains to the point that they could not be excavated without causing further breakages. Additionally, the fossils had been subject to some compressional distortion forces.

Because of these factors, only two of the best-preserved individuals were described in the research that introduced

Yimenosaurus. Taken together, they represent a relatively complete picture of the animal, with the exception of the forelimbs. (Only the skull and jaw elements were actually figured in the article.)

The original phylogenetic analysis of *Yimenosaurus* placed it within **Plateosauridae**, although other taxonomic aspects from this 1990 study have not aged well. To date, *Yimenosaurus* has not been included in any modern cladistic analyses; one set of researchers was apparently even denied permission to reexamine the bones (Peyre de Fabrègues et al., 2020). Thus, its actual position on the sauropodomorph family tree currently remains uncertain.

The generic name *Yimenosaurus* refers to the Chinese county of Yimen, where the animal's fossils were first discovered. The specific name *youngi* honors paleontologist Yang Zhongjian, also known as C. C. Young.

CLASSIFICATION

Dinosauria
  Saurischia
    Sauropodomorpha
      Plateosauria
        Plateosauridae (?)

LOCATION

China

KNOWN REMAINS

Skull and partial skeleton

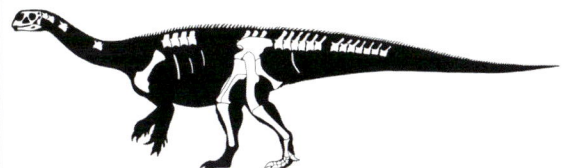

A. pereirabdalorum

(Yates et al., 2011)
Length: 2.3 m (7.5 ft)
Height: 1 m (3.3 ft)
Reconstruction: ☐☐☐☐

190

| TRIASSIC | JURASSIC | CRETACEOUS |

Arcusaurus pereirabdalorum (meaning "Pereira's and Abdala's rainbow lizard") is an enigmatic sauropodomorph, one which is difficult to classify accurately because of the incompleteness of its known skeletal remains. With only some portions of the skull and a few other fragments, there is little information upon which to base a phylogenetic analysis. Further confounding the issue is that the individual was a juvenile at the time of its death—immature skeletal remains tend to skew phylogenetic results in the basal direction. According to the describers, *Arcusaurus* is most likely a late-surviving example of a basal species, although it does also share traits with more contemporary *Plateosaurus*.

The generic name *Arcusaurus* combines the Latin "arcus" (meaning "rainbow") and the Greek "sauros" (meaning "lizard"), alluding to South Africa being known as the "rainbow nation." The specific name *pereirabdalorum* honors fossil discoverers Lucille Pereira and Fernando Abdala.

N. roychowdhurii

(Novas et al., 2010)
Length: 3.7 m (12.1 ft)
Height: 1.2 m (3.9 ft)
Reconstruction: ☐☐☐☐

205

| TRIASSIC | JURASSIC | CRETACEOUS |

Nambalia roychowdhurii (meaning "Roy Chowdhuri's one from Nambal") is a little-studied genus that has thus far been closely examined only by the original study that described it. In this study, it was found to be a very basal sauropodomorph, closely related to *Thecodontosaurus* and *Efraasia*.

Nambalia is known from fossil elements that were gathered from a small erosion slope, representing the remains of at least three different individuals, judging by overlapping elements including foot bones. Its hands have been described as gracile and similar to those of *Herrerasaurus* and *Guaibasaurus*. The femur is discernible from that of *Alwalkeria*, one of the only other Triassic dinosaurs known from India.

The generic name *Nambalia* refers to the town of Nambal, India, near where the fossils were located. The specific name *roychowdhurii* honors paleontologist Roy Chowdhuri.

X. suni

(Sekiya, 2010)

Length: 4.3 m (14.1 ft)
Height: 1.3 m (4.3 ft)
Reconstruction: ▢▢▢▢

190

| TRIASSIC | JURASSIC | CRETACEOUS |

Xixiposaurus suni (meaning "Sun's lizard from Xixipo") is a little-studied sauropodomorph. It was briefly described in a primarily Chinese-language journal, *Global Geology*, and the holotype specimen (ZLJ 01018, consisting of a skull, jaw, and partial skeleton) has not been examined by any subsequent works or included in phylogenetic analyses. This original paper concluded that *Xixiposaurus* was the sister taxon of *Mussaurus*, both of which were considered to lie within Plateosauridae. The authors considered *Xixiposaurus* to be the "most derived taxon among Chinese prosauropod dinosaurs."

The generic name *Xixiposaurus* refers to the village of Xixipo in Lufeng County, China, where the holotype was discovered. The specific name *suni* honors Sun Ge, professor at Jilin University.

R. bedheimensis

(Galton, 2001)

Length: 7.6 m (25 ft)
Height: 3.5 m (11.5 ft)
Reconstruction: ▢▢▢▢

210

| TRIASSIC | JURASSIC | CRETACEOUS |

Ruehleia bedheimensis (meaning "Rühle's one from Bedheim"), a German sauropodomorph, was briefly described in the journal *Revue de Paléobiologie* in an addendum to an article about *Plateosaurus*. Differentiating the animal from *Plateosaurus*, which has 15 dorsal (back) vertebrae, *Ruehleia* has only 14. Additionally, the animal's sacrum (collection of fused hip vertebrae) is "dorsosacral" rather than "caudosacral," as seen in *Plateosaurus*.

According to *The Dinosauria* (Weishampel et al., 2004), the known remains of *Ruehleia* include one "nearly complete skeleton" as well as "2 incomplete skeletons, juvenile to adult." The holotype was originally referred to as an "unnumbered" specimen from the Berlin Museum of Nature; a later paper would refer to the remains as MB.R.4718–42 (Otero, 2018). The animal was named for the German paleontologist Hugo Rühle von Lilienstern of Bedheim.

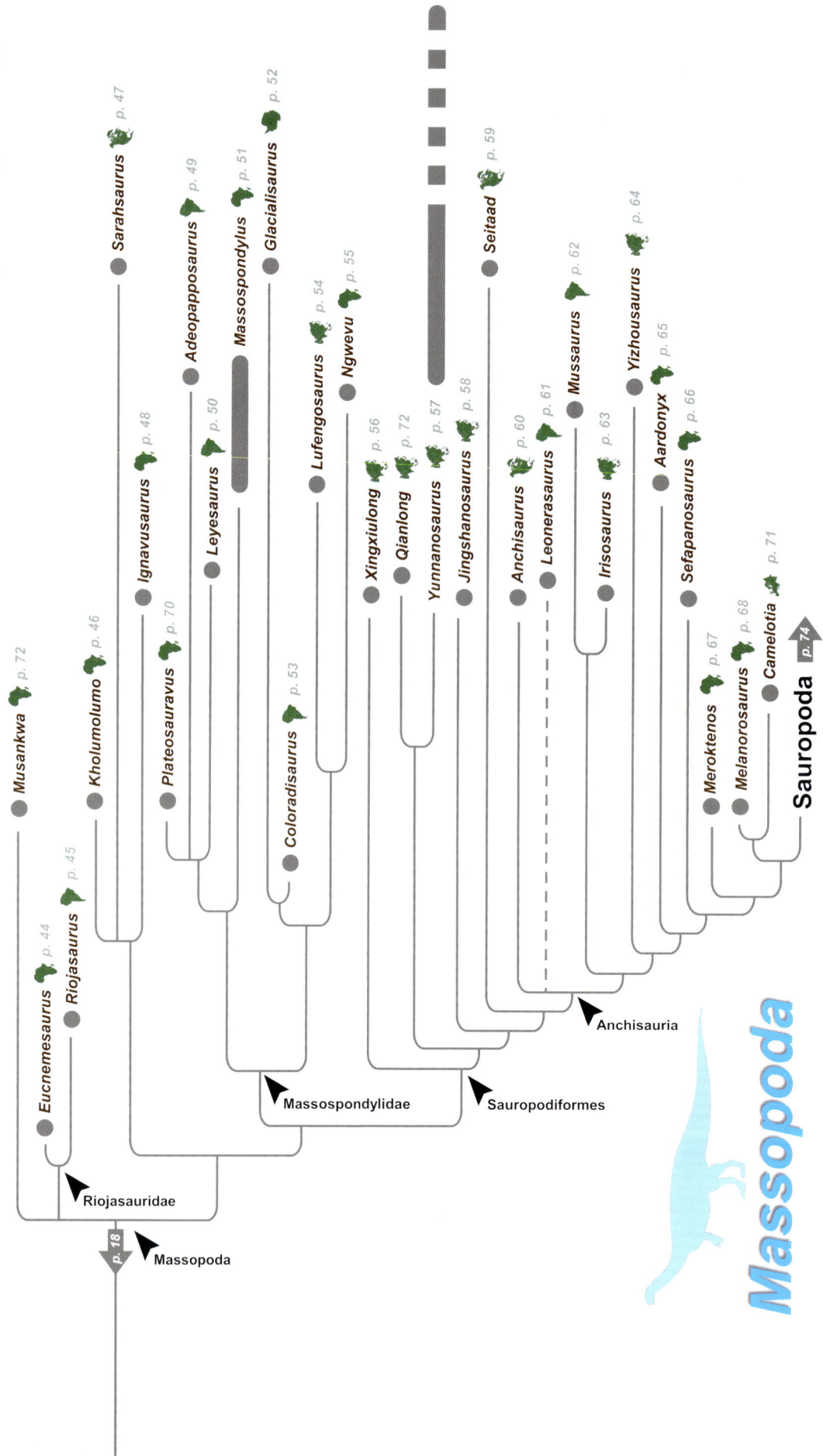

The name Massopoda comes from the Latin "massa" (meaning "lump") and from the Greek "podi" (meaning "foot"). It is also a contraction of the names Massospondylidae (a family within the group) and Sauropoda.

The massopods include many animals that once fell within the umbrella of the "prosauropod" group. In the past, it was thought that the various genera within Plateosauridae, Riojasauridae, and Massospondylidae were all members of a monophyletic group—in other words, that they shared their own singular branch on the family tree, known as Prosauropoda. However, this view began falling out of favor in the 2000s, as it became clear that these various families were paraphyletic with respect to sauropods, branching off the family tree at numerous points along the way.

There is a high degree of variability regarding the exact placements of most species near the **Massopoda** node on the sauropodomorph family tree. One common thread, though, places *Eucnemesaurus* and *Riojasaurus* as sister taxa within their own group, **Riojasauridae**, at the very base of Massopoda (Wang et al., 2017b; Rauhut et al., 2020; Fernández and Werneburg, 2022), although they have been placed even more basally (Beccari et al., 2021) or in a more derived position (Peyre de Fabrègues et al., 2020; Zhang et al., 2020). The describers of *Musankwa* placed it as the basalmost massopod (Barrett et al., 2024).

Sarahsaurus has variably been placed basal to Massospondylidae (Peyre de Fabrègues and Allain, 2020; Rauhut et al., 2020; Fernández and Werneburg, 2022), within that family (Chapelle et al., 2019), or more derived than it (Beccari et al., 2021). The description of *Kholumolumo* placed it as the sister taxon to *Sarahsaurus* (Peyre de Fabrègues and Allain, 2020). *Ignavusaurus* has not been included in as many analyses, but those that have included it tend to find it to be very closely related to *Sarahsaurus* as well (Apaldetti et al., 2011; Chapelle et al., 2019). **Massospondylidae** is currently accepted as being its own family and thus not being directly ancestral to the true sauropods; which species belong in the group, though, is unsurprisingly ambiguous. In addition to *Massospondylus*, *Adeopapposaurus* and *Leyesaurus* are regarded as definite members and as sister taxa (Peyre de Fabrègues et al., 2020; Zhang et al., 2020).

Additionally, *Glacialisaurus*, *Coloradisaurus*, and *Lufengosaurus* are most often grouped into their own clade within Massospondylidae (Rauhut et al., 2020; Beccari et al., 2021; Fernández and Werneburg, 2022). The description of the newer genus *Ngwevu* placed it as the sister taxa to *Lufengosaurus* (Chapelle et al., 2019). Although *Plateosauravus* has been depicted in almost any position imaginable among the early sauropodomorphs, the most recent comprehensive study on the genus identified it as a massospondylid (Krupandan, 2019).

Yunnanosaurus, *Jingshanosaurus*, and *Seitaad* have sometimes been placed as closely related genera within Massospondylidae (Rauhut et al., 2020; Beccari et al., 2021) but are more commonly shown (in various configurations) as being among the basalmost **Sauropodiformes** (Wang et al., 2017b; Zhang et al., 2020; Fernández and Werneburg, 2022). *Xingxiulong* has, at times, been placed as the very basalmost sauropodiform (Peyre de Fabrègues et al., 2020; Rauhut et al., 2020; Zhang et al., 2020), although it has also been placed as a massopod (Peyre de Fabrègues and Allain, 2020) and basally to the massospondylids (Fernández and Werneburg, 2022). *Qianlong* was placed by its describers as sister to *Yunnanosaurus* (Han et al., 2024).

Some analyses have placed *Anchisaurus* (and thus **Anchisauria**) basal to (and thus including) Massospondylidae (Peyre de Fabrègues et al., 2020; Zhang et al., 2020); however, this is not generally favored. *Mussaurus* and *Aardonyx* are often placed in the subsequent derived positions, along with *Sefapanosaurus* (Otero et al., 2015; Rauhut et al., 2020; Fernández and Werneburg, 2022). *Yizhousaurus* has been similarly placed by its describers (Zhang et al., 2020), and the describers of *Irisosaurus* placed it as the sister taxon of *Mussaurus* (Peyre de Fabrègues et al., 2020). Placement of the fragmentary *Leonerasaurus* is quite inconsistent between studies.

Hovering on the boundary of Sauropoda are *Melanorosaurus* and the closely related *Meroktenos*, along with the enigmatic *Camelotia* (Apaldetti et al., 2018; Chapelle et al., 2019; Fernández and Werneburg, 2022).

Massospondylus carinatus

2 m   4 m   6 m   8 m

2 m

E. fortis

(van Hoepen, 1920)
Length: 7.8 m (25.6 ft)
Height: 2.7 m (8.9 ft)
Hip height: 2 m (6.6 ft)
Body mass: 1,200 kg (1.3 t)
Reconstruction: ☐☐☐☐

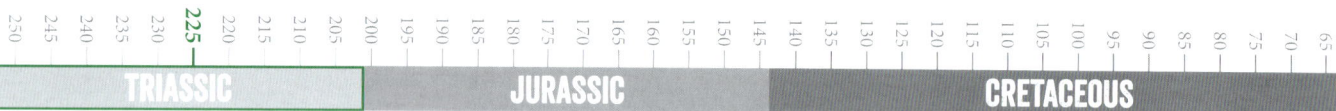

250 245 240 235 230 **225** 220 215 210 205 200 195 190 185 180 175 170 165 160 155 150 145 140 135 130 125 120 115 110 105 100 95 90 85 80 75 70 65

TRIASSIC	JURASSIC	CRETACEOUS

Eucnemesaurus fortis (meaning "strong good tibia lizard") is a long-dormant genus that was recently revitalized.

Eucnemesaurus was originally described in 1920 by Egbert van Hoepen based on the specimen TrM 119, consisting of fragmentary vertebrae, leg bones, and hip bones. After that time, the genus was largely forgotten.

Starting in the 1860s, collections of fossils—mostly sauropodomorph bones—were sent by Alfred Brown to several European institutions. A distinctive femur (NMW 1889-XV-39) was identified by Friedrich von Huene in 1906 as belonging to the now-dubious "prosauropod" *Euskelosaurus browni*, although this was not to last. In 1985, the femur, along with a carnivorous upper-jaw bone, was reinterpreted by Peter Galton as the remains of a herrerasaurid-type theropod. Together, these fossils formed the holotype of the now-debunked chimera *Aliwalia rex*,

which, owing to the large size of the femur, was interpreted as a gargantuan *Herrerasaurus*-like carnivore.

Finally, in 2003, a new femur (BP/1/6111) was unearthed among other sauropodomorph remains, clarifying the nature of the aforementioned specimens: *Eucnemesaurus* was reestablished as a valid genus while *Aliwalia rex* was invalidated (Yates, 2007). This analysis also established that the genus was a close relation of *Riojasaurus*, which together form the family **Riojasauridae**.

In 2015, a new articulated set of remains (BP/1/6234) was identified as a second species, within the genus, *E. entaxonis* (McPhee et al., 2015).

The generic name *Eucnemesaurus* combines the Greek "eu" (meaning "good" or "true"), "kneme" (meaning "tibia"), and "sauros" (meaning "lizard"). The specific name *fortis* is Latin for "strong"; "entaxonic" is an anatomical term referring to the weight-bearing nature of the foot bones.

CLASSIFICATION
Dinosauria
 Sauropodomorpha
  Bagualosauria
   Plateosauria
    Massopoda
     Riojasauridae

LOCATION
South Africa

KNOWN REMAINS
Leg, hip, and fragments

2 m   4 m   6 m

2 m

R. incertus

(Bonaparte, 1969)
Length: 6.8 m (22.3 ft)
Height: 2.3 m (7.5 ft)
Hip height: 1.8 m (5.9 ft)
Body mass: 800 kg (1,760 lb)
Reconstruction: ☐☐☐☐

250 245 240 235 230 225 **220** 215 210 205 200 195 190 185 180 175 170 165 160 155 150 145 140 135 130 125 120 115 110 105 100 95 90 85 80 75 70 65

TRIASSIC   **JURASSIC**   **CRETACEOUS**

Riojasaurus incertus (meaning "uncertain lizard from Rioja") was quite large for an early sauropodomorph; it also possessed dense leg bones, contrasted by its partially hollow, lightened vertebrae. This combination of traits could potentially indicate that *Riojasaurus* was something of a transitional form between the early, bipedal sauropodomorphs and the later, quadrupedal sauropods.

Whether or not *Riojasaurus* could actually walk on all four legs is a matter that is open for debate. On the one hand, the animal's forelimbs are longer, proportionately, than those of more basal sauropodomorphs, making it easier for those appendages to touch the ground. *Riojasaurus* also had four sacral (hip) vertebrae, similar to later sauropods and unlike the three sacral vertebrae found in bipedal sauropodomorphs. However, researcher Scott Hartman has pointed out aspects of the animal's shoulders and spine that disfavor the quadrupedal interpretation, and no detailed analyses of these options have been carried out in the last two decades.

Some sources from the twentieth century suggested that *Riojasaurus* was most closely related to *Melanorosaurus*, but most modern analyses instead place *Riojasaurus* as one of the basalmost massopods (Müller et al., 2018), with one pushing its position even further back than that (Beccari et al., 2021). *Riojasaurus* is now thought to be most closely related to *Eucnemesaurus*, with these two genera being the only members of the family **Riojasauridae**.

A study of *Riojasaurus*'s scleral eye rings suggests that it was active at both day and night (Schmitz and Motani, 2011).

The generic name *Riojasaurus* refers to La Rioja Province in Argentina. The specific name *incertus* is Latin for "uncertain." The genus *Strenusaurus* (meaning "vigorous lizard") (Bonaparte, 1969) has been synonymized with *Riojasaurus* (Galton, 1985).

CLASSIFICATION
Dinosauria
 Sauropodomorpha
  Bagualosauria
   Plateosauria
    Massopoda
     Riojasauridae

LOCATION
Argentina

KNOWN REMAINS
Nearly complete

2 m   4 m   6 m   8 m   10 m   12 m

4 m

2 m

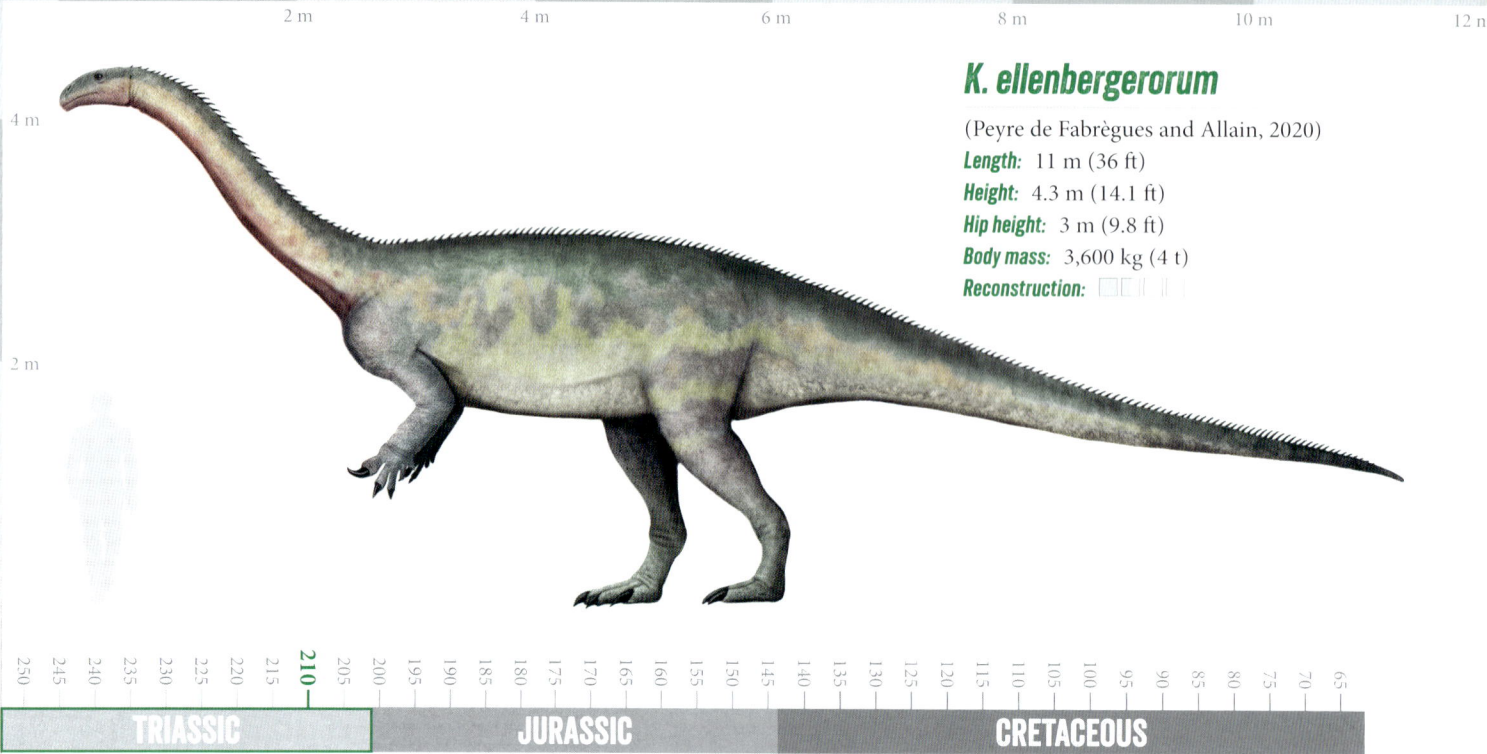

K. ellenbergerorum

(Peyre de Fabrègues and Allain, 2020)

Length: 11 m (36 ft)
Height: 4.3 m (14.1 ft)
Hip height: 3 m (9.8 ft)
Body mass: 3,600 kg (4 t)
Reconstruction: ☐☐❙❙

250 245 240 235 230 225 220 215 **210** 205 200 195 190 185 180 175 170 165 160 155 150 145 140 135 130 125 120 115 110 105 100 95 90 85 80 75 70 65

TRIASSIC   **JURASSIC**   **CRETACEOUS**

Kholumolumo ellenbergerorum (meaning "Ellenbergers' dragon") is among the largest animals known to have lived during the Late Triassic. Despite its size, it was clearly a bipedal creature, not a quadruped.

Between the years of 1955 and 1970, a large quantity of dinosaur bones and trackways were discovered in Lesotho, in southern Africa. These discoveries were made adjacent to a large trash pile, known as a "thotobolo" in the indigenous Sotho language. A number of the fossils were initially attributed to *Euskelosaurus browni*; in 1970, the name "*Thotobolosaurus mabeatae*" was mentioned in literature as a suggested name but was never formally published. Similarly, an unpublished doctoral dissertation referred to the specimen by the name "*Kholumolumosaurus ellenbergerorum*" (Gauffre, 1996), but this too remained invalid. Finally, the total sum of the species' known remains (210 bones that

originate from at least five separate individuals) were described in 2020. Not every bone was thoroughly analyzed, though, and more material remains in museum collections.

The phylogenetic examination conducted on the *Kholumolumo* remains concluded that the animal was a member of the earliest-branching clade within **Massopoda**, sandwiched between Plateosauridae and Massospondylidae. The analysis showed that the creature's closest relatives (*Xingxiulong*, from China, and *Sarahsaurus*, from North America) were geographically widespread. This could suggest this particular lineage of sauropodomorphs originated in Gondwana before geographically dispersing.

The generic name *Kholumolumo* refers to a mythological dragon-like creature of indigenous Sotho folklore. The specific name *ellenbergerorum* honors paleontologists Paul and François Ellenberger.

CLASSIFICATION
Dinosauria
  Sauropodomorpha
    Bagualosauria
      Plateosauria
        Massopoda

LOCATION
Lesotho

KNOWN REMAINS
Partial limbs, other fragments

1 m 2 m 3 m 4 m

2 m

1 m

S. aurifontanalis

(Rowe et al., 2011)
Length: 4.3 m (14.1 ft)
Height: 1.9 m (6.2 ft)
Hip height: 1 m (3.3 ft)
Body mass: 190 kg (419 lb)
Reconstruction: ☐☐☐☐

250 245 240 235 230 225 220 215 210 205 200 195 190 185 180 175 170 165 160 155 150 145 140 135 130 125 120 115 110 105 100 95 90 85 80 75 70 65

TRIASSIC **JURASSIC** **CRETACEOUS**

Sarahsaurus aurifontanalis (meaning "Sarah's lizard from Gold Spring") is a possibly omnivorous North American sauropodomorph.

Sarahsaurus is known primarily from two specimens that were found at the same location (along with a juvenile *Dilophosaurus* and a handful of unidentified bones) in the 1970s. These remains include the majority of the skeleton but little skull material.

A third specimen, MCZ 8893 (consisting of a much more complete skull and jaw, but very little other skeletal material), was discovered in 1978 less than a kilometer away and was initially referred to *Massospondylus* (Attridge et al., 1985). Later works questioned this identification and referred to the specimen as the "undescribed Kayenta prosauropod" (Yates, 2003). When *Sarahsaurus* was first described, this specimen was referred to the new genus, based, in part, on the location of its discovery and that no characteristics that were shared between the specimens

could clearly differentiate them. That being said, only one of the numerous tested characteristics could be used to unambiguously show their shared identity. The skull was also from an individual that was less mature than the other two specimens.

The exact phylogenetic positioning of *Sarahsaurus* depends on whether or not this third specimen is included in the analysis, as well as with which other genera it is being compared. A number of tested variations favored the interpretation of *Sarahsaurus* being a member of Massospondylidae, although some results show it placed in a more basal position (Apaldetti et al., 2011; Marsh and Rowe, 2018; Peyre de Fabrègues and Allain, 2020).

The generic name *Sarahsaurus* honors philanthropist Sarah "Mrs. Ernest" Butler. The specific name *aurifontanalis* combines the Latin "aurum" (meaning "gold") and "fontinalis" (meaning "of the spring"), in reference to where the holotype was discovered: Gold Spring, Arizona.

CLASSIFICATION
Dinosauria
 Sauropodomorpha
  Bagualosauria
   Plateosauria
    Massopoda

LOCATION
Arizona, USA

KNOWN REMAINS
Nearly complete

47

MASSOPODA

1 m   2 m   3 m   4 m

1 m

I. rachelis

(Knoll, 2010)
Length: 4 m (13.1 ft)
Height: 1.4 m (4.6 ft)
Hip height: 1 m (3.3 ft)
Body mass: 190 kg (419 lb)
Reconstruction: ☐☐☐☐

250 245 240 235 230 225 220 215 210 205 200 195 190 185 180 175 170 165 160 155 150 145 140 135 130 125 120 115 110 105 100 95 90 85 80 75 70 65

| TRIASSIC | JURASSIC | CRETACEOUS |

Ignavusaurus rachelis (meaning "Raquel's coward lizard") is known primarily from a partial yet well-preserved and mostly articulated specimen. Although the skull and jaw were fragmented into more than 100 pieces, numerous preserved teeth have shown that *Ignavusaurus* was likely more of a generalist or opportunist, as opposed to engaging primarily in herbivory. It had teeth of various shapes, with some being more pointed and less serrated than others, and the teeth generally lacked any of the overlapping placement seen in other early sauropodomorphs.

The holotype specimen (BM HR 20) is that of a juvenile, with an estimated body length measuring only 1.5 meters. This age determination was made based on the internal features of both the femur and humerus, which revealed that the individual was no more than one year of age. As such, the full size of an adult *Ignavusaurus* can only be estimated.

The immature nature of the skeletal remains has complicated the matter of determining the phylogenetic placement of *Ignavusaurus*. The original description placed the animal quite basally, outside Massopoda, although a lack of certainty was emphatically expressed. Soon thereafter, another paper questioned the validity of the genus, suggesting that the specimen could actually be a juvenile *Massospondylus* (Yates et al., 2011). This notion, however, has not been universally accepted, with the two genera continuing to be considered separate and distinct by some studies. One more recent analysis placed the genus within Massospondylidae (Chapelle et al., 2019).

The generic name *Ignavusaurus* combines the Latin "ignavus" (meaning "coward") and the Greek "sauros" (meaning "lizard"); this refers to the name of the locality of the specimen's discovery, Ha Ralekoala, which literally translates as "the place of the father of the coward." The specific name *rachelis* honors paleontologist Raquel Lopez-Antonanzas.

CLASSIFICATION

Dinosauria
  Sauropodomorpha
    Bagualosauria
      Plateosauria
        Massopoda

LOCATION

Lesotho

KNOWN REMAINS

Partial skeleton (juvenile)

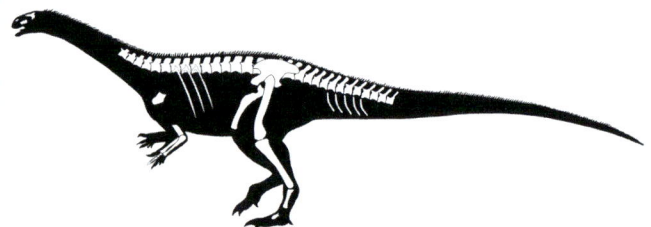

50 cm   100 cm   150 cm   200 cm   250 cm

100 cm

50 cm

A. mognai

(Martínez, 2009)
Length: 2.9 m (9.7 ft)
Height: 1.2 m (4 ft)
Hip height: 0.8 m (2.6 ft)
Body mass: 60 kg (132 lb)
Reconstruction: ☐☐☐☐

250 245 240 235 230 225 220 215 210 205 200 195 190 185 180 175 170 165 160 155 150 145 140 135 130 125 120 115 110 105 100 95 90 85 80 75 70 65

TRIASSIC | **JURASSIC** | **CRETACEOUS**

Adeopapposaurus mognai (meaning "far-eating lizard from Mogna") is based on several partial sets of remains that were initially speculated to represent *Massospondylus* (Martínez, 1999). In comparison with related species, the foremost dorsal vertebra of *Adeopapposaurus* has actually transitioned into being a cervical (neck) vertebra.

The researchers who initially described *Adeopapposaurus* proposed the idea that the animal actually had a bony, keratinous beak. This notion is based on a combination of distinctive traits, such as the animal's jaw being actually slightly shorter than its skull. The sides of the snout also host a pronounced "bony platform," and both the snout and jaw feature an increased number of openings meant for the passage of blood vessels and nerve connections.

Taken altogether, it would seem that *Adeopapposaurus* likely had some form of enhanced structure on its face.

Given that limited archosaurian facial musculature would eliminate the possibility of *Adeopapposaurus* having fleshy, horse-like lips, the next most likely conclusion would be a beak. This beak would have widened the animal's selection of potential food sources by giving it the ability to snip and prune tougher vegetation, perhaps compensating for the animal's smaller size in comparison with other sauropodomorphs present at the time.

Phylogenetic analyses tend to agree that *Adeopapposaurus* is a close relative of *Massospondylus* (Rauhut et al., 2020).

The generic name *Adeopapposaurus* combines the Latin "adeo" (meaning "far"), "pappo" (meaning "eating"), and the Greek "saurus" (meaning "lizard"); this is in reference to the animal's long neck. The specific name *mognai* refers to the location, Mogna, in San Juan Province, Argentina.

CLASSIFICATION
Dinosauria
 Sauropodomorpha
  Bagualosauria
   Plateosauria
    Massopoda
     Massospondylidae

LOCATION
Argentina

KNOWN REMAINS
Complete

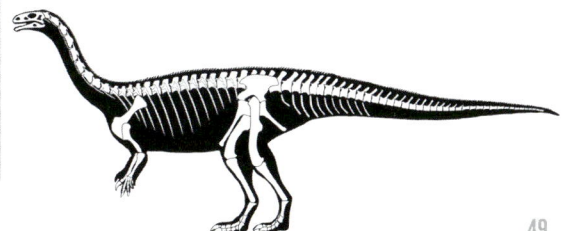

49

1 m   2 m   3 m

L. marayensis

(Apaldetti et al., 2011)

Length: 3.1 m (10.3 ft)
Height: 1.1 m (3.6 ft)
Hip height: 0.7 m (2.3 ft)
Body mass: 70 kg (154 lb)
Reconstruction: ☐☐☐☐

| 250 | 245 | 240 | 235 | 230 | 225 | 220 | 215 | 210 | 205 | 199 | 195 | 190 | 185 | 180 | 175 | 170 | 165 | 160 | 155 | 150 | 145 | 140 | 135 | 130 | 125 | 120 | 115 | 110 | 105 | 100 | 95 | 90 | 85 | 80 | 75 | 70 | 65 |

TRIASSIC   **JURASSIC**   **CRETACEOUS**

Leyesaurus marayensis (meaning "Leyes's lizard from Marayes") represents an intermediate state within the sauropodomorph lineage. While it possessed an extended neck and leaf-shaped teeth that were suitable for an herbivorous diet, it lacked the huge sauropods' column-like extremities and likely walked on just two legs.

Leyesaurus was not an ancestor of the enormous quadrupedal sauropods that lived in the later portions of the Mesozoic. Instead, throughout the Late Triassic and Early Jurassic, the **massospondylids** were just one of several sauropodomorph lineages that dispersed over the globe (Rauhut et al., 2020).

The only known *Leyesaurus* specimen (PVSJ 706) was recovered from the Quebrada del Barro geological formation; the exact age of this strata has not been accurately determined and could range from the Late Triassic all the way through the Early Jurassic. Other fragmentary sauropodomorph fossils that have been discovered from this formation have previously been speculated to be *Riojasaurus* remains, although this identification is ambiguous. The best-preserved portions of the *Leyesaurus* specimen are the skull and the first several neck vertebrae, although several other skeletal fragments are also present.

The incompleteness of the remains makes the animal's size difficult to gauge, but estimates put the animal's length somewhere near the 3 meter mark. This makes *Leyesaurus* one of the smallest sauropodomorphs of its time, as some genera—such as *Kholumolumo*—had already reached lengths greater than 10 meters.

The generic name *Leyesaurus* honors the finders of the fossils, the Leyes family from the town of Balde de Leyes. The specific name *marayensis* refers to Marayes-El Carrizal Basin, where the fossils were unearthed.

CLASSIFICATION

Dinosauria
 Sauropodomorpha
  Bagualosauria
   Plateosauria
    Massopoda
     Massospondylidae

LOCATION

Argentina

KNOWN REMAINS

Partial skull and skeleton

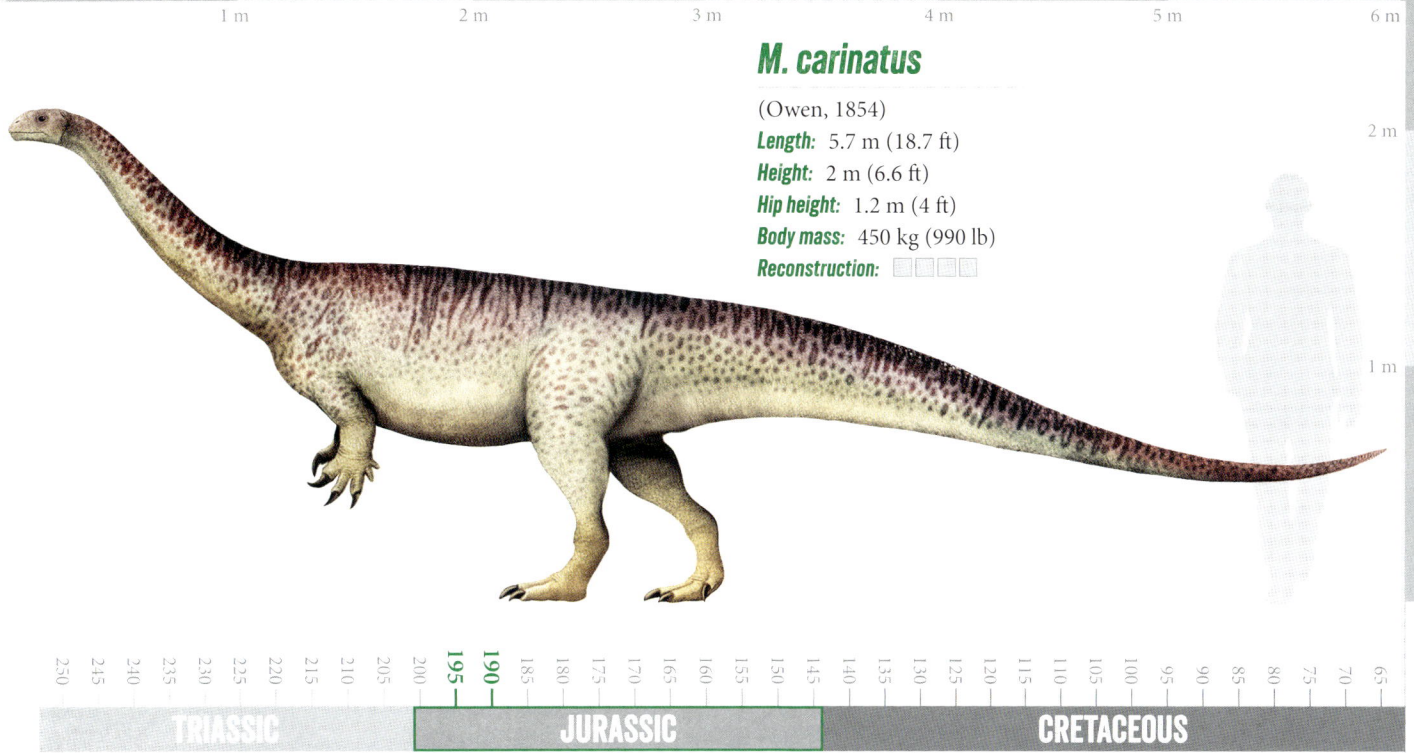

| 1 m | 2 m | 3 m | 4 m | 5 m | 6 m |

M. carinatus

(Owen, 1854)
Length: 5.7 m (18.7 ft)
Height: 2 m (6.6 ft)
Hip height: 1.2 m (4 ft)
Body mass: 450 kg (990 lb)
Reconstruction: ☐☐☐☐

| 250 | 245 | 240 | 235 | 230 | 225 | 220 | 215 | 210 | 205 | 200 | 195 | 190 | 185 | 180 | 175 | 170 | 165 | 160 | 155 | 150 | 145 | 140 | 135 | 130 | 125 | 120 | 115 | 110 | 105 | 100 | 95 | 90 | 85 | 80 | 75 | 70 | 65 |

| TRIASSIC | JURASSIC | CRETACEOUS |

Massospondylus carinatus (meaning "longer-keeled vertebrae") is a well-known, midsized sauropodomorph that had a proportionately small head. The presence of relatively large openings for blood vessels on the animal's jaws has been interpreted as evidence that *Massospondylus* had fleshy cheeks and, thus, chewed its food (Galton and Upchurch, 2004).

The maximum size of *Massospondylus* adults seems to have been variable, with some reaching nearly 6 meters in length, while others only grew to about 4 meters. Its thumb claw was proportionately large and could have been used for defense or foliage manipulation. Although *Massospondylus* was long thought to be partially quadrupedal, it is now known to have been an obligate biped, as it was incapable of rotating its wrists enough to have walked quadrupedally (Bonnan and Senter, 2007).

Numerous *Massospondylus* specimens of varying age and completeness are known from locations across southern Africa. The holotype remains were destroyed in a World War II bombing, necessitating the designation of a neotype, BP/1/4934 (Yates and Barrett, 2010). In the century and a half since its initial description, numerous *Massospondylus* species have been named, but only *M. carinatus* and *M. kaalae* are typically considered to be valid by modern researchers (Barrett, 2009). Various obsolete genera are now considered to be synonymous with *Massospondylus*, including *Leptospondylus*, *Pachyspondylus*, *Aristosaurus*, *Dromicosaurus*, and *Hortalotarsus*. Conversely, several specimens previously considered to be *Massospondylus* remains have since been reclassified as new genera, such as *Sarahsaurus* and *Ngwevu*.

The generic name *Massospondylus* combines the Greek "masson" (meaning "longer") and "spondylos" (meaning "vertebra"). The specific name *carinatus* is Latin for "keeled"; *kaalae* honors museum worker Sheena Kaal.

CLASSIFICATION
Dinosauria
  Sauropodomorpha
   Bagualosauria
    Plateosauria
     Massopoda
      Massospondylidae

LOCATION
South Africa, Lesotho, Zimbabwe

KNOWN REMAINS
Complete

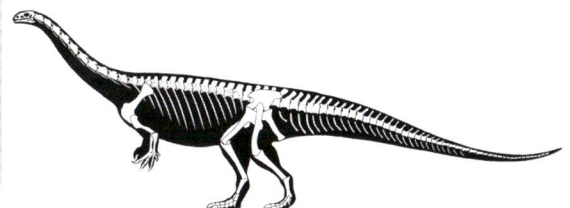

1 m   2 m   3 m   4 m   5 m   6 m

G. hammeri

(Smith and Pol, 2007)
Length: 6.3 m (20.7 ft)
Height: 2.2 m (7.2 ft)
Hip height: 1.6 m (5.2 ft)
Body mass: 600 kg (1,300 lb)
Reconstruction: ☐☐☐☐

250 245 240 235 230 225 220 215 210 205 200 195 190 185 180 175 170 165 160 155 150 145 140 135 130 125 120 115 110 105 100 95 90 85 80 75 70 65

| TRIASSIC | JURASSIC | CRETACEOUS |

Glacialisaurus hammeri (meaning "Hammer's icy lizard") was the first sauropodomorph to be discovered from Antarctica. At the time of its literature debut, Antarctica represented the sixth continent on which sauropodomorph remains had been found (leaving out only Australia). This helped to show the near-global distribution that sauropodomorphs had achieved by the early stages of the Jurassic. During this time, Antarctica was located farther north, and its coastal regions were likely quite mild in temperature.

Only two fragmentary specimens of *Glacialisaurus* are currently known; they were found near to one another, although they are believed to have originated from two separate individuals. These fossils were recovered in 1990 and 1991, during the same expedition that discovered the theropod *Cryolophosaurus*. The *Glacialisaurus* material consists of fragments of the foot, ankle, and leg; some

vertebrae initially suspected to belong to the same animal were later attributed to *Cryolophosaurus* instead.

During a later expedition in 2003–2004, fossil material was discovered from the same geological formation that is thought to have belonged to a true sauropod dinosaur. If so, this would indicate that at least some Early Jurassic ecosystems hosted both the giant, derived sauropods, and the smaller, "primitive" sauropodomorphs at the same time. This state of affairs has been borne out by various phylogenetic analyses, which have shown that the group **Massospondylidae** was not directly ancestral to the true sauropods but rather was an earlier-diverging offshoot (Rauhut et al., 2020).

The generic name *Glacialisaurus* is derived from the Latin "glacialis" (meaning "icy" or "frozen"). The specific name *hammeri* honors paleontologist William R. Hammer, who led the expedition to Mt. Kirkpatrick that unearthed the fossils.

CLASSIFICATION
Dinosauria
 Sauropodomorpha
  Bagualosauria
   Plateosauria
    Massopoda
     Massospondylidae

LOCATION
Antarctica

KNOWN REMAINS
Leg fragments

1 m 2 m 3 m 4 m 5 m 6 m

2 m

1 m

C. brevis

(Bonaparte, 1978)
Length: 6.3 m (20.7 ft)
Height: 2.2 m (7.2 ft)
Hip height: 1.6 m (5.2 ft)
Body mass: 600 kg (1,300 lb)
Reconstruction: ☐☐☐☐

250 245 240 235 230 225 220 213 210 205 200 195 190 185 180 175 170 165 160 155 150 145 140 135 130 125 120 115 110 105 100 95 90 85 80 75 70 65

TRIASSIC   **JURASSIC**   **CRETACEOUS**

Coloradisaurus brevis (meaning "short lizard from Colorados") is known from two partial specimens that were unearthed in 1971 by José Bonaparte. For many years, the skull of the holotype (PVL 3967) was the only portion of the remains to be well described. Later works would reexamine this material (Apaldetti et al., 2014) and finally provide adequate detail and analysis on the remainder of the skeleton (Apaldetti et al., 2013).

The rather gracile *Coloradisaurus* was only one of several sauropodomorphs to apparently share an ecosystem with one another. The presence of the robust *Riojasaurus* and the larger *Lessemsaurus*, along with at least one as-yet unidentified species (PULR 136; Ezcurra and Apaldetti, 2012), potentially demonstrates a case of niche partitioning.

Along with the animal's closest known relative, *Lufengosaurus*, *Coloradisaurus* is considered to be a member of the family **Massospondylidae** (Müller, 2020).

It has also been noted, though, that *Coloradisaurus* shares some traits in common with the more "primitive" plateosaurids. These traits are thought to have evolved independently and is one of several examples of sauropodomorph traits that can be difficult to interpret and can easily muddy the waters of numerous phylogenetic interpretations (Apaldetti et al., 2014).

The name originally intended for the genus was *Coloradia*, but it was later found that this name was preoccupied by a type of moth. David Lambert, after communication with José Bonaparte, used the name *Coloradisaurus* in 1983, but this was not an official name until its proper use by Peter Galton in 1990. The generic name *Coloradisaurus* refers to the Los Colorados geological formation. The specific name *brevis* is Latin for "short," apparently in reference to the length of the animal's skull in relation to its height and width.

CLASSIFICATION
Dinosauria
 Sauropodomorpha
  Bagualosauria
   Plateosauria
    Massopoda
     Massospondylidae

LOCATION
Argentina

KNOWN REMAINS
Nearly complete

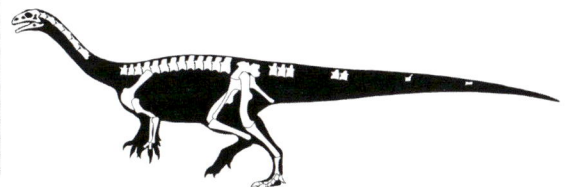

2 m   4 m   6 m   8 m

L. huenei

(Young, 1940)
Length: 8.9 m (29.2 ft)
Height: 2.7 m (8.9 ft)
Hip height: 2.1 m (6.9 ft)
Body mass: 1,750 kg (1.9 t)
Reconstruction: ☐☐☐☐

2 m

250 245 240 235 230 225 220 215 210 205 200 **195** 190 185 180 175 170 165 160 155 150 145 140 135 130 125 120 115 110 105 100 95 90 85 80 75 70 65

TRIASSIC   **JURASSIC**   **CRETACEOUS**

Lufengosaurus huenei (meaning "Huene's lizard from Lufeng") was described by a pioneer of Chinese paleontology, Yang Zhongjian (also known as "C. C. Young"). He designated two species, first *L. huenei*, and then the larger *L. magnus* in 1947, although many modern sources consider the two to be one and the same. Judging only from specimens historically attributed to *L. huenei*, the animal's average adult length was once considered to be approximately 6 meters, but when *L. magnus* specimens are included in the calculations, this length increases to nearly 9 meters.

Numerous specimens of *Lufengosaurus* are currently known, many of which were cataloged more than half a century ago. The first dinosaur skeleton to ever be mounted and displayed in China was that of *Lufengosaurus*. These fossils were originally considered to be of Late Triassic origin but now are known to be from the Early Jurassic.

Known specimens include the remains of individuals of differing ages, allowing for the analysis of growth patterns and ontogenetic changes (Sekiya and Dong, 2010). The genus *Tawasaurus* and at least one species of *Gyposaurus* are often considered to be synonymous with *Lufengosaurus*.

Several bony bumps are present on the skull of *Lufengosaurus*, as well as a ridge of bone that has been interpreted as the anchoring location of substantial cheek muscles (Barrett et al., 2005).

In 2015, researchers identified soft-tissue collagen proteins from a fragment of a *Lufengosaurus* rib bone, breaking the record for the oldest such discovery by more than 100 million years (Lee et al., 2017).

The generic name *Lufengosaurus* refers to the city of Lufeng, China. The specific name *huenei* honors paleontologist Friedrich von Huene, while the name *magnus* means "large one" in Latin.

CLASSIFICATION
Dinosauria
 Sauropodomorpha
  Bagualosauria
   Plateosauria
    Massopoda
     Massospondylidae

LOCATION
China

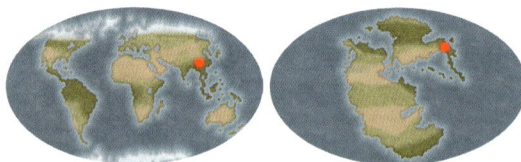

KNOWN REMAINS
Complete

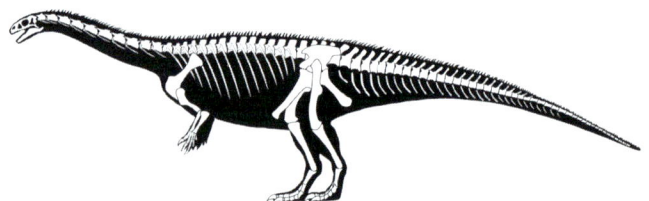

N. intloko

(Chapelle et al., 2019)
Length: 3.5 m (11.5 ft)
Height: 1.2 m (3.9 ft)
Hip height: 0.8 m (2.6 ft)
Body mass: 100 kg (220 lb)
Reconstruction: ☐☐☐☐

1 m

191

TRIASSIC   **JURASSIC**   **CRETACEOUS**

Ngwevu intloko (meaning "gray skull," pronounced Ng-g'where-voo) is known from a single specimen (BP/1/4779) that was discovered 1978 and was long believed to represent *Massospondylus*. A more detailed analysis, though, would eventually reveal it to be its own, unique type of creature.

The skull of *Ngwevu* has quite a different shape than that of *Massospondylus*, but this was initially put down to taphonomic distortion—in other words, deformation of a geologic nature that had warped the shape of the fossil. It was also assumed that the individual had been a juvenile, which could possibly account for the shorter dimensions of the skull.

The researchers who described the genus, though, found that histological growth patterns within the specimen's bones indicated that the individual was nearly fully grown. Thus, different physical characteristics of the skull could not

be attributed to age-related development. Further, the skull bones were actually found to be quite intact and not nearly so distorted as originally reported. Analysis of the specimen's inner-ear structure also revealed marked differences between itself and specimens of *Massospondylus*.

Consequently, *Ngwevu* was erected as a new genus. With a shorter and more robust skull, it was proposed that *Ngwevu* was able to develop tougher jaw musculature than *Massospondylus* and was thus able to base its diet upon sturdier types of foliage. The longer, narrower snout of *Massospondylus* would have been better suited for targeted, selective browsing. These differing feeding strategies could have allowed for niche partitioning, letting *Massospondylus* and *Ngwevu* share the same habitat without competition.

The binomial name is taken from the indigenous Xhosa language, meaning "gray skull".

CLASSIFICATION
Dinosauria
 Sauropodomorpha
  Baqualosauria
   Plateosauria
    Massopoda
     Massospondylidae

LOCATION
South Africa

KNOWN REMAINS
Skull and partial skeleton

1 m   2 m   3 m   4 m   5 m   6 m

2 m

1 m

X. chengi

(Wang et al., 2017b)

Length: 6 m (19.7 ft)
Height: 2.6 m (8.5 ft)
Hip height: 2 m (6.6 ft)
Body mass: 460 kg (1,010 lb)
Reconstruction: ☐☐☐☐

250 245 240 235 230 225 220 215 210 205 **200** 195 190 185 180 175 170 165 160 155 150 145 140 135 130 125 120 115 110 105 100 95 90 85 80 75 70 65

| TRIASSIC | JURASSIC | CRETACEOUS |

Xingxiulong chengi (meaning "Cheng's bridge dragon") is just one of numerous sauropodomorph species that inhabited what is now the Lufeng geological formation. The remains of *Lufengosaurus*, *Yunnanosaurus*, *Jingshanosaurus*, and *Chuxiongosaurus* have also been unearthed from this Early Jurassic layer. These latter three, along with *Xingxiulong*, are among the most basal genera of **Sauropodiformes** currently known, perhaps suggesting that the clade first originated in China.

Despite its "primitive" positioning on the sauropod family tree, *Xingxiulong* possesses some traits that were common in the more derived "true" sauropods and less common in the basal sauropodomorph species. For instance, *Xingxiulong* had four sacral (hip) vertebrae, rather than three, and had comparatively robust leg bones. These adaptations are thought to have facilitated the development of an increased body mass and, in particular, a larger digestive system, which was physically supported by a very sauropod-like pubis bone.

Despite its heavy weight, and despite its rather strong shoulder blades, *Xingxiulong* was likely a bipedal creature and was not partially quadrupedal. This conclusion is supported by the available range of motion and the relative proportions of the forearms. *Xingxiulong* was also not particularly long; two of the three known fossil specimens are known to have been fully grown adults, based on the fused nature of their cranial and vertebral bones, so the animal's maximum size is confidently known.

The generic name *Xingxiulong* refers to the ancient "Xingxiu Bridge" located in Lufeng County, China; the term "xingxiu" translates literally as "constellation," while "long" means "dragon." The specific name *chengi* honors Zheng-Wu Cheng.

CLASSIFICATION

Dinosauria
 Sauropodomorpha
  Bagualosauria
   Plateosauria
    Massopoda
     Sauropodiformes

LOCATION

China

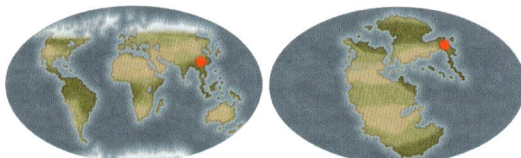

KNOWN REMAINS

Nearly complete

1 m 2 m 3 m 4 m 5 m 6 m 7 m

2 m

1 m

Y. huangi

(Young, 1942)
Length: 7 m (23 ft)
Height: 2.4 m (7.9 ft)
Hip height: 1.8 m (5.9 ft)
Body mass: 600 kg (1,300 lb)
Reconstruction: ☐☐☐☐

250 245 240 235 230 225 220 215 210 205 200 195 **190 185** 180 175 170 165 160 155 150 145 140 135 130 125 120 115 110 105 100 95 90 85 80 75 70 65

TRIASSIC | **JURASSIC** | **CRETACEOUS**

Yunnanosaurus huangi (meaning "lizard from Huangchiatien, Yunnan") is a long-studied animal known from numerous sets of fossil remains that include several skulls and individuals of varying ages and, thus, varying developmental states.

The *Y. huangi* holotype specimen was discovered by Yang Zhongjian (i.e., C. C. Young) in 1939, then fully described in 1942. In 1951, he erected a second species, *Y. robustus*; many subsequent analyses have concluded that the two species are synonymous, although the study of a juvenile specimen has led at least one group to continue differentiating the two, citing minor differences such as the presence of serrations on some of the animal's teeth (Sekiya et al., 2014).

A third species, *Y. youngi*, was described in 2007 based on material that was excavated in 2000. Whereas the adult length of *Y. huangi* is approximately 7 meters, the new *Y.*

youngi was significantly larger, reaching lengths of 13 meters. *Yunnanosaurus youngi* also lived several million years later than the previously known species, during the beginning stages of the mid-Jurassic (Lü et al., 2007).

Since 1942, *Yunnanosaurus* has been placed all over the messy and ever-changing "prosauropod" family tree. Most modern analyses, though, place the genus near the base of the **Sauropodiformes** (Wang et al., 2017b; Zhang et al., 2020). Although its teeth were very much akin to those of the true sauropods, this is likely a result of convergent evolution.

The animal's binomial name refers to the holotype's palace of discovery, the village of Huangchiatien in Yunnan Province, China. The specific name *robustus* refers to the comparative robustness of the specimen; *youngi* honors paleontologist C. C. Young.

CLASSIFICATION
Dinosauria
 Sauropodomorpha
  Bagualosauria
   Plateosauria
    Massopoda
     Sauropodiformes

LOCATION
China

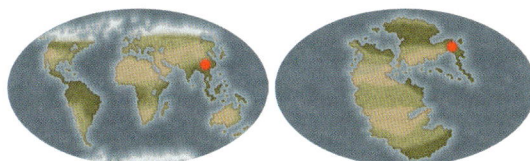

KNOWN REMAINS
Nearly complete

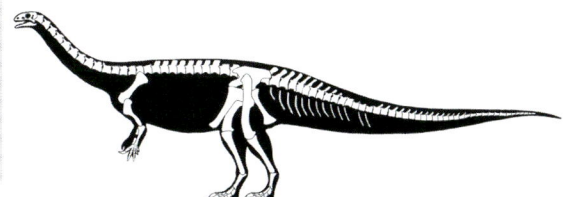

JINGSHANOSAURUS

J. xinwaensis

(Zhang and Yang, 1995)
Length: 9.2 m (30.2 ft)
Height: 4.2 m (13.8 ft)
Hip height: 2.3 m (7.5 ft)
Body mass: 1,600 kg (1.8 t)
Reconstruction: ☐☐☐☐

TRIASSIC | **JURASSIC** | **CRETACEOUS**

Jingshanosaurus xinwaensis (meaning "lizard from Xinwa, Jinghsan") is the largest example out of numerous sauropodomorph species that inhabited the area that is now the Lufeng geological formation, outsizing *Lufengosaurus*, *Yunnanosaurus*, *Xingxiulong*, and *Chuxiongosaurus*.

The nearly complete holotype specimen of *Jingshanosaurus* (LFGT-ZLJ0113) was discovered in 1988 by paleontologist Zheng-Ju Wang in the area that has since been designated the Lufeng Dinosaur National Geopark. In 1995, *Jingshanosaurus* was described in the Chinese-language book *A New Complete Osteology of Prosauropoda in Lufeng Basin, Yunnan, China*. After this publication, *Jingshanosaurus* was included in various phylogenetic classification studies based on its stated measurements, but no other research material was produced that took any closer of a look at the remains until 2019 (Zhang et al., 2020).

This new analysis focused on the animal's skull, updating and clarifying the unique anatomical details of the cranium. This area of focus was chosen because the phylogenetic positions of several underdescribed early sauropodomorphs were very dependent on skull morphology; small changes in the stated characteristics of an animal's skull could lead to significantly different outcomes on models of the family tree. Thus, getting these exact details right is crucial to piecing together the precise nature of sauropodomorph evolution.

Although it was originally placed within Plateosauridae, the newer reanalysis of the skull has determined that *Jingshanosaurus* was one of the earliest-branching members of **Sauropodiformes**.

The creature's binomial name references the village of Xinwa and the town of Jinghsan in Yunnan Province, China.

CLASSIFICATION
Dinosauria
 Sauropodomorpha
  Bagualosauria
   Plateosauria
    Massopoda
     Sauropodiformes

LOCATION
China

KNOWN REMAINS
Nearly complete

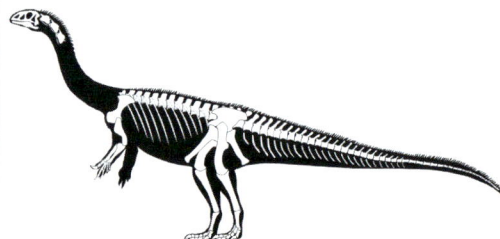

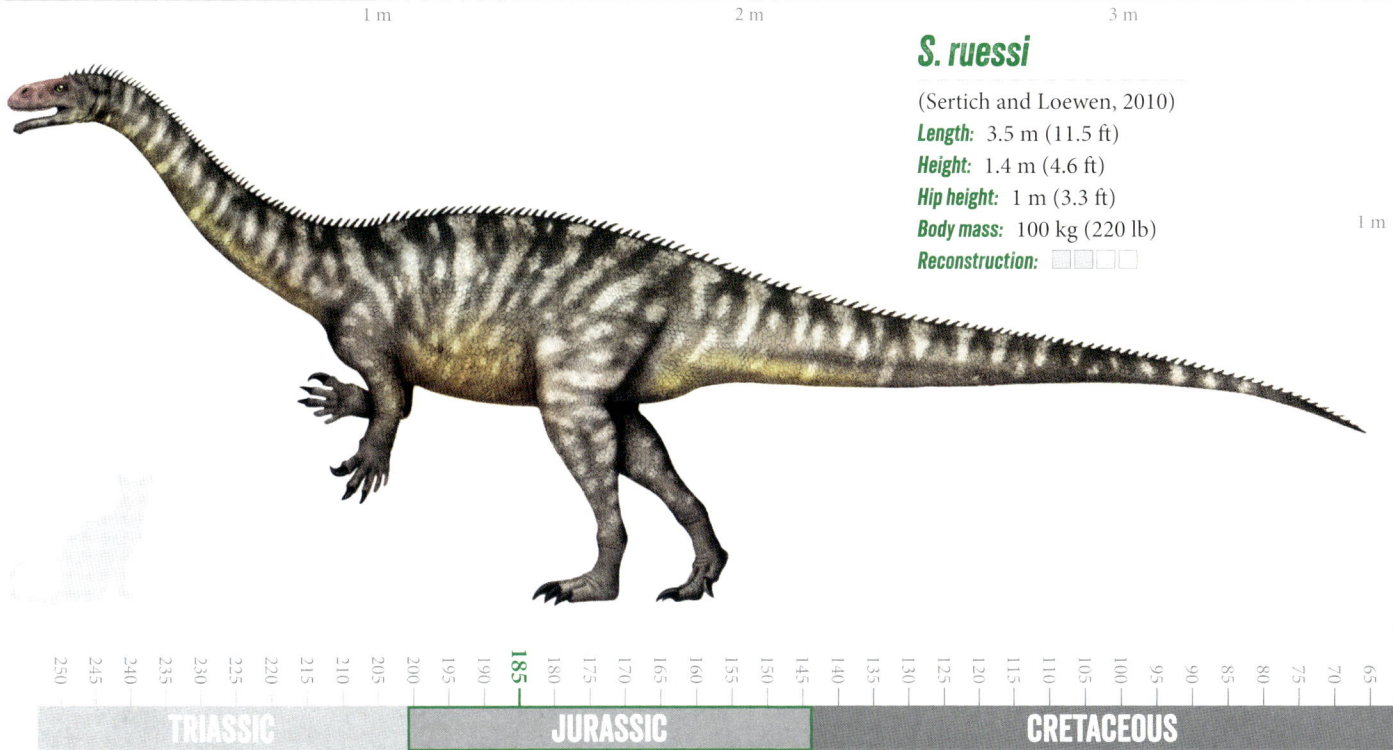

S. ruessi

(Sertich and Loewen, 2010)
Length: 3.5 m (11.5 ft)
Height: 1.4 m (4.6 ft)
Hip height: 1 m (3.3 ft)
Body mass: 100 kg (220 lb)
Reconstruction: ☐☐☐☐

1 m   2 m   3 m

1 m

250 245 240 235 230 225 220 215 210 205 200 195 **185** 180 175 170 165 160 155 150 145 140 135 130 125 120 115 110 105 100 95 90 85 80 75 70 65

TRIASSIC   **JURASSIC**   **CRETACEOUS**

Seitaad ruessi (meaning "Ruess's sand-monster") is one of the few sauropodomorphs to be described from North America. Although sauropodomorphs from the Late Triassic and Early Jurassic were common and widespread, thanks to the connections that existed between the modern continents, North American remains have proven to be curiously elusive, especially from the American West. Although scattered fragments from the region had been previously found, the remains of *Seitaad* proved to be among the first that were complete enough to diagnose and identify accurately.

The only known specimen (UMNH VP 18040) appears to have been buried in sand, likely by a collapsing dune, after the individual was already dead. Thus, the recovered remains (which consisted of portions of the animal's trunk and limbs) were preserved three-dimensionally and in articulation. With no skull, neck, or tail being preserved, the length of

Seitaad is difficult to accurately surmise, but estimates place the value between 3 and 4 meters total. It is also not known for certain whether or not the individual was a juvenile or a fully grown adult.

The original description of the genus was unable to determine the genus's phylogenetic position confidently, suggesting several possibilities and leaning toward the conclusion that *Seitaad* was a massospondylid. However, subsequent studies have favored a more derived position, with *Seitaad* being among the **Sauropodiformes** (Apaldetti et al., 2011; McPhee et al., 2015).

The generic name *Seitaad* is taken from the Navajo term "séít'áád," the name of a mythological creature from Diné folklore that is said to bury its victims in sand dunes, just as the dinosaur specimen appears to have been. The specific name *ruessi* honors the American artist and naturalist Everett Ruess, who went missing in Utah in 1934.

CLASSIFICATION
Dinosauria
 Sauropodomorpha
  Bagualosauria
   Plateosauria
    Massopoda
     Sauropodiformes

LOCATION
Utah, USA

KNOWN REMAINS
Partial skeleton

1 m   2 m   3 m

1 m

A. polyzelus

(Hitchcock, 1865)
Length: 3.2 m (10.3 ft)
Height: 0.9 m (3 ft)
Hip height: 0.6 m (2 ft)
Body mass: 55 kg (121 lb)
Reconstruction: ☐☐☐☐

| 250 | 245 | 240 | 235 | 230 | 225 | 220 | 215 | 210 | 205 | 200 | 195 | 190 | 185 | 180 | 175 | 170 | 165 | 160 | 155 | 150 | 145 | 140 | 135 | 130 | 125 | 120 | 115 | 110 | 105 | 100 | 95 | 90 | 85 | 80 | 75 | 70 | 65 |

TRIASSIC	JURASSIC	CRETACEOUS

Anchisaurus polyzelus (meaning "much sought for near-lizard") was once considered to be the "smallest sauropod" but is now thought to be more basally placed on the family tree, outside Sauropoda. It had narrow feet and an elongated midsection, which would have allowed for more effective digestion of plants.

Anchisaurus is known from four established sets of partial remains, discovered throughout the 1800s. The original type specimen (ACM 41/109) was unearthed during a blast excavation, and was quite damaged as a result; in 2012, it was proposed that a more complete and diagnostic specimen be given neotype status (YPM 1883; Galton, 2012). Additional partial remains from Arizona and Canada have been tentatively suggested but remain unconfirmed (Galton, 1971; Fedak, 2007).

The various specimens now attributed to *A. polyzelus* have been known by many names over time, including *A. colurus*, *A. major*, *Yaleosaurus*, *Ammosaurus*, *Amphisaurus*, and *Megadactylus*. The latter two names were replaced because they turned out to already be occupied by other organisms.

Some of the specimens were not recognized as being attributable to the same species because of their different states of growth. Some are clearly of juvenile status, while at least one is thought to have been fully grown, being approximately 3.2 meters in length. However, the size of some footprints suggests individuals of at least 6 meters in length (Weems, 2019).

The generic name *Anchisaurus* combines the Greek "anchi" (meaning "near") and "sauros" (meaning "lizard"). The specific name *polyzelus* means "much sought for" in Greek; Edward Hitchcock Jr. gave the animal this name because his father had spent much time and effort in finding the identity of the mysterious reptile that had created certain fossilized trackways.

CLASSIFICATION

Dinosauria
 Sauropodomorpha
  Bagualosauria
   Plateosauria
    Massopoda
     Sauropodiformes
      Anchisauria

LOCATION

Eastern United States

KNOWN REMAINS

Skull and majority of skeleton

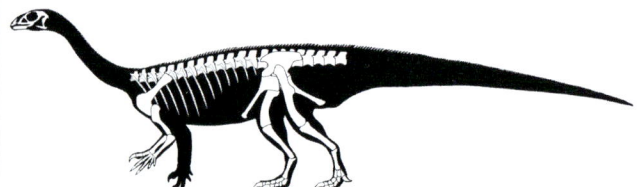

1 m   2 m   3 m

1 m

L. taquetrensis

(Pol et al., 2011)
Length: 3 m (9.8 ft)
Height: 1 m (3.2 ft)
Hip height: 0.7 m (2.3 ft)
Body mass: 70 kg (154 lb)
Reconstruction: ▣☐☐☐

250 245 240 235 230 225 220 215 210 205 **199** 195 190 185 180 175 170 165 160 155 150 145 140 135 130 125 120 115 110 105 100 95 90 85 80 75 70 65

TRIASSIC   **JURASSIC**   **CRETACEOUS**

Leonerasaurus taquetrensis (meaning "lizard from Leoneras, Taquetrén") is a relatively small animal that has many skeletal traits that mark it as a non-sauropod sauropodomorph. However, one very sauropod-like trait stands out as significant, which is that *Leonerasaurus* possesses four sacral (i.e., pelvic) vertebrae rather than just three.

Paleontologists have long debated how and why the four-element sacrum developed among the sauropods. Many had hypothesized that incorporating a fourth vertebra into the sacrum was necessary to help support the increased weight of the huge sauropod species as well as their increased gut volume; as a consequence, the presence of four vertebrae in the sacrum was often considered a diagnostic trait of Sauropoda.

However, the discovery of certain *Melanorosaurus* specimens cast doubt onto the notion that the larger sacrum was present only within Sauropoda, and *Leonerasaurus* has further muddied the waters by showing that this condition evolved even earlier in the sauropodomorph lineage, counterintuitively within small species that did not have enormous bodies to support.

The lone *Leonerasaurus* specimen (MPEF-PV 1663) is quite partial, and comes from an individual that was not fully grown. The animal is thought to have been roughly six years of age at the time of its death, which means it was not particularly young, nor was it an adult. Some of the remains were found in articulation, and all were unearthed from a small area. The lower limbs were almost entirely missing, but based on its phylogenetic position, *Leonerasaurus* was probably bipedal, at least to some degree.

The generic name *Leonerasaurus* refers to the Las Leoneras geological formation from which the fossils were unearthed. The specific name *taquetrensis* refers to the Sierras de Taquetrén, the region where the excavation took place.

CLASSIFICATION
Dinosauria
  Sauropodomorpha
    Bagualosauria
      Plateosauria
        Massopoda
          Sauropodiformes
            Anchisauria

LOCATION
Argentina

KNOWN REMAINS
Partial skeleton and jaw

2 m   4 m   6 m   8 m

2 m

M. patagonicus

(Bonaparte and Martin, 1979)
Length: 8 m (26.2 ft)
Height: 2.7 m (8.8 ft)
Hip height: 2.3 m (7.5 ft)
Body mass: 1,350 kg (1.5 t)
Reconstruction: ☐☐☐☐

250 245 240 235 230 225 220 215 210 205 200 195 **192** 185 180 175 170 165 160 155 150 145 140 135 130 125 120 115 110 105 100 95 90 85 80 75 70 65

TRIASSIC | **JURASSIC** | **CRETACEOUS**

Mussaurus patagonicus (meaning "mouse lizard from Patagonia") was originally known only from the remains of hatchlings, but more recent finds have been able to illuminate the animal's entire lifespan.

Several adult specimens, which had originally been misidentified as representing *Plateosaurus*, were reinterpreted in 2013; this analysis provided the first true glimpse of how non-sauropod sauropodiforms developed as they aged by allowing the comparison of the two age groups (Otero and Pol, 2013).

In 2021, an analysis of several rich fossil sites (from the original type locality where the first hatchlings were found) described dozens of substantial, articulated sets of remains representing six different ontogenetic stages (i.e., developmental growth states) that illuminated two key pieces of information. First, the *Mussaurus* individuals were mostly clustered in groups of similarly aged individuals;

this grouping was interpreted as clear evidence of herd-forming social behavior being present 40 million years prior to the next-oldest evidence of such an occurrence (Pol et al., 2021).

Second, it became clear that as hatchlings, *Mussaurus* were quadrupedal, but that as they aged, they shifted their stance until they were entirely bipedal as adults; this adds another piece to the puzzle of how the sauropods transitioned into their later, giant forms (Otero et al., 2019).

The site of these discoveries, the Laguna Colorada Formation, was once thought to be Late Triassic in age but is now known to be Early Jurassic (Pol et al., 2021).

The generic name *Mussaurus* combines the Latin "mus" (meaning "mouse") with the Greek "sauros" (meaning "lizard"), referring to the initial hatchlings' size. The specific name *patagonicus* refers to the animal's discovery in the Patagonia region of Argentina.

CLASSIFICATION
Dinosauria
  Sauropodomorpha
    Bagualosauria
      Plateosauria
        Massopoda
          Sauropodiformes
          Anchisauria

LOCATION
Argentina

KNOWN REMAINS
Complete

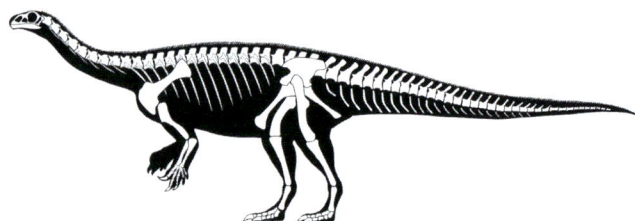

1 m   2 m   3 m   4 m   5 m

2 m

1 m

I. yimenensis

(Peyre de Fabrègues et al., 2020)
Length: 5.7 m (18.7 ft)
Height: 2 m (6.6 ft)
Hip height: 1.6 m (5.2 ft)
Body mass: 500 kg (1,100 lb)
Reconstruction: ☐☐☐☐

250 245 240 235 230 225 220 215 210 205 **200** 195 190 185 180 175 170 165 160 155 150 145 140 135 130 125 120 115 110 105 100 95 90 85 80 75 70 65

TRIASSIC   **JURASSIC**   **CRETACEOUS**

Irisosaurus yimenensis (meaning "iridescent lizard from Yimen") is a medium-sized sauropodomorph that can be distinguished from other species by a unique combination of character traits that, in and of themselves, are not individually unique.

The partial holotype skeleton (CVEB 21901) was discovered in 2018 in the Fengjiahe geological formation, near the village of Zhanmatian, China. This Chinese formation, and others of a similar age (such as the Lufeng Formation) have proven to be remarkably diverse in non-sauropodan sauropodomorph genera; indeed, over half of the described Laurasian species have come from Chinese strata.

The phylogenetic analysis conducted by the describers placed *Irisosaurus* within **Sauropodiformes**. This placement was an unexpectedly derived result, as the animal shares several traits in common with more "primitive" sauropodomorphs (such as having elongated cervical vertebrae) and also appears to have been entirely bipedal, with the forelimbs being fairly gracile and not well adapted for locomotion. The claw on the first finger was definitely quite mobile and is speculated to have been important for the browsing of vegetation.

The analysis also concluded that *Irisosaurus* was the sister taxon of *Mussaurus*, even though the two genera have some significant differences: the latter hails from Gondwana rather than from Laurasia, and *Irisosaurus* is approximately 15–20 million years more recent. The two animals also have more than a few anatomical differences.

The generic name *Irisosaurus* refers to the famous "iridescent clouds" of its discovery location. The specific name *yimenensis* refers to Yimen County, China.

CLASSIFICATION
Dinosauria
  Sauropodomorpha
    Bagualosauria
      Plateosauria
        Massopoda
          Sauropodiformes
            Anchisauria

LOCATION
China

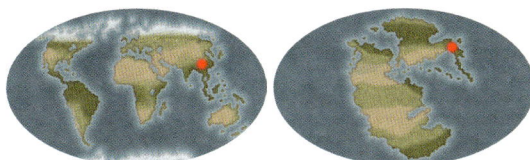

KNOWN REMAINS
Partial skeleton and skull fragments

2 m   4 m   6 m   8 m

2 m

Y. sunae

(Zhang et al., 2018)
Length: 8.2 m (27 ft)
Height: 2.8 m (9.2 ft)
Hip height: 2.3 m (7.5 ft)
Body mass: 1,500 kg (1.7 t)
Reconstruction: ☐☐☐☐☐

250 245 240 235 230 225 220 215 210 205 200 195 190 185 180 175 170 165 160 155 150 145 140 135 130 125 120 115 110 105 100 95 90 85 80 75 70 65

TRIASSIC   **JURASSIC**   **CRETACEOUS**

Yizhousaurus sunae (meaning "Sun's lizard from Chuxiong Yi") is possibly a transitional form of sauropodomorph, as it bears a curious mixture of traits—some that resemble ancestral forms (particularly in the skeleton) and others that are similar to more derived species (especially in the skull).

Skeletally, *Yizhousaurus* does not differ greatly from related species: the animal's forearm and femur suggest that it was bipedal, like the more basal sauropodomorphs were. However, the skull of *Yizhousaurus* is notably short and wide as opposed to the moderately elongated forms that were common for plateosaurids and massospondylids; this condition would instead be common for the true sauropods, which would soon come to dominate the biosphere. Additionally, the skull was also relatively small in comparison with the animal's body and had developed more robust bones and reinforcement structures that are consistent with those found in true sauropods.

These traits might make it seem possible that *Yizhousaurus* was a direct ancestor of the true sauropods;

however, cladistic analysis suggests that it was more likely that *Yizhousaurus* was on a side branch of the family tree, among the **Sauropodiformes**. This would mean that its sauropod-like features were a result of convergent evolution rather than being the progenitor of directly inherited traits.

The remains of *Yizhousaurus* (consisting of a three-dimensionally preserved skull and the majority of the skeleton, lacking just the lower legs and tail) were first discovered in 2002 and were introduced to the scientific community in a preliminary fashion at a conference in 2010 (Chatterjee et al., 2010). One noteworthy aspect of the remains is simply that the skeleton and skull were found together, as this allowed for comparisons to be made between both the cranial and skeletal traits of other related species.

The generic name *Yizhousaurus* makes reference to the Chinese region Chuxiong Yi. The specific name *sunae* honors paleontologist Ai-Ling Sun.

CLASSIFICATION

Dinosauria
  Sauropodomorpha
    Bagualosauria
      Plateosauria
        Massopoda
          Sauropodiformes
            Anchisauria

LOCATION

China

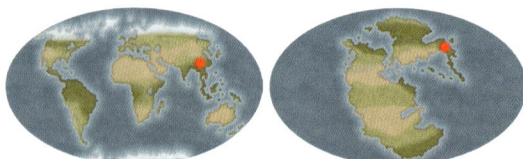

KNOWN REMAINS

Skull and partial skeleton

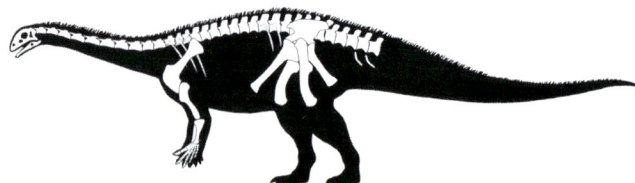

2 m   4 m   6 m   8 m

2 m

A. celestae

(Yates et al., 2010)
Length: 8.7 m (28.5 ft)
Height: 2.8 m (9.2 ft)
Hip height: 2 m (6.6 ft)
Body mass: 1,700 kg (1.9 t)
Reconstruction: ☐ ☐ ☐ ☐

250 245 240 235 230 225 220 215 210 205 200 **195** 190 185 180 175 170 165 160 155 150 145 140 135 130 125 120 115 110 105 100 95 90 85 80 75 70 65

TRIASSIC   **JURASSIC**   **CRETACEOUS**

Aardonyx celestae (meaning "Celeste's Earth-claw") is notable because its anatomy seems to be truly intermediate between bipedality and quadrupedality. This transitionary form of locomotion is a valuable clue for understanding how the bipedal sauropodomorphs gave rise to the quadrupedal sauropods.

Full quadrupedality was beyond the ability of *Aardonyx*, as its hands could not fully rotate to the degree that would allow them to be placed flat on the ground. The convex shape of the animal's femur is also a trait seen in bipedal sauropodomorphs. However, regions for muscle attachment along the femur indicate that *Aardonyx* was stronger and slower than other bipedal sauropodomorphs, and the arrangement of the radius and ulna show that they could support one another in order to bear increased weight.

Taken altogether, it seems that *Aardonyx* was a habitual, but not an obligatory, quadruped—an animal that spent some time on all fours but still relied on its hindlimbs for the majority of its locomotory movement. Having such traits appear so early in the Jurassic was surprising to the fossil's describers, as these characteristics had been thought to have appeared only much later in the Sauropodiformes' lineage.

The jaws of *Aardonyx* also bear an intriguing mixture of traits. Its jaws are narrow, as opposed to the broad mandibles of the later true sauropods. However, *Aardonyx* seems to have lacked fleshy cheeks, which would help its mouth open wide, enabling the broad-grazing style of herbivory that was ubiquitous among the colossal sauropods.

The most studied *Aardonyx* specimens are two subadults that imply an animal roughly 10 meters in length. Certain isolated elements, though, tantalizingly suggest a potential maximum size of roughly twice that (Yates et al., 2012).

The generic name *Aardonyx* combines the Afrikaans "aard" (meaning "Earth") and the Greek "onyx" (meaning "claw"). The specific name *celestae* honors fossil preparator Celeste Yates.

CLASSIFICATION
Dinosauria
  Sauropodomorpha
    Bagualosauria
      Plateosauria
        Massopoda
          Sauropodiformes
            Anchisauria

LOCATION
South Africa

KNOWN REMAINS
Partial skull and skeleton

1 m   2 m   3 m   4 m   5 m   6 m

2 m

1 m

S. zastronensis

(Otero et al., 2015)

Length: 6 m (19.7 ft)
Height: 1.7 m (5.6 ft)
Hip height: 1.5 m (4.9 ft)
Body mass: 550 kg (1,210 lb)
Reconstruction:

250 245 240 235 230 225 220 215 210 205 200 195 190 185 180 175 170 165 160 155 150 145 140 135 130 125 120 115 110 105 100 95 90 85 80 75 70 65

TRIASSIC	JURASSIC	CRETACEOUS

Sefapanosaurus zastronensis (meaning "cross lizard from Zastron") is another potential "transitional form" of sauropodomorph, displaying some traits that seem intermediate compared with the smaller and clearly bipedal early sauropodomorphs and the quadrupedal true sauropods.

The remains now attributed to *Sefapanosaurus* were collected by A. W. Keyser in the late 1930s; they were designated *Euskelosaurus* and essentially remained hidden and unstudied among many half-forgotten university specimens until 2010, when they were attributed to the closely related sauropodomorph *Aardonyx* (Yates et al., 2010). Closer scrutiny by researcher Emil Krupandan, however, would reveal that some of these fragments—originating from at least four individual animals—were from a separate species entirely. In particular, it was the subtle

differences in the bones of the ankle and upper foot that first became apparent.

It has become clear, particularly with the addition of *Sefapanosaurus*, that southern Gondwanan localities were hotspots for sauropodiform diversity. To date, the vast majority of Sauropodiformes species originating near the Triassic-Jurassic boundary have been unearthed from either southern Africa or South America. In order to gain a clearer understanding of the evolutionary trends which gave rise to the true sauropods, further discoveries from these regions will continue to be paramount.

The generic name *Sefapanosaurus* incorporates the Sesotho word "sefapano" (meaning "cross"), which refers to the distinctive cross-shaped structure of the animal's astragalus bone. The specific name *zastronensis* refers to the Zastron locality where the specimen was discovered.

CLASSIFICATION

Dinosauria
 Sauropodomorpha
  Bagualosauria
   Plateosauria
    Massopoda
     Sauropodiformes
      Anchisauria

LOCATION

Lesotho

KNOWN REMAINS

Fragments

1 m   2 m   3 m   4 m

1 m

M. thabanensis

(Peyre de Fabrègues and Allain, 2016)

Length: 4.8 m (15.7 ft)
Height: 1.4 m (4.6 ft)
Hip height: 1.2 m (3.9 ft)
Body mass: 300 kg (660 lb)
Reconstruction: ☐☐☐☐☐

250 245 240 235 230 225 220 215 210 205 200 195 190 185 180 175 170 165 160 155 150 145 140 135 130 125 120 115 110 105 100 95 90 85 80 75 70 65

| TRIASSIC | JURASSIC | CRETACEOUS |

Meroktenos thabanensis (meaning "femur beast from Thaban") is based on a very partial set of remains that were originally attributed to *Melanorosaurus*, specifically to the non–type species *M. thabanensis*. These remains were unearthed in 1959 and briefly described in 1962, although it was not until 1993 that they were properly assigned to a genus and species (Gauffre, 1993). Further description came just a few years later (van Heerden and Galton, 1997).

The exact location from which these remains originated was never properly documented. For a time, it was thought that they came from the Upper Elliot geological formation, which would have made them 20 million years younger than the type species of *Melanorosaurus, M. readi*. In part, for this reason, they were assigned to a new, second species in the genus. However, later work would suggest that they had instead come from the Lower Elliot (Gauffre, 1996).

In 2016, the specimen (MNHN.F.LES16) was reexamined and compared with all other known specimens of *Melanorosaurus*; it was found to be sufficiently unique to justify its reassignment to a new genus, *Meroktenos*. Additional remains, which were previously overlooked but are suspected to have come from the same animal, were also included in this analysis. The specimen's relatively small size suggests that it might not have been fully grown, but its ontogenetic status has not actually been determined.

The femur of *Meroktenos* is shaped similarly to those of the true sauropods, despite the creature's small size and Triassic provenance. This means that it could have been among the first sauropodomorphs to develop this particular trait, which the later, larger sauropods would retain.

The generic name *Meroktenos* combines the Greek "meros" (meaning "femur") and "ktenos" (meaning "animal" or "beast"). The specific name *thabanensis* refers to the village near where the specimens were found, Thabana-Morena.

CLASSIFICATION

Dinosauria
  Sauropodomorpha
    Bagualosauria
      Plateosauria
        Massopoda
          Sauropodiformes
            Anchisauria

LOCATION

Lesotho

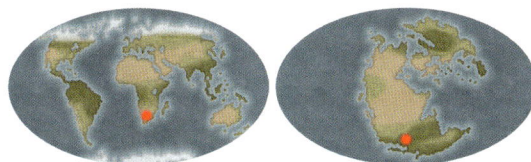

KNOWN REMAINS

Fragments

67

2 m   4 m   6 m

M. readi

(Haughton, 1924)
Length: 6.5 m (21.3 ft)
Height: 1.6 m (5.2 ft)
Hip height: 1.4 m (4.6 ft)
Body mass: 700 kg (1,543 lb)
Reconstruction: ☐☐☐☐

2 m

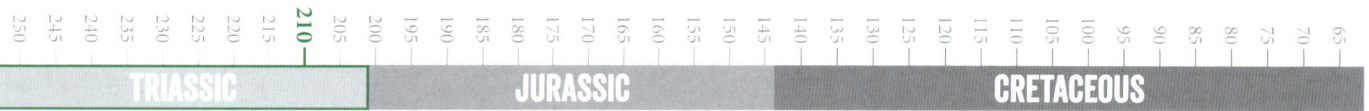

250 245 240 235 230 225 220 215 **210** 205 200 195 190 185 180 175 170 165 160 155 150 145 140 135 130 125 120 115 110 105 100 95 90 85 80 75 70 65

TRIASSIC	JURASSIC	CRETACEOUS

Melanorosaurus readi (meaning "Read's Black Mountain lizard") was one of the largest animals of its time, and despite its early appearance, seems to have already been at least partially quadrupedal.

The original syntype specimens (SAM 3449 and SAM 3450) were discovered and described in 1924 by Sidney H. Haughton. Regrettably, even then it was unclear whether or not all of these bones actually belonged to the same creature, with Haughton noting that many were "lying isolated" and that the femur, in particular, was "in doubtful association with the other remains." Many of these fossils were later reassigned to *Euskelosaurus* instead (van Heerden, 1979), but that genus is now largely considered a *nomen dubium* wastebasket taxon. Currently, it is believed that selected material from the syntype specimens is indeed diagnostically relevant and is attributable to *Melanorosaurus* instead (Galton et al., 2005).

Some bones from SAM 3532 were also referred to the genus by Haughton but have thus far remained relatively unstudied. The most complete sets of remains often referred to *Melanorosaurus* are NM QR3314 (Welman, 1998; Bonnan and Yates, 2007) and NM QR1551 (Galton et al., 2005). However, a later review cast doubt on whether or not these specimens were of the same species as one another (citing differences in the feet and sacral vertebrae) or even attributable to *M. readi* at all (noting a lack of overlapping, comparable elements with the syntype specimens) (McPhee et al., 2015, 2017).

The generic name *Melanorosaurus* combines the Greek "melas" (meaning "back"), "oros" (meaning "mountain"), and "sauros" (meaning "lizard"). This refers to the location of the holotype's discovery, on the slope of Thaba 'Nyama (meaning "Black Mountain" in the indigenous Nyanja language) in South Africa. The specific name *readi* is meant to honor B. Read of the Bensonvale Training School.

CLASSIFICATION
Dinosauria
 Sauropodomorpha
  Bagualosauria
   Plateosauria
    Massopoda
     Sauropodiformes
      Anchisauria

LOCATION
South Africa

KNOWN REMAINS
Nearly complete

P. gracilis

(Kutty et al., 2007)
Length: 4.6 m (15.1 ft)
Height: 1.6 m (5.2 ft)
Reconstruction: ☐☐☐☐

0.5 m

195—

TRIASSIC | JURASSIC | CRETACEOUS

Pradhania gracilis (meaning "Pradhan's slender one") is known from a single set of very fragmentary remains (ISI R265) that were unearthed in India. Its most distinguishing characteristic, which prompted the erection of this genus, is the presence of a prominent ridge along the inside of the upper jaw bone.

Owing to the paucity of the remains, the original description of the genus was unable to determine its phylogenetic placement accurately; a later study concluded that *Pradhania* was likely a **massospondylid** (Novas et al., 2010). The generic name honors fossil collector Dhuiya Pradhan.

C. lufengensis

(Lü et al., 2010)
Length: 9.2 m (30.2 ft)
Height: 3.1 m (10.2 ft)
Reconstruction: ☐☐☐☐

1 m

195—

TRIASSIC | JURASSIC | CRETACEOUS

Chuxiongosaurus lufengensis (meaning "lizard from Chuxion, Lufeng") was described on the basis of a single specimen (CMY LT9401) consisting of a mostly complete skull and jaw. It was compared with *Thecodontosaurus* and was phylogenetically placed as a basal sauropodiform. It was distinguished from similar animals, such as *Jingshanosaurus*, based on several characteristics, including the number of teeth it had.

This same Chinese specimen had been previously reported as representing *Jingshanosaurus*. After the description of *Chuxiongosaurus*, a reanalysis of *Jingshanosaurus* disputed the validity of the genus and questioned the accuracy of the criteria that had been used to erect it; it was then concluded that *Chuxiongosaurus* was merely a synonym of *Jingshanosaurus* (Zhang et al., 2020).

50 cm

"G. sinensis"

(Young, 1941)
Length: 2.9 m (9.5 ft)
Height: 1 m (3.3 ft)
Reconstruction: ☐☐☐☐

195

| TRIASSIC | JURASSIC | CRETACEOUS |

The original (and, technically, current) type species of the genus **Gyposaurus** (meaning "vulture lizard") is *G. capensis*, named in 1911 from South Africa. However, this species was later determined to be the same animal as *Massospondylus* (Barrett et al., 2007), rendering the generic name invalid and undescriptive. But it seems that a second species within the genus, "*G. sinensis*", may actually be distinct from *Massospondylus* (Cooper, 1981). This Chinese species was named by C. C. Young in 1941.

If "*G. sinensis*" is indeed distinct from other known genera, it would not be proper to refer to it as *Gyposaurus*, since that name was invalidated. It could be renamed to a new genus, or the type species of the genus could formally be recategorized. More analyses are needed in order to determine the uniqueness of the species, as some researchers have suggested a synonymy with *Lufengosaurus* instead (Wang et al., 2017a).

PLATEOSAURAVUS

1 m

P. cullingworthi

(von Huene, 1932)
Length: 8.3 m (27 ft)
Height: 3.6 m (11.8 ft)
Reconstruction: ☐☐☐☐

210?

| TRIASSIC | JURASSIC | CRETACEOUS |

In 1924, a smattering of dinosaur bones discovered in South Africa were named to a new species of *Plateosaurus*, *P. cullingworthi*, by Sidney Haughton. In 1932, though, Friedrich von Huene reassessed the remains and decided that they actually represented an all-new genus, which he dubbed **Plateosauravus cullingworthi** (meaning "Cullingworth's grandfather of *Plateosaurus*"). In 1979, the fossils were once again reassigned by Jacques van Heerden, being lumped together with other remains as belonging to

the genus *Euskelosaurus*. However, this genus has now largely been recognized as a wastebasket taxon and is often treated as *nomen dubium*. Throughout the 2000s, paleontologist Adam Yates advocated that the name *Plateosauravus* once again be adopted. In 2019, work done by Emil Darius Krupandan reassessed each of the supposed *Plateosauravus* fossils, reevaluating which fossils could actually be confidently assigned as coming from the same animal and revalidating the establishment of the genus.

C. borealis

(Galton, 1985)
Length: 10.2 m (33.5 ft)
Height: 3.5 m (11.5 ft)
Reconstruction: ▢▢▢▢▢

205

| TRIASSIC | JURASSIC | CRETACEOUS |

Camelotia borealis (meaning "northern one from Camelot") is based on a very fragmentary set of remains found in the Westbury geological formation of Somerset, England. The publication that named the dinosaur was very brief, providing no analysis apart from a basic diagnostic definition. A later analysis would go on to examine the remains in greater detail and further solidified *Camelotia*'s status as a valid genus by highlighting the anatomical differences between it and the much more common *Plateosaurus* (Galton, 1998). This analysis grouped *Camelotia* into a family within Anchisauria known as **Melanorosauridae**, along with *Melanorosaurus* and *Lessemsaurus*. However, most subsequent studies have not found similar phylogenetic results, with many placing *Lessemsaurus* within Sauropoda; at most, *Melanorosaurus* and *Camelotia* may be sister taxa (Fernández and Werneburg, 2022).

AMYGDALODON

A. patagonicus

(Cabrera, 1947)
Length: 12 m (39.4 ft)
Height: 4.2 m (13.8 ft)
Reconstruction: ▢▢▢▢▢

175

| TRIASSIC | JURASSIC | CRETACEOUS |

Amygdalodon patagonicus (meaning "almond tooth from Patagonia") is notable for being the first sauropod species discovered from South America, and to this day it remains one of the few Jurassic dinosaurs known from the continent. The disparate remains were first discovered in 1936, consisting mostly of teeth and vertebral fragments.

With so little skeletal material to go on, *Amygdalodon* has occasionally been seen as *nomen dubium*, although a number of modern analyses have treated it as valid in their attempts to place the genus phylogenetically. Although its exact position is unclear, what seems fairly certain is that *Amygdalodon* sits somewhere near the base of Sauropoda (Pol et al., 2022).

1 m

Q. shouhu

(Han et al., 2024)
Length: 8.1 m (26.6 ft)
Height: 2.6 m (8.5 ft)
Reconstruction: ☐☐☐☐

199

| TRIASSIC | JURASSIC | CRETACEOUS |

Qianlong shouhu (meaning "guarding dragon from Qian") is known from three partial adult specimens found among five clutches of eggs, indicating a colonial type of reproductive strategy. Some of the eggs preserved embryos, which showed that hatchlings were likely quadrupedal. The eggshells are thought to have been "leathery," as opposed to either "soft" or "hard."

The generic name *Qianlong* combines the Chinese "Qian" (an alternate name for Guizhou Province) and "long" (meaning "dragon"). The specific name *shouhu* is Chinese for "guarding."

50 cm

M. sanyatiensis

(Barrett et al., 2024)
Length: 5.7 m (18.7 ft)
Height: 2.1 m (6.9 ft)
Reconstruction: ☐☐☐☐

210

| TRIASSIC | JURASSIC | CRETACEOUS |

Musankwa sanyatiensis (meaning "Musankwa from Sanyati") is an early sauropodomorph from an undersampled region of Africa, hinting that the area could still hold many other potential discoveries. Its describers considered *Musankwa* to be the basalmost member of Massopoda.

The generic name reflects the name of the house-boat that housed the field paleontologists, the *Musankwa*. The specific name *sanyatiensis* refers to the Sanyati River.

| 2 m | 4 m | 6 m | 8 m | 10 m |

4 m

2 m

L. wangi

(Zhang et al., 2024)
Length: 10.9 m (35.8 ft)
Height: 4.2 m (13.8 ft)
Hip height: 2.3 m (7.5 ft)
Body mass: 2,900 kg (6.4 t)
Reconstruction: ▢▢▢▢

| 250 | 245 | 240 | 235 | 230 | 225 | 220 | 215 | 210 | 205 | 200 | 193 | 190 | 185 | 180 | 175 | 170 | 165 | 160 | 155 | 150 | 145 | 140 | 135 | 130 | 125 | 120 | 115 | 110 | 105 | 100 | 95 | 90 | 85 | 80 | 75 | 70 | 65 |

| TRIASSIC | JURASSIC | CRETACEOUS |

Lishulong wangi (meaning "Wang's chestnut tree dragon") is the eighth genus of sauropodomorph dinosaur to be described from the Lufeng Formation of China. Out of all of these, *Lishulong* possessed the largest cranium (40 cm in length); this likely reflects the maximum size of *Lishulong*, as the fused state of the specimen's skeletal elements indicates that it had reached adulthood.

The large number of sauropodomorph species has offered a glimpse into the origins of the first dinosaurs to inhabit this region of China. These various species, while being somewhat closely related, do not form a singular group on the family tree. This means that sauropodomorphs didn't just arrive once in the region and then diversify; rather, it indicates that multiple dispersal events must have occurred, signaling the great success that sauropodomorphs enjoyed in the Early Jurassic.

The phylogenetic analysis conducted by the describing authors placed *Lishulong* as the sister taxon of *Yunnanosaurus*.

The generic name *Lishulong* combines the Chinese "lishu" (meaning "chestnut tree," which is the name of the region where the fossils were found), combined with "long" (meaning "dragon"). The specific name *wangi* honors fossil-hunter Zheng-Ju Wang.

CLASSIFICATION
Dinosauria
 Sauropodomorpha
  Bagualosauria
   Plateosauria
    Massopoda
     Sauropodiformes

LOCATION
China

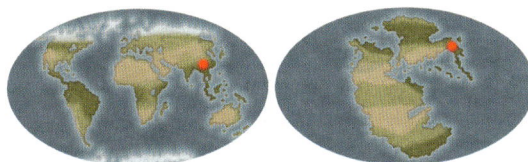

KNOWN REMAINS
Skull and neck

Sauropoda

Triassic

Late

Norian	Rhaetian
225	220	215	210	205

Jurassic

Early

Het.	Sinemurian	Pliensbachian	Toarcian
200	195	190	185	180

Middle

Aalenian	Baj.	Bath.	Callo.
175	170	165

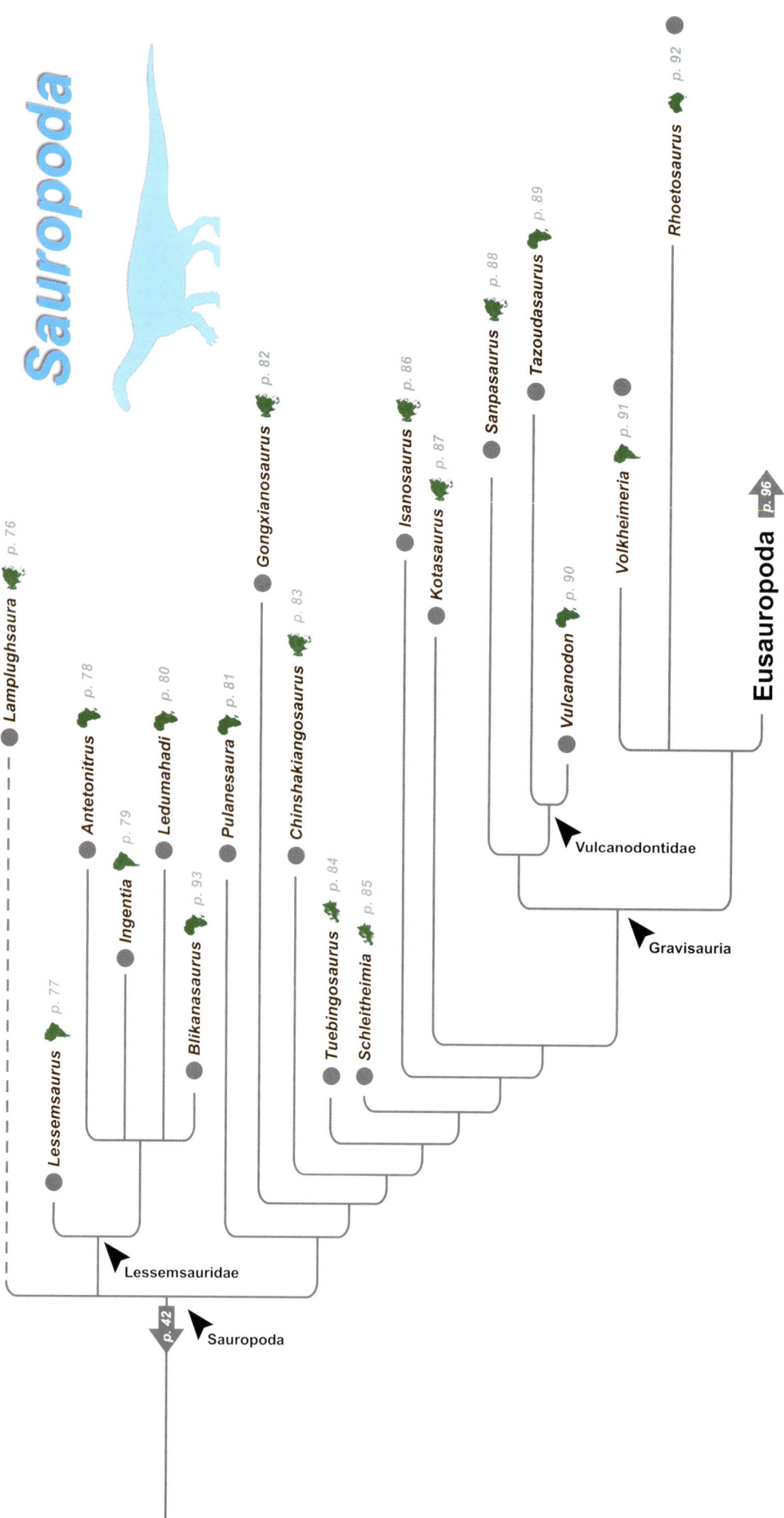

Late

Oxfo.

Lamplughsaura *p. 76*

Lessemsaurus *p. 77*

Antetonitrus *p. 78*

Ingentia *p. 79*

Ledumahadi *p. 80*

Blikanasaurus *p. 93*

Pulanesaura *p. 81*

Gongxianosaurus *p. 82*

Chinshakiangosaurus *p. 83*

Tuebingosaurus *p. 84*

Schleitheimia *p. 85*

Isanosaurus *p. 86*

Kotasaurus *p. 87*

Sanpasaurus *p. 88*

Tazoudasaurus *p. 89*

Vulcanodon *p. 90*

Volkheimeria *p. 91*

Rhoetosaurus *p. 92*

Lessemsauridae

Sauropoda *p. 42*

Vulcanodontidae

Gravisauria

Eusauropoda *p. 96*

What makes a sauropod? In recent history, there have been two main competing definitions for the group. In short, one definition put forth by Leonardo Salgado et al. in 1997 defined Sauropoda as beginning further up the family tree, whereas another definition suggested by Adam Yates in 2007 moved Sauropoda into a more basal position, essentially renaming the node previously known as "Sauropoda" to "Gravisauria." This second definition has definitely pulled ahead in recent years and is what is used in this book, although the 1997 definition is not without its proponents.

The earliest sauropods include a host of "transitional" species, those that display a number of characteristics of both the more basal sauropodomorphs and the more derived eusauropodan giants that would begin to dominate the Jurassic. Many of these intermediate lineages, though, did not survive the moderate extinction event that occurred between the Pliensbachian and Toarcian stages of the Jurassic, leaving the eusauropods to dominate (Pol et al., 2020).

Lamplughsaura could conceivably be either a sauropodomorph or a sauropod (Kutty et al., 2007). The basal subfamily **Lessemsauridae** includes *Lessemsaurus*, *Antetonitrus* (Apaldetti et al., 2018), *Ledumahadi* (McPhee et al., 2018), *Ingentia* (Pol et al., 2020), and possibly *Blikanasaurus* (Fernández and Werneburg, 2022).

Pulanesaura and *Gongxianosaurus* are often shown as the next most derived taxa (Zhang et al., 2018; Krupandan, 2019; Pol et al., 2020). The fragmentary *Chinshakiangosaurus*, when included in analyses, has also been depicted within this phylogenetic region (Upchurch et al., 2007; Bandyopadhyay et al., 2010). Following these are *Tuebingosaurus* (Fernández and Werneburg, 2022) and *Schleitheimia* (Rauhut et al., 2020).

Isanosaurus and (on the occasion when it is included) *Kotasaurus* are typically depicted as lying just outside **Gravisauria** (Bandyopadhyay et al., 2010; Pol et al., 2020; Rauhut et al., 2020). A thorough reanalysis of *Sanpasaurus* suggested that it was among the most derived non-eusauropod (McPhee et al., 2016), and a subsequent study placed it as the sister clade of Vulcanodontidae (Pol et al., 2020).

Vulcanodontidae has most recently been suggested to include *Vulcanodon* and *Tazoudasaurus* (Krupandan, 2019; Zhang et al., 2019; Pol et al., 2020; Rauhut et al., 2020). Modern reanalyzes of *Rhoetosaurus* suggest that it may be the most derived of all known non-eusauropods (Nair and Salisbury, 2012; Jannel et al., 2019). *Volkheimeria*'s affinities are highly uncertain but likely place it outside of Eusauropoda (Pol et al., 2020).

Pulanesaura eocollum

2 m   4 m   6 m   8 m

2 m

L. dharmaramensis

(Kutty et al., 2007)
Length: 7.8 m (25.6 ft)
Height: 2.5 m (8.2 ft)
Hip height: 2 m (6.6 ft)
Body mass: 1,200 kg (1.3 t)
Reconstruction:

250 245 240 235 230 225 220 215 210 205 200 195 190 185 180 175 170 165 160 155 150 145 140 135 130 125 120 115 110 105 100 95 90 85 80 75 70 65

TRIASSIC	JURASSIC	CRETACEOUS

Lamplughsaura dharmaramensis (meaning "Lamplugh's lizard from Dharmaram") is a transitional early sauropod featuring a heavy build and a quadrupedal gait.

Several sets of partial remains, which were discovered within the same locality and horizon as one another, constitute all of the known fossil material of *Lamplughsaura*. Based on size, at least two of the five individuals represented at the site were likely subadults. Taken together, approximately 80% of the animal's skeleton is accounted for, including significant portions of the skull.

The original description of *Lamplughsaura* could not definitively determine the animal's phylogenetic placement. Various methods of analysis placed it either among the earlier stem sauropodomorphs or as a basal member of **Sauropoda**. The authors' conclusion was that *Lamplughsaura* was most likely to be a true sauropod.

If *Lamplughsaura* is indeed a basal sauropod, it could potentially represent something of a transitional form, as it displays a mixture of both ancestral and derived characteristics. For instance, some traits of the animal's jaw are indicative of sauropods, while at least one trait had previously been considered unique to "prosauropod" sauropodomorphs. The presence of a claw that is only slightly curved, instead of greatly curved, on the first digit suggests a derived phylogenetic position, whereas the twisted shape of the first finger bone is indicative of a more basal state. The teeth of *Lamplughsaura* also bore a textured enamel surface indicative of Sauropoda.

The generic name *Lamplughsaura* honors Pamela Lamplugh, founder of the Indian Statistical Institute. The suffix "saura" is intended to be the feminine form of "saurus." The specific name *dharmaramensis* refers to the Dharmaram geological formation of India.

CLASSIFICATION
Dinosauria
  Sauropodomorpha
   Bagualosauria
    Plateosauria
     Massopoda
      Sauropodiformes
       Anchisauria
        Sauropoda

LOCATION
India

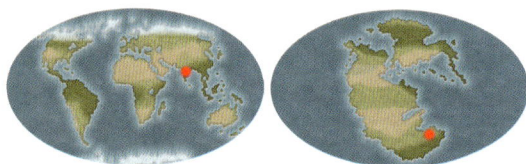

KNOWN REMAINS
Partial skull and skeleton

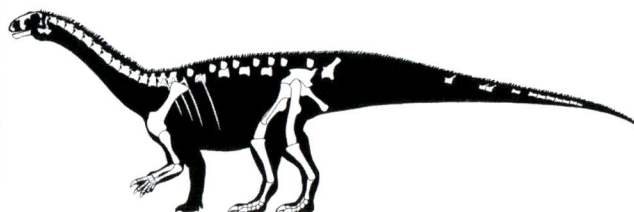

2 m   4 m   6 m   8 m   10 m   4 m

2 m

L. sauropoides

(Bonaparte, 1999)
Length: 10.3 m (33.8 ft)
Height: 3 m (9.8 ft)
Hip height: 2.5 m (8.2 ft)
Body mass: 2,900 kg (3.2 t)
Reconstruction: ▢▢▢▢

250 245 240 235 230 225 220 **215** 210 205 200 195 190 185 180 175 170 165 160 155 150 145 140 135 130 125 120 115 110 105 100 95 90 85 80 75 70 65

| TRIASSIC | JURASSIC | CRETACEOUS |

Lessemsaurus sauropoides (meaning "Lessem's sauropod-like lizard") is an early sauropod that shares numerous traits with the later, large-bodied, derived eusauropods, although at the same time it is also lacking in many of that group's typical characteristics.

Lessemsaurus is primarily known from an assemblage of various fossils unearthed in 1971, and catalogued as PVL 4822. In the animal's original description, only a collection of vertebral fragments were used as the basis of the genus; these exact bones are now specifically known as PVL 4822–1. A host of other remains, though, were found in association (but not articulation) with these vertebral bones, and a more recent analysis has identified them as also belonging to *Lessemsaurus* (and thus not to the other sauropodomorphs known from the Los Colorados geological formation, such as *Riojasaurus* or *Coloradisaurus*). These specimens come from more than one individual, as evidenced by repeated skeletal elements and bones of varying size (Pol and Powell, 2007).

The animal's original description placed it within the now-dubious subfamily known as Melanorosauridae. Later studies, though, found that *Lessemsaurus* bore a strong resemblance only to the South African genus *Antetonitrus*, suggesting that the two might form a clade of their own. The more recent descriptions of *Ledumahadi* and *Ingentia* have seemingly confirmed that a distinct branch of the sauropod family tree does indeed exist, namely **Lessemsauridae** (Apaldetti et al., 2018).

The generic name *Lessemsaurus* honors Donald Lessem, founder of the Dinosaur Society. The specific name *sauropoides* is meant to indicate Bonaparte's view that the species is comparable to sauropods but not actually within that clade.

CLASSIFICATION
Dinosauria
  Sauropodomorpha
    Sauropodiformes
      Anchisauria
        Sauropoda
          Lessemsauridae

LOCATION
Argentina

KNOWN REMAINS
Partial skeleton

2 m 4 m 6 m 8 m

2 m

A. ingenipes

(Yates and Kitching, 2003)
Length: 7.8 m (25.6 ft)
Height: 2.2 m (7.2 ft)
Hip height: 1.8 m (5.9 ft)
Body mass: 1,250 kg (1.4 t)
Reconstruction: ▢▢▢▢

250 245 240 235 230 225 220 215 210 205 200 195 190 185 180 175 170 165 160 155 150 145 140 135 130 125 120 115 110 105 100 95 90 85 80 75 70 65

TRIASSIC **JURASSIC** **CRETACEOUS**

In terms of forelimb morphology, *Antetonitrus ingenipes* (meaning "massive foot before thunder") is one of the clearest examples of an intermediate form between the bipedal sauropodomorphs and the quadrupedal sauropods. Although *Antetonitrus* engaged in some degree of quadrupedalism, its forelimbs were not well adapted for heavy weight bearing; rather, its thumb still retained a degree of mobility and could possibly be used to perform grasping actions. A strengthened hindfoot, along with the presence of certain muscle attachment sites on the animal's femur, suggest that it could have also been capable of some form of bipedalism (McPhee et al., 2014).

The first known remains of *Antetonitrus* were unearthed in 1981 and were misidentified as the now-dubious *Euskelosaurus*. In addition to the remains that were initially cataloged in the animal's description, further fragmentary remains from several individuals have subsequently been assigned to the genus (McPhee et al., 2014).

Multiple phylogenetic analyses have supported the position that *Antetonitrus* is a member of the subfamily **Lessemsauridae** (McPhee et al., 2014; Apaldetti et al., 2018; Rauhut et al., 2020). At the time of its description, the holotype material was believed to have been from the Lower Elliot Formation, which has a Late Triassic age; however, later studies have shown that an Early Jurassic designation from the Upper Elliot Formation is the actual origin (McPhee et al., 2017).

The generic name *Antetonitrus* combines the Latin "ante" (meaning "before") and "tonitrus" (meaning "thunder"). This is in reference to the fact that the animal existed before well-known sauropods such as *Brontosaurus*, the "thunder lizard." The specific name *ingenipes* combines the Latin "ingens" (meaning "massive") and "pes" (meaning "foot").

CLASSIFICATION
Dinosauria
  Sauropodomorpha
   Sauropodiformes
   Anchisauria
   Sauropoda
    Lessemsauridae

LOCATION
South Africa

KNOWN REMAINS
Partial skeleton

2 m   4 m   6 m

I. prima

(Apaldetti et al., 2018)
Length: 6.8 m (22.3 ft)
Height: 2.3 m (7.5 ft)
Hip height: 1.9 m (6.2 ft)
Body mass: 850 kg (1,870 lb)
Reconstruction: ▢▢▢▢

2 m

| 250 | 245 | 240 | 235 | 230 | 225 | 220 | 215 | 210 | **205** | 200 | 195 | 190 | 185 | 180 | 175 | 170 | 165 | 160 | 155 | 150 | 145 | 140 | 135 | 130 | 125 | 120 | 115 | 110 | 105 | 100 | 95 | 90 | 85 | 80 | 75 | 70 | 65 |

TRIASSIC   **JURASSIC**   **CRETACEOUS**

Ingentia prima (meaning "first huge one") broke the mold for what it took for sauropods to become truly "big."

The genuinely gargantuan titanosauriforms, such as the famous *Brachiosaurus*, had a well-defined suite of characteristics that seemed to be inexorably tied to their massive stature: sturdy, column-like legs to best support their massive weight; thermoregulating air sacs present along their elongated necks; and a rapid, continuous growth phase that allowed them to get big quickly.

Ingentia, however, had none of these characteristics, despite being one of the biggest living creatures of its time. Its forelimbs were bent and relatively flexible, yet were somehow still sufficiently sturdy enough to support *Ingentia*'s increased mass. *Ingentia* also possessed pneumatic structures that intertwined with their vertebrae, similar to

the respiratory system seen in modern birds, which served to help regulate the animal's body temperature and lighten the weight of the massive bones.

Perhaps most notably, *Ingentia* had a peculiar pattern of bone growth as it aged. Like many dinosaurs—but unlike the titanic later sauropods—*Ingentia* went through regular cycles of growth spurts. Unusually, though, these spurts were turbocharged, occurring at two to three times the growth rate of the colossal titanosaurs. This difference shows that the path toward gigantism is not linear but can, in fact, be achieved through differing suites of adaptations. Further, these traits were concluded by researchers to be present in all members of the early sauropod subfamily **Lessemsauridae**.

The binomial name *Ingentia prima* literally translates to "huge first" from Latin, using the feminine word forms.

CLASSIFICATION

Dinosauria
  Sauropodomorpha
    Sauropodiformes
      Anchisauria
        Sauropoda
          Lessemsauridae

LOCATION

Argentina

KNOWN REMAINS

Fragments

2 m 4 m 6 m 8 m 10 m 12 m

4 m

2 m

L. mafube

(McPhee et al., 2018)
Length: 10.8 m (35.4 ft)
Height: 3.3 m (10.8 ft)
Hip height: 2.9 m (9.5 ft)
Body mass: 3,400 kg (3.7 t)
Reconstruction:

250 245 240 235 230 225 220 215 210 205 200 195 190 185 180 175 170 165 160 155 150 145 140 135 130 125 120 115 110 105 100 95 90 85 80 75 70 65

| TRIASSIC | JURASSIC | CRETACEOUS |

Ledumahadi mafube (meaning "giant thunderclap at dawn") is an early sauropod that displayed quadrupedal traits.

Unlike the later giant sauropods, such as *Diplodocus*, *Ledumahadi* did not have column-like limbs that were composed of relatively narrow bones. Rather, *Ledumahadi* likely had a "flexed" or "crouched" limb posture, as evidenced by the shape and robustness of its ulna and femur, which greatly resemble those of earlier sauropodomorphs. This posture is not the sort of "sprawl" that one might observe in a crocodile but rather is more akin to the way that some smaller mammals hold themselves up.

This combination of traits indicates that the sauropod lineage did not simply follow a single path toward the evolution of gigantism—namely, by developing column-like limbs. Rather, *Ledumahadi* highlights how multiple evolutionary pathways can arrive at the same destination.

Additionally, the describers extrapolate that many earlier sauropodomorphs that had previously been classified as bipedal might actually have been quadrupedal, adopting a stance similar to that of *Ledumahadi*. If true, then the shift in the mode of locomotion must have evolved very rapidly.

Like other **lessemsaurids**, *Ledumahadi* grew quite rapidly in annual spurts, as evidenced by the microstructure observed within the animal's bones. The only known specimen of *Ledumahadi* (BP/1/7120) had ended its growth by the time it died, indicating that it was a fully grown adult.

The generic name *Ledumahadi* translates from the indigenous Sotho language as "giant thunderclap," which is meant to refer to the animal's large size. The specific name *mafube* means "dawn" in the same language, referring to the creature's early appearance among the sauropods.

CLASSIFICATION
Dinosauria
  Sauropodomorpha
    Sauropodiformes
      Anchisauria
        Sauropoda
          Lessemsauridae

LOCATION
South Africa

KNOWN REMAINS
Fragments

P. eocollum

(McPhee et al., 2015)
Length: 8 m (26.2 ft)
Height: 2.5 m (8.2 ft)
Hip height: 2.2 m (7.2 ft)
Body mass: 1,100 kg (1.2 t)
Reconstruction: ☐☐☐☐

250	245	240	235	230	225	220	215	210	205	200	195	190	185	180	175	170	165	160	155	150	145	140	135	130	125	120	115	110	105	100	95	90	85	80	75	70	65

TRIASSIC | **JURASSIC** | **CRETACEOUS**

Pulanesaura eocollum (meaning "rain-bringing dawn-neck lizard") is thought to have been a "low browser," feeding off vegetation that was at or close to ground level. This conclusion was reached partially based on the structure and orientation of the animal's vertebrae, as well as the enamel structure of its teeth.

What makes this conclusion significant is that this trait clearly differentiates *Pulanesaura* from several other sauropodomorph species that have been discovered from the same geological formation, the Upper Elliot. These differences likely indicate niche partitioning, by which similar species avoid competition by engaging in different behaviors or by favoring different foods.

Citing uncertainties in which criteria definitively define Sauropoda, the describers acknowledged that *Pulanesaura* may be the basalmost sauropod but favored the view that *Pulanesaura* was actually sister to Sauropoda. Although

several analyses have supported the conclusion that *Pulanesaura* falls well within **Sauropoda** (Krupandan, 2019; Fernández and Werneburg, 2022), numerous others have placed *Pulanesaura* in a more basal position among the group (Chapelle et al., 2019; Zhang et al., 2019; Peyre de Fabrègues et al., 2020; Rauhut et al., 2020).

As a consequence of this disagreement, and the fragmentary nature of the fossils, it is unclear whether *Pulanesaura* was bipedal or quadrupedal.

The generic name *Pulanesaura* combines the indigenous Sesotho term "pulane" (meaning "rain-bringer," referring to the weather conditions with which the fossil excavators dealt) with the Greek "saura" (a feminine form of "sauros," meaning "lizard"). The specific name *eocollum* combines the Greek "eo" (meaning "dawn") with the Latin "collum" (meaning "neck"), referring to the animal's relatively short (and, thus, "primitive") neck.

CLASSIFICATION

Dinosauria
  Sauropodomorpha
    Bagualosauria
      Plateosauria
        Massopoda
          Sauropodiformes
            Anchisauria
              Sauropoda

LOCATION

South Africa

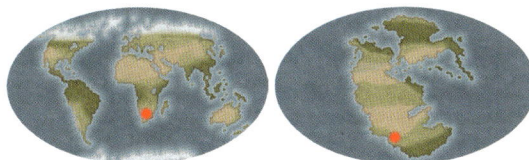

KNOWN REMAINS

Fragments

| 2 m | 4 m | 6 m | 8 m | 10 m | 12 m |

4 m

2 m

G. shibeiensis

(He et al., 1998)
Length: 12.5 m (41 ft)
Height: 4 m (13.1 ft)
Hip height: 3 m (9.8 ft)
Body mass: 4,000 kg (4.4 t)
Reconstruction: ☐☐☐☐☐

| 250 | 245 | 240 | 235 | 230 | 225 | 220 | 215 | 210 | 205 | 200 | 195 | 190 | 187 | 180 | 175 | 170 | 165 | 160 | 155 | 150 | 145 | 140 | 135 | 130 | 125 | 120 | 115 | 110 | 105 | 100 | 95 | 90 | 85 | 80 | 75 | 70 | 65 |

| TRIASSIC | JURASSIC | CRETACEOUS |

Although *Gongxianosaurus shibeiensis* (meaning "lizard from Shibei, Gongxian") has not been described in great detail, there are nevertheless several known characteristics of the animal that are noteworthy and can be analyzed.

Gongxianosaurus has several skeletal traits that are more indicative of ancestral sauropodomorphs, as opposed to later sauropods. In particular, its limbs are unusually ossified in comparison to the vast majority of sauropods, which normally replace certain areas of bone with shock-absorbing cartilage; in contrast, *Gongxianosaurus* is among the only known sauropods with bony, ossified distal ankle elements. Additionally, only three fused vertebrae make up the sacrum portion of its hip, which is fewer than in the later sauropods, and its vertebrae lacked certain weight-reducing grooves that are common in sauropods.

Still, some traits of *Gongxianosaurus* are clearly sauropod in nature: its forelimbs are elongated in comparison to its hindlimbs, its foot bones are broadened, its toe claws are stout, and its overall body size is large. Taken altogether, this combination of traits likely indicates that *Gongxianosaurus* is an "intermediate" form of sauropod, caught in transition between "primitive" and "derived" forms.

Gongxianosaurus is known from four individual specimens of varying levels of completeness, including at least one younger individual. Taken altogether, the majority of the body is accounted for, except for most of the skull and the hand. These specimens were all discovered in 1997 from the Ziliujing geological formation. It has been suggested, though, that some of the sauropod remains found at the location might actually belong to a second species within the genus (Yaonan and Changsheng, 2000), or even potentially to another genus entirely, unofficially dubbed "*Yibinosaurus zhoui*" (Ouyang, 2003).

Owing to the lack of detailed specimen descriptions or measurements, the phylogenetic relationships of *Gongxianosaurus* are uncertain. What data exists, though, suggests that it is among the basalmost **sauropods** (Apaldetti et al., 2011).

CLASSIFICATION
Dinosauria
　Sauropodomorpha
　　Bagualosauria
　　　Plateosauria
　　　　Massopoda
　　　　　Sauropodiformes
　　　　　　Anchisauria
　　　　　　　Sauropoda

LOCATION
China

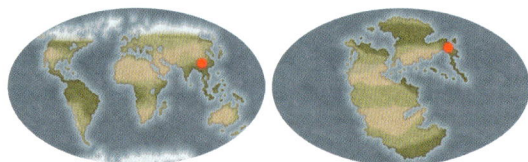

KNOWN REMAINS
Nearly complete

2 m   4 m   6 m   8 m   10 m

4 m

C. chunghoensis

(Dong, 1992)

Length: 10.1 m (33.1 ft)
Height: 3.6 m (11.8 ft)
Hip height: 3 m (9.8 ft)
Body mass: 2,750 kg (3 t)
Reconstruction: ☐☐☐☐

2 m

250 245 240 235 230 225 220 215 210 205 **200** 195 190 185 180 175 170 165 160 155 150 145 140 135 130 125 120 115 110 105 100 95 90 85 80 75 70 65

| TRIASSIC | JURASSIC | CRETACEOUS |

Chinshakiangosaurus chunghoensis (meaning "Chinshakian lizard from Chungho") is notable because of the features of its jaw, although much of its fossil material remains almost entirely unstudied. At the time of its description, this jaw was the only example known from an Early Jurassic sauropod, which provided integral clues regarding the evolution of this lineage's characteristic feeding habits.

The various bones that were cataloged as specimen IVPP V14474 were unearthed in 1970 from the lower reaches of the Fengjiahe geological formation. The name *Chinshakiangosaurus chunghoensis* was proposed first (Ye, 1975), while the alternative translation *C. zhonghoensis* was later used (Zhao, 1985). A lack of sufficient description rendered these *nomina nuda*, or "bare names." This situation was partially rectified in 1992 by Dong Zhiming, who officially penned a brief description of the animal's

mandible, femur, and partial neck vertebra. However, the specimen also includes "dorsal vertebrae, several cranial caudal vertebrae, a pair of scapulae, an incomplete pelvic girdle, and the hindlimbs," which have never been figured or described in any way.

During a more recent attempt to study *Chinshakiangosaurus*, these other fossils still remained "in storage" and "could not be accessed." Still, the reevaluation of the jaw confirmed that the animal retained a fleshy "cheek," in much the same way the earlier sauropodomorphs did. Moreover, *Chinshakiangosaurus* was phylogenetically placed among the basalmost **sauropods** (Upchurch et al., 2007), a position upheld by later studies (Bandyopadhyay et al., 2010).

The animal's binomial name refers to the Yangtze River and the village of Zhonghe in China.

CLASSIFICATION
Dinosauria
  Sauropodomorpha
    Bagualosauria
      Plateosauria
        Massopoda
          Sauropodiformes
            Anchisauria
              Sauropoda

LOCATION
China

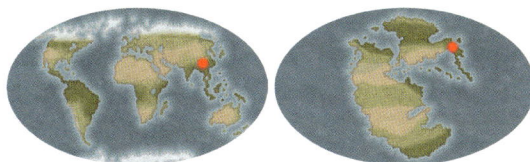

KNOWN REMAINS
Jaw and skeletal elements

2 m

2 m | 4 m | 6 m | 8 m

T. maierfritzorum

(Fernández and Werneburg, 2022)
Length: 7.8 m (25.6 ft)
Height: 3.2 m (10.5 ft)
Hip height: 2.6 m (8.5 ft)
Body mass: 1,000 kg (1.1 t)
Reconstruction:

250 — 245 — 240 — 235 — 230 — 225 — 220 — 215 — **211** — 205 — 200 — 195 — 190 — 185 — 180 — 175 — 170 — 165 — 160 — 155 — 150 — 145 — 140 — 135 — 130 — 125 — 120 — 115 — 110 — 105 — 100 — 95 — 90 — 85 — 80 — 75 — 70 — 65

TRIASSIC | **JURASSIC** | **CRETACEOUS**

A full century after its discovery, the fossil specimen often known as "GPIT IV" was reexamined and identified as a new genus and species, ***Tuebingosaurus maierfritzorum*** (meaning "Maiers and Fritz's lizard from Tübingen").

Unearthed in 1922, the specimen now identified as GPIT-PV-30787 was long thought to be the remains of a *Plateosaurus*. Indeed, the specimen was often used as an example reference for *Plateosaurus* anatomy. Even though hundreds of *Plateosaurus* specimens have been "identified" in the last century, there is no definitive reference of all known fossils, and relatively few have undergone thorough analysis. Oftentimes, similar-looking fossils (especially those from Germany) have been lumped into the genus without detailed examination.

Hence, a recent reexamination of this particular specimen has concluded that it belongs to an animal that is discernibly different from *Plateosaurus*. Dubbed *Tuebingosaurus*, it appears to have not been all that closely related to *Plateosaurus* at all. Rather, *Tuebingosaurus* has certain traits that suggest that it is of a more derived evolutionary form, being more closely related to true **sauropods** than the early sauropodomorphs.

For example, although the available fossil material lacks any trace of the animal's forelimbs, the describing researchers have concluded that *Tuebingosaurus* was likely quadrupedal rather than being bipedal like the more "primitive" sauropodomorphs. This determination was made, in part, by comparing features of the animal's ankle with other related species.

The generic name *Tuebingosaurus* refers to Tübingen, Germany. The specific name *maierfritzorum* honors Wolfgang Maier along with editor Uwe Fritz.

CLASSIFICATION
Dinosauria
  Sauropodomorpha
    Bagualosauria
      Plateosauria
        Massopoda
          Sauropodiformes
            Anchisauria
              Sauropoda

LOCATION
Germany

KNOWN REMAINS
Leg and hip

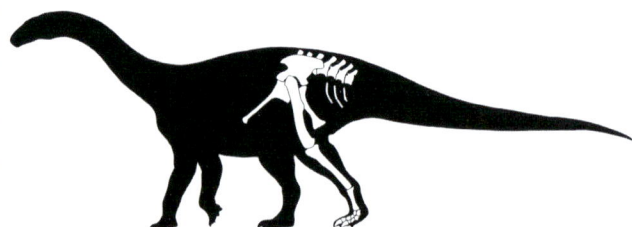

2 m   4 m   6 m

S. schutzi

(Rauhut et al., 2020)
Length: 6.5 m (21.3 ft)
Height: 2.3 m (7.5 ft)
Hip height: 1.6 m (5.2 ft)
Body mass: 550 kg (1,200 lb)
Reconstruction: ▢▢▢▢

2 m

| 250 | 245 | 240 | 235 | 230 | 225 | 220 | 215 | **210** | 205 | 200 | 195 | 190 | 185 | 180 | 175 | 170 | 165 | 160 | 155 | 150 | 145 | 140 | 135 | 130 | 125 | 120 | 115 | 110 | 105 | 100 | 95 | 90 | 85 | 80 | 75 | 70 | 65 |

| TRIASSIC | JURASSIC | CRETACEOUS |

Schleitheimia schutzi (meaning "Schutz's one from Schleitheim") is the first known European species that straddles the line between the sauropodomorphs and the sauropods.

The first fossils now assigned to *Schleitheimia* were collected between 1952 and 1954 and were subsequently donated to the University of Zürich. Many of these bones were examined by Peter Galton in 1986, who then referred them to *Plateosaurus engelhardti*. But a 2015 reevaluation of the material revealed distinct traits that were contradictory with their assignment to *Plateosaurus*; in particular, the robustness of many of the skeletal elements was not characteristic of *Plateosaurus*, and while *Plateosaurus* was primarily bipedal, this new animal seemed to have been quadrupedal. Thus, the new genus *Schleitheimia* was erected.

Despite their anatomical differences, *Schleitheimia* and *Plateosaurus* did indeed share a habitat. This shows that "primitive" bipedal sauropodomorphs shared a habitat with "advanced" quadrupedal sauropods for quite some time. While one lineage of sauropodomorphs (including *Schleitheimia*) developed their bodies to achieve increased size (eventually becoming the **sauropods**), others survived contemporaneously while maintaining their smaller sizes.

It remains unclear why the sauropods eventually outcompeted the bipedal sauropodomorphs such as *Plateosaurus*. What is clear, though, is that the end-Triassic mass extinction had little effect on the overall diversity or evolution of the sauropod lineages.

The generic name *Schleitheimia* refers to the locality of Schleitheim in Switzerland, where the holotype fossil was unearthed. The specific name *schutzi* honors the collector of the type material, Emil Schutz.

CLASSIFICATION

Dinosauria
 Sauropodomorpha
  Bagualosauria
   Plateosauria
    Massopoda
     Sauropodiformes
      Anchisauria
       Sauropoda

LOCATION

Switzerland

KNOWN REMAINS

Fragments

2 m 4 m 6 m 8 m 10 m 12 m 14 m

I. attavipachi

(Buffetaut et al., 2000)
Length: 13 m (43 ft)
Height: 4.6 m (15.1 ft)
Hip height: 3.6 m (9.8 ft)
Body mass: 7,000 kg (7.7 t)
Reconstruction:

4 m

2 m

250 245 240 235 230 225 220 215 210 205 200 195 190 **187** 185 180 175 170 165 160 155 150 145 140 135 130 125 120 115 110 105 100 95 90 85 80 75 70 65

| TRIASSIC | JURASSIC | CRETACEOUS |

Isanosaurus attavipachi (meaning "Attavipach's lizard from Isan") is a somewhat enigmatic sauropod that was once thought to have been much more ancient than it actually was.

The holotype specimen of *Isanosaurus* (CH4) was discovered in 1998, and consisted of a handful of vertebral fragments as well as most of the scapula and femur. This individual was presumed to be a fully grown adult and was estimated to have been about 8.3 meters in length. Shortly thereafter, though, a second, heavily eroded and fragmentary specimen (MH 350) was discovered nearby—one that was significantly larger. Based on the most intact bone (the humerus), this individual was estimated to have been at least 13 meters in length. (Since there was little to no skeletal overlap between the two specimens, MH 350 could not confidently be assigned to *I. attavipachi*, and thus was only referred to as *Isanosaurus* sp. [Buffetaut et al., 2002]).

Both of these specimens were originally thought to have been from the lower reaches of the Nam Phong geological formation, which would have meant that truly massive sauropods had already evolved by the Late Triassic—a revolutionary claim that would have pushed back the known **Sauropoda** timeline by tens of millions of years. However, newer research (including the discovery of another specimen, CH8–66, the partial humerus of a young adult) has shown that the various *Isanosaurus* remains have actually come from the upper reaches of the formation, making them solidly Jurassic in age, likely from the Pliensbachian epoch (Laojumpon et al., 2017; Jentgen-Ceschino et al., 2020).

The generic name *Isanosaurus* is derived from "Isan," the local name for the northeastern reaches of Thailand. The specific name *attavipachi* honors P. Attavipach, former director general of the Thai Department of Mineral Resources.

CLASSIFICATION

Dinosauria
  Sauropodomorpha
    Bagualosauria
      Plateosauria
        Massopoda
          Sauropodiformes
            Anchisauria
              Sauropoda

LOCATION

Thailand

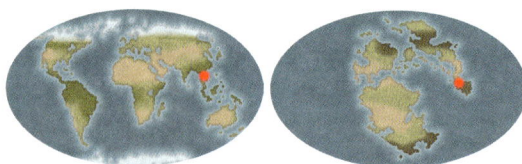

KNOWN REMAINS

Fragments

K. yamanpalliensis

(Yadagiri, 1988)
Length: 12.4 m (40.7 ft)
Height: 5 m (16.4 ft)
Hip height: 3 m (9.8 ft)
Body mass: 4,500 kg (5 t)
Reconstruction: ☐☐☐☐

| TRIASSIC | JURASSIC | CRETACEOUS |

Kotasaurus yamanpalliensis (meaning "Kota lizard from Yamanpally") is clearly of the sauropodan lineage, as the animal possessed four fused sacral (hip) vertebrae, a relatively narrow and straight femur, and spoon-shaped teeth, among numerous other telltale qualities. Yet, it also retained a few ancestral sauropodomorph traits (such as the shape and proportionality of its humerus), which mark *Kotasaurus* as one of the very earliest of sauropods (Yadagiri, 2001).

In the late 1970s, a large collection of fossils was unearthed amidst what had once been a riverbed during the Early Jurassic. It has been hypothesized that these individuals perished together as a herd during a flash flood and that their bodies were deposited collectively.

These scattered remains, which once belonged to at least a dozen individuals, together constitute nearly the entire skeleton of *Kotasaurus*, with only the skull and jaw being absent (with the exception of two teeth). The lack of skull material is standard for sauropods, as the connection between the head and the neck of these animals was apparently very weak.

Phylogenetic studies consistently place *Kotasaurus* among the most basal of **sauropods**, but its exact placement in relation to closely related genera, such as *Vulcanodon*, often differs between analyses (Bandyopadhyay et al., 2010; Sekiya, 2011; Mocho et al., 2014).

A preliminary study has suggested the possibility that *Kotasaurus* had a small club on the end of its tail, similar to its close relative *Omeisaurus*, based on the remains of 12 different individuals (Kareem et al., 2024).

The generic name *Kotasaurus* refers to the Kota geological formation. The specific name *yamanpalliensis* refers to the village of Yamanpally.

CLASSIFICATION

Dinosauria
  Sauropodomorpha
    Sauropodiformes
      Anchisauria
        Sauropoda

LOCATION

India

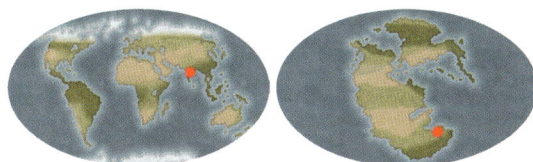

KNOWN REMAINS

Skeleton

2 m   4 m   6 m   8 m

2 m

S. yaoi

(Young, 1944)
Length: 7.2 m (23.6 ft)
Height: 4 m (13.1 ft)
Hip height: 2 m (6.6 ft)
Body mass: 1,100 kg (1.2 t)
Reconstruction: ☐☐☐☐

250 245 240 235 230 225 220 215 210 205 200 195 190 **183** 180 175 170 165 160 155 150 145 140 135 130 125 120 115 110 105 100 95 90 85 80 75 70 65

TRIASSIC   **JURASSIC**   **CRETACEOUS**

Sanpasaurus yaoi (meaning "Yao's lizard from Sanpa") was long held to be *nomen dubium* but has since been validated as being a legitimate taxon.

When the specimens were first described by the famous Chinese paleontologist C. C. Young, he misidentified *Sanpasaurus* as an iguanodontid. One of several reasons for this misidentification is that the pedal claws of *Sanpasaurus* are unusually compressed, a trait that was unknown in sauropods at the time but which has since been identified in African taxa, such as *Vulcanodon*. In 1967, A. K. Rozhdestvensky correctly reinterpreted *Sanpasaurus* as a sauropod.

Efforts to determine the animal's phylogenetic placement within Sauropoda are hampered both by the incompleteness of the remains and the fact that the records regarding where, exactly, each bone was excavated are nonexistent; this means

that, despite researchers' best efforts, it is possible that not all of these bones (which undoubtedly came from several individuals of varying ages) are actually from the same species.

Even so, the reevaluation that relegitimized *Sanpasaurus* as a genus proposed that it was likely a highly derived non-eusauropodan sauropod, and subsequent studies have come to similar conclusions. A close relationship with certain African taxa, such as *Vulcanodon*, shows that at least some sauropodan lineages had achieved global coverage by this point in the Jurassic (McPhee et al., 2016; Pol et al., 2020).

The generic name *Sanpasaurus* refers to the ancient name for Szechuan Province, China. The specific name *yaoi* honors H. H. Yao, of the National Geological Survey of China, who discovered the remains in 1939.

CLASSIFICATION
Dinosauria
  Sauropodomorpha
   Sauropodiformes
    Anchisauria
     Sauropoda
      Gravisauria

LOCATION
China

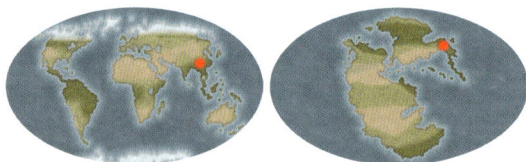

KNOWN REMAINS
Partial skeleton

2 m 4 m 6 m 8 m 10 m 12 m 14 m

6 m

T. naimi

(Allain et al., 2004)
Length: 12.4 m (40.7 ft)
Height: 5.8 m (19 ft)
Hip height: 2.7 m (8.9 ft)
Body mass: 6,000 kg (6.6 t)
Reconstruction: ▢▢▢▢

4 m

2 m

250 245 240 235 230 225 220 215 210 205 200 195 190 185 **181** 175 170 165 160 155 150 145 140 135 130 125 120 115 110 105 100 95 90 85 80 75 70 65

TRIASSIC **JURASSIC** **CRETACEOUS**

Tazoudasaurus naimi (meaning "slender lizard from Tazouda") is one of the only Northern Hemisphere sauropods known from the Toarcian stage of the Jurassic period.

Several different dig sites at nearby locations have yielded multiple fossils of *Tazoudasaurus*. Some of these come from very young juveniles, while others are clearly adults. Excavations at these sites are expected to yield more fossil material in the future. A few theropod bones, from a basal abelisauroid known as *Berberosaurus*, mark the only other vertebrate fossils of any kind to be found at the sites.

Given that none of the bones show any signs of weathering and these fossils have all been excavated from a single layer of mudstone suggests that a deadly mudslide is what spelled the doom of these creatures. That so many *Tazoudasaurus* are clustered so close to one another suggests that the species displayed some level of herding behavior (Allain and Aquesbi, 2008).

One adult specimen of *Tazoudasaurus* is notable for preserving a nearly intact mandible as well as a partial cranium—a rarity for sauropods. By examining the jaw and tooth wear patterns, it has been determined that *Tazoudasaurus* likely chewed their food, as opposed to many later sauropods that simply stripped vegetation from branches to swallow whole (Peyer and Allain, 2010).

Analyses generally agree that *Tazoudasaurus* is the sister taxon to *Vulcanodon*, placing both into the subfamily **Vulcanodontidae**, which lies outside Eusauropoda (Pol et al., 2020; Rauhut et al., 2020).

The generic name *Tazoudasaurus* refers to the locality where the first fossils were collected, Tazouda. The specific name *naimi* is the Berber word for "slender," owing to the animal's relatively small size.

CLASSIFICATION
Dinosauria
  Sauropodomorpha
   Sauropodiformes
    Anchisauria
     Sauropoda
      Gravisauria
       Vulcanodontidae

LOCATION
Morocco

KNOWN REMAINS
Mostly complete

89

2 m   4 m   6 m   8 m   10 m   12 m

V. karibaensis

(Raath, 1972)
Length: 11 m (36.1 ft)
Height: 4 m (13.1 ft)
Hip height: 3 m (9.8 ft)
Body mass: 3,500 kg (3.9 t)
Reconstruction: ▢▢▢▢▢

4 m

2 m

250 245 240 235 230 225 220 215 210 205 200 **195** 190 185 180 175 170 165 160 155 150 145 140 135 130 125 120 115 110 105 100 95 90 85 80 75 70 65

TRIASSIC   **JURASSIC**   **CRETACEOUS**

Vulcanodon karibaensis (meaning "volcano tooth from Kariba") was once believed to be the earliest sauropod on record. However, more accurate dates have determined that it actually lived much later in time than was originally thought (Yates et al., 2004).

The describers of *Vulcanodon* originally classified the animal as an early sauropodomorph, based on features of its pelvis, and the observation that some of its teeth were sharp, suggesting omnivory. However, just a few years later, other researchers noticed that aspects of the animal's foot matched those of sauropods, not sauropodomorphs (Cruickshank, 1975). Additionally, its supposed teeth are now known to have actually come from a theropod, one that was likely scavenging the carcass of *Vulcanodon* (Cooper, 1984).

It is well established that *Vulcanodon* is a very basal member of Sauropoda, but who its closest relatives are is still a matter of debate. The subfamily **Vulcanodontidae** has, at various points in time, included numerous species, but its validity has occasionally been called into question.

The estimated size and mass of *Vulcanodon* have varied greatly between different studies as a consequence of the fragmentary nature of the skeleton. Many sources cite a length of 6.5 meters for the animal, but more recent works cite larger numbers (Paul, 2016; McPhee et al., 2018).

The generic name *Vulcanodon* combines the name Vulcanus (the Roman god of fire) with the Greek "odon" (meaning "tooth"). This partially refers to the fossil-bearing layer being in between two igneous layers. The specific name *karibaensis* refers to Lake Kariba, as the specimen was discovered on a small island of the lake.

CLASSIFICATION

Dinosauria
  Sauropodomorpha
   Sauropodiformes
    Anchisauria
     Sauropoda
      Gravisauria
       Vulcanodontidae

LOCATION

Zimbabwe

KNOWN REMAINS

Partial skeleton

2 m   4 m   6 m   8 m

V. chubutensis

(Bonaparte, 1979)
Length: 8.7 m (28.5 ft)
Height: 4.2 m (13.8 ft)
Hip height: 2 m (6.6 ft)
Body mass: 4,000 kg (4.4 t)
Reconstruction: ▣☐☐☐

4 m

2 m

250 245 240 235 230 225 220 215 210 205 200 195 190 185 **179** 175 170 165 160 155 150 145 140 135 130 125 120 115 110 105 100 95 90 85 80 75 70 65

TRIASSIC   **JURASSIC**   **CRETACEOUS**

Volkheimeria chubutensis (meaning "Volkheimer's one from Chubut") is known from a single set of remains (PVL 4077) consisting of fragmentary vertebrae, a femur, a tibia, and a nearly complete pelvis. Nearby, the remains of the theropod *Piatnitzkysaurus* were first discovered around the same time, as were the fossils of at least five individuals pertaining to another sauropod, *Patagosaurus*. *Volkheimeria* was clearly distinct from this genus, in part due to having an ilium that was proportionately much smaller. The overall size of the specimen was also noticeably smaller; the *Volkheimeria* specimen is thus thought to have been less than fully grown at the time of its death, although it was likely past the juvenile stage (Cerda et al., 2017).

These fossils come from the Cañadón Asfalto geological formation, which is composed of river- and lake-based sedimentary deposits. These strata have yielded numerous dinosaurian finds. The age of the layers that held

Volkheimeria were once believed to have been approximately 165 million years old, but more recent and precise measurements have placed the age at 179–178 million years (Pol et al., 2022).

The taxonomic affinities of *Volkheimeria* have varied considerably over the years, with different results usually being ascribed to the specimen's incompleteness. Some analyses have retrieved it as a brachiosaurid (McIntosh, 1990), an early eusauropod (Pol et al., 2011), a vulcanodontid (Cerda et al., 2017), or as a **gravisaurian** non-eusauropod (Becerra et al., 2017; Holwerda and Pol 2018; Rauhut et al., 2020; Pol et al., 2022).

The generic name *Volkheimeria* honors Wolfgang Volkheimer (1928–2018), a German-born Argentine paleontologist who was instrumental in expanding Argentine paleontological pursuits. The specific name *chubutensis* refers to the province of Chubut, Argentina.

CLASSIFICATION
Dinosauria
  Sauropodomorpha
    Sauropodiformes
      Anchisauria
        Sauropoda
          Gravisauria

LOCATION
Argentina

KNOWN REMAINS
Partial skeleton

3 m   6 m   9 m   12 m   15 m

6 m

3 m

R. brownei

(Longman, 1926)
Length: 15 m (49.2 ft)
Height: 6.5 m (21.3 ft)
Hip height: 3.2 m (10.5 ft)
Body mass: 9,000 kg (9.9 t)
Reconstruction:

250 245 240 235 230 225 220 215 210 205 200 195 190 185 180 175 170 **163** 160 155 150 145 140 135 130 125 120 115 110 105 100 95 90 85 80 75 70 65

TRIASSIC | **JURASSIC** | **CRETACEOUS**

Rhoetosaurus brownei (meaning "Browne's Rhoetus lizard") is not just the only pre-Cretaceous sauropod to be named from Australia, but it is also the most complete pre-Cretaceous dinosaur specimen to be discovered from the entire continent.

Rhoetosaurus was originally described on the basis of the posterior vertebral column, which was found in 1924. On the suspicion that further remains from the same animal might be present as well, further digs uncovered a nearly complete leg in 1976, in addition to various other fragments in later years; additional, unprepared specimens, still encased in rock, could potentially still reveal significant findings in the future (Nair and Salisbury, 2012).

The nearly complete, articulated foot was a fortuitous find, as relatively few sauropod feet have been unearthed in their natural configuration, and none before from the Gondwanan Jurassic. By studying the foot of *Rhoetosaurus*, a comprehensive analysis of all sauropod feet was undertaken that helped to reveal just how, exactly, these titans managed to support their immense weight.

The bones of sauropods' hind feet were arranged in an almost "tiptoe" or "high-heeled" arrangement, which would have been insufficient to hold up the animal's weight on their own. Instead, the bones rested upon a thick, fleshy pad of soft tissue, somewhat similar to the arrangement found in modern elephants. This reinforcement was a vital component in understanding the locomotion of the giant sauropods (Jannel et al., 2019).

The generic name *Rhoetosaurus* refers to Rhoetus, an enormous Titan from Greek mythology. The specific name *brownei* honors Arthur Browne, the manager of the Durham Downs cattle station who was responsible for bringing the fossils to the attention of paleontologist Heber Longman.

CLASSIFICATION
Dinosauria
 Sauropodomorpha
  Sauropodiformes
   Anchisauria
    Sauropoda
     Gravisauria

LOCATION
Australia

KNOWN REMAINS
Partial rear skeleton

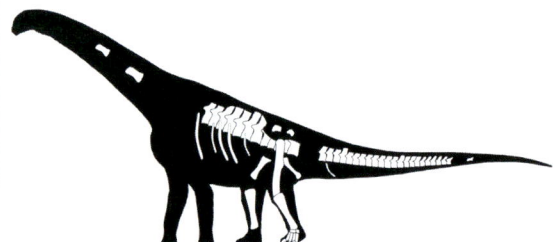

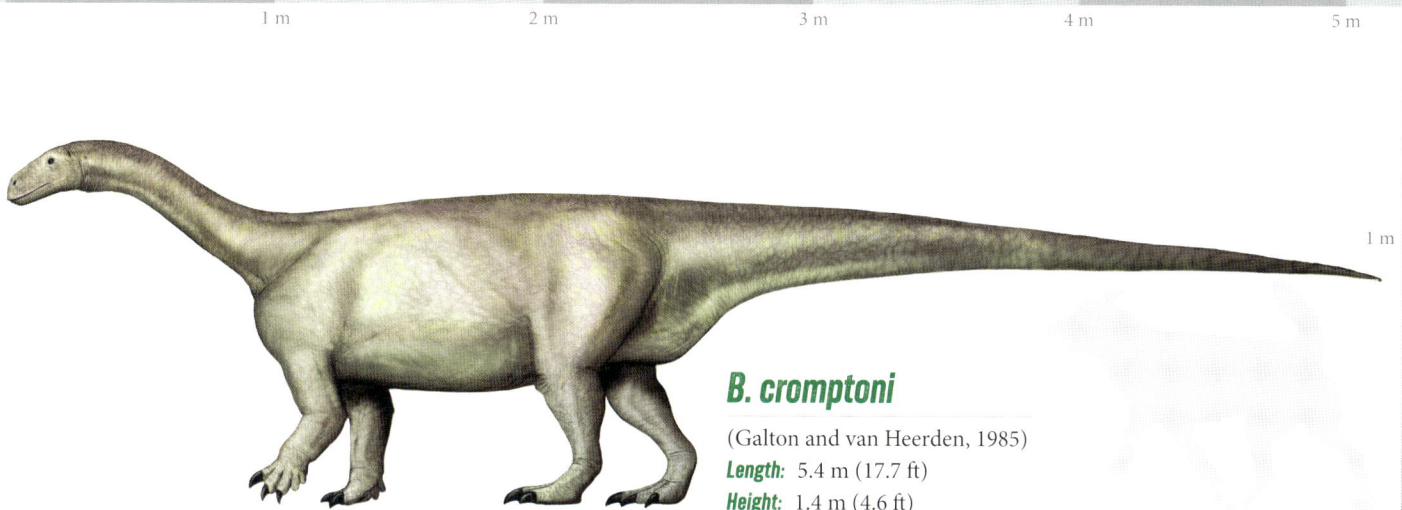

| 1 m | 2 m | 3 m | 4 m | 5 m |

B. cromptoni

(Galton and van Heerden, 1985)
Length: 5.4 m (17.7 ft)
Height: 1.4 m (4.6 ft)
Hip height: 1 m (3.3 ft)
Body mass: 420 kg (925 lb)
Reconstruction: ☐☐☐☐

| 250 | 245 | 240 | 235 | 230 | 225 | 220 | 215 | 210 | 205 | 200 | 195 | 190 | 185 | 180 | 175 | 170 | 165 | 160 | 155 | 150 | 145 | 140 | 135 | 130 | 125 | 120 | 115 | 110 | 105 | 100 | 95 | 90 | 85 | 80 | 75 | 70 | 65 |

TRIASSIC — **JURASSIC** — **CRETACEOUS**

Blikanasaurus cromptoni (meaning "Crompton's lizard from Blikana") is a South African sauropodomorph of uncertain affinity. It is known only from the bones of the lower leg and foot (discovered in 1965), severely limiting the data that is available to make any sort of phylogenetic determination. A second specimen, consisting of a single metatarsal bone, was described in 2008 (Yates, 2008). Other isolated bones, such as an ilium, could possibly belong to the genus as well (McPhee and Choiniere, 2016).

The original descriptive publication postulated that *Blikanasaurus* was a "prosauropod" that was actually not directly ancestral to the lineage of Sauropoda, being an "early experiment" in heavy, load-bearing, quadrupedal locomotion. As of yet, the specimens have not received a modern, detailed reanalysis.

At least one study, though, noted some similarities among *Blikanasaurus*, *Vulcanodon*, and *Antetonitrus* (McPhee et al., 2016), while another phylogenetically placed the taxon as sister to Sauropoda (Bandyopadhyay et al., 2010). Most recently, placement within Lessemsauridae has also been proposed (Regalado Fernández and Werneburg, 2022).

Blikanasaurus was found in the Lower Elliot geological formation, which is a famous fossil-bearing locality. The generic name *Blikanasaurus* refers to Blikana, the place where the holotype fossil was discovered. The specific name *cromptoni* honors field paleontologist A. W. "Fuzz" Crompton "in recognition of his research on fossil vertebrates from South Africa."

CLASSIFICATION
Dinosauria
　Sauropodomorpha
　　Sauropodiformes
　　　Anchisauria
　　　　Sauropoda
　　　　　Lessemsauridae

LOCATION
South Africa

KNOWN REMAINS
Rear foot

2 m 4 m 6 m 8 m

2 m

O. liasicus

(Wild, 1978)
Length: 7 m (23 ft)
Height: 4 m (13.1 ft)
Hip height: 2 m (6.6 ft)
Body mass: 1,100 kg (2.4 t)
Reconstruction: ▢▢▢▢

| 250 | 245 | 240 | 235 | 230 | 225 | 220 | 215 | 210 | 205 | 200 | 195 | 190 | 185 | **182** | 175 | 170 | 165 | 160 | 155 | 150 | 145 | 140 | 135 | 130 | 125 | 120 | 115 | 110 | 105 | 100 | 95 | 90 | 85 | 80 | 75 | 70 | 65 |

TRIASSIC   **JURASSIC**   **CRETACEOUS**

The only known fossils of **Ohmdenosaurus liasicus** (meaning "Lias lizard from Ohmden") were discovered in the famous fossil-bearing location called the Posidonia Shale. As this region of Germany would have been in the middle of a shallow inland sea while the animal was alive, the bones were originally misidentified as belonging to an aquatic plesiosaur.

No records exist regarding exactly where or when the fossils were recovered, although the museum in which they are housed was specifically opened in 1936 to house fossils discovered in the area. Pieces of rock attached to the bones at least narrow down the area of discovery as being part of the "lower slate."

Ohmdenosaurus is known only from a single tibia and two ankle bones. Weathering patterns on the bones suggest that they had already been exposed once prior to their final burial. This suggests that either ocean currents or predators carried the bones to their resting place.

With so few bones to study, it is difficult to determine exactly what type of sauropod *Ohmdenosaurus* was. The tibia is relatively slender, which is a feature usually associated with sauropods that lived later in time, making *Ohmdenosaurus* among the earliest to have this trait (Nair and Salisbury, 2012). One analysis suggested *Ohmdenosaurus* to be a very basal eusauropod, closely related to *Cetiosaurus* (Fernández and Werneburg, 2022).

The generic name *Ohmdenosaurus* refers to the town of Ohmden. The specific name *liasicus* refers to the Lias, a term for the Lower Jurassic layers of Europe.

CLASSIFICATION
Dinosauria
  Sauropodomorpha
   Sauropodiformes
    Anchisauria
    Sauropoda

LOCATION
Germany

KNOWN REMAINS
Partial rear leg

PROTOGNATHOSAURUS

P. oxyodon

(Zhang, 1988)

Length: 14 m (45.9 ft)
Height: 4.8 m (15.7 ft)
Reconstruction: ☐☐☐☐

2 m

168

| TRIASSIC | JURASSIC | CRETACEOUS |

Protognathosaurus oxyodon (meaning "sharp-toothed early jawed lizard") is known from a single bone, a fragment of the animal's left lower jaw that was unearthed in China. The jaw still held some replacement teeth, which were considered to be sharp for a sauropod—hence the animal's name.

Since this fragmentary specimen has only been briefly analyzed in Chinese-language journals, most researchers tend to overlook *Protognathosaurus* or ignore the genus on the basis that it lacks diagnostically relevant information, essentially classifying the genus as *nomen dubium*.

The generic name combines the Greek "protos" (meaning "first" or "early"), "gnathos" (meaning "jaw"), and "sauros" (meaning "lizard"). The specific name combines the Greek "oxy" (meaning "sharp") and "odon" (meaning "tooth"). The animal was originally called *Protognathus* before it was realized that the name was previously occupied by a species of beetle.

ZIZHONGOSAURUS

Z. chuanchengensis

(Dong et al., 1983)

Length: 8.4 m (27.6 ft)
Height: 3.1 m (10.2 ft)
Reconstruction: ☐☐☐☐

1 m

176

| TRIASSIC | JURASSIC | CRETACEOUS |

Zizhongosaurus chuanchengensis (meaning "lizard from Chuancheng, Zizhong") is a basal sauropod from China, known only from three partial bones: a vertebra, humerus, and pubis. It was described only very briefly, and only the vertebral element was even depicted. In and of themselves, the fossils are not usually considered to be diagnostic, so the genus is often considered to be *nomen dubium*. One source suggests a synonymy with *Omeisaurus* (Weishampel et al., 2004). A second species was briefly named in a subsequent work, but as no details or description regarding the specimen ever appeared, the species *Z. huangshibanensis* is *nomen nudum* (Li et al., 1999).

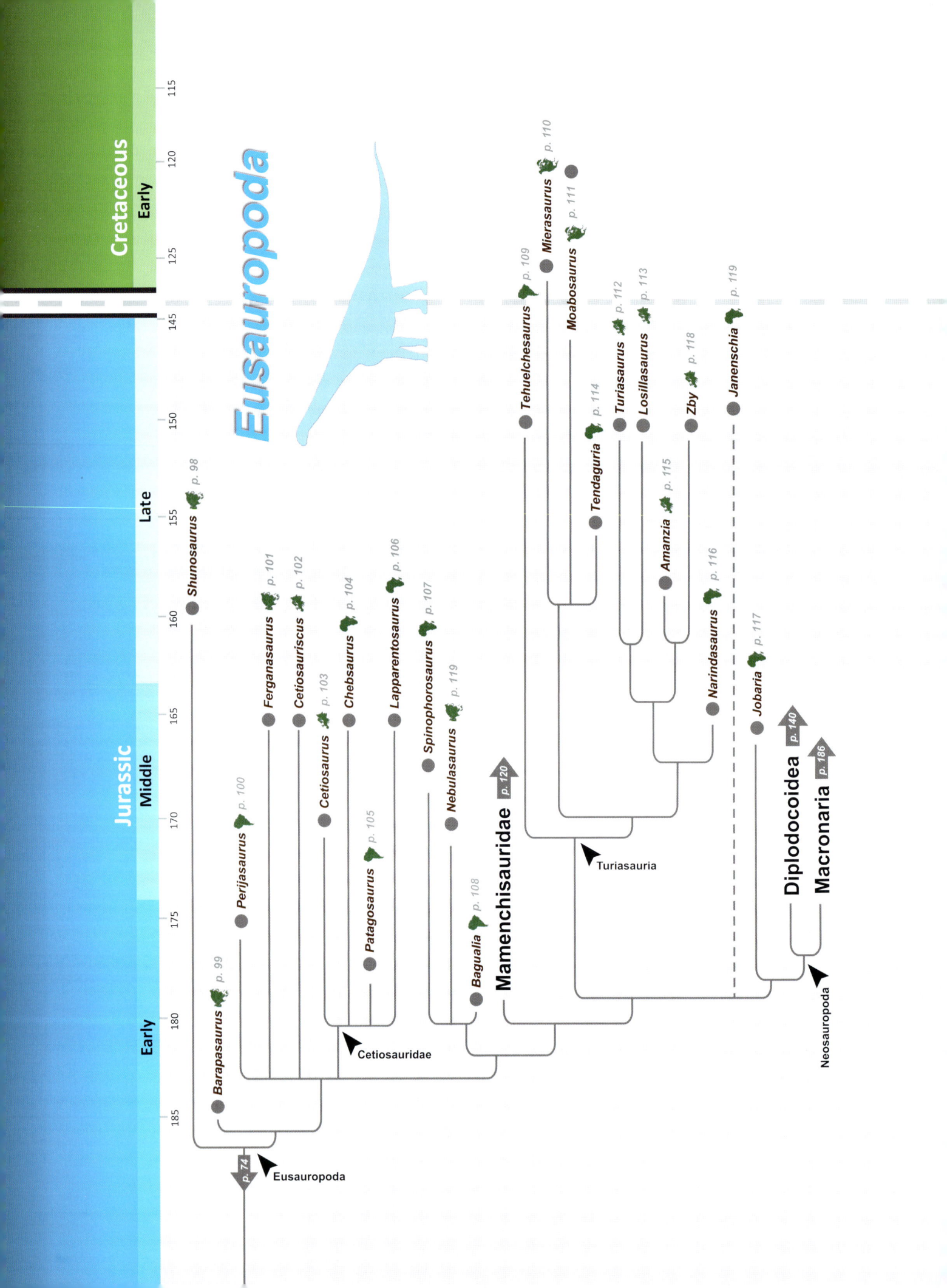

Eusauropoda

p. 74

Cretaceous — Early

Jurassic — Early / Middle / Late

115 120 125 145 150 155 160 165 170 175 180 185

Shunosaurus p. 98

Mierasaurus p. 110

Moabosaurus p. 111

Tehuelchesaurus p. 109

Turiasaurus p. 112

Losillasaurus p. 113

Tendaguria p. 114

Amanzia p. 115

Zby p. 118

Janenschia p. 119

Narindasaurus p. 116

Jobaria p. 117

Ferganasaurus p. 101

Cetiosauriscus p. 102

Cetiosaurus p. 103

Chebsaurus p. 104

Lapparentosaurus p. 106

Spinophorosaurus p. 107

Nebulasaurus p. 119

Patagosaurus p. 105

Bagualia p. 108

Perijasaurus p. 100

Barapasaurus p. 99

Mamenchisauridae p. 120

Turiasauria

Cetiosauridae

Eusauropoda

Diplodocoidea p. 140

Macronaria p. 186

Neosauropoda

The species within Eusauropoda (meaning "true sauropods") are obligate quadrupeds, unlike most earlier forms, which were at least partially bipedal. They developed anatomical features that enabled feeding via "bulk-browsing," maximizing the amount of food that could be consumed; these features included vegetation-stripping teeth, a wide jaw, and the ability to open their mouths wider by not having restrictive fleshy cheeks. The position of their nostrils also became retracted, moving toward the back of the skull.

The derived eusauropods will eventually split into two main categories, the Diplodocoidea and the Macronaria. The basal forms are represented by a smattering of smaller clades, such as Mamenchisauridae, Cetiosauridae, and Turiasauria. Turiasauria was initially suspected to be a uniquely European group but is now thought to have been much more widespread.

Placing *Shunosaurus* and *Barapasaurus* as basalmost members of **Eusauropoda** is uncontroversial, although *Barapasaurus* has, at times, been found to be just barely a non-eusauropod (Holwerda et al., 2021) or, contrastingly, to be more derived than the mamenchisaurids (Ren et al., 2018).

The validity or usefulness of the family **Cetiosauridae** has been debated but, if used, obviously includes *Cetiosaurus*. *Patagosaurus* has been considered a member (Ren at al., 2020, 2023; Holwerda et al., 2021; although not always [Pol et al., 2020]), as has *Chebsaurus* (Taquet, 2010) and *Lapparentosaurus* (Raveloson et al., 2019; Royo-Torres et al., 2020). Alternatively, *Lapparentosaurus* has also been placed within Turiasauria (Moore et al., 2023). Cetiosauridae is most often placed basally to Mamenchisauridae but has also been calculated as being more derived instead (Pol et al., 2020; Royo-Torres et al., 2020).

The oft-neglected *Cetiosauriscus* and *Ferganasaurus* have sometimes been found to be basal eusauropods closely related to *Cetiosaurus* (Ksepka and Norell, 2010; Upchurch et al., 2021; Moore et al., 2023). The describers of *Perijasaurus* place it in a similar position (Rincón et al., 2022).

The describers of *Bagualia* place it within a clade along with *Nebulasaurus* and *Spinophorosaurus* (Pol et al., 2020), with previous studies often showing the latter pair as sister taxa (Xing et al., 2015a; Ren et al., 2018). Differing versions of analyses place this *Bagualia*

clade either before, or after, Mamenchisauridae (Pol et al., 2020).

The placement of *Janenschia* is very uncertain, variously being placed within basal Macronaria (Ren et al., 2023), within Turiasauria (Mannion et al., 2019), or as one of the most derived non-neosauropods (Royo-Torres et al., 2020). The placement of *Jobaria* is nearly as unclear; it has often been found as the most derived non-neosauropod (Pol et al., 2020; Holwerda et al., 2021; Rincón et al., 2022; Ren et al., 2023), but it has also been placed within Turiasauria (Moore et al., 2023) or even basal to Turiasauria (Mannion et al., 2019; Royo-Torres et al., 2020).

Aside from *Turiasaurus*, **Turiasauria** confidently includes *Losillasaurus*, *Zby*, *Mierasaurus*, and *Moabosaurus* (Moore et al., 2023), as well as *Amanzia* (Royo-Torres et al., 2020; Schwarz et al., 2020). *Tehuelchesaurus* has traditionally been viewed as a camarasauromorph, but has also been suggested as a turiasaurian (Mannion et al., 2019; Schwarz et al., 2020). Multiple studies have also placed *Tendaguria* in the group (Pol et al., 2020). The describers of *Narindasaurus* likewise find it within this clade (Royo-Torres et al., 2020).

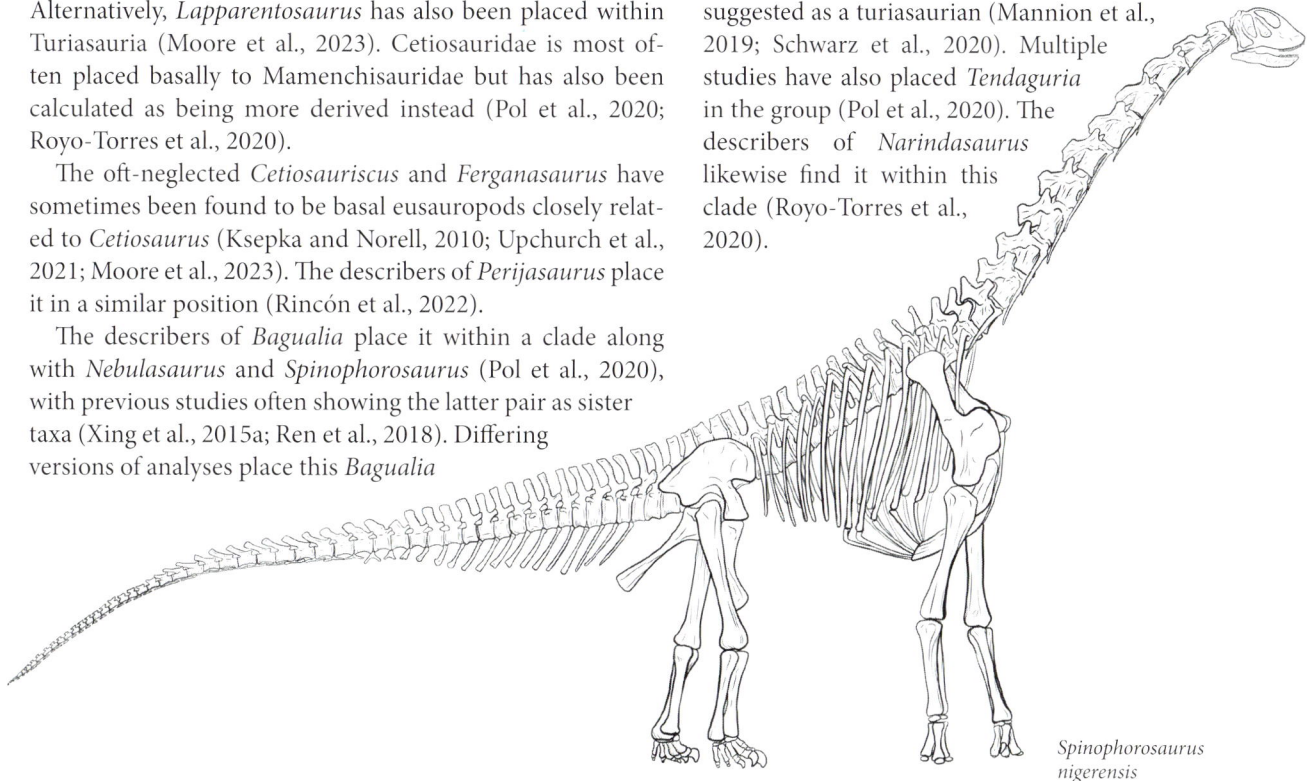

Spinophorosaurus nigerensis

SHUNOSAURUS

S. lii

(Dong et al., 1983)
Length: 12.5 m (41 ft)
Height: 4.6 m (15.1 ft)
Hip height: 3.5 m (11.5 ft)
Body mass: 7,000 kg (7.7 t)
Reconstruction: ☐☐☐☐

2 m — 4 m — 6 m — 8 m — 10 m — 12 m

4 m

2 m

250 245 240 235 230 225 220 215 210 205 200 195 190 185 180 175 170 165 **159** 155 150 145 140 135 130 125 120 115 110 105 100 95 90 85 80 75 70 65

TRIASSIC | **JURASSIC** | **CRETACEOUS**

Shunosaurus lii (meaning "Li's lizard from Shu") has two very distinctive (although not quite unique) physical traits.

First, *Shunosaurus* possesses an offensive tail club—the sort that one would normally associate with the armored ankylosaurs, not the sauropods. This bony protuberance consists of the last few vertebrae of the tail as well as at least two short spikes, formed by projecting osteoderms. Whether this appendage was used primarily for defense against predators or for competition between males is unknown (Dong et al., 1989).

Second, the neck of *Shunosaurus* is proportionally among the shortest of all sauropods. This indicates that *Shunosaurus* was a low browser, subsisting either on ground-level plants or the lower tree branches. This positioning could indicate a case of niche-partitioning, as *Shunosaurus* is known to have shared an ecosystem with larger sauropods; by having a different primary food source, competition for resources could thus have been largely avoided.

Numerous specimens of *Shunosaurus* have been discovered, including both adults and juveniles. Altogether, the vast majority of the animal's skeleton is known, including—crucially—the cranium. That being said, the various skull fossils unearthed have been either distorted or disarticulated, which means that the animal's head has been depicted in several different configurations over time as interpretations of its anatomy have evolved. The upper and lower jaws both curve upward, much like the shape of common garden shears, which would have made *Shunosaurus* very efficient at shearing off plant matter (Chatterjee and Zheng, 2002).

The generic name *Shunosaurus* refers to Shu, an old name for the Sichuan region of China. The specific name *lii* honors local Chinese historical figure Li Bing.

CLASSIFICATION

Dinosauria
  Sauropodomorpha
  Sauropodiformes
  Anchisauria
  Sauropoda
  Gravisauria
  Eusauropoda

LOCATION

China

KNOWN REMAINS

Nearly complete

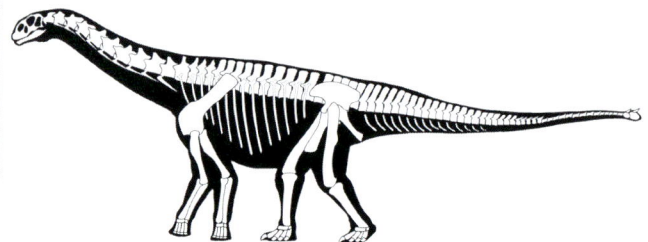

| 2 m | 4 m | 6 m | 8 m | 10 m | 12 m | 14 m |

B. tagorei

(Jain et al., 1975)
Length: 14 m (45.9 ft)
Height: 6.1 m (20 ft)
Hip height: 3.3 m (10.8 ft)
Body mass: 8,500 kg (9.4 t)
Reconstruction: ▢▢▢▢

| 250 | 245 | 240 | 235 | 230 | 225 | 220 | 215 | 210 | 205 | 200 | 195 | 190 | 185 | 180 | 175 | 170 | 165 | 160 | 155 | 150 | 145 | 140 | 135 | 130 | 125 | 120 | 115 | 110 | 105 | 100 | 95 | 90 | 85 | 80 | 75 | 70 | 65 |

TRIASSIC | **JURASSIC** | **CRETACEOUS**

Barapasaurus tagorei (meaning "Tagore's big leg lizard") displays some early indications of weight-reducing anatomical traits that would become highly adapted and specialized in the later eusauropods, in the form of some hollow areas and grooves that are present in the vertebral column. Its vertebrae also bear stabilizing projections that help to link them securely to one another.

Additional traits that hint at the later important developments of eusauropodan anatomy include teeth with wrinkled enamel, a strengthened sacrum, and slender, column-like limbs.

The totality of known *Barapasaurus* specimens come from a single locality. Over 300 individual bones, from at least six different individuals, have been unearthed from a series of three excavations. In total, nearly the entire skeleton of the animal is accounted for, lacking just the skull and jaw. Most of these bones were mounted and put on display by the Indian Statistical Institute, becoming in 1977 the first mounted dinosaur specimen in Asia. Sadly, while technically housed within a public museum, the mounted specimen is currently under restricted access, and insufficient museum funds put the specimen in danger of disrepair (Sarkar, 2019).

Phylogenetic assessments disagree regarding the exact placement of *Barapasaurus*. Numerous evaluations have placed it as a non-eusauropod (Bandyopadhyay et al., 2010), while several others have placed it among the basalmost **eusauropods** (Gomez et al., 2021; Ren et al., 2023).

The generic name *Barapasaurus* combines the Bengali words "bara" (meaning "big") and "pa" (meaning "leg"), with the Greek "sauros" (meaning "lizard"). The specific name *tagorei* honors Indian poet Rabindranath Tagore, as the original excavation took place on what would have been his 100th birthday.

CLASSIFICATION
Dinosauria
 Sauropodomorpha
  Sauropodiformes
   Anchisauria
    Sauropoda
     Gravisauria
      Eusauropoda

LOCATION
India

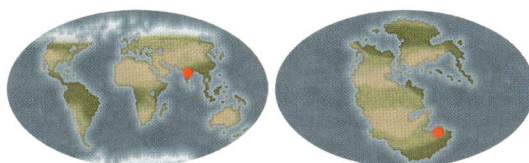

KNOWN REMAINS
Nearly complete, lacking skull

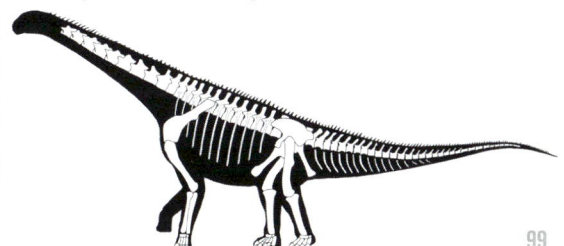

99

EUSAUROPODA

6 m 2 m 4 m 6 m 8 m 10 m 12 m 14 m

4 m

2 m

P. lapaz

(Rincón et al., 2022)
Length: 13 m (42.6 ft)
Height: 5 m (16.4 ft)
Hip height: 2.5 m (8.2 ft)
Body mass: 6,000 kg (6.6 t)
Reconstruction: ☐☐☐☐

250 245 240 235 230 225 220 215 210 205 200 195 190 185 180 175 170 165 160 155 150 145 140 135 130 125 120 115 110 105 100 95 90 85 80 75 70 65

TRIASSIC **JURASSIC** **CRETACEOUS**

Perijasaurus lapaz (meaning "lizard from La Paz, Perija") is known from a single bone: a mostly complete backbone, likely the ninth dorsal vertebra.

This specimen was discovered in 1943 by a geologist who was working for the Tropical Oil Company. It was then sent to the University of California, Berkeley, and was briefly described in the *Journal of Paleontology* in 1955. It was still partially encased in rocky matrix and plaster when paleontologist Jeffrey Wilson first saw it in 1997 and remained so until he finally got the opportunity to properly prepare the specimen in 2018.

With the revelation of newly visible details, it became clear that the bone represented a type of sauropod that was new to science. Based on an original, hand-drawn map, researchers were able to pinpoint the original dig site that had yielded the fossil by using the sediment that had encased the bone to positively match it to the area. Fossilized plant debris in the vicinity confirmed that *Perijasaurus* lived in a wooded area, near a body of water.

The vertebra reveals that the spinal column of *Perijasaurus* was only partially pneumatized. The later neosauropods evolved a system of bone-lightening air cavities that reduced the skeleton's weight by at least 50%, whereas the limited openings present in *Perijasaurus* removed only about a quarter of the possible volume.

The generic name *Perijasaurus* refers to the Serranía del Perijá mountain range. The specific name *lapaz* refers to the town of La Paz, close to where the fossil was found. Additionally, "la paz" is Spanish for "the peace," which references the 2016 peace agreement between the Colombian government and the FARC guerrilla group, which made it possible for paleontologists to access geographic areas that had long been unsafe to visit.

CLASSIFICATION

Dinosauria
  Sauropodomorpha
   Sauropodiformes
    Anchisauria
     Sauropoda
      Gravisauria
       Eusauropoda

LOCATION

Columbia

KNOWN REMAINS

Single vertebra

3 m   6 m   9 m   12 m   15 m   18 m

F. verzilini

(Alifanov and Averianov, 2003)

Length: 18 m (59 ft)
Height: 6.1 m (20 ft)
Hip height: 4 m (13 ft)
Body mass: 15,000 kg (16.5 t)
Reconstruction: ☐☐☐☐☐

6 m

3 m

250 245 240 235 230 225 220 215 210 205 200 195 190 185 180 175 170 **165** 160 155 150 145 140 135 130 125 120 115 110 105 100 95 90 85 80 75 70 65

TRIASSIC	JURASSIC	CRETACEOUS

Ferganasaurus verzilini (meaning "Verzilin's lizard from Fergana") is an enigmatic and poorly understood animal.

The first fossils of *Ferganasaurus* were spotted in July 1966 by a team from Leningrad State University led by N. N. Verzilin. They were subsequently excavated by a group from the USSR Academy of Sciences, under the leadership of A. K. Rozhdestvensky. Although he was reportedly preparing a description of the fossils, Rozhdestvensky passed away before its completion, and no draft was ever recovered.

Poor record keeping made it a challenge for modern researchers to return to the original site, but T. Martin managed to do so in 2001 and recovered a tooth and a foot bone that possibly belonged to *Ferganasaurus*.

Unfortunately, when researchers set out to finally describe the remains of the creature, they found that they were "not able to locate all elements in the collection" of the Paleontological Institute of the Russian Academy of Sciences. The dorsal vertebrae, distal caudal (tail) vertebrae, and the entire manus (front foot) were missing. (These lost elements have been colored gray in the diagram below.) One consolation is that original drawings of the bones were available for reference, as well as some photographs.

The description placed *Ferganasaurus* as one of the basalmost neosauropods, but a subsequent analysis placed it outside that group (Ksepka and Norell, 2010) as a **eusauropod**. Very few phylogenetic analyses have subsequently included *Ferganasaurus*.

The generic name *Ferganasaurus* refers to the Fergana Valley. The specific name *verzilini* honors the fossil's discoverer, Nikita N. Verzilin.

CLASSIFICATION
Dinosauria
  Sauropodomorpha
  Sauropodiformes
    Anchisauria
    Sauropoda
    Gravisauria
    Eusauropoda

LOCATION
Kyrgyzstan

KNOWN REMAINS
Partial skeleton

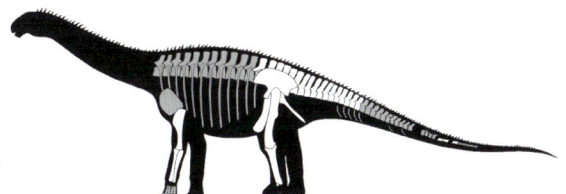

8 m   4 m   8 m   12 m   16 m

4 m

C. stewarti

(Charig, 1980)
Length: 16.8 m (55.1 ft)
Height: 8 m (26.2 ft)
Hip height: 3.3 m (10.8 ft)
Body mass: 11,000 kg (12.1 t)
Reconstruction: ☐☐☐☐

250 245 240 235 230 225 220 215 210 205 200 195 190 185 180 175 170 **165** 160 155 150 145 140 135 130 125 120 115 110 105 100 95 90 85 80 75 70 65

TRIASSIC   **JURASSIC**   **CRETACEOUS**

As is the case with many taxa described during this time period, the issues of naming and bookkeeping surrounding **Cetiosauriscus stewarti** are tedious and tumultuous. The partial skeleton known as BMNH R.3078 was excavated in May of 1898; a separate fragmentary pelvis (NHMUK R1988) was used as the holotype for the description of *Cetiosaurus leedsii* by Friedrich von Huene in 1927. However, in 1980, it was determined that the holotype was too indistinct and of too poor quality to be used as a means of identification; thus, *C. leedsii* (as well as the dubious *C. greppini*) was rendered invalid, and the more complete specimen BMNH R.3078 was named to the new species *Cetiosauriscus stewarti*, which therefore became the type species for the genus (Charig, 1980). Although some isolated and fragmentary bones have been potentially ascribed to the genus over the years, BMNH R.3078 remains the only substantial specimen known thus far.

Cetiosauriscus was long suspected to be a diplodocid, due in part to its rather elongated tail (Berman and McIntosh,

1978), though it was reinterpreted as a mamenchisaurid in some later studies (Naish and Martill, 2007). However, a recent study regarding the subfamily Mamenchisauridae concluded that *Cetiosauriscus* fell outside that clade, as a basal **eusauropod** (Moore et al., 2023).

During the Middle Jurassic, the area where *Cetiosauriscus* was found would have been a coastal environment adjacent to a shallow sea that was dotted with many islands or small landmasses. The Oxford Clay geological formation preserves numerous marine and dinosaurian remains.

The generic name *Cetiosauriscus* is meant to mean "like *Cetiosaurus*." *Cetiosaurus* is a sauropod genus named in 1842, whose name itself means "whale lizard." The specific name *leedsii* honors fossil collector Alfred Nicholson Leeds, whereas the name *stewarti* honors Sir Ronald Stewart, the chairman of the company that owned the location where the fossils had been found.

CLASSIFICATION
Dinosauria
  Sauropodomorpha
   Sauropodiformes
    Anchisauria
     Sauropoda
      Gravisauria
       Eusauropoda

LOCATION
England

KNOWN REMAINS
Partial skeleton

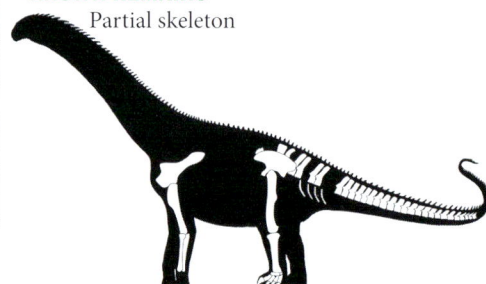

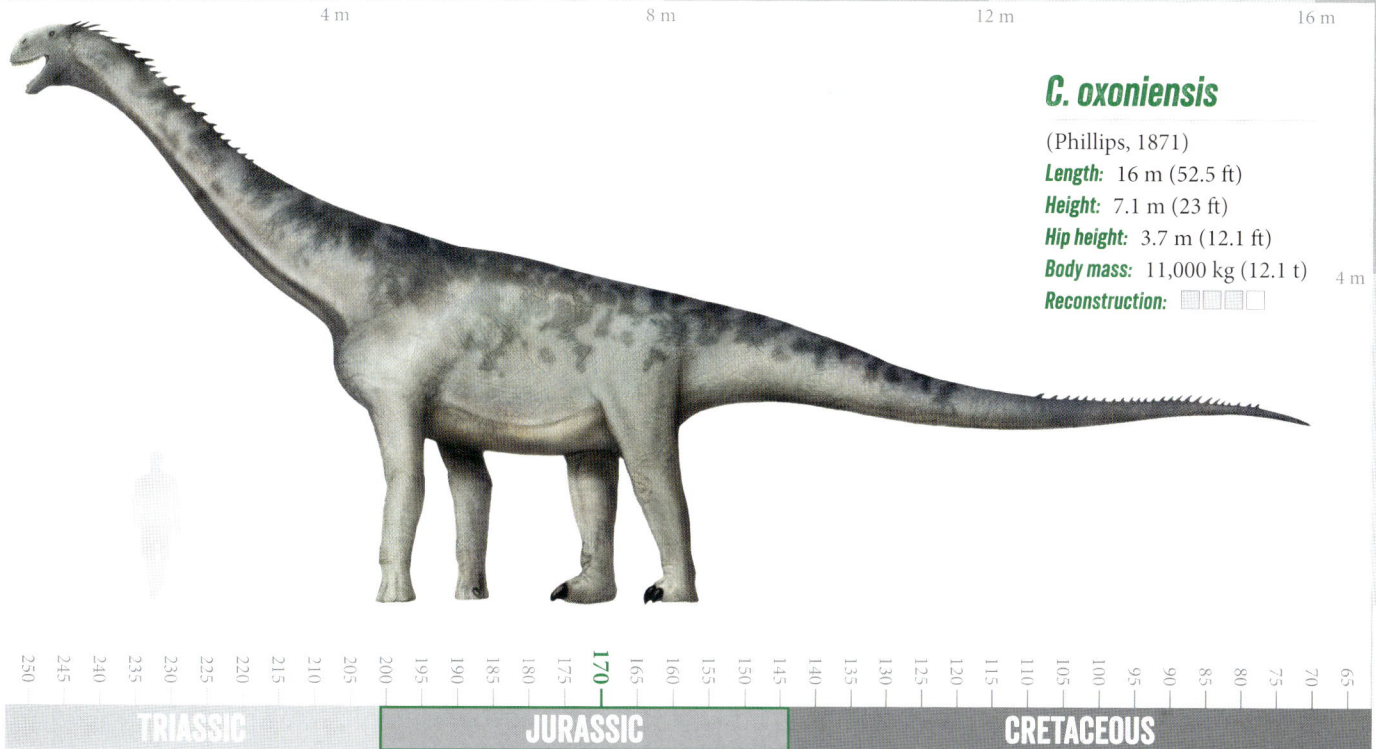

4 m   8 m   12 m   16 m

C. oxoniensis

(Phillips, 1871)

Length: 16 m (52.5 ft)
Height: 7.1 m (23 ft)
Hip height: 3.7 m (12.1 ft)
Body mass: 11,000 kg (12.1 t)
Reconstruction: ▢▢▢▢

4 m

250 245 240 235 230 225 220 215 210 205 200 195 190 185 180 175 **170** 165 160 155 150 145 140 135 130 125 120 115 110 105 100 95 90 85 80 75 70 65

TRIASSIC   **JURASSIC**   **CRETACEOUS**

Cetiosaurus oxoniensis (meaning "whale-like lizard from Oxfordshire") is currently the only valid species recognized within the genus, although more than a dozen have been named at one time or another. As is common for dinosaurs named in this era, the genus *Cetiosaurus* became something of a wastebasket taxon. A comprehensive reanalysis found numerous species to be *nomina nuda*, *nomina dubia*, synonyms, from sauropods that were not actually closely related to *C. oxoniensis*, or, in the case of *C. brevis*, reclassified using a preexisting alternative name, *Pelorosaurus* (Upchurch and Martin, 2003). *Cetiosaurus oxoniensis* officially became the new type species in 2014.

Cetiosaurus has a "typical" sauropod body, without any obvious specialized adaptations, although it is notable in some subtle ways. Its tail is proportionately long, as are its forearms, which are just as long as its upper arms.

Specimens originating from four distinct individuals are recognized. Three were described by John Phillips, including an adult, subadult, and juvenile, each of which is quite fragmentary. The fourth, known as the "Rutland" specimen is the most complete, containing about 40% of the total animal. This specimen was discovered in 1968, although it went unprepared until the early 1980s and did not receive a thorough description until 2002. Of its cranium, only a braincase and tooth are provisionally assigned to the species, so any adaptations of its feeding apparatus are thus far unknown (Upchurch and Martin, 2002).

The generic name *Cetiosaurus* is derived from the Greek "keteios" (meaning "whale-like"); Sir Richard Owen named the genus in 1841, believing the remains represented a marine reptile. This makes *Cetiosaurus* the first sauropod to be formally named. The specific name *oxoniensis* refers to Oxfordshire County, England, where the type specimen was found.

CLASSIFICATION

Dinosauria
  Sauropodomorpha
    Sauropodiformes
      Anchisauria
        Sauropoda
          Gravisauria
            Eusauropoda
              Cetiosauridae

LOCATION

Western Europe

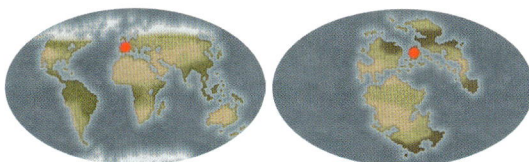

KNOWN REMAINS

Nearly complete

EUSAUROPODA

2 m   4 m   6 m   8 m

2 m

C. algeriensis

(Mahammed et al., 2005)
Length: 8.2 m (27 ft)
Height: 3 m (10 ft)
Hip height: 2 m (6.6 ft)
Body mass: 1,500 kg (1.7 t)
Reconstruction: ☐☐☐☐

250 245 240 235 230 225 220 215 210 205 200 195 190 185 180 175 170 **165** 160 155 150 145 140 135 130 125 120 115 110 105 100 95 90 85 80 75 70 65

TRIASSIC	JURASSIC	CRETACEOUS

Chebsaurus algeriensis (meaning "teenager lizard from Algeria") was discovered during an expedition that specifically set out to unearth fossils for a museum being constructed by an oil company, Sonatrach. The discovery of *Chebsaurus* would spur the launch of digs at more than 50 individual sites.

The holotype specimen is that of a juvenile, based on the condition of unfused elements of the vertebrae. It is very fragmentary and yet still contains pieces of many body parts, including the cranium. It was unearthed alongside bones of another indeterminate sauropod, as well as theropod teeth. When discovered, it was also the most complete sauropod skeleton to ever be discovered from Algeria.

During subsequent return trips to the same site, more material belonging to the same specimen was uncovered, along with remnants of a second specimen, which was also a juvenile. At the time these remains were described, still more fossils awaited final preparation and cleaning (Läng and Mahammed, 2009).

Chebsaurus was found to be a close relative of *Cetiosaurus*. If the subfamily **Cetiosauridae** is indeed valid, *Chebsaurus* would likely be a part of that group (Taquet, 2010). As a non-neosauropod, the animal's vertebrae are mostly solid, lacking extensive pneumatization. Like many Middle Jurassic sauropods, *Chebsaurus* bears some similarities in common with both derived sauropods (such as a lack of certain tooth grooves) and with basal sauropods.

The generic name *Chebsaurus* derives from the Arab "chab" (meaning "teenager"), as the holotype specimen is that of a juvenile. The specific name *algeriensis* refers to the North African nation of Algeria. Prior to its description, the holotype specimen had gained the nickname "the Giant of Ksour," after the Ksour Mountain range.

CLASSIFICATION

Dinosauria
  Sauropodomorpha
    Sauropodiformes
      Anchisauria
        Sauropoda
          Gravisauria
            Eusauropoda
              Cetiosauridae

LOCATION

Algeria

KNOWN REMAINS

Partial skull and skeleton

3 m   6 m   9 m   12 m   15 m   18 m

P. fariasi

(Bonaparte, 1979)
Length: 18.6 m (61 ft)
Height: 7.7 m (25 ft)
Hip height: 3.9 m (12.8 ft)
Body mass: 12,300 kg (13.6 t)
Reconstruction: ☐☐☐☐

6 m

3 m

250 245 240 235 230 225 220 215 210 205 200 195 190 185 178 175 170 165 160 155 150 145 140 135 130 125 120 115 110 105 100 95 90 85 80 75 70 65

TRIASSIC   **JURASSIC**   **CRETACEOUS**

Patagosaurus fariasi (meaning "Farias's lizard from Patagonia") was briefly described in 1979 but was soon after given a more extensive write-up, which assigned numerous partial specimens to the genus, including those of subadults and at least one juvenile. Taken together, nearly the entire axial skeleton of the animal is known, along with some cranial fragments (Bonaparte, 1986).

Later examinations have called into question whether or not each of these specimens is actually referable to *Patagosaurus*, suggesting that at least some of these remains could possibly be attributed to separate, as-yet-unidentified species (Rauhut, 2003). More thorough and modern examinations of the holotype have also been conducted (Holwerda et al., 2021).

Although *Patagosaurus* represents a fairly basal, Jurassic member of Sauropoda, its teeth were similar to those of some much later sauropods, having one concave surface and pronounced crowns. Histological analysis suggests that *Patagosaurus* would have replaced a worn-out tooth every 58 days (D'Emic et al., 2013). Some aspects of its shoulders and pelvis are comparatively robust, while its forelimbs seem to have been slightly slender.

Patagosaurus was first attributed to the sauropod group **Cetiosauridae**, and since that time, nearly all studies have also found *Patagosaurus* and *Cetiosaurus* to be very closely related (Holwerda et al., 2021). Still, some studies have concluded that the two are not technically a part of the exact same family (Wilson and Upchurch, 2009; Pol et al., 2020).

The generic name *Patagosaurus* refers to the South American region of Patagonia. The specific name *fariasi* honors Ricardo Farias, the landowner of the site of the first fossil discoveries. Other discoveries made around the same time and place include the first known remains of the sauropod *Volkheimeria* and the theropod *Piatnitzkysaurus*.

CLASSIFICATION
Dinosauria
 Sauropodomorpha
  Sauropodiformes
   Anchisauria
    Sauropoda
     Gravisauria
      Eusauropoda
       Cetiosauridae

LOCATION
Argentina

KNOWN REMAINS
Nearly complete

EUSAUROPODA

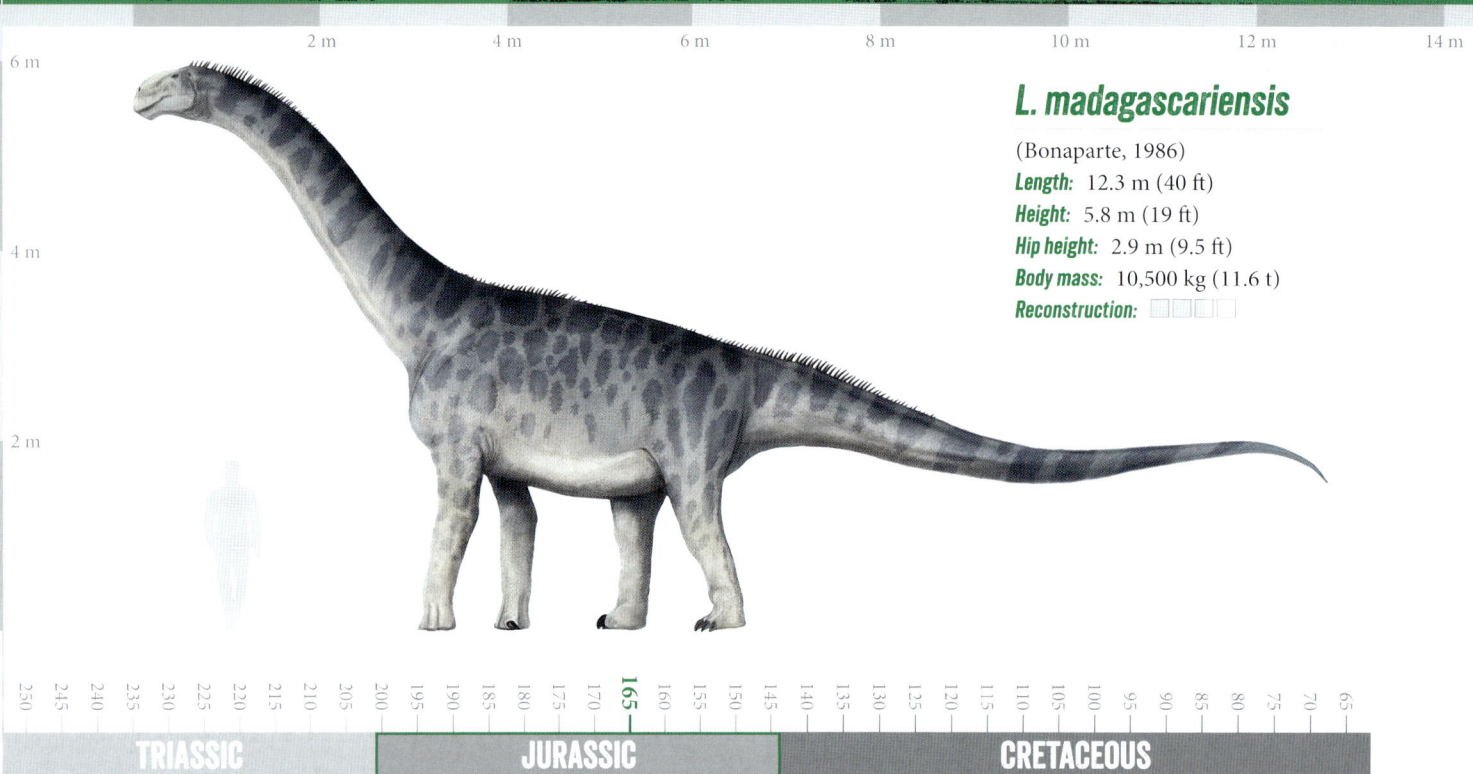

L. madagascariensis

(Bonaparte, 1986)
Length: 12.3 m (40 ft)
Height: 5.8 m (19 ft)
Hip height: 2.9 m (9.5 ft)
Body mass: 10,500 kg (11.6 t)
Reconstruction: ☐☐☐☐☐

TRIASSIC   JURASSIC   CRETACEOUS

Lapparentosaurus madagascariensis (meaning "Lapparent's lizard from Madagascar") is an ill-described sauropod genus with a convoluted naming history.

In 1875, Sir Richard Owen named several species to the now-dubious genus *Bothriospondylus*, based on remains found in England. Over the years, numerous fossils and species were assigned to the genus, and it became something of a wastebasket taxon. One such species was *B. madagascariensis*—a name that, itself, has been used to describe the remains of what are now known to be several different species.

Deciding that an animal from Madagascar was probably not in the same genus as a species from England, José Bonaparte erected a new genus for many of the country's sauropod fossils: *Lapparentosaurus*. These remains were individually quite incomplete but came from numerous individuals (the exact number remains indeterminate) of varying growth stages, representing juveniles, subadults, and adults. The limbs, hips, and shoulders are best represented, with relatively few vertebral fragments and no cranial material. Histological sampling suggests that it took *Lapparentosaurus* 31–45 years to reach its maximum size (Ricqlès, 1983).

Lapparentosaurus has remained a little-studied sauropod. It was long assumed to be a brachiosaur, but newer studies began to suggest a position within **Cetiosauridae** (Läng, 2008; Royo-Torres et al., 2020). This conclusion has been further solidified by the examination of some newly described skeletal fragments, which found unique correlations between *Lapparentosaurus* and *Cetiosaurus* with regard to their vertebral anatomy (Raveloson et al., 2019).

The generic name *Lapparentosaurus* honors paleontologist Albert-Félix de Lapparent. The specific name *madagascariensis* refers to the nation of Madagascar.

CLASSIFICATION
Dinosauria
 Sauropodomorpha
  Sauropodiformes
   Anchisauria
    Sauropoda
     Gravisauria
      Eusauropoda
       Cetiosauridae

LOCATION
Madagascar

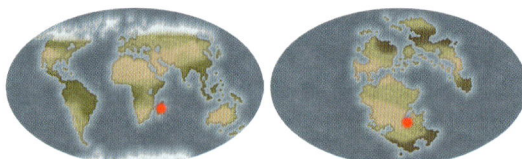

KNOWN REMAINS
Partial skeleton

S. nigerensis

(Remes et al., 2009)
Length: 13.8 m (45 ft)
Height: 5.6 m (18.3 ft)
Hip height: 3.5 m (11.5 ft)
Body mass: 8,500 t (9.4 t)
Reconstruction: ☐☐☐☐

TRIASSIC	JURASSIC	CRETACEOUS

Spinophorosaurus nigerensis (meaning "spine-bearing lizard from Niger") is known from the substantial remains of two adults and one juvenile. With this nearly complete set of information, tests regarding the animal's neck flexibility have shown that it would have had a similar range of motion to that of a giraffe (Vidal et al., 2020). Despite this skeletal completeness, studies disagree on whether *Spinophorosaurus* was a member of **Eusauropoda** (Xing et al., 2015a; Ren et al., 2023) or was instead more basally placed (Holwerda and Pol, 2018).

Initially, *Spinophorosaurus* was believed to possess a feature unique among sauropods—a set of short spikes situated at the end of its tail, without the presence of any sort of bulbous club. The bones originally interpreted as these osteoderm-spikes were not discovered attached to the tail, but the tail's stiffness was similar to other dinosaurs known to possess tail weaponry, and so the hypothesis seemed plausible. However, later work would show that these bones lacked the surface morphology associated with osteoderm, and that they were most likely broken bits of the clavicles (Tschopp and Mateus, 2013; Mocho et al., 2018).

Another initial inaccuracy regarding *Spinophorosaurus* involves its posture. It was originally interpreted as a low browser, eating vegetation close to the ground, and as a consequence having a quite horizontal stance. However, after digitally reconstructing the skeleton, it was found that *Spinophorosaurus* would instead have a much more upright posture. One key trait enabling this feature is that the vertebrae in its hip and lower back are slightly wedge shaped, which alters the spine's overall trajectory.

The generic name *Spinophorosaurus* combines the Latin "spina" (meaning "spine") with the Greek "phora" (meaning "to bear") and "sauros" (meaning "lizard"). The specific name *nigerensis* refers to the African nation of Niger.

CLASSIFICATION
Dinosauria
  Sauropodomorpha
    Sauropodiformes
      Anchisauria
        Sauropoda
          Gravisauria
            Eusauropoda

LOCATION
Niger

KNOWN REMAINS
Nearly complete

2 m   4 m   6 m   8 m   10 m

4 m

2 m

B. alba

(Pol et al., 2020)
Length: 9.5 m (31.2 ft)
Height: 3.3 m (10.8 ft)
Hip height: 2.9 m (9.8 ft)
Body mass: 2,800 kg (3.1 t)
Reconstruction: ☐☐☐☐☐

250 245 240 235 230 225 220 215 210 205 200 195 190 185 **179** 175 170 165 160 155 150 145 140 135 130 125 120 115 110 105 100 95 90 85 80 75 70 65

TRIASSIC　　**JURASSIC**　　**CRETACEOUS**

Bagualia alba (meaning "wild horse of dawn") is among the earliest eusauropods known in the fossil record.

Between the Pliensbachian and Toarcian stages of the Early Jurassic, Earth faced a climate crisis, spurred on by ongoing supervolcanic activity. Before this time, the presence of a diverse array of plant fossils suggests a wet, humid ecosystem; by contrast, the reduced overall diversity of plants, leaving mainly conifers to dominate the ecosystem, indicates a much drier and seasonal climate afterward. In order to survive, it seems that sauropod lineages needed to adapt to eat these tougher, hardier flora.

Bagualia, one of the first sauropods known to have appeared after this event, had developed a number of attributes in order to do just that. Its increased body mass and enlarged gut capacity meant that it could digest more resistant plant matter; it procured this material thanks to especially tough teeth, which boasted an enamel layer that

was seven times thicker than that of more primitive sauropods. These spoon-shaped teeth display telltale wear patterns, which further bolsters the idea that *Bagualia* was subsisting on very hardy vegetation.

A proportionally elongated neck meant that *Bagualia* could acquire more calories without expending the energy necessary to move its massive body. Its neck had also begun to develop an increased pneumaticity, which would both lighten its weight and, potentially, aid in the dissipation of heat being generated within its massive body. These traits would prove to be vital for **eusauropods**, as members of their lineage continued to balloon in size.

The generic name *Bagualia* comes from the Spanish term "bagual" (meaning "wild horse"), referring to Bagual Canyon, where the fossils were discovered. The specific name *alba* is Spanish for "dawn," referring to the specimen's early place on the sauropod family tree.

CLASSIFICATION

Dinosauria
　Sauropodomorpha
　　Sauropodiformes
　　　Anchisauria
　　　　Sauropoda
　　　　　Gravisauria
　　　　　　Eusauropoda

LOCATION

Argentina

KNOWN REMAINS

Nearly complete

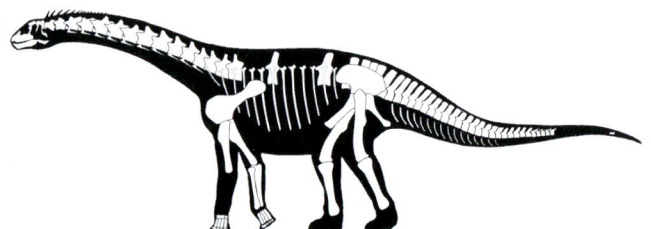

4 m 8 m 12 m 16 m 8 m

4 m

T. benitezii

(Rich et al., 1999)
Length: 16 m (52 ft)
Height: 8 m (26.2 ft)
Hip height: 3.3 m (10.8 ft)
Body mass: 16,000 kg (17.6 t)
Reconstruction: ☐☐☐☐☐

250 — 245 — 240 — 235 — 230 — 225 — 220 — 215 — 210 — 205 — 200 — 195 — 190 — 185 — 180 — 175 — 170 — 165 — 160 — 155 — **150** — 145 — 140 — 135 — 130 — 125 — 120 — 115 — 110 — 105 — 100 — 95 — 90 — 85 — 80 — 75 — 70 — 65 —

TRIASSIC	JURASSIC	CRETACEOUS

Tehuelchesaurus benitezii (meaning "Benitez's lizard of the Tehuelche") is known from a single partial specimen (MPEF-PV 1125), consisting of the forelimb, shoulder, femur, partial pelvis, and the midvertebral column. The vertebrae and pelvic elements were found in articulation, with the other bones found in close association.

Skin impressions were also found at the site, corresponding to various body regions, such as the forelimbs, shoulders, and thorax. These revealed hexagonal- and rhomboidal-shaped scales of varying sizes (del Valle Giménez, 2007).

When the genus was first described (which was in brief), many of the vertebral elements had not yet been fully prepared. A later, more complete examination was conducted once all of the material was free of obstruction. Unique skeletal traits had been revealed during the process, such as the arrangement of "accessory laminae"—sheet-like

bony support structures—on the vertebrae. Other details of the vertebrae show some similarities to disparate lineages, such as to primitive sauropodomorphs or to derived titanosaurs (Carballido et al., 2011).

Tehuelchesaurus was originally considered to be a camarasauromorph, and several studies have subsequently agreed with this phylogenetic positioning (Carballido et al., 2020). However, a taxonomic study that specifically incorporated updated data found *Tehuelchesaurus* to more likely be a member of **Turiasauria** (Mannion et al., 2019).

The generic name *Tehuelchesaurus* refers to the Indigenous Tehuelche people of the area. The specific name *benitezii* honors the fossil's discoverer, Aldino Benitez. The Cañadón Calcáreo Formation, which yielded the bones, contains many fish fossils, and at least one crocodylomorph, but comparatively few dinosaurian remains.

CLASSIFICATION
Dinosauria
  Sauropodomorpha
   Sauropodiformes
    Anchisauria
     Sauropoda
      Gravisauria
       Eusauropoda
        Turiasauria

LOCATION
Argentina

KNOWN REMAINS
Partial skeleton

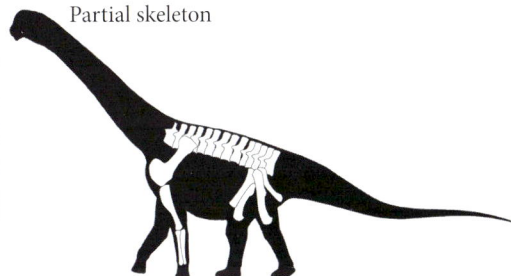

109

MIERASAURUS

3 m · **6 m** · **9 m** · **12 m**

3 m

M. bobyoungi
(Royo-Torres et al., 2017)
Length: 11.5 m (38 ft)
Height: 5.4 m (17.7 ft)
Hip height: 2.7 m (8.8 ft)
Body mass: 4,800 kg (5.3 t)
Reconstruction: ☐☐☐☐

250 245 240 235 230 225 220 215 210 205 200 195 190 185 180 175 170 165 160 155 150 145 140 135 130 125 120 115 110 105 100 95 90 85 80 75 70 65

TRIASSIC **JURASSIC** **CRETACEOUS**

Mierasaurus bobyoungi (meaning "Miera and Robert Young's lizard") is known from a nearly complete, mostly disarticulated skeleton that was unearthed in 2010. One of the few articulated elements was a foot, which was situated lower than the rest of the body in mudstone; this placement suggests that the animal perhaps died after becoming mired in mud. A few bones of a juvenile sauropod, likely also *Mierasaurus*, were found at the site as well as iguanodontian and ankylosaurian fragments and theropod teeth. The local environment during this time was wet and lush.

The species' identification as a **turiasaur** was somewhat surprising because prior to this point in time, turiasaurs were largely considered to be limited to the Jurassic of Europe. Finding an example of a North American turiasaur that survived well into the Cretaceous expands the overall picture of the sauropod family tree. As continental bridges are known to have existed between Europe and North America around this time, this example of "faunal interchange" is entirely plausible.

The existence of *Mierasaurus* also raises the question of why other, similar turiasaur remains had never before been found. It is possible that, based on this knowledge, future reviews of previously discovered museum-housed specimens will lead to some being reclassified in a new light. One potential candidate for such a specimen is FMNH PR 487, a sauropod humerus from South Dakota of similar age. Unfortunately, the humerus is one of the few bones not present in the *Mierasaurus* holotype, so a direct comparison of the two is currently impossible.

The generic name *Mierasaurus* honors historic Spanish cartographer Bernardo de Miera y Pacheco for his early work in mapping Utah. The specific name *bobyoungi* honors paleontologist Robert Young for his work regarding the Cretaceous outcrops of Utah.

CLASSIFICATION
Dinosauria
 Sauropodomorpha
  Sauropodiformes
   Anchisauria
    Sauropoda
     Gravisauria
      Eusauropoda
       Turiasauria

LOCATION
Utah, USA

KNOWN REMAINS
Nearly complete

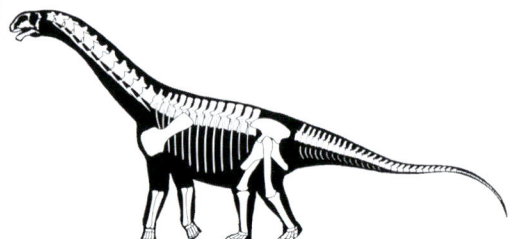

3 m   6 m   9 m   12 m

M. utahensis

(Britt et al., 2017)
Length: 12 m (39 ft)
Height: 5.5 m (18 ft)
Hip height: 3 m (10 ft)
Body mass: 6,000 kg (6.6 t)
Reconstruction: ☐☐☐☐

3 m

250 245 240 235 230 225 220 215 210 205 200 195 190 185 180 175 170 165 160 155 150 145 140 135 130 125 120 115 110 105 100 95 90 85 80 75 70 65

| TRIASSIC | JURASSIC | CRETACEOUS |

Moabosaurus utahensis (meaning "lizard from Moab, Utah") is known from the disarticulated remains of multiple individuals that were uncovered together at the Dalton Wells Quarry. Since 18 separate braincases were recovered, at least that many individual *Moabosaurus* contributed to the bone bed. Other dinosaurian remains at the site include the brachiosaurid *Venenosaurus* and theropods such as *Utahraptor*.

The remains at the site (which first underwent excavation in 1975) show signs of having been trampled on at least two occasions (as evidenced by distinctive markings and breakage), having been consumed and burrowed into by insect larvae, and having been transported some distance by flowing water. It is surmised that this motion washed away most of the small bones and flat bones, leaving behind only the heavy or irregularly shaped bones. As a consequence,

there is a preservation bias at the quarry that favors bones like the vertebrae of *Moabosaurus*; comparatively rare are the remains of plants, invertebrates, or lightweight creatures such as pterosaurs.

The initial description of *Moabosaurus* suggested that it was a basal titanosauriform, yet a slew of subsequent papers have recovered the genus as a member of **Turiasauria**, often being very closely related to *Mierasaurus*, which was also discovered in Utah (Royo-Torres et al., 2017; Mannion et al., 2019; Moore et al., 2023).

The reconstructed *Moabosaurus* presented by researchers was nearly 10 meters in length, but since many of the recovered fossils seem to be from juveniles or subadults, the adult size of *Moabosaurus* would have been larger.

The binomial name of the animal refers to the city of Moab, Utah, which is near the Dalton Wells Quarry.

CLASSIFICATION
Dinosauria
 Sauropodomorpha
  Sauropodiformes
   Anchisauria
    Sauropoda
     Gravisauria
      Eusauropoda
       Turiasauria

LOCATION
Utah, USA

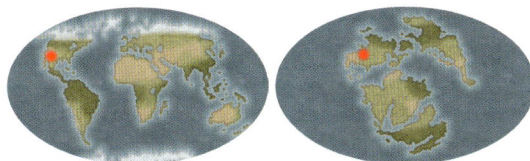

KNOWN REMAINS
Partial skull and skeleton

4 m   8 m   12 m   16 m   20 m   24 m

8 m

4 m

T. riodevensis

(Royo-Torres et al., 2006)
Length: 24 m (78 ft)
Height: 10.2 m (33.5 ft)
Hip height: 6.7 m (22 ft)
Body mass: 30,000 kg (33.1 t)
Reconstruction: ☐☐☐☐

| 250 | 245 | 240 | 235 | 230 | 225 | 220 | 215 | 210 | 205 | 200 | 195 | 190 | 185 | 180 | 175 | 170 | 165 | 160 | 155 | **150** | 145 | 140 | 135 | 130 | 125 | 120 | 115 | 110 | 105 | 100 | 95 | 90 | 85 | 80 | 75 | 70 | 65 |

| TRIASSIC | JURASSIC | CRETACEOUS |

Turiasaurus riodevensis (meaning "lizard from Riodeva, Turia") is known primarily from two partial specimens. The holotype was first discovered in 2003, with excavations continuing for several years. It consisted of mostly disarticulated, yet associated, remains that likely belonged to a single individual. Elements of the skull were preserved and revealed that *Turiasaurus* had a cranium similar in appearance to the famous *Camarasaurus*, being relatively short and rounded (Royo-Torres and Upchurch, 2012). A second, less complete specimen was briefly described in subsequent works (Royo-Torres et al., 2009).

The original, brief description of *Turiasaurus* estimated it to have an enormous body mass of 40,000 to 48,000 kilograms (44–53 tons), and compared it to some of the largest known sauropods, those with body lengths exceeding 30 meters. Thus, some publications have quoted some truly earth-shattering figures regarding the animal's size. However, these early estimates are now often considered to be overly grandiose; a more likely size range for *Turiasaurus* is between 21 and 24 meters in length, which still places it among the leading behemoths of the non-neosauropods.

The discoveries of the related genus *Zby* (whose holotype was originally attributed to *Turiasaurus*) and new specimens of *Losillasaurus* have revealed the existence of a distinct eusauropodan clade, **Turiasauria**. The details of this distinct group are only now beginning to come into focus (Royo-Torres et al., 2020); one unique feature of these sauropods is their "heart-shaped" teeth (Mocho et al., 2016).

The generic name *Turiasaurus* is derived from the term "Turia," which historically has been used to refer to the Teruel Province of Spain. The specific name *riodevensis* refers to the village of Riodeva, Spain.

CLASSIFICATION
Dinosauria
 Sauropodomorpha
  Sauropodiformes
   Anchisauria
    Sauropoda
     Gravisauria
      Eusauropoda
       Turiasauria

LOCATION
Spain

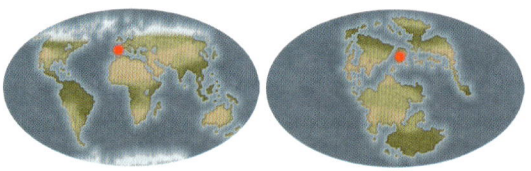

KNOWN REMAINS
Partial skull and skeleton

L. giganteus

(Casanovas et al., 2001)
Length: 20 m (65 ft)
Height: 8.25 m (27 ft)
Hip height: 5 m (16.4 ft)
Body mass: 26,000 kg (28.7 t)
Reconstruction:

3 m | 6 m | 9 m | 12 m | 15 m | 18 m

250 245 240 235 230 225 220 215 210 205 200 195 190 185 180 175 170 165 160 155 150 145 140 135 130 125 120 115 110 105 100 95 90 85 80 75 70 65

TRIASSIC **JURASSIC** **CRETACEOUS**

Losillasaurus giganteus (meaning "giant lizard from Losilla") is known from two primary specimens. The original, consisting of fragmentary cranial and postcranial material, was unearthed in the early 1990s. The specimen is that of a subadult, based on features of the pelvis. Encased in a solid sandstone matrix that was difficult to remove, preparation of the specimen took several years to complete.

A second substantial specimen, discovered from the same geological formation and also consisting of fragmentary cranial and postcranial remains, was excavated in 2010 and 2011 and was described in 2020. Other fragments have also been referred to the genus, such as a solitary tail vertebra (Royo-Torres et al., 2020).

The description of the original specimen included an analysis that placed the animal as a basal member of Diplodocoidea, and the maxilla of *Losillasaurus* does indeed possess some traits that had been previously recorded only in diplodocoids. However, the subsequent description of the second *Losillasaurus* specimen reached the conclusion that *Losillasaurus* was the sister genus of *Turiasaurus*, placing it within **Turiasauria**.

This work helped to solidify the affinities of turiasaurians in general, adding more certainty to the accuracy of phylogenetically placing eusauropod species; the study was conducted using firsthand observations and new measurements of nearly every turiasaur specimen in Europe, Argentina, and the United States (Royo-Torres et al., 2020).

The generic name *Losillasaurus* refers to the village of Losilla in Valencia, Spain. The specific name *giganteus* is Latin, meaning "like that of giants."

CLASSIFICATION
Dinosauria
 Sauropodomorpha
 Sauropodiformes
 Anchisauria
 Sauropoda
 Gravisauria
 Eusauropoda
 Turiasauria

LOCATION
Spain

KNOWN REMAINS
Partial skull and skeleton

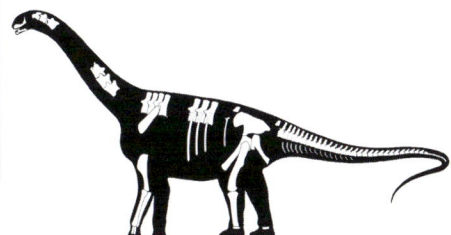

113

EUSAUROPODA

4 m   8 m   12 m   16 m

4 m

T. tanzaniensis

(Bonaparte et al., 2000)
Length: 16 m (52 ft)
Height: 6.7 m (22 ft)
Hip height: 4 m (13.1 ft)
Body mass: 12,000 kg (13.2 t)
Reconstruction: ☐☐☐☐

250 245 240 235 230 225 220 215 210 205 200 195 190 185 180 175 170 165 160 155 150 145 140 135 130 125 120 115 110 105 100 95 90 85 80 75 70 65

TRIASSIC   **JURASSIC**   **CRETACEOUS**

Tendaguria tanzaniensis (meaning "Tendaguru one from Tanzania") was described on the basis of only a few vertebrae.

The two front-dorsal vertebrae that now form the defining syntypes of *Tendaguria* were unearthed during a series of paleontological expeditions that took place between 1909 and 1912. Carried out by the German geologist Wilhelm Bornhardt, one such 1911 dig found these two bones only a short distance away from one another. On this basis, they are presumed to have come from the same animal.

The bones were first described in 1929 by German paleontologist Werner Janensch but were not assigned to a particular species, in part because of their unique features (such as extremely reduced neural spines) that defied categorization. Their classification would have to wait until 2000, when they were included in a review of material that had been potentially assigned to the enigmatic sauropod genus *Janenschia*. During this process, the bones were reevaluated and subsequently assigned to their own distinct genus. Another bone from the original series of expeditions, a cervical vertebra that had once been assigned to the now-dubious genus *Gigantosaurus*, was also tentatively referred to *Tendaguria*.

This 2000 analysis failed to identify the phylogenetic affinities of *Tendaguria*. A subsequent study found it to be a titanosauriform (Poropat et al., 2015). However, a number of newer studies have all found *Tendaguria* to be a member of **Turiasauria** (Moore et al., 2020; Schwarz et al., 2020; Royo-Torres et al., 2020).

The generic name *Tendaguria* refers to the Tendaguru Beds, a geological formation known for the numerous fossils it contains. The specific name *tanzaniensis* refers to the African nation of Tanzania.

CLASSIFICATION
Dinosauria
  Sauropodomorpha
  Sauropodiformes
  Anchisauria
  Sauropoda
  Gravisauria
  Eusauropoda
  Turiasauria

LOCATION
Tanzania

KNOWN REMAINS
Three vertebrae

2 m   4 m   6 m

2 m

A. greppini

(Schwarz et al., 2020)
Length: 6.3 m (20.7 ft)
Height: 2.6 m (8.5 ft)
Hip height: 1.6 m (5.2 ft)
Body mass: 1,000 kg (1.1 t)
Reconstruction: ☐ ☐ ☐ ☐

250 245 240 235 230 225 220 215 210 205 200 195 190 185 180 175 170 165 160 **157** 150 145 140 135 130 125 120 115 110 105 100 95 90 85 80 75 70 65

TRIASSIC	JURASSIC	CRETACEOUS

Amanzia greppini (meaning "Amanz and Greppin's one") is the first sauropod to be identified from Switzerland. Typically, Europe is thought to have consisted of numerous small islands during the Jurassic and Cretaceous, but the discovery of *Amanzia* is one of several clues to suggest that, at least part of the time, more of Europe consisted of dry land.

A set of disarticulated, highly deformed fossils was discovered in a limestone quarry in the 1860s, along with the tooth of a ceratosaurian theropod. Most of the bones were sold into private collections but were subsequently reassembled by Jean-Baptiste Greppin. All of the remains were assumed to have come from a theropod and were dubbed "*Megalosaurus meriani*" in 1870. It would not be until the 1920s that the sauropod affinities of most of the bones were realized; they were subsequently known as *Ornithopsis greppini*, and then *Cetiosauriscus greppini*.

The fossils were largely overlooked until 2003, when they were at last cleaned and prepared using modern methods.

Having come from at least four individual animals, the bones were revealed to have preserved the remnants of cartilage located in a limb joint—a rare example of soft tissue preservation. It was also noted that the species to which the bones belonged was not likely to have actually been all that closely related to *Cetiosauriscus* after all. Eventually, a comprehensive redescription of the specimen would lead to the erection of a new genus, *Amanzia*. Despite the bones' small size, indications are that the animal was fully grown.

The phylogenetic component of this analysis suggested that *Amanzia* was either a member of the **Turiasauria** clade or sat just outside that group, in a slightly more derived position.

The generic name *Amanzia* honors Swiss geologist Amanz Gressly, who discovered the first dinosaur fossil from Switzerland in 1856. The specific name *greppini* honors geologist Jean-Baptiste Greppin.

CLASSIFICATION
Dinosauria
 Sauropodomorpha
  Sauropodiformes
   Anchisauria
    Sauropoda
     Gravisauria
      Eusauropoda
       Turiasauria

LOCATION
Switzerland

KNOWN REMAINS
Partial skeleton

115

EUSAUROPODA

N. thevenini

(Royo-Torres et al., 2020)
Length: 13.5 m (44 ft)
Height: 6.2 m (20 ft)
Hip height: 3.7 m (12.1 ft)
Body mass: 8,000 kg (8.8 t)
Reconstruction: ☐☐☐☐

TRIASSIC   **JURASSIC**   **CRETACEOUS**

Narindasaurus thevenini (meaning "Thevenin's lizard from Narinda") is based on scant remains that were found by naturalist Joseph Thomas Last in Madagascar, sometime prior to 1894. They were deposited in the French National Museum of Natural History in 1907, where they remain to this day. The remains consist of a tooth, scattered tail elements, lower limb bones, and the pubis portion of the pelvic bones.

In 1907, paleontologist Armand Thevenin assigned the specimen to "*Bothriospondylus madagascariensis*," which has since been reclassified as *Lapparentosaurus* (Bonaparte, 1986). The specimen received very little attention until it was evaluated during work that reexamined various materials formerly attributed to "*Bothriospondylus*" (Mannion, 2010). The solitary tooth attributed to the specimen was identified as **turiasaurian** shortly thereafter; a distinguishing

characteristic of turiasaurians is having teeth that are "heart-shaped" (Mocho et al., 2016).

The fossils were finally described and assigned in 2020 to a new genus, *Narindasaurus*. This was justified not only by the specimen's Madagascar origins, which differ from what is normally thought of as a predominately European clade, but from a unique combination of skeletal features unknown from other sauropods. Specific attention was paid to other taxa from Madagascar and Africa in order to definitively exclude them from consideration. This taxonomic analysis found *Narindasaurus* to be the basalmost known turiasaurian.

The generic name *Narindasaurus* refers to Narinda Bay, near the "Ankinganivalaka site" where the fossils were discovered. The specific name *thevenini* honors French paleontologist Armand Thevenin.

CLASSIFICATION

Dinosauria
  Sauropodomorpha
    Sauropodiformes
      Anchisauria
        Sauropoda
          Gravisauria
            Eusauropoda
              Turiasauria

LOCATION

Madagascar

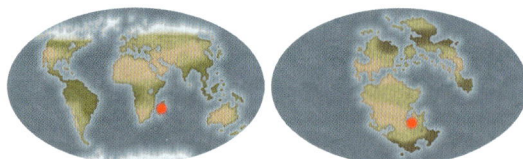

KNOWN REMAINS

Fragments

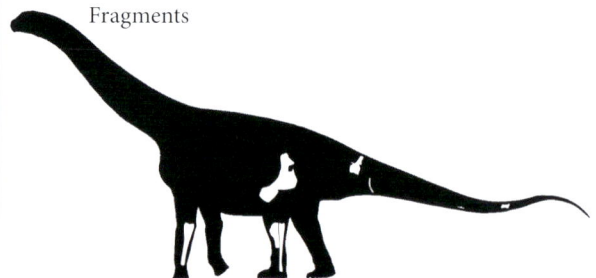

2 m   4 m   6 m   8 m   10 m   12 m   14 m   16 m

J. tiguidensis

(Sereno et al., 1999)
Length: 15.2 m (50 ft)
Height: 6 m (19.7 ft)
Hip height: 3.3 m (10.8 ft)
Body mass: 17,000 kg (18.7 t)
Reconstruction: ☐☐☐☐

6 m

4 m

2 m

250 245 240 235 230 225 220 215 210 205 200 195 190 185 180 175 170 165 160 155 150 145 140 135 130 125 120 115 110 105 100 95 90 85 80 75 70 65

TRIASSIC **JURASSIC** **CRETACEOUS**

Jobaria tiguidensis (meaning "Jobar's one from Tiguidi") was originally interpreted as being capable of rearing upright on its hindlegs. Its describers calculated that its center of gravity was near the animal's hips and compared the measured ratios of its limb bones with that of a modern elephant, which is similarly capable of rearing up. Currently, there is no consensus on whether or not this was actually likely.

Several specimens of varying levels of completeness were recovered from the same area, suggesting that perhaps *Jobaria* engaged in herding behavior. The fossil site was interpreted as a flash-flood induced burial, although predatory tooth marks (speculated to have been inflicted by *Afrovenator*) were noticed on the ribs of one of the specimens, which happened to be a juvenile. The remains also included the majority of the animal's skull, a rarity for sauropods.

The majority of modern analyses phylogenetically place *Jobaria* as a non-neosauropod **eusauropod**, one that is very closely related to Turiasauria (Jannel et al., 2019; Mannion et al., 2019; Ren et al., 2023), although at least one study goes as far as to actually place it within Turiasauria (Moore et al., 2023), while one suggests Cetiosauridae (Royo-Torres et al., 2020).

When it was first described, the sediments that bore *Jobaria* were thought to have been Early Cretaceous in age, which made the animal's "primitive" position doubly complexing; now, though, the strata in question are known to have been Middle Jurassic in origin, which does help it "fit in" the family tree a bit more reasonably.

The generic name *Jobaria* refers to the name of a local mythical giant beast, Jobar, which is thought to have been directly inspired by the presence of large dinosaur bones. The specific name *tiguidensis* refers to the cliff of Tiguidi, the specimens' site of discovery.

CLASSIFICATION

Dinosauria
 Sauropodomorpha
  Sauropodiformes
   Anchisauria
    Sauropoda
     Gravisauria
      Eusauropoda

LOCATION

Niger

KNOWN REMAINS

Nearly complete

117

ZBY

Z. atlanticus

(Mateus et al., 2014)
Length: 18 m (59 ft)
Height: 7.5 m (24.6 ft)
Reconstruction: ☐☐☐☐☐

2 m

150

TRIASSIC | JURASSIC | CRETACEOUS

Zby atlanticus (meaning "Zby's Atlantic one," pronounced "zee-bee") is known from a single specimen consisting of a nearly complete forelimb as well as a few other scattered fragments. The bones were found in close association during excavations in the late 1990s and early 2000s in Portugal. They were preliminarily identified as belonging to *Turiasaurus* but later evaluation would conclude that

they represented a new and distinct genus within **Turiasauria**.

The generic name *Zby* honors Russian-French paleontologist Georges Zbyszewski, who had a large impact on the study of Portuguese paleontology. The specific name *atlanticus* refers to the scenic view of the Atlantic Ocean near the fossil dig site.

HAESTASAURUS

H. becklesii

(Mantell, 1852)
Length: 9 m (30 ft)
Height: 5 m (16.4 ft)
Reconstruction: ☐☐☐☐☐

1 m

140

TRIASSIC | JURASSIC | CRETACEOUS

Haestasaurus becklesii (meaning "Haesta and Beckles's lizard") is known from a single forelimb that was long identified as belonging to the genus *Pelorosaurus*, although many researchers had noted that the specimen differed considerably from the type species, *P. brevis*. It was finally assigned its own genus in 2015 (Upchurch et al., 2015).

What is still uncertain is what kind of sauropod *Haestasaurus* exactly was. It has variously been recovered as

a basal titanosaur (Upchurch et al., 2015), a turiasaurian (Mannion et al., 2019), a derived eusauropod outside Neosauropoda (Royo-Torres et al., 2020), and as a basal macronarian (Moore et al., 2020).

The generic name *Haestasaurus* refers to Haesta, the original historic leader of the people who first settled Hastings, England. The specific name *becklesii* honors the fossil's discoverer, Samuel Husband Beckles.

J. robusta

(Wild, 1991)
Length: 15.3 m (50 ft)
Height: 5.5 m (18 ft)
Reconstruction: ☐☐☐☐

2 m

150

| TRIASSIC | JURASSIC | CRETACEOUS |

Janenschia robusta (meaning "Janensch's strong one") has an especially complicated naming history. Originally named *Gigantosaurus robustus* by Eberhard Fraasin in 1908, it was eventually renamed and reclassified. A number of bones once attributed to the genus have been reassigned to their own, distinct species. Currently, only bones from the limbs and hips are assigned to *Janenschia* (Mannion et al., 2019).

Phylogenetic placements of the genus often vary by a significant amount. Most recently, it has been placed within Turiasauria (Mannion et al., 2019) or basal Macronaria (Ren et al., 2023). The animal's generic name honors paleontologist Werner Janensch. The species was discovered in Tanzania.

N. taito

(Xing et al., 2015a)
Length: 8 m (26 ft)
Height: 3.5 m (11.5 ft)
Reconstruction: ☐☐☐☐

1 m

170

| TRIASSIC | JURASSIC | CRETACEOUS |

Nebulasaurus taito (meaning "misty cloud lizard of Taito") is known from only a single braincase. It is distinct in that it represents a clade of eusauropods that had previously been unknown in Asia and thus expands the known sauropodan diversity of the time period.

The generic name *Nebulasaurus* derives from the Latin word "nebulae" (meaning "misty cloud," a reference to Yunnan Province in China); "Yunnan" translates as "southern misty cloudy province." The specific name honors the Japanese gaming company Taito, which funded the recovery field project.

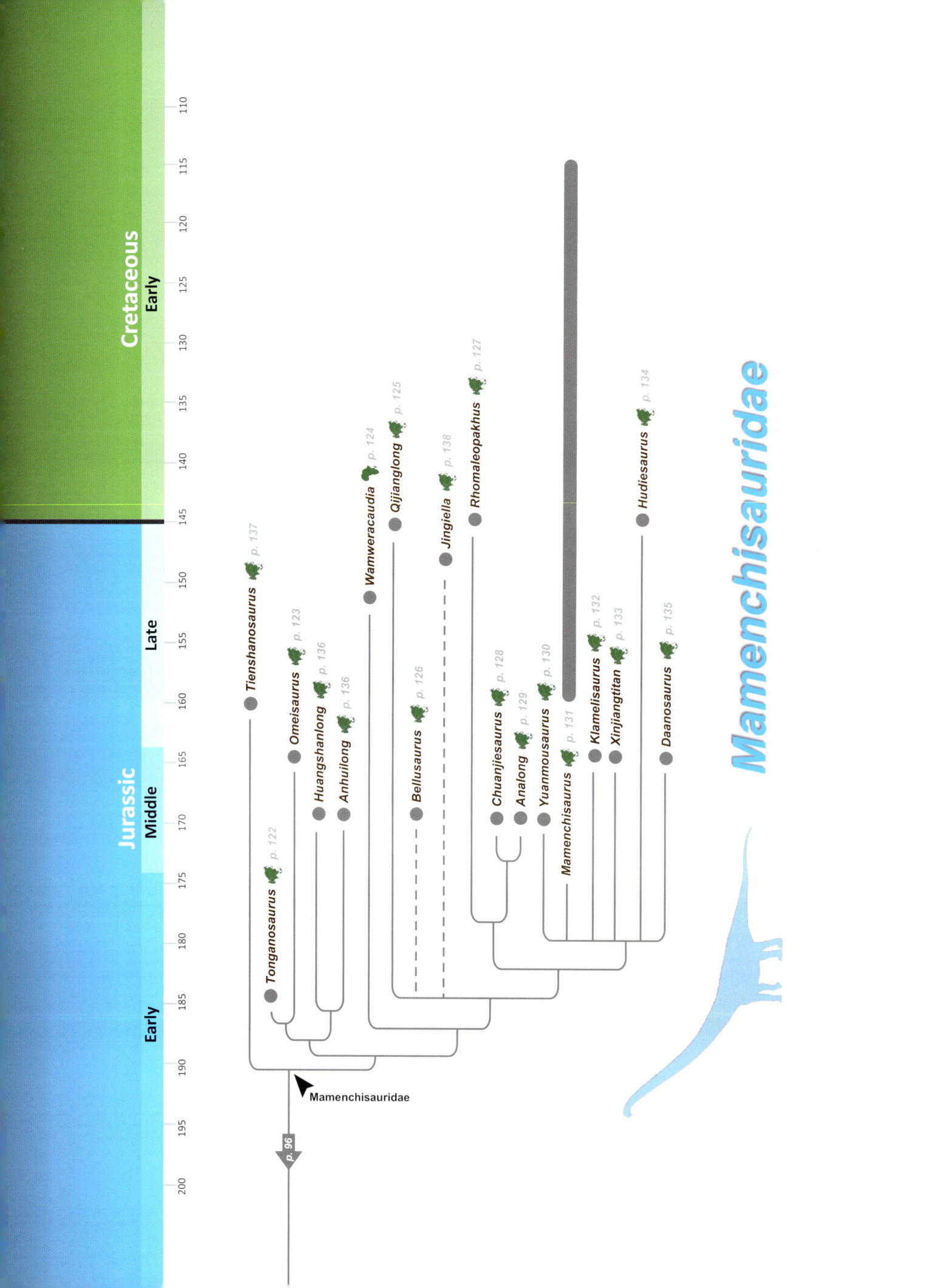

Mamenchisauridae

Cretaceous		Jurassic		
Early		Late	Middle	Early

Tienshanosaurus *p. 137*

Omeisaurus *p. 123*

Huangshanlong *p. 136*

Anhuilong *p. 136*

Wamweracaudia *p. 124*

Qijianglong *p. 125*

Jingiella *p. 138*

Rhomaleopakhus *p. 127*

Bellusaurus *p. 126*

Chuanjiesaurus *p. 128*

Analong *p. 129*

Yuanmousaurus *p. 130*

Mamenchisaurus *p. 131*

Klamelisaurus *p. 132*

Xinjiangtitan *p. 133*

Hudiesaurus *p. 134*

Daanosaurus *p. 135*

Tonganosaurus *p. 122*

Mamenchisauridae

p. 96

The mamenchisaurids are famous for having the longest necks in the animal kingdom—both in proportional and absolute terms. The vast majority of specimens have been discovered from Jurassic deposits in central Asia, but Cretaceous and African specimens have shown that the group's range is greater in scope than was once believed.

The relationships between the mamenchisaurid taxa are far from certain. With a fair degree of uncertainty even regarding which taxa are, or are not, part of the clade, phylogenetic analyses tend to limit the clade's inclusion to contain just a small number of core taxa. Many researchers consider the task of creating a more detailed and accurate reckoning of Mamenchisauridae to be impossible without first subjecting each of the distinct *Mamenchisaurus* species to thorough, modern reanalysis. Until such work is completed, there are currently limited sources to draw upon that attempt to lay out Mamenchisauridae as a whole phylogenetically. As such, the arrangement here will be largely based on the work of Moore et al. (2023), which itself conducted multiple versions of analyses utilizing differing datasets and parameters.

The oft-neglected *Tonganosaurus* was considered by its describers to be most closely related to *Omeisaurus* (Li et al., 2010), so here it has been placed as the sister taxa of that genus for lack of alternatives. The describers of *Anhuilong* found it to be the sister taxon of *Huangshanlong*, both of which are known from only forelimb material (Ren et al., 2018). Some results have also placed *Yuanmousaurus* somewhere within the family (Xing et al., 2015b; Ren et al., 2023). The describers of *Jingiella* determined it to be a mamenchisaurid of some sort (Ren et al., 2024).

Mamenchisaurus constructus

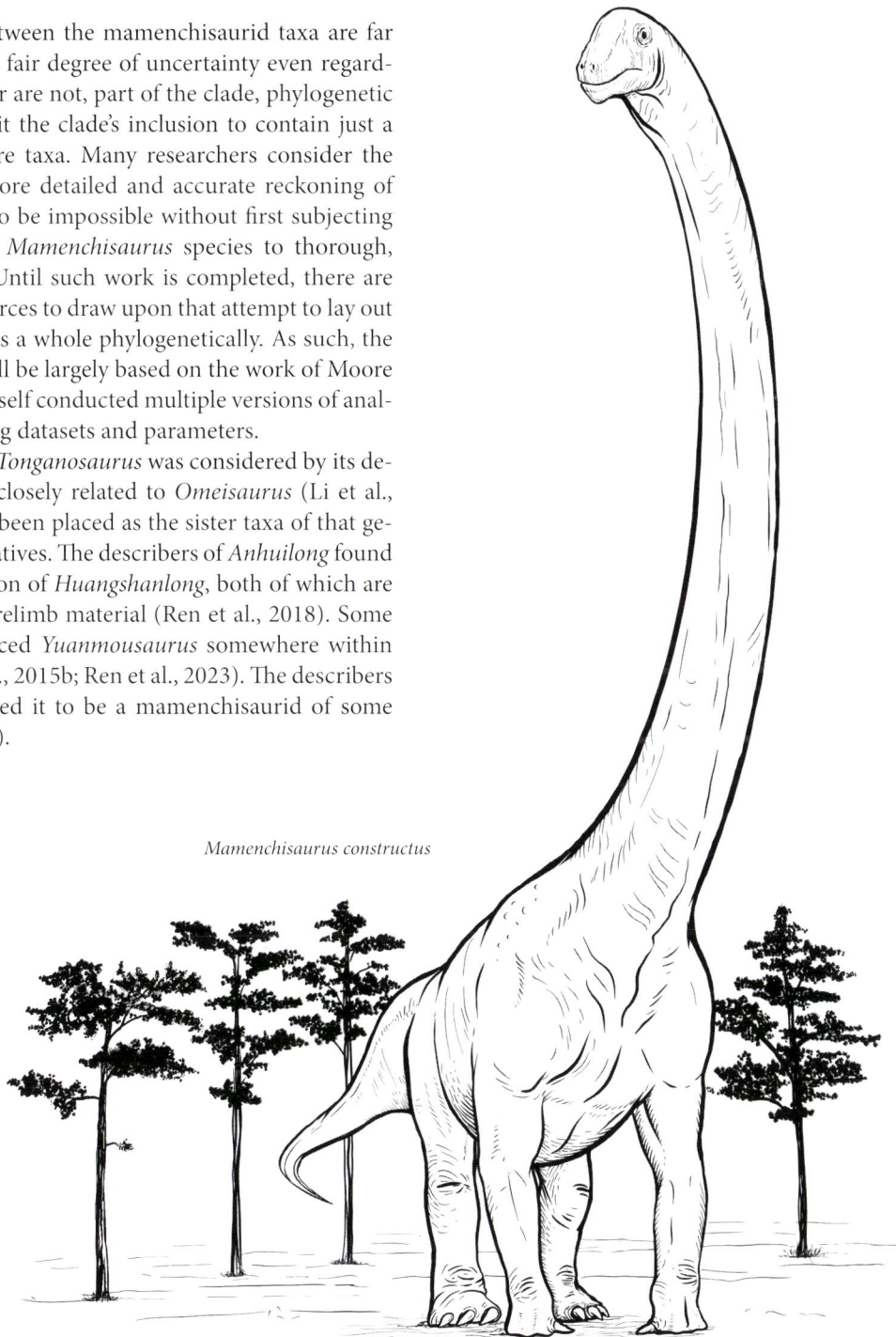

2 m 4 m 6 m 8 m 10 m 12 m 14 m

6 m

T. hei

(Li et al., 2010)
Length: 11 m (36 ft)
Height: 7.2 m (23.6 ft)
Hip height: 2.9 m (9.5 ft)
Body mass: 2,000 kg (2.2 t)
Reconstruction:

4 m

2 m

250 245 240 235 230 225 220 215 210 205 200 195 190 **184** 180 175 170 165 160 155 150 145 140 135 130 125 120 115 110 105 100 95 90 85 80 75 70 65

TRIASSIC	JURASSIC	CRETACEOUS

Tonganosaurus hei (meaning "He's lizard from Tong'an") is a sauropod from the Early Jurassic of China. Prior to its discovery, only three sauropod genera were known from Early Jurassic deposits within the country. Also present in the same geological formation as the fossils themselves are sauropod trackways, known by the ichnogenus *Brontopodus*; the trackways could very well have been made by *Tonganosaurus* (Xing et al., 2016). The Yimen geological formation primarily contains plant and invertebrate fossils that indicate a floodplain or lake-based environment. Other fragmentary sauropodomorph remains have also been uncovered from the formation.

Acting on information about the presence of fossils that was provided by citizen scientists, paleontologists from the Chengdu University of Technology Museum found and excavated the disarticulated bones that would go on to form

the holotype specimen of *Tonganosaurus*. Based on their location and size, the bones were believed to have come from a single individual.

The description of the genus indicates that *Tonganosaurus* belongs to **Mamenchisauridae**, being most closely related to *Omeisaurus*. Whereas *Omeisaurus* possesses "complex" structures in the centra of its dorsal vertebrae, *Tonganosaurus* possesses those with a "simple" structure, which the researchers conclude makes it more basally placed on the family tree. If true, this would make it the earliest known example of Mamenchisauridae. However, no other taxonomic studies have, to date, tested this result or offered any alternatives.

The generic name *Tonganosaurus* refers to the town of Tong'an, in Sichuan Province, China. The specific name *hei* honors researcher He Xinlu.

CLASSIFICATION
Dinosauria
  Sauropodomorpha
  Sauropodiformes
  Anchisauria
  Sauropoda
  Gravisauria
  Eusauropoda
  Mamenchisauridae

LOCATION
China

KNOWN REMAINS
Partial skeleton

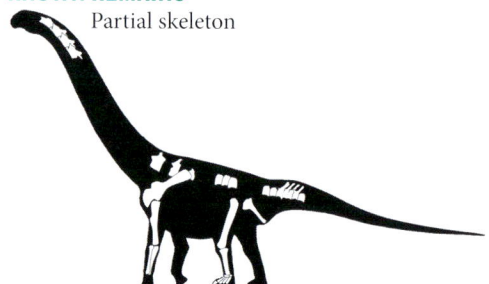

4 m   8 m   12 m   16 m   20 m

8 m

4 m

O. junghsiensis

(Young, 1939)
Length: 16.5 m (54 ft)
Height: 7.7 m (25 ft)
Hip height: 3 m (9.8 ft)
Body mass: 6,000 kg (6.6 t)
Reconstruction: ☐☐☐☐

250 245 240 235 230 225 220 215 210 205 200 195 190 185 180 175 170 165 160 155 150 145 140 135 130 125 120 115 110 105 100 95 90 85 80 75 70 65

TRIASSIC　　　　**JURASSIC**　　　　**CRETACEOUS**

Omeisaurus junghsiensis (meaning "Omei lizard from Junghsien") was described based on a partial skeleton that was unearthed in 1936, then lost during WW II. Like many dinosaur genera described from the early 1900s and earlier, many separate species have been placed into the genus over time. Unlike most examples, though, the majority of these species have not been rendered dubious and are by and large still considered to be valid. The most recent species, *O. puxiani*, was named in 2021 (Tan et al., 2021). Some studies, though, have found these species to not actually be very closely related to one another (Upchurch et al., 2021).

One recent analysis sought to clarify the relatedness of various *Omeisaurus* species and found *O. junghsiensis*, *O. tianfuensis*, *O. jiaoi*, and *O. puxiani* to indeed form a distinct clade, which at the very least suggests a very close relationship (Chao et al., 2020). *Omeisaurus maoianus*, however, was found to be separate from the others; given this result had also been found by other researchers

(Ren et al., 2018; Moore et al., 2023), it likely means that this particular species should be reassigned to its own genus (D'Angelo, 2021).

Other attributed species include *O. changshouensis*, *O. fuxiensis*, and *O. luoquanensis*. The largest species, *O. tianfuensis*, measured between 18 and 20 meters in length. Taken altogether, the dozens of separate specimens account for the genus's entire skeleton, including numerous examples of relatively intact skull and jaw material.

Based on disarticulated remains found nearby, *O. tianfuensis* has been hypothesized to have possessed a small club on the end of its tail (Dong et al., 1989; Lida et al., 2009), although the possibility remains that the fossil came from a separate genus, such as *Shunosaurus*.

The generic name *Omeisaurus* refers to the sacred mountain Omeishan. The specific name *junghsiensis* refers to the city of Junghsien.

CLASSIFICATION
Dinosauria
　Sauropodomorpha
　Sauropodiformes
　　Anchisauria
　　Sauropoda
　　　Gravisauria
　　　Eusauropoda
　　　　Mamenchisauridae

LOCATION
China

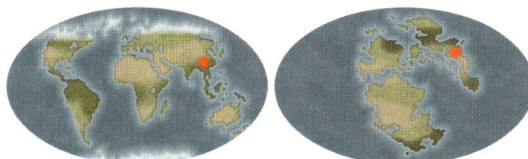

KNOWN REMAINS
Complete

MAMENCHISAURIDAE

| 4 m | 8 m | 12 m | 16 m | 20 m |

W. keranjei

(Mannion et al., 2019)
Length: 16.7 m (55 ft)
Height: 9.8 m (32.1 ft)
Hip height: 3.3 m (10.8 ft)
Body mass: 7,000 kg (7.7 t)
Reconstruction: ☐☐☐☐☐

| TRIASSIC | JURASSIC | CRETACEOUS |

Wamweracaudia keranjei (meaning "Keranje's tail lizard of the Wamwera") is described on the basis of a single, nearly complete tail.

This set of caudal (tail) vertebrae was unearthed between 1909 and 1912 during a series of digs orchestrated by German paleontologist Werner Janensch. In 1929, he referred these fossils to the now-dubious genus *Gigantosaurus robustus*, apparently because he believed that the bones bore a resemblance to another *Gigantosaurus* specimen that he had previously observed. This other set of remains, however, is no longer available for comparison because they were destroyed during World War II.

Gigantosaurus was later rebranded as *Janenschia* (Wild, 1991). However, it was subsequently concluded that this specimen could not be attributed to *Janenschia*, partially because the remains were not comparable to any overlapping material present in the *Janenschia* type specimen (Bonaparte et al., 2000; Mannion et al., 2013).

A thorough redescription of the specimen in 2019 led to the erection of the new genus, *Wamweracaudia*. This study also concluded that, rather than having titanosaur affinities, the owner of the vertebrae was a **mamenchisaurid**. This interpretation was very noteworthy, because before this analysis, mamenchisaurids had been unknown outside Asia. The presence of mamenchisaurids in southern Gondwana indicates that the clade was more widespread and diversified than was previously realized.

The generic name *Wamweracaudia* honors the Wamwera, an Indigenous tribe in the Lindi District, near the area of the fossil's discovery; "caudia" is Greek for "tail." The specific name *keranjei* honors excavation supervisor Mohammadi Keranje.

CLASSIFICATION
Dinosauria
  Sauropodomorpha
    Sauropodiformes
      Anchisauria
        Sauropoda
          Gravisauria
            Eusauropoda
              Mamenchisauridae

LOCATION
Tanzania

KNOWN REMAINS
Tail

3 m   6 m   9 m   12 m   15 m

6 m

3 m

Q. guokr

(Xing et al., 2015b)
Length: 13.2 m (43 ft)
Height: 9.5 m (31.1 ft)
Hip height: 3.1 m (10.2 ft)
Body mass: 5,200 kg (5.7 t)
Reconstruction: ☐☐☐☐

250 245 240 235 230 225 220 215 210 205 200 195 190 185 180 175 170 165 160 155 150 **145** 140 135 130 125 120 115 110 105 100 95 90 85 80 75 70 65

TRIASSIC   **JURASSIC**   **CRETACEOUS**

Qijianglong guokr (meaning "Guokr's dragon from Qijiang," pronounced "chee-gee-ang-long") is a mamenchisaurid known from the Suining geological formation. The strata's geology has long been considered to be Late Jurassic in origin, occurring near the Jurassic-Cretaceous boundary. One study made some waves in the literature by suggesting that the formation was actually 30 million years younger than this (Wang et al., 2019), but other assertions have disagreed with this claim (Huang, 2018).

The first signs of sauropod remains in the area came from a local farmer who uncovered some vertebrae in the early 1990s. Further study of the location would not commence until 2006, when a construction project unveiled a Jurassic fish fossil. Subsequent paleontological excavations revealed the rest of the remains, which are now attributed to *Qijianglong*. These include the rear portions of the skull—a particularly valuable find.

The individual in question was likely not fully grown, as the preserved elements of the skull had not yet completely fused together. Despite its immaturity, it had already achieved a length of approximately 15 meters. The size of a mature individual remains unknown. A unique feature of *Qijianglong* is that its cervical (neck) vertebrae are highly pneumatized, meaning that they possess hollow spaces that would be occupied by open air sacs. These features are quite common among the more derived eusauropods, such as the diplodocoids, but are never-before-seen in **mamenchisaurids**. This naturally highlights how previously known taxonomic groups can contain more variety and diversity than initially assumed.

The generic name *Qijianglong* refers to the Qijiang District in China; "long" is Chinese for "dragon." The specific name refers to the Guokr Chinese scientific social network; the term itself translates as "nutshell."

CLASSIFICATION
Dinosauria
 Sauropodomorpha
  Sauropodiformes
   Anchisauria
    Sauropoda
     Gravisauria
      Eusauropoda
       Mamenchisauridae

LOCATION
China

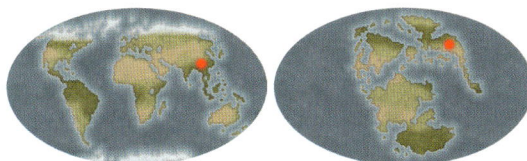

KNOWN REMAINS
Partial skull and skeleton

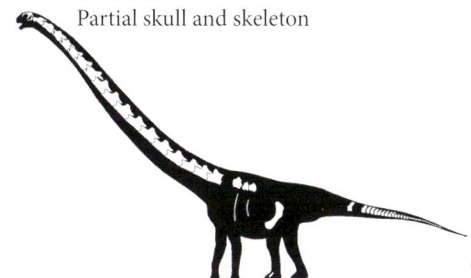

125

BELLUSAURUS

B. sui

(Dong, 1990)
Length: 12 m (39 ft)
Height: 6.8 m (22.3 ft)
Hip height: 2.1 m (6.9 ft)
Body mass: 8,000 kg (8.8 t)
Reconstruction: ☐☐☐☐

| TRIASSIC | JURASSIC | CRETACEOUS |

Bellusaurus sui (meaning "Sui's pretty lizard") has proven to be very tricky to describe and classify, despite having a great many specimens to work with—this is because all of the specimens are from juvenile individuals.

A bone bed that has yielded only bones pertaining to *Bellusaurus* was first excavated in 1983 and again in 2003. Of the hundreds of bones procured, it can be deduced that the remains originate from at least 24 separate individuals based on the number of shoulder blades discovered. This collection also includes a number of skull fragments (Moore et al., 2018).

Phylogenetically classifying a juvenile specimen is complicated by the fact that skeletal morphology can vary greatly with an animal's growth stage. All studies agree that *Bellusaurus* fits in somewhere in the vicinity of the base of Neosauropoda, but more specific diagnoses differ greatly. Two popular interpretations place the genus either just outside of Neosauropoda (Mannion et al., 2019) or at the base of Macronaria, a view favored by a modern study that thoroughly reexamined the combined cranial anatomy of *Bellusaurus* (Moore et al., 2018).

However, the same lead author of the aforementioned paper would later postulate that, in actuality, *Bellusaurus* (as well as *Daanosaurus*) is in fact just a juvenile form of a *Mamenchisaurus*-like genus and, thus, is invalid as its own taxon. This determination was made based on certain uniquely shared anatomical traits, as well as the propensity of **mamenchisaurid** species known from the region (Moore et al., 2023).

The generic name *Bellusaurus* incorporates the Latin "bellus" (meaning "pretty" or "delicate"), as the known specimens were so interpreted by Dong. The specific name *sui* honors the senior preparator, Youling Sui.

CLASSIFICATION

Dinosauria
  Sauropodomorpha
   Sauropodiformes
    Anchisauria
     Sauropoda
      Gravisauria
       Eusauropoda
        Mamenchisauridae

LOCATION

China

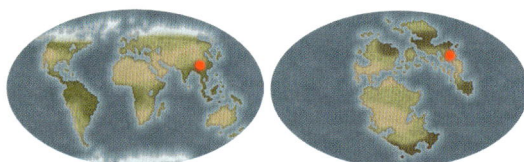

KNOWN REMAINS

Nearly complete (juvenile)

126

3 m 6 m 9 m 12 m 15 m 18 m 21 m 24 m

12 m

R. turpanensis

(Upchurch et al., 2021)
Length: 20 m (65 ft)
Height: 11.7 m (38.4 ft)
Hip height: 3.6 m (11.8 ft)
Body mass: 18,000 kg (19.8 t)
Reconstruction: ◻◻◻◻

9 m

6 m

3 m

| 250 245 240 235 230 225 220 215 210 205 200 195 190 185 180 175 170 165 160 155 150 **145** 140 135 130 125 120 115 110 105 100 95 90 85 80 75 70 65 |
| TRIASSIC | JURASSIC | CRETACEOUS |

Rhomaleopakhus turpanensis (meaning "robust forearm from Turpan") is distinguished by possessing a very stout forelimb (the ulna and radius). Sauropod forelimbs range from "slender" to "hyperrobust" across lineages, and it is thought that extreme stoutness evolved independently at least five different times.

There are several possibilities regarding why such a feature would evolve. One hypothesis regards a sauropod's possible ability to rear up onto its hindlegs; both pushing off of the ground and landing would create considerable stress forces that would need to be distributed.

Alternatively, the simple fact that a **mamenchisaurid's** center of mass lies closer to the shoulders than it does to the hips could account for the development of a greater load-bearing capability in the forelimbs.

Other aspects of *Rhomaleopakhus*'s forelimbs suggest that they were capable of an enhanced range of motion and thus greater stride length. This lengthened stride could potentially indicate a developed specialty for more efficient movement between distant feeding grounds

The *Rhomaleopakhus* holotype specimen was discovered on the same 1993 expedition that found the remains of the closely related *Hudiesaurus*. Until 2004, these bones continued to be identified as *Hudiesaurus*; however, as the two sets of remains contained no overlapping elements, it was concluded that they could not be attributed to the same species with any level of confidence (Upchurch et al., 2004). Thus, these separate limb bones were later used as the basis for describing *Rhomaleopakhus*.

The generic name *Rhomaleopakhus* combines the Greek "rhomaleos" (meaning "robust") and "pakhus" (meaning "forearm"). The specific name *turpanensis* refers to the Turpan Basin, China.

CLASSIFICATION

Dinosauria
 Sauropodomorpha
  Sauropodiformes
   Anchisauria
    Sauropoda
     Gravisauria
      Eusauropoda
       Mamenchisauridae

LOCATION

China

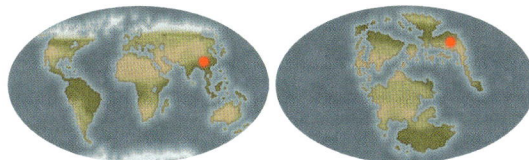

KNOWN REMAINS

Forelimb

C. anaensis

(Fang et al., 2000)
Length: 17 m (56 ft)
Height: 9 m (29.5 ft)
Hip height: 3.9 m (12.8 ft)
Body mass: 11,000 kg (12.1 t)
Reconstruction: ▢▢▢▢

| 250 | 245 | 240 | 235 | 230 | 225 | 220 | 215 | 210 | 205 | 200 | 195 | 190 | 185 | 180 | 175 | 170 | 165 | 160 | 155 | 150 | 145 | 140 | 135 | 130 | 125 | 120 | 115 | 110 | 105 | 100 | 95 | 90 | 85 | 80 | 75 | 70 | 65 |

| TRIASSIC | JURASSIC | CRETACEOUS |

Chuanjiesaurus anaensis (meaning "lizard from Chuanjie and Ana") was discovered by Tao Wang in 1995, and the specimen has been preserved in situ at the location ever since.

In 2000, a bone bed of material was described from the Lufeng World Dinosaur Valley; the bones in this area were described as the new genus *Chuanjiesaurus* (Fang et al., 2000). Later work deduced that the assemblage contained the remains of at least two distinct individuals, more or less side by side. Both were then labeled as *Chuanjiesaurus* (Sekiya, 2011). However, a more detailed examination would later reveal numerous subtle differences between the two specimens, with the second specimen being used to erect the new genus *Analong*.

Chuanjiesaurus can be differentiated from *Analong* based on the sites of muscle attachment in the animal's hips and forelimbs, which are comparatively enlarged or otherwise altered. It is possible these differences relate to the size and posture of the animal's neck and, thus, its feeding strategy, but the overall lack of remains renders this uncertain. The lack of cervical (neck) vertebrae necessitates some guesswork when it comes to the animal's total length.

The initial description of *Chuanjiesaurus* was brief and did not provide a detailed analysis of the remains. It speculated that *Chuanjiesaurus* was a member of Cetiosauridae. When a more thorough investigation was eventually conducted, however, it was concluded that *Chuanjiesaurus* in fact belonged to **Mamenchisauridae**. This determination was made, in part, due to the number of sacral vertebrae, the lack of pneumaticity in the dorsal vertebrae, and rear-dorsal vertebrae which are fused together (Sekiya, 2011).

The animal's binomial name refers to the villages of Ana and Chuanjie located in Yunnan Province, China.

CLASSIFICATION

Dinosauria
 Sauropodomorpha
  Sauropodiformes
   Anchisauria
    Sauropoda
     Gravisauria
      Eusauropoda
       Mamenchisauridae

LOCATION

China

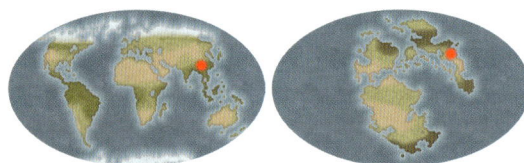

KNOWN REMAINS

Partial skeleton

4 m　　　　　8 m　　　　　12 m　　　　　16 m

A. chuanjieensis

(Ren et al., 2018)
Length: 16.7 m (55 ft)
Height: 9.4 m (30.8 ft)
Hip height: 3.7 m (12.1 ft)
Body mass: 10,000 kg (11 t)
Reconstruction: ☐☐☐☐

8 m

4 m

| 250 | 245 | 240 | 235 | 230 | 225 | 220 | 215 | 210 | 205 | 200 | 195 | 190 | 185 | 180 | 175 | 170 | 165 | 160 | 155 | 150 | 145 | 140 | 135 | 130 | 125 | 120 | 115 | 110 | 105 | 100 | 95 | 90 | 85 | 80 | 75 | 70 | 65 |

| TRIASSIC | JURASSIC | CRETACEOUS |

Analong chuanjieensis (meaning "Dragon from Ana, Chuanjie") is based on remains that were originally identified as the mamenchisaurid *Chuanjiesaurus*.

In 2000, a bone bed of material was described from the Lufeng World Dinosaur Valley; the bones in this area were described as the new genus *Chuanjiesaurus* (Fang et al., 2000). Later work deduced that the assemblage contained the remains of at least two distinct individuals, more or less side by side. Both were then labeled as *Chuanjiesaurus* (Sekiya, 2011). However, a more detailed examination would later reveal numerous subtle differences between the two specimens, with the second specimen being used to erect the new genus *Analong*.

Analong can be differentiated from other mamenchisaurids based on the sites of muscle attachment in the animal's hips and forelimbs, which have been reduced or otherwise altered. It is possible these differences relate to the size and posture of the animal's neck and, thus, its feeding strategy, but the overall lack of remains renders this uncertain.

The description of *Analong* found it to be quite taxonomically different from *Chuanjiesaurus*, despite the two both falling within **Mamenchisauridae**. *Analong* was found to be the basalmost member of the group, whereas *Chuanjiesaurus* was found to be quite derived. This result was repeated in a subsequent study by the same lead author (Ren et al., 2023). Consequently, the two genera being contemporaries suggests intriguing and complex evolutionary pathways. However, a different study contrasted this finding by calculating the two genera to be very closely related within the family (Moore et al., 2023).

The animal's binomial name refers to the villages of Ana and Chuanjie located in Yunnan Province, China.

CLASSIFICATION
Dinosauria
　Sauropodomorpha
　　Sauropodiformes
　　　Anchisauria
　　　　Sauropoda
　　　　　Gravisauria
　　　　　　Eusauropoda
　　　　　　　Mamenchisauridae

LOCATION
China

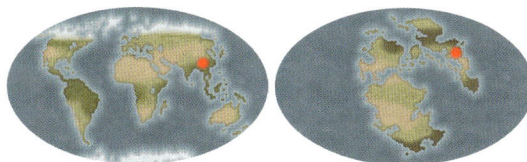

KNOWN REMAINS
Partial skeleton

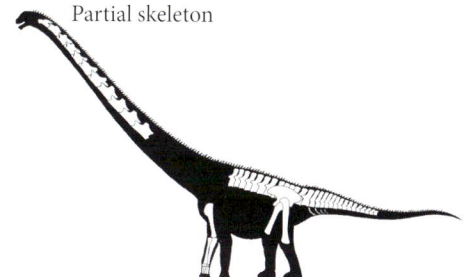

| | 4 m | 8 m | 12 m | 16 m | 20 m |

Y. jiangyiensis

(Lü et al., 2006)
Length: 17 m (58 ft)
Height: 9.5 m (31.2 ft)
Hip height: 3 m (9.8 ft)
Body mass: 11,000 kg (12.1 t)
Reconstruction: ☐☐☐☐

| 250 | 245 | 240 | 235 | 230 | 225 | 220 | 215 | 210 | 205 | 200 | 195 | 190 | 185 | 180 | 175 | **170** | 165 | 160 | 155 | 150 | 145 | 140 | 135 | 130 | 125 | 120 | 115 | 110 | 105 | 100 | 95 | 90 | 85 | 80 | 75 | 70 | 65 |

| TRIASSIC | JURASSIC | CRETACEOUS |

Yuanmousaurus jiangyiensis (meaning "lizard from Jiangyi, Yuanmou") is known from a single set of partial remains that were unearthed from the Zhanghe geological formation in May of 2000. The excavation was conducted by the Yunnan Provincial Institute of Cultural Relics and Archaeology, the Yuanmou Museum, and the Chuxiong Museum.

The set of remains consists primarily of a postcervical vertebral column, shoulder blade, bones of the upper and lower forelimb, and bones of the upper and lower hindlimb. A single fragment of a cervical (neck) vertebra led the fossil's describers to compare it to the mamenchisaurid *Omeisaurus*, in part because of its elongated form.

Yuanmousaurus was also compared to mamenchisaurids because of a similar length ratio between its femur and humerus. The physical structure of the "neural arches of [the] dorsal vertebrae" was said to be "more complex" than

those of *Omeisaurus* (Lü et al., 2006), with the implication being that *Yuanmousaurus* was more derived in comparison. Comparisons were also made with *Patagosaurus*, suggesting that the two genera might share a close relationship. Altogether, these aspects led the describers to place *Yuanmousaurus* within the family Euhelopodidae.

Despite the partial nature of the remains, several newer studies have included the genus in their taxonomic analyses and have found *Yuanmousaurus* to be placed within **Mamenchisauridae** (Sekiya, 2011; Ren et al., 2018, 2023). However, there has not been any modern, detailed analysis completed that describes the holotype set of remains in any more detail, hampering attempts to more accurately pin down its taxonomic position.

The binomial name refers to the Jiangyi region in Yuanmou County, China.

CLASSIFICATION
Dinosauria
 Sauropodomorpha
  Sauropodiformes
   Anchisauria
    Sauropoda
     Gravisauria
      Eusauropoda
       Mamenchisauridae

LOCATION
China

KNOWN REMAINS
Partial skeleton

M. constructus

(Young, 1954)

Length: 20 m (65.6 ft)
Height: 8.3 m (27.2 ft)
Hip height: 3.1 m (10.2 ft)
Body mass: 12,000 kg (13.2 t)
Reconstruction: ☐☐☐☐

TRIASSIC	JURASSIC	CRETACEOUS

Mamenchisaurus (meaning "horse gate brook") is famous for having the longest neck of any known animal.

The first remains of *Mamenchisaurus* were discovered at a construction site in 1952, which inspired the name of the type species, *M. constructus*. The type specimen was fragmentary, disarticulated, and excavated in a hurry—the lack of detailed analysis regarding the specimen has been detrimental in understanding the evolutionary relationships among the multitude of species that are now assigned to the genus. Researchers have long called for a comprehensive reevaluation of the bloated genus.

Mamenchisaurus sinocanadorum is the largest of the recognized species. Although its remains are incomplete, it has been calculated to have the longest neck of any animal, measuring 15 meters (50 feet) in length (Moore et al. 2023).

Other widely recognized species include *M. hochuanensis*, *M. youngi*, *M. anyuensis*, and *M. jingyanensis*. However, many scientists have raised serious doubts as to whether each of these animals should actually be classified within the same genus. Rather, it seems likely that future comparative studies will lead to many of these species being reassigned to new, distinct genera. Multiple studies have already found evidence that the various species are actually widely spread across the **Mamenchisauridae** family tree (Moore et al. 2023). Additionally, the strata that bore *M. anyuensis* have been dated to an age that is 30 million years younger than the age of the other supposed species; this difference makes it even less likely to belong to the same genus (Wang et al. 2019).

The generic name *Mamenchisaurus* combines the Chinese words "ma," "men," and "xi" (meaning "horse gate brook"). This was a miscommunication, as "men" was meant to be "ming" (meaning "neighing").

CLASSIFICATION

Dinosauria
 Sauropodomorpha
 Sauropodiformes
 Anchisauria
 Sauropoda
 Gravisauria
 Eusauropoda
 Mamenchisauridae

LOCATION

China

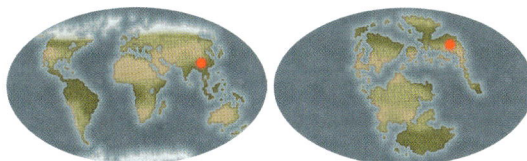

KNOWN REMAINS

Complete

K. gobiensis

(Zhao, 1993)

Length: 14.5 m (48 ft)
Height: 9.9 m (32.5 ft)
Hip height: 3.3 m (10.8 ft)
Body mass: 6,000 kg (6.6 t)
Reconstruction:

TRIASSIC | JURASSIC | CRETACEOUS

Klamelisaurus gobiensis (meaning "lizard from Kelameili, Gobi") is known from a single specimen that was excavated in 1984 by the Institute of Vertebrate Paleontology and Paleoanthropology. Exposed parts of the specimen had already been weathered, and further deterioration occurred during the fossil's subsequent preparation. Many of the bones were subjected to extensive "reconstruction" using paint and plaster, which consequently obscured portions of the fossils. Zhao Xijin's original paper for the genus was limited to a brief "simple description."

After its description, the genus was largely overlooked, until an exhaustive reexamination was completed in 2020. During this review, a few of the bones that were originally reported could not be located, although conversely, some fragments were examined that had never before been mentioned. Based on certain skeletal details, such as the complete fusion of vertebral elements, the specimen was deemed to be that of an adult. The original assignment of individual teeth to the genus was called into question (Moore et al., 2020).

This redescription conducted a phylogenetic study that incorporated several different methods of calculation. (This analysis was sorely needed, as the original study was hopelessly out of date.) In all variants, *Klamelisaurus* was found to lie within **Mamenchisauridae** (or "core *Mamenchisaurus*-like taxa").

It was once hypothesized that the genus *Bellusaurus* could actually just represent the juvenile form of *Klamelisaurus* (Paul, 2016), but this is considered to be unlikely by some.

The animal's binomial name refers to the Kelameili Mountains (also known as the Klameli) within the Gobi Desert.

CLASSIFICATION

Dinosauria
 Sauropodomorpha
 Sauropodiformes
 Anchisauria
 Sauropoda
 Gravisauria
 Eusauropoda
 Mamenchisauridae

LOCATION

China

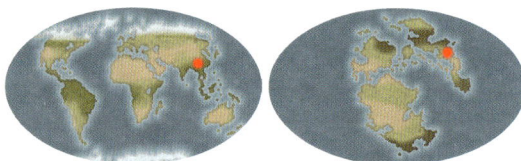

KNOWN REMAINS

Skeleton

4 m   8 m   12 m   16 m   20 m   24 m   28 m

X. shanshanesis

(Wu et al., 2013)
Length: 27 m (88 ft)
Height: 15.3 m (50 ft)
Hip height: 4.2 m (13.7 ft)
Body mass: 25,000 kg (27.6 t)
Reconstruction: ☐☐☐☐

12 m

8 m

4 m

250 245 240 235 230 225 220 215 210 205 200 195 190 185 180 175 170 165 160 155 150 145 140 135 130 125 120 115 110 105 100 95 90 85 80 75 70 65

| TRIASSIC | JURASSIC | CRETACEOUS |

Xinjiangtitan shanshanesis (meaning "giant from Xinjiang, Shanshan") is among the largest of sauropods ever discovered.

First unearthed in 2012, the specimen SSV12001 had been excavated to the point at which its dorsal and sacral (back and hip) vertebrae were exposed, along with some other elements. On the basis of these visible remains, the first brief description of the genus was carried out.

Further excavations of the area in 2014 and 2015 would go on to expose the animal's entire neck and nearly complete tail. These bones, preserved in articulation, make up the longest specimen of a *complete* neck ever discovered, for any animal. Ten of the neck vertebrae were, individually, more than a meter long. Depending on how much cartilage and soft tissue existed between the bones when the animal was alive, the neck of *Xinjiangtitan* would have been between 13.5 and 15 meters long.

Both the cervical series and the dorsal series have received their own separate, detailed write-ups. In addition to the vertebral column, elements of the pelvis and rear leg have also been preserved and studied. A "partial cranium" has been mentioned but still remains unfigured (Zhang et al., 2020, 2022).

Each analysis has concluded that *Xinjiangtitan* is a **mamenchisaurid**, although it possesses unique characteristics that are reminiscent of other derived sauropod groups, such as diplodocids.

The skeletal description states that the remains were found within the Qigu geological formation, which has a Late Jurassic age, but a separate study has suggested that the region is actually part of the Qiketai Formation of the Middle Jurassic (Maisch and Matzke, 2019).

The binomial name refers to the Shanshan region of Xinjiang Province, China.

CLASSIFICATION
Dinosauria
 Sauropodomorpha
  Sauropodiformes
   Anchisauria
    Sauropoda
     Gravisauria
      Eusauropoda
       Mamenchisauridae

LOCATION
China

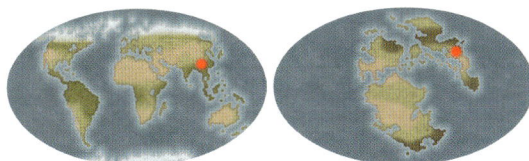

KNOWN REMAINS
Partial skeleton

3 m 6 m 9 m 12 m 15 m 18 m 21 m 24 m

12 m

9 m

6 m

3 m

H. sinojapanorum

(Dong, 1997)
Length: 25 m (78.7 ft)
Height: 9.1 m (30 ft)
Hip height: 3.7 m (12 ft)
Body mass: 18,000 kg (19.8 t)
Reconstruction: ☐ ☐ ☐ ☐

250 245 240 235 230 225 220 215 210 205 200 195 190 185 180 175 170 165 160 155 150 145 140 135 130 125 120 115 110 105 100 95 90 85 80 75 70 65

TRIASSIC　　　　**JURASSIC**　　　　**CRETACEOUS**

Hudiesaurus sinojapanorum (meaning "Sino-Japan's butterfly lizard") is currently known only from a single vertebra.

Originally, the bone in question was listed by Dong Zhiming as being the first of the dorsal (back) vertebra, being located immediately after the last of the cervical (neck) vertebrae. Comparing its size with the corresponding bone in the largest Chinese sauropod known at the time, *Mamenchisaurus hochuanensis*, estimates were made for the overall length of *Hudiesaurus* being anywhere from 30 meters to a colossal 48 meters.

However, later research would conclude that the bone was likely one of the rearmost (but not *the* last) of the cervical vertebrae instead. This recalculation brings the estimates of the animal's size down to more reasonable proportions (Upchurch et al., 2021).

Another fossil specimen, that of a front leg, was discovered approximately a kilometer away from the holotype and was also originally assigned to *Hudiesaurus*. However, the argument was made that there was no way to definitively link the two specimens as belonging to the same species (Upchurch et al., 2004), and these separate limb bones were later used as the basis for describing *Rhomaleopakhus* (Upchurch et al., 2021).

The generic name *Hudiesaurus* derives from the Mandarin "hudie" (meaning "butterfly"), referring to the butterfly-shaped protrusion on the front of the vertebra. The specific name *sinojapanorum* refers to the members of the Sino-Japan Silk Road Dinosaur Expedition of 1993. Additionally, the specific name can also be read as "central part" in Chinese, which corresponds to the Japanese name of an organization that financially assisted the fossil expedition, Chunichi Shinbun (or "central part").

CLASSIFICATION

Dinosauria
　Sauropodomorpha
　Sauropodiformes
　　Anchisauria
　　Sauropoda
　　　Gravisauria
　　　Eusauropoda
　　　　Mamenchisauridae

LOCATION

China

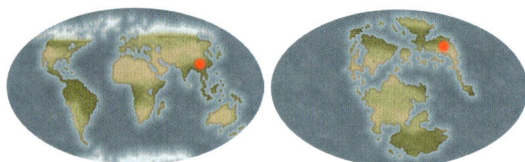

KNOWN REMAINS

Neck vertebra

2 m 4 m 6 m 8 m 10 m 12 m

6 m

D. zhangi

(Ye et al., 2005)
Length: 10 m (32.8 ft)
Height: 7 m (23 ft)
Hip height: 3.3 m (10.8 ft)
Body mass: 1,200 kg (1.3 t)
Reconstruction: ☐☐☐☐

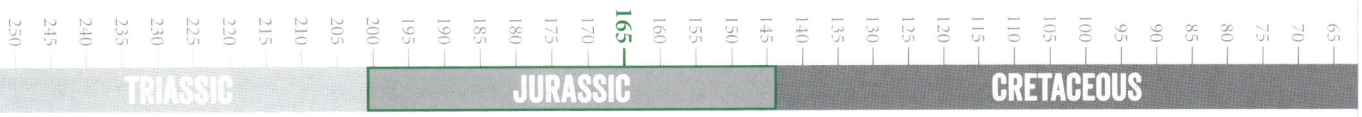

4 m

2 m

250 245 240 235 230 225 220 215 210 205 200 195 190 185 180 175 170 165 160 155 150 145 140 135 130 125 120 115 110 105 100 95 90 85 80 75 70 65

TRIASSIC **JURASSIC** **CRETACEOUS**

Daanosaurus zhangi (meaning "Zhang's lizard from Da'an") is known from a single set of partial remains. These bones belonged to a juvenile, as evidenced by unfused elements in the vertebrae, the specimen's small size, and other factors.

Phylogenetically classifying a juvenile specimen is complicated by the fact that skeletal morphology can vary greatly throughout an animal's growth stages. All studies agree that *Daanosaurus* fits in somewhere in the vicinity of the base of Neosauropoda, but more specific diagnoses differ greatly.

Two popular interpretations place the genus either just outside Neosauropoda (Xing et al., 2009) or at the base of Macronaria within Brachiosauridae—a view favored by the original describers of *Daanosaurus*, among others (D'Emic, 2012).

However, later authors would postulate that, in actuality, *Daanosaurus* (as well as *Bellusaurus*) is, in fact, just a

juvenile form of a *Mamenchisaurus*-like genus and thus is invalid as its own taxon. This determination was made based on certain uniquely shared anatomical traits, as well as the propensity of **mamenchisaurid** species known from the region; in fact, as far as sauropod remains are concerned, the upper Shaximiao Formation has only ever produced the mamenchisaurid *Omeisaurus* (Liao et al., 2021; Moore et al., 2023).

Accepting *Daanosaurus* as a juvenile *Mamenchisaurus*-like genus also helps to show that bifurcation of vertebral neural spines may be an ontogenetically acquired trait in sauropods.

The generic name *Daanosaurus* refers to the Da'an District in Sichuan Province, China. The specific name *zhangi* honors Chinese paleontologist Fucheng Zhang.

CLASSIFICATION

Dinosauria
  Sauropodomorpha
   Sauropodiformes
    Anchisauria
     Sauropoda
      Gravisauria
       Eusauropoda
        Mamenchisauridae

LOCATION

China

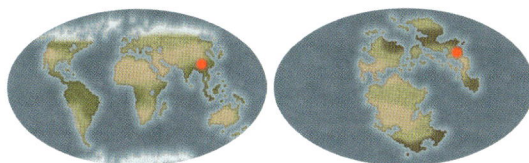

KNOWN REMAINS

Partial skull and skeleton (juvenile)

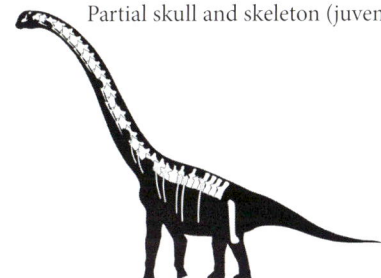

135

ANHUILONG

Anhuilong diboensis (meaning "dragon from Anhui, Dibo") is known only from a humerus, radius, and ulna. Although it was unearthed from the same geological formation that yielded the closely related *Huangshanlong*, the animal's bones were distinct enough for researchers to ascribe them to their own new genus. Specifically, the humerus of *Anhuilong* was found to be noticeably less robust than that of *Huangshanlong*, and its radius and ulna are quite short, proportionally, when compared with the humerus. *Anhuilong* and *Huangshanlong* were found to be sister taxa within the clade **Mamenchisauridae**.

A. diboensis

(Ren et al., 2018)
Length: 17.5 m (57.4 ft)
Height: 10.3 m (34 ft)
Reconstruction: ▢▢▢▢

The animal's binomial name refers to the Dibo region in the Anhui Province of China; "long" is Chinese for "dragon."

170

| TRIASSIC | JURASSIC | CRETACEOUS |

HUANGSHANLONG

Huangshanlong anhuiensis (meaning "dragon from Huangshan, Anhui") is known only from the humerus, radius, and ulna, which come from a single individual. The fossils were found in 2002 during the construction of the Huihang Expressway. The brief Chinese-language description of the animal determined that the genus was representative of **Mamenchisauridae**. A subsequent analysis confirmed the taxon's identity, determining it to be the sister taxon of *Anhuilong* (Ren et al., 2018). The binomial name refers to Huangshan City in Anhui Province, China. *Huangshanlong* was the first Jurassic animal named and discovered in the province.

H. anhuiensis

(Huang et al., 2014)
Length: 16 m (52 ft)
Height: 9.5 m (31.1 ft)
Reconstruction: ▢▢▢▢

170

| TRIASSIC | JURASSIC | CRETACEOUS |

EOMAMENCHISAURUS

Eomamenchisaurus yuanmouensis (meaning "dawn-*Mamenchisaurus* from Yuanmou") is a questionably valid genus of mamenchisaurid from China. Aside from the very brief paper that described it, the fragmentary specimen has not been the subject of any other analysis, and the genus itself has rarely been mentioned since. One of the few subsequent comments on the taxon called its defining characteristics "ambiguous" (Suteethorn et al., 2012).

Researcher Mike Taylor called its diagnostic identifiers "not entirely convincing" ("Sauropods of 2008: *Eomamenchisaurus*," 2009). Author Gregory S. Paul suggests that the specimen could simply be a juvenile form of *Yuanmousaurus* (Paul, 2016).

2 m

E. yuanmouensis

(Lü et al., 2008)
Length: 15.5 m (51 ft)
Height: 7.1 m (23.3 ft)
Reconstruction: ☐☐☐☐

170

TRIASSIC | JURASSIC | CRETACEOUS

TIENSHANOSAURUS

Tienshanosaurus chitaiensis (meaning "lizard from Tianshan, Chiatia") is a sauropod that has not been thoroughly described or examined since its 1937 write-up by Yang Zhongjian (also known as "C. C. Young"). As such, there is little that can be definitely said about it. At times, it has been considered to be synonymous with *Euhelopus* or to represent a new species within that genus (Molnar, 1991). Various phylogenetic studies have placed it all over the map, but newer analyses seem to generally agree on a **mamenchisaurid** placement (Sekiya, 2011; Moore et al., 2023). The "Tian Shan" are also known as the "heavenly mountains."

1 m

T. chitaiensis

(Young, 1937)
Length: 10 m (33 ft)
Height: 7.6 m (25 ft)
Reconstruction: ☐☐☐☐

160

TRIASSIC | JURASSIC | CRETACEOUS

MAMENCHISAURIDAE

J. dongxingensis

(Ren et al., 2024)
Length: 16 m (52.5 ft)
Height: 9.3 m (30.5 ft)
Reconstruction: ▨☐☐☐

2 m

147?

TRIASSIC	JURASSIC	CRETACEOUS

Jingiella dongxingensis (meaning "the Jing's one from Dongxing") is a mamenchisaurid, based on fragments of a few vertebrae, the ulna, and the femur.

The generic name *Jingiella* honors "Chinese people of the Jing Nationality who emigrated from Vietnam." The name was originally printed as "Jingia," but it turned out that this name was preoccupied by a type of moth. The specific name *dongxingensis* refers to Dongxing City in Guangxi Province, China.

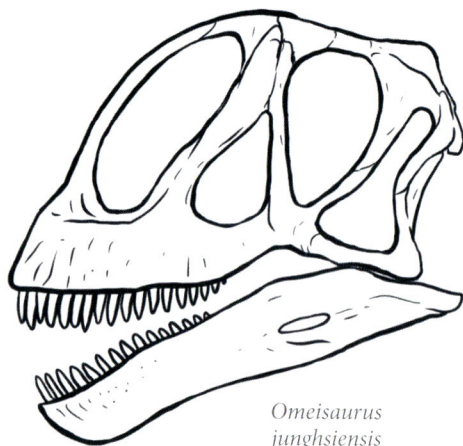

Omeisaurus junghsiensis

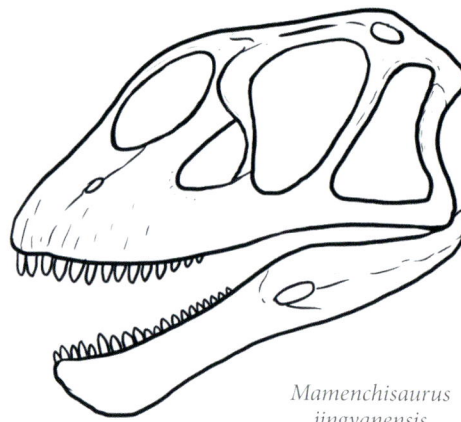

Mamenchisaurus jingyanensis

Mamenchisaurus youngi

Bajadasaurus

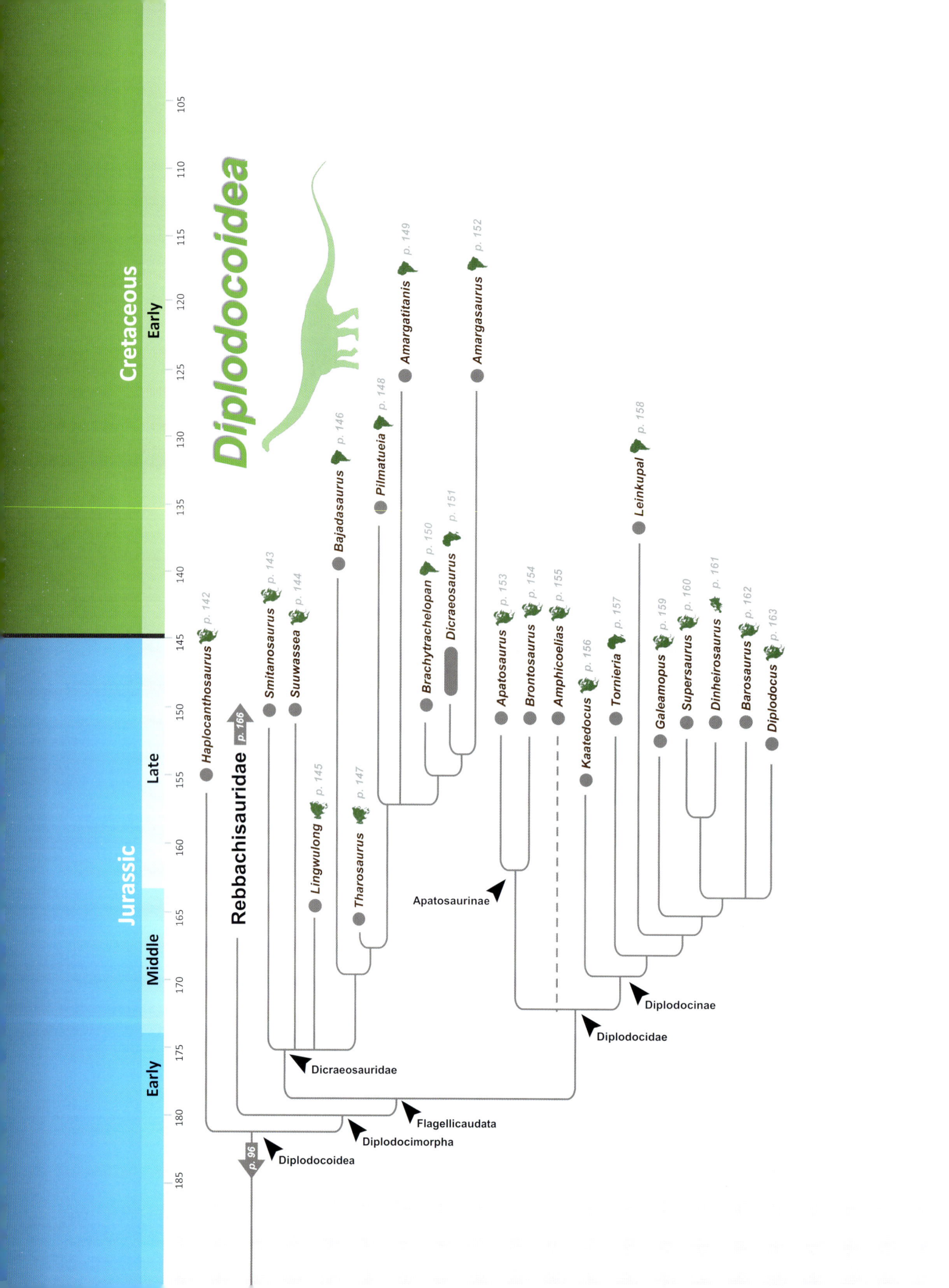

Diplodocoidea

105 110 115 120 125 130 135 140 145 150 155 160 165 170 175 180 185

● *Haplocanthosaurus* p. 142

Rebbachisauridae p. 166

● *Smitanosaurus* p. 143
● *Suuwassea* p. 144
● *Lingwulong* p. 145
● *Tharosaurus* p. 147

● *Bajadasaurus* p. 146

● *Pilmatueia* p. 148

● *Amargatitanis* p. 149

● *Brachytrachelopan* p. 150
● *Dicraeosaurus* p. 151

● *Amargasaurus* p. 152

● *Apatosaurus* p. 153
● *Brontosaurus* p. 154
● *Amphicoelias* p. 155
● *Kaatedocus* p. 156

● *Leinkupal* p. 158

● *Tornieria* p. 157
● *Galeamopus* p. 159
● *Supersaurus* p. 160
● *Dinheirosaurus* p. 161
● *Barosaurus* p. 162
● *Diplodocus* p. 163

◄ Apatosaurinae

◄ Dicraeosauridae

◄ Diplodocinae

◄ Diplodocidae

◄ Flagellicaudata

◄ Diplodocimorpha

p. 96 ◄ Diplodocoidea

The diplodocoids are a diverse group, encompassing the uniquely mandibled rebbachisaurids, the high-spined dicraeosaurids, and the extremely long-bodied diplodocids.

The group Flagellicaudata are known for their long, thin tails, deriving their name from the Latin "flagellum" (meaning "whip") and "cauda" (meaning "tail"). The idea that they might have used their tails as literal whips has become a popular one, although from a biomechanical standpoint this idea is seeming less and less likely. One alternative idea suggests that their tails were used as feeling, tactile organs.

The diplodocids themselves include some of the longest animals ever to walk the Earth, such as Supersaurus. Their jaws were specialized to easily strip vegetation whole from fronds or branches. Some evidence exists of iguana-looking spiny protuberances along the top of their tails.

Among the **diplodocoids**, *Haplocanthosaurus* is widely considered by most studies to be the basalmost genus and to be the sister taxon to Diplodocimorpha, a group that splits into Rebbachisauridae (see the next chapter) and **Flagellicaudata**; the latter, in turn, splits into Dicraeosauridae and Diplodocidae.

Among the **dicraeosaurids**, Whitlock and Mantilla (2020) have named *Smitanosaurus* to be among the most ancestral. Studies generally agree on the order of the next most derived members, *Suuwassea* and *Lingwulong* (Royo-Torres et al., 2020; Windholz et al., 2022), although *Suuwassea* is shown being rather more derived by Whitlock and Mantilla (2020). The placement of *Dicraeosaurus* is a bit less certain, with (Windholz et al., 2022) showing it to be fairly basally placed, whereas other sources show it to be one of the most derived dicraeosaurids (Ren et al., 2018; Xu et al., 2018). The describers of *Tharosaurus* place it in this clade (Bajpai et al., 2023).

Among the derived dicraeosaurids, although their exact placement shifts somewhat from study to study, it is well agreed that their ranks include *Amargasaurus, Brachytrachelopan*, and *Pilmatueia* (Whitlock and Mantilla, 2020; Windholz et al., 2022). *Amargatitanis* is less often included in taxonomic studies but has been recently placed among the most derived dicraeosaurids (Windholz et al., 2022). *Bajadasaurus* has contrarily been shown as being quite ancestral (Gallina et al., 2019; Whitlock and Mantilla, 2020) or quite derived (Windholz et al., 2022).

As for the **diplodocids**, *Amphicoelias* has variously been shown as the basalmost example, lying outside the two main clades of Apatosaurinae and Diplodocinae (Mannion et al., 2021), or as an apatosaurine, along with *Apatosaurus* and *Brontosaurus* (Tschopp and Mateus, 2017; Windholz et al., 2022).

While Tschopp and Mateus (2017) and Windholz et al. (2022) both found *Kaatedocus* to be the basalmost **diplodocine**, Whitlock and Mantilla (2020) instead placed it as the basalmost dicraeosaurid. Although the details of placement change slightly between studies, well-agreed-upon members of the clade include *Tornieria, Leinkupal, Galeamopus, Supersaurus, Barosaurus, Dinheirosaurus*, and *Diplodocus* (Tschopp and Mateus, 2017; Windholz et al., 2022).

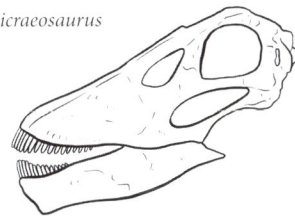
Dicraeosaurus

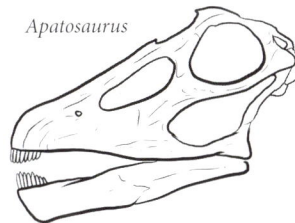
Apatosaurus

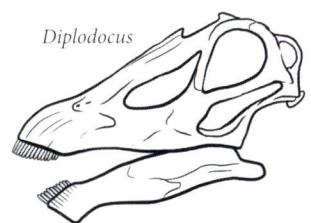
Diplodocus

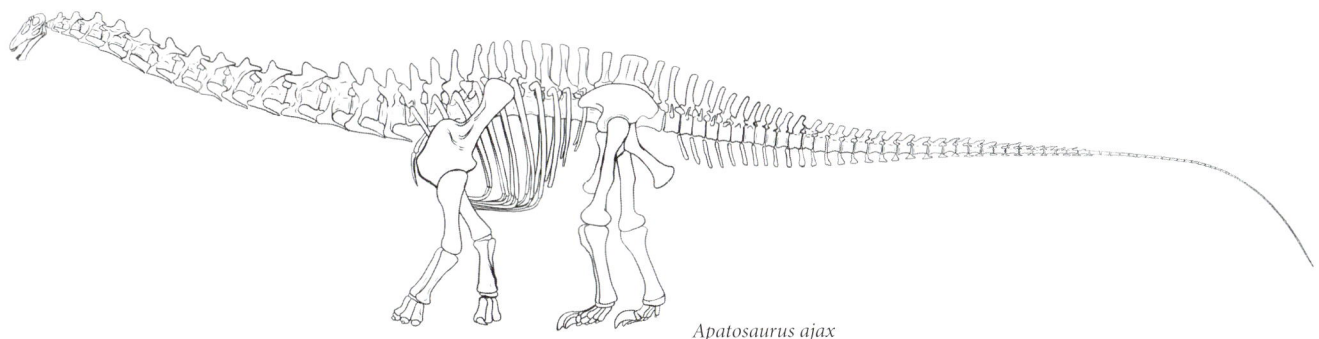
Apatosaurus ajax

4 m   8 m   12 m   16 m

4 m

H. priscus

(Hatcher, 1903)
Length: 16 m (55 ft)
Height: 4.6 m (15.1 ft)
Hip height: 3.5 m (11.5 ft)
Body mass: 13,000 kg (14.3 t)
Reconstruction: ☐☐☐☐

250 245 240 235 230 225 220 215 210 205 200 195 190 185 180 175 170 165 160 155 150 145 140 135 130 125 120 115 110 105 100 95 90 85 80 75 70 65

| TRIASSIC | JURASSIC | CRETACEOUS |

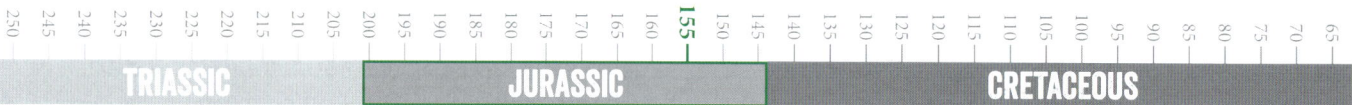

Haplocanthosaurus priscus (meaning "ancient simple-spined lizard") is known from four substantive (yet still quite partial) specimens, which together constitute a good portion of the spinal column and shoulder girdle. Three of these are examples of the type species (although one was once known as *H. utterbacki*), while one is of a separate species, *H. delfsi*.

Two additional specimens, which include many elements of the limbs, have also been described; however, their identification has been given only as "indeterminate Haplocanthosaurid" or "*Haplocanthosaurus?*", and thus are not included in the diagram of known remains below. The scarcity of *Haplocanthosaurus* fossils has led researchers to consider it to be the rarest sauropod of the dinosaur-rich Morrison geological formation (Forster and Wedel, 2014).

Another tentative specimen nicknamed "Big Monty" is currently in the hands of private collectors and purportedly measures 34 meters in length (Ronson, 2016).

Studies most often place *Haplocanthosaurus* as the basalmost member of **Diplodocoidea** (Royo-Torres et al., 2020; Ren et al., 2023). One analysis pushed its position one step back, placing it as a non-diplodocoid basal neosauropod (Ren et al., 2018).

The generic name *Haplocanthosaurus* combines the Greek "haplos" (meaning "simple" or "single"), "akantha" (meaning "spine"), and "sauros" (meaning "lizard"). The animal was originally to be named *Haplocanthus*, but that name was mistakenly believed to be preoccupied, and thus it was hastily amended.

The specific name *priscus* is Greek for "ancient"; *delfsi* honors fossil discoverer Edwin Delfs; *utterbacki* honors fossil discoverer W. H. Utterback.

CLASSIFICATION
Dinosauria
 Sauropoda
  Gravisauria
   Eusauropoda
    Neosauropoda
     Diplodocoidea

LOCATION
Colorado, USA

KNOWN REMAINS
Partial skeleton

4 m   8 m   12 m   16 m   20 m

8 m

S. agilis

(Whitlock and Mantilla, 2020)
Length: 22 m (72 ft)
Height: 5.5 m (18 ft)
Hip height: 4.5 m (13.1 ft)
Body mass: 17,000 kg (18.7 t)
Reconstruction: ☐☐☐☐

4 m

250 245 240 235 230 225 220 215 210 205 200 195 190 185 180 175 170 165 160 155 **150** 145 140 135 130 125 120 115 110 105 100 95 90 85 80 75 70 65

TRIASSIC   **JURASSIC**   **CRETACEOUS**

The naming of **Smitanosaurus agilis** (meaning "Smith's agile reptile") resulted from the reexamination of a set of old fossils that had long suffered an ambiguity of identity.

The braincase and first several neck vertebrae of a sauropod were discovered in a Morrison Formation quarry in 1883, beneath a large femur. Othniel Charles Marsh named them to a new species of *Morosaurus*, *M. agilis*. As is the case with many dinosaur genera erected in the nineteenth century, though, the case surrounding *Morosaurus* is rather convoluted. The type species, *M. impar*, was created by Marsh in 1878, and four additional species were swiftly described thereafter, including *M. agilis*.

Needless to say, these identifications were mostly made using insufficient, undiagnostic material, and by 1907 some researchers (such as C. W. Gilmore) had begun to suspect synonymy between *Morosaurus* and *Camarasaurus*—except for *M. agilis*, which seemed to be distinct. Even after

Morosaurus was recognized as being defunct, though, the identity of "*M.*" *agilis* remained a mystery for more than a century. Despite being well preserved, the remains were an ill-fitting match for any other Morrison Formation sauropod and left researchers stumped.

A careful reexamination of the fossil material, as well as original journals and notes, revealed key fragments of the fossil that had been previously left unidentified. With these clues, the animal was finally identified as a juvenile **dicraeosaurid** and was described as a new genus.

The generic name *Smitanosaurus* is derived from the Old Saxon "smitan," which means "smith." This is meant to reference both J. August Smith, who excavated the fossil, as well as the Smithsonian Institution, where the fossil has been kept for more than a century. The specific name *agilis* is Latin for "agile."

CLASSIFICATION

Dinosauria
 Sauropoda
  Eusauropoda
   Neosauropoda
    Diplodocoidea
     Flagellicaudata
      Dicraeosauridae

LOCATION

Colorado, USA

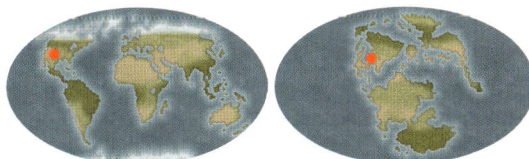

KNOWN REMAINS

Partial skull and vertebrae

2 m   4 m   6 m   8 m   10 m   12 m   14 m

S. emilieae

(Harris and Dodson, 2004)
Length: 16.2 m (53 ft)
Height: 4 m (13.1 ft)
Hip height: 3 m (9.8 ft)
Body mass: 7,000 kg (7.7 t)
Reconstruction: ☐☐☐☐

| TRIASSIC | JURASSIC | CRETACEOUS |

250 245 240 235 230 225 220 215 210 205 200 195 190 185 180 175 170 165 160 155 **150** 145 140 135 130 125 120 115 110 105 100 95 90 85 80 75 70 65

Suuwassea emilieae (meaning "Emilie's first thunder heard in spring") is known from a partial skeleton that was first spotted by Peter Dodson during a horseback ride. One unusual feature of the animal is that its skull possessed an extra opening, a second "postparietal foramen" that, at the time of discovery, had only been seen in three sauropod species, none of which originated in North America.

The description of the genus placed it somewhere within the clade Flagellicaudata but was unable to determine a more precise positioning, owing to an unusual mixture of traits, some of which were deemed "primitive" while others were "derived." More recent works are quite unified in their placement of *Suuwassea* within **Dicraeosauridae**, a determination aided by the discovery of a jaw bone that was not initially part of the known specimen. Within the group, it is often found to be the most basal of members, which could explain its "primitive" appearance, lacking the distinctively tall neural spines seen in many dicraeosaurids (Ren et al., 2023). Typically viewed as a Gondwanan clade, the presence of *Suuwassea* in Laurasia suggests that this is possibly where the group originated (Whitlock and Harris, 2010).

After the initial publication that briefly summarized *Suuwassea* as a whole, a series of publications quickly followed that examined in more detail the axial skeleton, appendicular skeleton, dentary, cranium, and the animal's evolutionary significance (Harris, 2006a, b, 2007; Whitlock and Harris, 2010).

The generic name *Suuwassea* is a combination of Crow Indigenous American terms meaning "ancient thunder heard in spring," alluding to the "thunder lizard" moniker of sauropods such as *Brontosaurus*. The specific name *emilieae* honors Emilie de Hellebranth, who helped fund the discovery expedition.

CLASSIFICATION

Dinosauria
  Sauropoda
    Eusauropoda
      Neosauropoda
        Diplodocoidea
          Flagellicaudata
            Dicraeosauridae

LOCATION

Montana, USA

KNOWN REMAINS

Partial skull and skeleton

L. shenqi

(Xu et al., 2018)
Length: 17 m (56 ft)
Height: 4.2 m (13.8 ft)
Hip height: 3.7 m (12.1 ft)
Body mass: 8,000 kg (8.8 t)
Reconstruction: ▨▨▨▢

2 m   4 m   6 m   8 m   10 m   12 m

4 m

2 m

| 250 | 245 | 240 | 235 | 230 | 225 | 220 | 215 | 210 | 205 | 200 | 195 | 190 | 185 | 180 | 175 | 170 | **164** | 160 | 155 | 150 | 145 | 140 | 135 | 130 | 125 | 120 | 115 | 110 | 105 | 100 | 95 | 90 | 85 | 80 | 75 | 70 | 65 |

TRIASSIC | **JURASSIC** | **CRETACEOUS**

Lingwulong shenqi (meaning "amazing dragon from Lingwu") is one of the earliest known examples of a neosauropod and, additionally, among the earliest known diplodocoids. It was also the first definitive example of a **diplodocoid** known from eastern Asia. This is especially significant because, prior to the discovery of *Lingwulong*, it was thought that diplodocoids had never been present in East Asia.

The East Asian Isolation Hypothesis had surmised that a shallow sea, known as the Turgai Strait, had cut off access to the continent around this time; this separation would explain the local absence of diplodocoid sauropods, as well as why mamenchisaurid sauropods are almost universally found in Asia but not elsewhere. The very existence of *Lingwulong* could cast this notion into doubt; according to its describers, *Lingwulong* also pushes back the origin of the neosauropod lineages by at least 15 million years. To explain why no Asian diplodocoids had been found previously, the initial researchers suggest a sampling bias: mamenchisaurids could be most common in the southwest of China, while *Lingwulong* was found in the northwest.

However, later works have challenged the assertion that the fossils were found in the Yanan Formation, which is aged from the Toarcian-Bajocian (approximately 174 MYA), instead giving an age of Bathonian-Callovian (approximately 164 MYA), from the Zhiluo Formation (Bajpai et al., 2023; Ren et al., 2023).

Lingwulong is known from the remains of seven to ten individuals of various ontogenetic ages, ranging from juveniles to adults. The first specimens were spotted by a local sheep herder named Ma Yun in 2004, and several years' worth of excavations began in 2005.

The generic name *Lingwulong* refers to the city of Lingwu, in Ningxia, China; "long" is Mandarin for "dragon." The specific name *shenqi* is Mandarin for "amazing," alluding to the highly unexpected discovery of a diplodocoid from this region and time period.

CLASSIFICATION

Dinosauria
  Sauropoda
    Eusauropoda
      Neosauropoda
        Diplodocoidea
          Flagellicaudata
            Dicraeosauridae

LOCATION

China

KNOWN REMAINS

Nearly complete

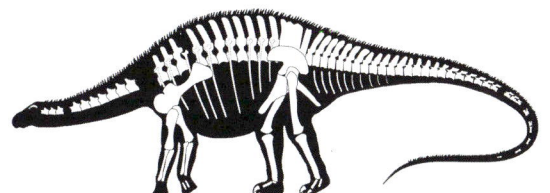

2 m   4 m   6 m   8 m   10 m

4 m

2 m

B. pronuspinax

(Gallina et al., 2019)
Length: 11.5 m (37.7 ft)
Height: 3 m (9.8 ft)
Hip height: 2.5 m (8.2 ft)
Body mass: 2,200 kg (2.4 t)
Reconstruction: ☐☐☐☐

250–245–240–235–230–225–220–215–210–205–200–195–190–185–180–175–170–165–160–155–150–145–**139**–135–130–125–120–115–110–105–100–95–90–85–80–75–70–65

TRIASSIC	JURASSIC	CRETACEOUS

Bajadasaurus pronuspinax (meaning "forward-bending spine from Bajada") is the second sauropod ever discovered to have definitively had long, spikey projections along its neck. Like its cousin *Amargasaurus*, it had bifurcated, narrow extensions of its neck vertebrae; unlike *Amargasaurus*'s rear-slanting spines, *Bajadasaurus* instead features forward-sweeping projections. Additionally, as *Bajadasaurus* lived approximately 15 million years prior to *Amargasaurus*, this earlier appearance shows that this evolutionary adaptation persisted for a fairly significant period of time among the dicraeosaurid sauropods.

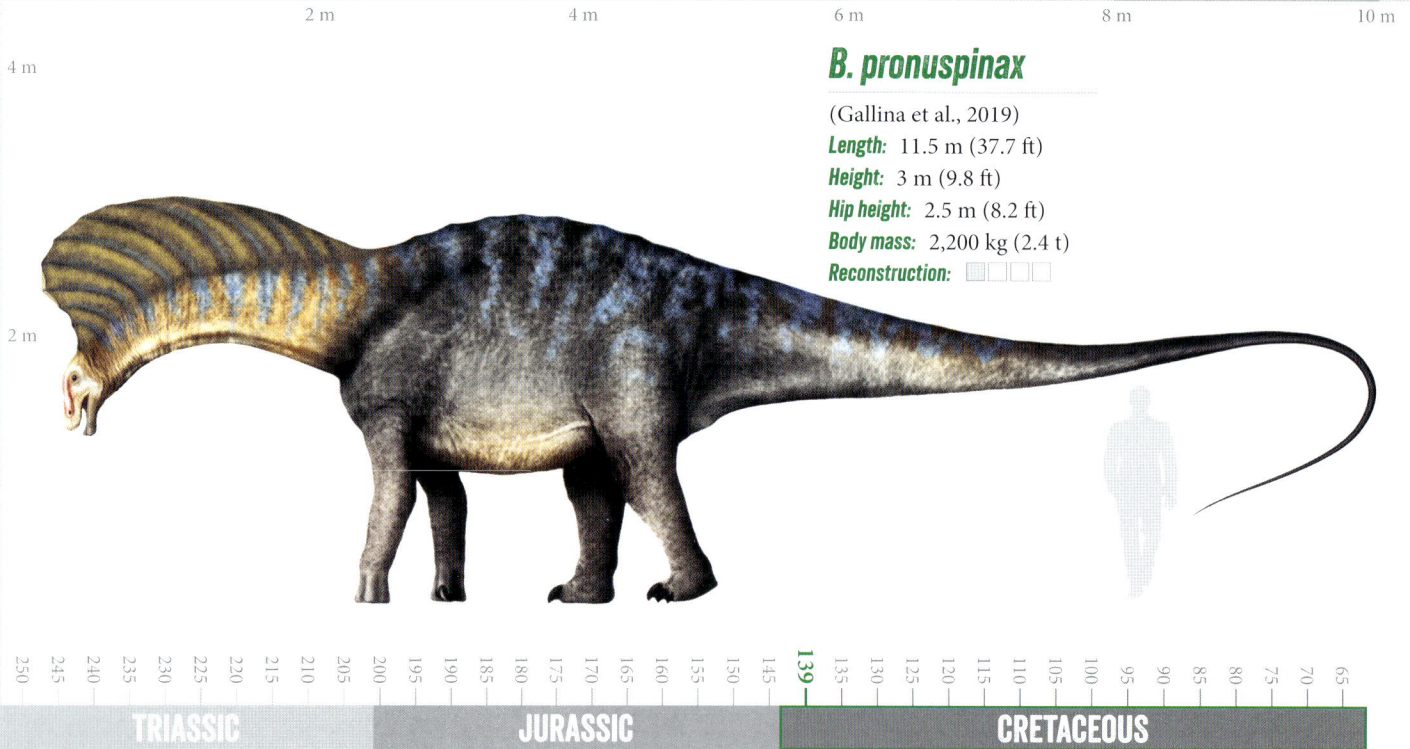

The describing authors propose that the spines of *Bajadasaurus* were horn-like in appearance, being covered and reinforced with a keratin sheath, and that their primary purpose was defensive in nature. Many studies have proposed similar ideas regarding *Amargasaurus* as well. However, a study regarding the microscopic internal bone structure of *Amargasaurus* has proposed a soft-tissue covering, forming a sail-like structure instead. And if this were the case for *Amargasaurus*, there would be a high likelihood of the same being true of *Bajadasaurus* as well (Cerda et al., 2022).

The fossils of *Bajadasaurus* were not extracted in the field but in the lab. When an exposed set of teeth was spotted in the ground in 2010 by the paleontologists of CONICET, the Argentine government's science agency, the entire surrounding area was extracted in a single mass. This is not an uncommon practice and was especially necessary in this case, as fossils from the area were known to often be quite fragile.

The generic name *Bajadasaurus* refers to the location Bajada Colorada, found in Patagonia, Argentina. The specific name *pronuspinax* combines the Latin "pronus" (meaning "bent over forward") and the Greek "spinax" (meaning "spine").

CLASSIFICATION
Dinosauria
 Sauropoda
  Eusauropoda
   Neosauropoda
    Diplodocoidea
     Flagellicaudata
     Dicraeosauridae

LOCATION
Argentina

KNOWN REMAINS
Partial skull and neck vertebrae

2 m   4 m   6 m   8 m   10 m

T. indicus

(Bajpai et al., 2023)
Length: 12 m (39 ft)
Height: 3 m (9.8 ft)
Hip height: 2 m (6.5 ft)
Body mass: 3,000 kg (3.3 t)
Reconstruction: ☐ ☐ ☐ ☐

4 m

2 m

250 245 240 235 230 225 220 215 210 205 200 195 190 185 180 175 170 **167** 160 155 150 145 140 135 130 125 120 115 110 105 100 95 90 85 80 75 70 65

TRIASSIC   **JURASSIC**   **CRETACEOUS**

Tharosaurus indicus (meaning "Indian lizard from Thar") is the first diplodocoid to be unearthed and described from India and is potentially the oldest diplodocoid ever to be discovered.

After local excavations in 2018 revealed the fossilized remains of sharks and other fishes, continued work in 2019 yielded the disarticulated and fragmentary remains of *Tharosaurus*, which were spread over an area of approximately 25 square meters. Various fragments of vertebrae from multiple regions of the spinal column were found, along with part of one rib.

The considerable age of *Tharosaurus* and its ensuing status as "oldest diplodocoid" show that the region which is now India was a central location for the evolution and dispersement of the clade as a whole. As India was still attached to the larger landmass of Gondwana at the time, it does make sense that diplodocoids would have had a presence in the area, given their prevalence in other Gondwanan regions such as Africa and South America. The describers of *Tharosaurus* therefore highlight the need for enhanced paleontological efforts focusing on India.

The describers' phylogenetic analysis recovered *Tharosaurus* as lying roughly in the middle of the clade **Dicraeosauridae**. Hence, it is plausible to assume that *Tharosaurus* had some sort of tall neural spines along its neck, similar to those found in related genera such as *Amargasaurus*, even though this sort of structure is only hinted at in the recovered fossil material.

The generic name *Tharosaurus* refers to the Thar Desert of western India; during the "middle" Jurassic, this was a coastal area adjacent to the Tethys Sea. The specific name *indicus* refers to the country of India.

CLASSIFICATION

Dinosauria
 Sauropoda
  Eusauropoda
   Neosauropoda
    Diplodocoidea
     Flagellicaudata
      Dicraeosauridae

LOCATION

India

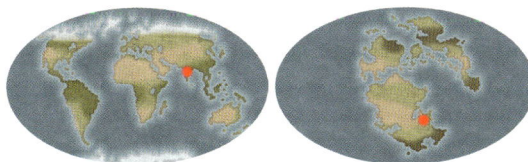

KNOWN REMAINS

Fragments

147

6 m 2 m 4 m 6 m 8 m 10 m 12 m 14 m

4 m

2 m

P. faundezi

(Coria et al., 2019)
Length: 14.5 m (48 ft)
Height: 3.7 m (12.1 ft)
Hip height: 3.1 m (10.2 ft)
Body mass: 4,200 kg (4.6 t)
Reconstruction: ☐☐☐☐

250 245 240 235 230 225 220 215 210 205 200 195 190 185 180 175 170 165 160 155 150 145 140 135 130 125 120 115 110 105 100 95 90 85 80 75 70 65

TRIASSIC **JURASSIC** **CRETACEOUS**

Pilmatueia faundezi (meaning "Faúndez's one from Pilmatué") differs from other known dicraeosaurids in that its cervical (neck) vertebrae contained pneumatic, air-filled chambers. This feature is common among sauropods in general, but the dicraeosaurids had previously been seen as outliers that were relatively lacking in this characteristic.

Fossil excavations that began in 2009 at the Mulichinco Formation would eventually produce the fossils now assigned to *Pilmatueia*. The genus was described on the basis of half a dozen disparate vertebrae and a femur. The lone neck vertebra was lacking its neural spines, but since the genus was found to be the sister taxon of *Amargasaurus*, it seemed very likely that *Pilmatueia* would have boasted similar spiny protrusions along its neck.

This notion was somewhat confirmed when additional fossils, discovered from the same locality, were later described and assigned to *Pilmatueia*. Among these were

some of the frontmost neck bones, which showed clear indications that these spines were forward-facing, such as they were in *Bajadasaurus*, rather than rear-facing like those of *Amargasaurus*. Still unknown, though, is the overall extent and height of the neck's external spines (Windholz et al., 2022).

It is possible that all the South American **dicraeosaurids** were part of a single subfamily that exclusively inhabited this one continent. One version of a phylogenetic analysis recovered such a result, and the researchers found it to be logical. However, other versions of the analysis utilizing various datasets came to differing conclusions (Windholz et al., 2022).

The generic name *Pilmatueia* refers to the region of Pilmatué in Neuquen, Argentina. The specific name *faundezi* honors Ramón Faúndez, manager of the Museo Municipal de Las Lajas.

CLASSIFICATION
Dinosauria
 Sauropoda
  Eusauropoda
   Neosauropoda
    Diplodocoidea
     Flagellicaudata
      Dicraeosauridae

LOCATION
Argentina

KNOWN REMAINS
Partial skeleton

2 m   4 m   6 m   8 m   10 m   12 m

A. macni

4 m

(Apesteguía, 2007)
Length: 12 m (40 ft)
Height: 3.1 m (10.2 ft)
Hip height: 2.7 m (8.9 ft)
Body mass: 3,100 kg (3.4 t)
Reconstruction: ☐ ☐ ☐ ☐

2 m

| 250 | 245 | 240 | 235 | 230 | 225 | 220 | 215 | 210 | 205 | 200 | 195 | 190 | 185 | 180 | 175 | 170 | 165 | 160 | 155 | 150 | 145 | 140 | 135 | 130 | 125 | 120 | 115 | 110 | 105 | 100 | 95 | 90 | 85 | 80 | 75 | 70 | 65 |

TRIASSIC   **JURASSIC**   **CRETACEOUS**

Amargatitanis macni (meaning "MACN's giant from Amarga") is known only from a handful of specimens that were originally collected in 1983 by the famous Argentine paleontologist, José Bonaparte.

Amargatitanis was originally described on the basis of three sets of remains from the Museo Argentino de Ciencias Naturales series of fossil specimens: N53, N51, and N34. However, examination of Bonaparte's original notes revealed that N34 (a scapula) and N51 (six tail vertebrae) actually originated from separate dig sites, as there had been four sites located in close proximity to one another. Since the remains could not have come from the same animal, their identification could not be verified, and so they are no longer considered definitive *Amargatitanis* fossils.

Conversely, material that had originally been excluded from the group, including lower leg and hip elements, was confirmed to have come from the same location and was able to be lumped in with collection N53 (Gallina, 2016).

A histological examination of growth patterns within the femur concluded that the specimen had been an adult at the time of its death. It was also determined that dicraeosaurid sauropods underwent similar patterns of growth as other groups of neosauropods (Windholz and Cerda, 2021).

The original taxonomic analysis included with the description of *Amargatitanis* concluded that the animal was a titanosaur. However, it is now believed to be a member of **Dicraeosauridae** (Gallina, 2016).

The generic name *Amargatitanis* refers to La Amarga Canyon, a Province in Neuquen, Argentina; this is combined with the Greek "titan" (meaning "giant"). The specific name *macni* honors the Museo Argentino de Ciencias Naturales (MACN).

CLASSIFICATION

Dinosauria
 Sauropoda
  Eusauropoda
   Neosauropoda
    Diplodocoidea
     Flagellicaudata
      Dicraeosauridae

LOCATION

Argentina

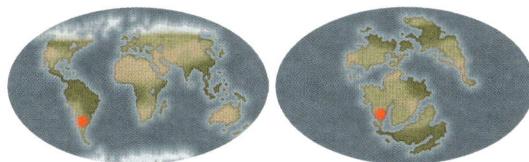

KNOWN REMAINS

Fragments

DIPLODOCOIDEA

BRACHYTRACHELOPAN

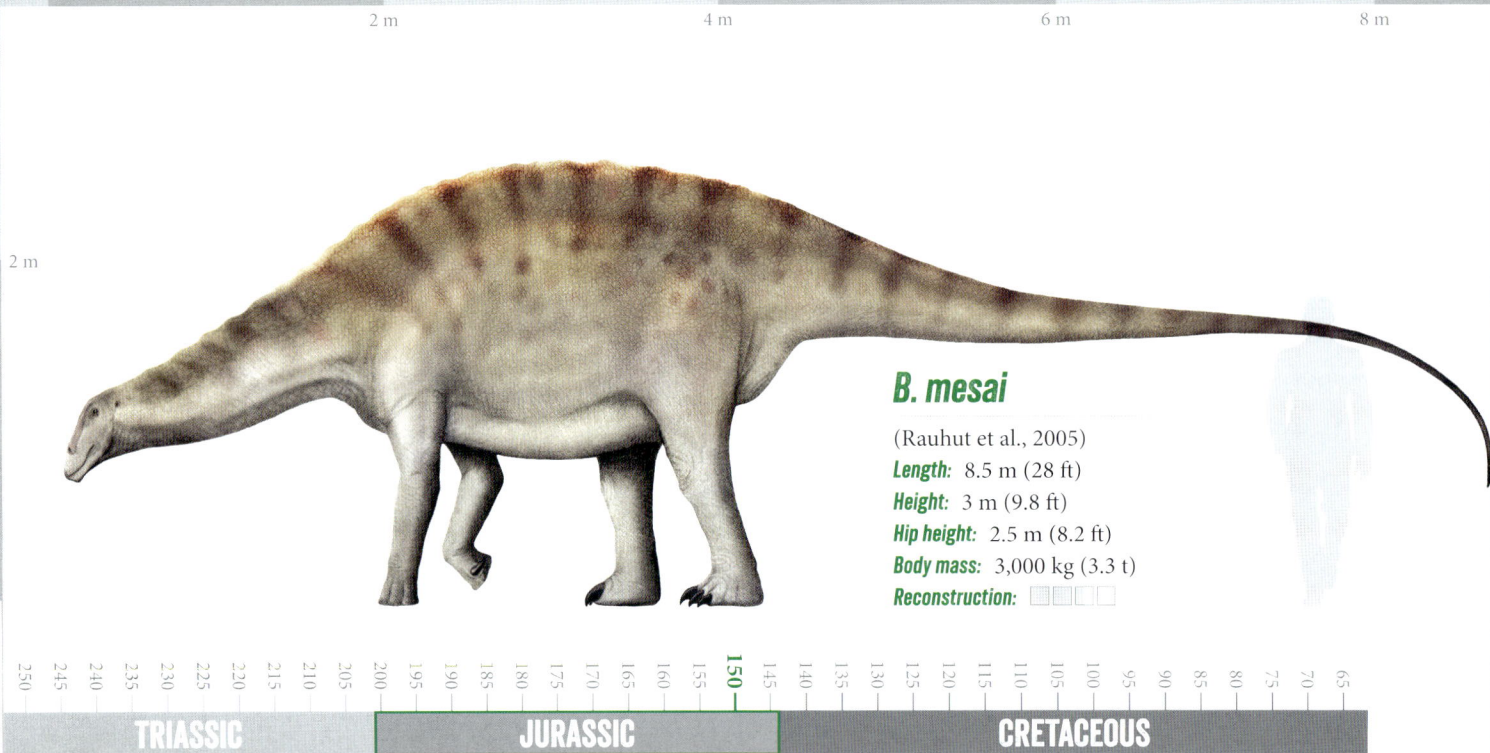

B. mesai

(Rauhut et al., 2005)
Length: 8.5 m (28 ft)
Height: 3 m (9.8 ft)
Hip height: 2.5 m (8.2 ft)
Body mass: 3,000 kg (3.3 t)
Reconstruction: ☐☐☐☐☐

| TRIASSIC | JURASSIC | CRETACEOUS |

Brachytrachelopan mesai (meaning "Mesa's short-necked shepherd") has the shortest neck of any known sauropod, a group that is obviously famous for their generally long necks. Proportionally, the average sauropod has a neck that is approximately 125–135% as long as its dorsal (back) vertebral column, but the neck of *Brachytrachelopan* is only 75% of that length.

In addition to the short length of the neck, the anatomy of the cervical (neck) vertebra reveals that the animal would not have been able to raise its head very high at all because of a low degree of joint freedom. Thus, it seems that *Brachytrachelopan* was specialized for low browsing, consuming vegetation only one or 2 meters above the ground. As such, it could very well have filled a similar ecological niche as some other kinds of herbivorous dinosaurs, such as the iguanodontians.

The original description of *Brachytrachelopan* concluded that the specimen had been an adult at the time of its death,

citing a high degree of fusion between vertebral elements. The small size of the animal, although unusual, would not be entirely unexpected, as many of its cousins within **Dicraeosauridae** were also on the smaller end of the sauropod scale. However, a newer study performed a histological analysis of the specimen's internal bone structure and concluded that the animal was not, in fact, fully grown. This determination was made, in part, by observing a high degree of vascularization—a sign that the bones were still in a mode of rapid growth—and a lack of a fundamental external layer (Windholz et al., 2023).

The generic name *Brachytrachelopan* derives from the Greek "brachytrachelos," meaning "short-necked"; this is combined with the name of the mythological Pan, the ancient Greek god of shepherds, because the specimen was discovered by a shepherd while searching for a stray sheep. The specific name *mesai* honors this shepherd, Daniel Mesa.

CLASSIFICATION
Dinosauria
 Sauropoda
  Eusauropoda
   Neosauropoda
    Diplodocoidea
     Flagellicaudata
      Dicraeosauridae

LOCATION
Argentina

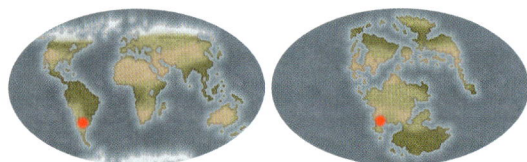

KNOWN REMAINS
Partial skeleton

150

DIPLODOCOIDEA

2 m   4 m   6 m   8 m   10 m   12 m   14 m

4 m

2 m

D. hansemanni

(Janensch, 1914)
Length: 15.5 m (51 ft)
Height: 4.1 m (13.5 ft)
Hip height: 3.8 m (12.5 ft)
Body mass: 6,000 kg (6.6 t)
Reconstruction: ▢▢▢▢

250 245 240 235 230 225 220 215 210 205 200 195 190 185 180 175 170 165 160 155 **150 145** 140 135 130 125 120 115 110 105 100 95 90 85 80 75 70 65

| TRIASSIC | JURASSIC | CRETACEOUS |

Dicraeosaurus hansemanni (meaning "Hansemann's bifurcated lizard") is named for the upward-projecting spines of the vertebra along the front half of its body being forked. Its neck was relatively short, making *Dicraeosaurus* a dedicated low browser.

From 1909 through 1912, an expedition from Berlin's Humboldt University excavated a great many fossils in the area around Tendaguru Hill in Tanzania. From these specimens, two species were soon described, *D. hansemanni* and *D. sattleri*. One of the reasons for erecting two separate species was that various specimens were unearthed from two different stratigraphic layers: *D. hansemanni* (consisting of three partial skeletons, one of which, known as skeleton "m," featured a nearly complete vertebral column preserved in articulation as well as various other scattered elements, including cranial material) was found in the lower layers of the Tendaguru

Formation, whereas *D. sattleri* (consisting of two partial skeletons, the most complete of which being known as skeleton "M") was found in the upper layers. Thus, the two species did not live alongside one another but were rather separated by approximately five million years.

Additionally, there are slight morphological differences that can be used to distinguish the two species. Comparing the humerus and the femur of the two species against those from the closely related *Amargasaurus* shows that *D. hansemanni* and *D. sattleri* are indeed more closely related to each other than either is to *Amargasaurus*, justifying their placement within a singular genus (Schwarz-Wings and Böhm, 2012).

The generic name *Dicraeosaurus* derives from the Greek "dikraios" (meaning "bifurcated"). The specific name *hansemanni* honors David von Hansemann; *sattleri* honors expedition promoter W. B. Sattler.

CLASSIFICATION
Dinosauria
 Sauropoda
  Eusauropoda
   Neosauropoda
    Diplodocoidea
     Flagellicaudata
      Dicraeosauridae

LOCATION
Tanzania

KNOWN REMAINS
Nearly complete

| 2 m | 4 m | 6 m | 8 m | 10 m | 12 m |

4 m

2 m

A. cazaui

(Salgado and Bonaparte, 1991)
Length: 13.5 m (44 ft)
Height: 3.3 m (10.8 ft)
Hip height: 2.5 m (8.2 ft)
Body mass: 3,500 kg (3.9 t)
Reconstruction: ☐☐☐☐

| 250 | 245 | 240 | 235 | 230 | 225 | 220 | 215 | 210 | 205 | 200 | 195 | 190 | 185 | 180 | 175 | 170 | 165 | 160 | 155 | 150 | 145 | 140 | 135 | 130 | 125 | 120 | 115 | 110 | 105 | 100 | 95 | 90 | 85 | 80 | 75 | 70 | 65 |

| TRIASSIC | JURASSIC | CRETACEOUS |

Amargasaurus cazaui (meaning "Cazau's lizard from Amarga") is among the most unique of all sauropods. Discovered in 1984 during one of José Bonaparte's expeditions, the holotype specimen remains the only substantial example of *Amargasaurus* material.

Amargasaurus possesses extremely long, bifurcated neural spines, which protrude at a rear-facing angle from the vertebrae of the neck. Numerous hypotheses have been proposed as to what these structures were for, and what they would have looked like in life, but so far the matter remains unresolved.

The original description of *Amargasaurus* championed the idea that the spines functioned as a sort of weaponry, and numerous studies have also concluded that this is a likely interpretation. In 1994, Gregory Paul posited that the bony spines would have been covered in a keratinous sheath and would have been longer and sturdier than the bone

alone would be. In 2007, Schwarz and colleagues compared striations on the bone to those seen on modern horned bovids, such as antelopes. In 2016, Hallett and Wedel proposed that the weaponized spines might have interlocked during competition among rival males.

However, another popular interpretation is that the spines supported a fleshy sail instead. In 1997, Jack Bailey compared the structure of the spines to those found in the sail-bearing *Dimetrodon*. And in 2022, Cerda and colleagues conducted an osteohistological study, examining the microscopic structures inside the bones, and found evidence for a high degree of vascularization and connective ligaments between the spines—signs interpreted as evidence for a sail of soft tissue.

The generic name *Amargasaurus* refers to La Amarga Canyon, a province in Neuquen, Argentina. The specific name *cazaui* honors geologist Luis B. Cazau.

CLASSIFICATION

Dinosauria
  Sauropoda
    Eusauropoda
      Neosauropoda
        Diplodocoidea
          Flagellicaudata
            Dicraeosauridae

LOCATION

Argentina

KNOWN REMAINS

Partial skull and skeleton

4 m   8 m   12 m   16 m   20 m

A. ajax

(Marsh, 1877b)
Length: 23 m (75 ft)
Height: 4.3 m (14.1 ft)
Hip height: 4 m (13.1 ft)
Body mass: 20,000 kg (22 t)
Reconstruction: ☐☐☐☐

8 m

4 m

| 250 | 245 | 240 | 235 | 230 | 225 | 220 | 215 | 210 | 205 | 200 | 195 | 190 | 185 | 180 | 175 | 170 | 165 | 160 | 155 | **150** | 145 | 140 | 135 | 130 | 125 | 120 | 115 | 110 | 105 | 100 | 95 | 90 | 85 | 80 | 75 | 70 | 65 |

TRIASSIC   **JURASSIC**   **CRETACEOUS**

Apatosaurus ajax (meaning "Ajax's deceptive lizard") is one of the most famous of all dinosaurs, and was the first sauropod skeleton to ever be mounted for a museum display.

There have been a vast number of fossils attributed to numerous species of *Apatosaurus* over the years; its fossils are the second most common of any sauropod from the extensive Morrison Formation, which stretches across much of the US Midwest. Disentangling the exact history of each specimen's ever-changing identity would fill an entire textbook. Even the contents of the holotype specimen, YPM 1860, are ambiguous, as they were once intermingled and mixed up with certain elements of a similar sauropod, *Atlantosaurus* (Tschopp et al., 2015).

Currently, only two species are widely considered to be valid, *A. ajax*, and *A. louisae*, which was named by William H. Holland in 1916. Many partial specimens are simply identified as *Apatosaurus* sp., meaning that their species is indeterminate. One such fossil, OMNH 1670, an especially large vertebra, suggests that particularly large individuals may have reached lengths of 30 meters (Wedel, 2013).

Compared with its close relative, *Diplodocus*, *Apatosaurus* was a stockier animal with a more robust build. For a long period of time, the skull of *Apatosaurus* was assumed to be shorter, akin to that of *Camarasaurus*, but it is now known to be longer and narrower, much like that of *Diplodocus*. *Apatosaurus* bore a single large claw on the side of its front feet, while its hind feet sported three claws apiece (Lovelace et al., 2007).

The generic name *Apatosaurus* combines the Greek "apatao" (meaning "deceive") with "sauros" (meaning "lizard"), referring to the fact that the bones were originally misidentified, being thought to belong to a mosasaur. The specific name *ajax* refers to the Greek mythological hero; *louisae* honors Louise Carnegie, wife of philanthropist Andrew Carnegie.

CLASSIFICATION

Dinosauria
 Sauropoda
  Eusauropoda
   Neosauropoda
    Diplodocoidea
     Flagellicaudata
      Diplodocidae
       Apatosaurinae

LOCATION

United States

KNOWN REMAINS

Nearly complete

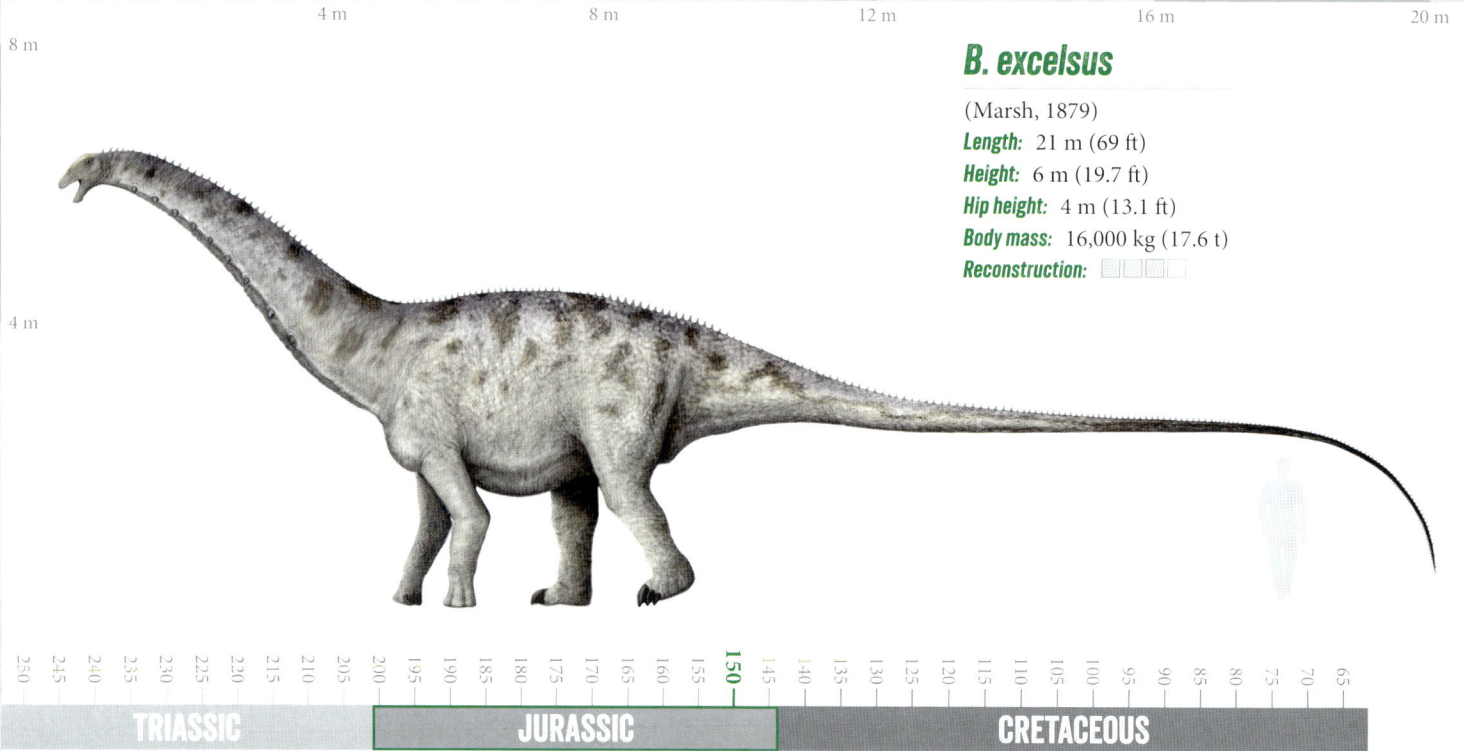

8 m

B. excelsus

(Marsh, 1879)
Length: 21 m (69 ft)
Height: 6 m (19.7 ft)
Hip height: 4 m (13.1 ft)
Body mass: 16,000 kg (17.6 t)
Reconstruction:

4 m

| 250 | 245 | 240 | 235 | 230 | 225 | 220 | 215 | 210 | 205 | 200 | 195 | 190 | 185 | 180 | 175 | 170 | 165 | 160 | 155 | **150** | 145 | 140 | 135 | 130 | 125 | 120 | 115 | 110 | 105 | 100 | 95 | 90 | 85 | 80 | 75 | 70 | 65 |

TRIASSIC	JURASSIC	CRETACEOUS

For more than a century, the infamous **Brontosaurus excelsus** (meaning "high thunder lizard") was not considered to be a valid taxon—and the genus is still not universally accepted.

Brontosaurus was first described on the basis of a nearly complete skeleton, YPM 1980, in 1879. However, in 1903, Elmer Riggs concluded that the animal was functionally indistinguishable from *Apatosaurus*, which had been named earlier and therefore had naming priority; thus, *B. excelsus* became *A. excelsus*. Nonetheless, when the American Museum of Natural History erected the first mounted sauropod skeleton in history (a composite skeleton made of the bones of numerous specimens), it was labeled as Brontosaurus—and from that point on, the name was firmly lodged into the public's consciousness, despite that the majority of paleontologists did not recognize the genus (Taylor, 2010).

In 2015, though, a comprehensive phylogenetic study rocked the boat. By comparing numerous diplodocid specimens, it was determined that *A. excelsus* actually did, in fact, have enough different traits to be considered a distinct genus, and the researchers proposed resurrecting the genus *Brontosaurus*. Further, they found that several specimens attributed to various *Apatosaurus* species formed their own branch on the family tree along with *B. excelsus*, and so the authors also proposed renaming them to the *Brontosaurus* genus, as well: *B. yahnahpin* and *B. parvus*. The genera *Elosaurus* and *Eobrontosaurus* were also synonymized with *Brontosaurus* (Tschopp et al., 2015).

The generic name *Brontosaurus* combines the Greek "bronte" (meaning "thunder") and "sauros" (meaning "lizard"). The specific name *excelsus* is Latin for "high" or "lofty."

CLASSIFICATION
Dinosauria
  Sauropoda
    Eusauropoda
      Neosauropoda
        Diplodocoidea
          Flagellicaudata
            Diplodocidae
              Apatosaurinae

LOCATION
United States

KNOWN REMAINS
Partial skeleton

4 m   8 m   12 m   16 m   20 m

8 m

A. altus

(Cope, 1878)
Length: 25 m (82 ft)
Height: 4.4 m (14.4 ft)
Hip height: 4 m (13.1 ft)
Body mass: 26,000 kg (28.7 t)
Reconstruction: ☐ ☐ ☐ ☐

4 m

250 245 240 235 230 225 220 215 210 205 200 195 190 185 180 175 170 165 160 155 150 145 140 135 130 125 120 115 110 105 100 95 90 85 80 75 70 65

TRIASSIC | **JURASSIC** | **CRETACEOUS**

Amphicoelias altus (meaning "high biconcave") is currently the only valid species within the genus, although others have been named in the past. The similarly named *A. latus* was later recognized as *Camarasaurus*, while *A. fragillimus* was reassigned to its own genus, *Maraapunisaurus*.

The holotype (and, currently, only accepted) specimen of *A. altus*, AMNH 5764, originally consisted of a tooth, two dorsal vertebrae, a pubis, and a femur. Later, in 1921, when Henry Osborn and Charles Mook cataloged the material for the American Museum of Natural History, they found that a scapula, coracoid, and an ulna had also been discovered in close proximity and included these elements with the rest of the specimens. However, modern examinations have cast doubt on the identity of some of these bones. The scapula and coracoid, in particular, have been described as having more features in common with *Camarasaurus* than with diplodocids (McIntosh, 1998; Tschopp et al., 2015), and the same is true for the tooth (Whitlock, 2011). The identity of the ulna remains more of an open question, pending more detailed analyses, but at the risk of inadvertently creating a chimera specimen, it is likely to be excluded from the genus going forward (Mannion et al., 2021).

Given the fragmentary nature of the remains, and the question as to which bones should actually be counted, *A. altus* is often excluded from phylogenetic studies. Various versions of analyses place it as either a basal **diplodocid**, or within the subfamily Apatosaurinae (Tschopp et al., 2015; Mannion et al., 2021).

The generic name *Amphicoelias* combines the Greek "amphi" (meaning "on both sides") and "koilos" (meaning "hollow" or "concave"), meant to be interpreted as "biconcave" in reference to the shape of the vertebrae. The specific name *altus* is Latin for "high," again referring to the vertebra—in this case, the neural spine.

CLASSIFICATION

Dinosauria
 Sauropoda
  Eusauropoda
   Neosauropoda
    Diplodocoidea
     Flagellicaudata
      Diplodocidae

LOCATION

Colorado, USA

KNOWN REMAINS

Fragments

| 2 m | 4 m | 6 m | 8 m | 10 m | 12 m |

K. siberi

(Tschopp and Mateus, 2013)
Length: 13 m (43 ft)
Height: 3.4 m (11.1 ft)
Hip height: 2 m (6.6 ft)
Body mass: 2,000 kg (2.2 t)
Reconstruction: ☐☐☐☐

| TRIASSIC | JURASSIC | CRETACEOUS |

Kaatedocus siberi (meaning "Siber's small beam") is known from material unearthed at the historic Howe Quarry, from the strata of the famous Morrison Formation.

Plentiful sauropod bones were first reported from Howe Quarry in 1934, but further excavations were canceled because of a dispute with the landowner. Most of the remains were lost in a fire that damaged the American Museum of Natural History in the 1940s; others had been irrevocably damaged due to poor storage conditions at a location near the dig site. All in all, roughly one-tenth of the original discoveries survived, and no new genera were described using these remains.

Much later, in 1989, Swiss researchers were able to revisit the site and excavated several hundred fossils. Among them were a partial skull and more than a dozen neck vertebrae, known together as SMA 0004. The remains were initially believed to belong to either *Barosaurus* or a juvenile *Diplodocus*. Closer examination, however, would reveal that they actually represented a new genus, *Kaatedocus*. The animal was a subadult at the time of death.

The description of *Kaatedocus* placed the animal well within **Diplodocidae**; a later study by the same lead author, specifically focusing on the phylogenetic relationships within Diplodocidae, came to the same conclusion (Tschopp et al., 2015). However, a different analysis calculated *Kaatedocus* to be the basalmost member of Dicraeosauridae, the sister clade to Diplodocidae (Whitlock and Mantilla, 2020).

The generic name *Kaatedocus* combines the indigenous Absaroka term "kaate" (meaning "small") with the Greek "docus" (meaning "beam"), which is an allusion to *Diplodocus*. The specific name *siberi* honors museum director Hans-Jakob "Kirby" Siber, who organized the excavation and preparation of the holotype material.

CLASSIFICATION

Dinosauria
 Sauropoda
  Eusauropoda
   Neosauropoda
    Diplodocoidea
     Flagellicaudata
      Diplodocidae
       Diplodocinae

LOCATION
Wyoming, USA

KNOWN REMAINS
Partial skull and neck

4 m 8 m 12 m 16 m 20 m

8 m

T. africana

(Fraas, 1908)
Length: 24 m (79 ft)
Height: 6.4 m (21 ft)
Hip height: 3.9 m (12.8 ft)
Body mass: 13,500 kg (14.9 t)
Reconstruction: ☐☐☐☐

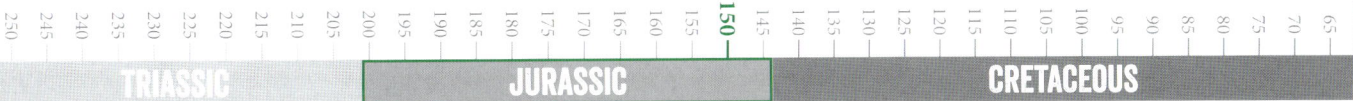

4 m

250 245 240 235 230 225 220 215 210 205 200 195 190 185 180 175 170 165 160 155 150 145 140 135 130 125 120 115 110 105 100 95 90 85 80 75 70 65

TRIASSIC   **JURASSIC**   **CRETACEOUS**

Tornieria africana (meaning "Tornier's one from Africa") is differentiated partially by the unique features of its robust frontal tail vertebrae and the stout proportions of its hindlimbs.

In 1907, two partial sauropod skeletons were discovered from the same site in what was then known as German East Africa. In 1908, Eberhard Fraas, a German paleontologist, named them to what he thought was a new genus, *Gigantosaurus*. However, unbeknownst to him, the name was already in use to describe a now-dubious European sauropod. To rectify this, paleontologist Richard Sternfeld renamed the sauropod genus *Tornieria* in 1911.

Various specimens once attributed to *Tornieria* have since been reassigned to *Malawisaurus* or *Janenschia*. For a time, it was thought that *Tornieria* might be synonymous with *Barosaurus* and that the genus name would be abandoned.

However, a modern reanalysis of the original two specimens, dubbed "skeleton A" and "skeleton k," confirmed *Tornieria* to be a distinct and valid genus (Remes, 2006).

Making this determination was hampered by some of these remains having been destroyed during an Allied bombing attack on Berlin during WWII. In addition to "A" and "k," the only other material reliably attributable to *Tornieria* are a collection of 26 tail vertebrae originating from a nearby site known as "trench dd", and coming from more than one individual animal. It is possible that similar sauropod fossils found around the same time and place, originally labeled as *Barosaurus*, might also represent *Tornieria*, but this determination cannot be made with confidence.

The generic name *Tornieria* honors German paleontologist Gustav Tornier. The specific name *africana* is in reference to Africa.

CLASSIFICATION
Dinosauria
  Sauropoda
   Eusauropoda
    Neosauropoda
     Diplodocoidea
      Flagellicaudata
       Diplodocidae
        Diplodocinae

LOCATION
Tanzania

KNOWN REMAINS
Partial skull and skeleton

157

L. laticauda

(Gallina et al., 2014)
Length: 11 m (36 ft)
Height: 2.7 m (8.9 ft)
Hip height: 1.6 m (5.2 ft)
Body mass: 1,700 t (1.9 t)
Reconstruction: ☐ ☐ ☐ ☐

TRIASSIC | JURASSIC | CRETACEOUS

Leinkupal laticauda (meaning "wide tail of the vanishing family") is the first diplodocid sauropod discovered from South America, and just the second confirmed diplodocid from what was once Gondwana. At the time of its naming, it was also the youngest recorded species of **Diplodocidae** known to exist.

Traditionally, it is typically thought that the majority of diplodocids went extinct around the Jurassic-Cretaceous boundary. The existence of *Leinkupal* several million years after this time suggests that the clade did not go extinct all at once, but that at least one pocket survived in South America for some time. Thus, the eventual diplodocid extinction did not occur all together but went through phases that were likely regional in nature.

The eight scattered vertebral elements from which *Leinkupal* was described were unearthed in 2010 and 2012, along with the remains of dicraeosaurid sauropods and some theropod elements. In a later conference paper, the researchers added additional vertebral elements referrable to the species, as well as a braincase and a portion of the breastplate that were found from the same location as the holotype material (Gallina et al., 2019).

The braincase was compared with the closely related dicraeosaurid sauropod *Bajadasaurus*, which was unearthed from the same site, and was determined to not belong to that genus. Given its diplodocid traits, it was assigned to *Leinkupal* (Garderes et al., 2022).

The generic name *Leinkupal* combines the indigenous Mapudungun "lein" (meaning "vanishing") and "kupal" (meaning "family"), referencing *Leinkupal* being the latest known species of Diplodocidae. The specific name *laticauda* combines the Latin "latus" (meaning "wide") and "cauda" (meaning "tail").

CLASSIFICATION
Dinosauria
 Sauropoda
  Eusauropoda
   Neosauropoda
    Diplodocoidea
     Flagellicaudata
      Diplodocidae
       Diplodocinae

LOCATION
Argentina

KNOWN REMAINS
Isolated vertebrae, braincase

4 m	8 m	12 m	16 m	20 m

8 m

4 m

G. hayi

(Tschopp et al., 2015)
Length: 23 m (75 ft)
Height: 5 m (16.4 ft)
Hip height: 3.9 m (12.8 ft)
Body mass: 10,000 kg (11 t)
Reconstruction: ☐☐☐☐

250	245	240	235	230	225	220	215	210	205	200	195	190	185	180	175	170	165	160	155	**153**	150	145	140	135	130	125	120	115	110	105	100	95	90	85	80	75	70	65

TRIASSIC	JURASSIC	CRETACEOUS

Galeamopus hayi (meaning "Hay's needed helmet") was long known as *Diplodocus hayi*, one of several species within that famous genus. A 2015 comparative study of numerous sets of diplodocid remains, however, determined that there were just barely enough differences in the specimen that it should actually constitute its own distinct genus (Tschopp et al., 2015).

Several different fossils that had long been assigned to *Diplodocus* were subsequently placed within the umbrella of *Galeamopus*, including the substantial holotype (HMNS 175) and the skull AMNH 969; skull material for sauropods is notoriously hard to come by.

Another substantial specimen, SMA 0011, which had been discovered in 1995, was subsequently described as a second species within the genus, *G. pabsti*. This specimen was able to highlight one of the key differences between *Galeamopus* and animals like *Diplodocus* or *Apatosaurus*, which has to do with where the neck vertebrae transition into back vertebrae; in *Galeamopus*, the length of the vertebra suddenly shortens by a considerable degree at this transition (Tschopp and Mateus, 2017).

The generic name *Galeamopus* combines the Latin "galeam" (meaning "helmet") and "opus" (meaning "need"). This is the literal translation of the German name "Wilhelm", which itself is the basis for the English name William. Thus, the name *Galeamopus* is meant to honor both William H. Utterback, who found the type specimen, and William J. Holland, who named "*D.*" *hayi* in 1924, while still recognizing that the animal could possibly be considered its own genus. The specific name *hayi* honors paleontologist Oliver Perry Hay; *pabsti* honors paleontologist Ben Pabst, who found the specimen.

CLASSIFICATION

Dinosauria
 Sauropoda
  Eusauropoda
   Neosauropoda
    Diplodocoidea
     Flagellicaudata
      Diplodocidae
       Diplodocinae

LOCATION

United States

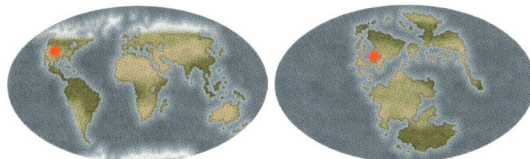

KNOWN REMAINS

Nearly complete

159

SUPERSAURUS

S. vivianae

(Jensen, 1985)
Length: 39 m (127 ft)
Height: 8.3 m (27.2 ft)
Hip height: 5 m (16.4 ft)
Body mass: 27,000 kg (29.8 t)
Reconstruction: ☐☐☐☐

TRIASSIC **JURASSIC** **CRETACEOUS**

Supersaurus vivianae (meaning "Vivian above lizard") is a contender for the longest dinosaur of all time, with some estimates putting the animal at a staggering 39 meters in length (Curtice, 2021). Even so, it would have been far from the heaviest of dinosaurs, as the titanosaur sauropods were built much more solidly.

In 1943, a large sauropod bone bed was discovered in Colorado by locals. Organized fossil collection efforts began in 1972. Described as a "bone salad," the area yielded disarticulated specimens that went on to be attributed to three new genera: *Supersaurus*, *Ultrasauros*, and *Dystylosaurus*. However, in 1996 it was determined that the majority of material attributed to *Ultrasauros*, which was found approximately 100 meters away from the rest of the remains, could actually be attributed to *Brachiosaurus*. Additionally, in 2001, it was determined that the material that had been labeled *Dystylosaurus* was part of the same

animal as *Supersaurus*, given that all the elements were found in the same small "pocket" and that there are no duplicate bones (Curtice and Stadtman, 2001). This left *Supersaurus* as the only valid genus of the original three.

In addition to these holotype remains (BYU 12962), a second substantial specimen from Wyoming was described in 2007 (WDC DMJ-021, aka "Jimbo"). Also from Wyoming, a third specimen, nicknamed "Goliath" is purportedly (as of 2021) being prepared at the Grandview Museum of Natural History.

Dinheirosaurus, a sauropod from Portugal, has been suggested by at least one study to be rebranded as a second species of *Supersaurus* (Tschopp et al., 2015).

The generic name *Supersaurus* combines the Latin "super" (meaning "above") with the Greek "saurus" (meaning "lizard"). The specific name *vivianae* honors Vivian Jones, who discovered the holotype.

CLASSIFICATION
Dinosauria
 Sauropoda
  Eusauropoda
   Neosauropoda
    Diplodocoidea
     Flagellicaudata
      Diplodocidae
       Diplodocinae

LOCATION
United States

KNOWN REMAINS
Partial skeleton

4 m   8 m   12 m   16 m   20 m

8 m

D. lourinhanensis

(Bonaparte and Mateus, 1999)
Length: 21 m (69 ft)
Height: 7.5 m (24.6 ft)
Hip height: 4.1 m (13.4 ft)
Body mass: 8,800 kg (9.7 t)
Reconstruction: ☐☐☐☐

4 m

| 250 | 245 | 240 | 235 | 230 | 225 | 220 | 215 | 210 | 205 | 200 | 195 | 190 | 185 | 180 | 175 | 170 | 165 | 160 | 155 | **150** | 145 | 140 | 135 | 130 | 125 | 120 | 115 | 110 | 105 | 100 | 95 | 90 | 85 | 80 | 75 | 70 | 65 |

TRIASSIC   **JURASSIC**   **CRETACEOUS**

Dinheirosaurus lourinhanensis (meaning "lizard from Dinheiro, Lourinhã") is known from only fragmentary remains, and its validity as a genus has also been called into question.

The only known specimen, ML 414, was first found in 1987 and continued to be excavated until 1992. It consisted primarily of the midvertebral column, along with some other fragments. In 1998, the specimen was attributed by Pedro Dantas as an example of the sauropod *Lourinhasaurus alenquerensis*. Shortly thereafter, though, an overview of *Lourinhasaurus* remains found that ML 414 was sufficiently different to justify the erection of a new genus (Mannion et al., 2012).

A 2015 study of numerous diplodocid specimens—the same study that proposed to reinstate *Brontosaurus* as a valid taxon—suggested that there were not enough differences between *Lourinhasaurus* and the sauropod *Supersaurus* to justify the distinction between the genera. The study concluded (somewhat arbitrarily) that 13 distinct anatomical differences were necessary to separate any two genera, and they only found 11 differences between *Lourinhasaurus* and *Supersaurus*. They suggested that ML 414 bc given the new combination *Supersaurus lourinhanensis* (Tschopp et al., 2015).

Contrastingly, a later study warned that further work needed to be done in order to justify or refute this distinction. It seems that the holotype specimen actually includes several caudal vertebrae that remain in an unprepared state, and the researchers believe that possibly hidden details of these bones should be taken into account with regard to the validity of the genus (Mocho et al., 2017).

The animal's binomial name refers to the Porto Dinheiro area located in the Lourinhã municipality, in Estremadura Province, Portugal.

CLASSIFICATION
Dinosauria
  Sauropoda
    Eusauropoda
      Neosauropoda
        Diplodocoidea
          Flagellicaudata
            Diplodocidae
              Diplodocinae

LOCATION
United States

KNOWN REMAINS
Nearly complete

| 4 m | 8 m | 12 m | 16 m | 20 m |

B. lentus

(Marsh, 1890)
Length: 27 m (88 ft)
Height: 8.7 m (28.5 ft)
Hip height: 3.9 m (12.8 ft)
Body mass: 10,000 kg (11 t)
Reconstruction: ▢▢▢▢

| TRIASSIC | JURASSIC | CRETACEOUS |

Barosaurus lentus (meaning "tough heavy lizard") differs from its cousin, the more famous *Diplodocus*, by having a much longer neck and a shorter tail; the limbs of the two animals are nearly identical.

The individual cervical (neck) vertebrae of *Barosaurus* can be up to 50% longer than those of *Diplodocus,* and are oriented in such a way that they would allow the animal a significant degree of lateral neck flexibility while simultaneously limiting its vertical flexibility. This differs from other diplodocids and suggests that *Barosaurus* made use of an altered feeding strategy, browsing low vegetation while sweeping its head along extended arcs (Taylor and Wedel, 2016).

Exactly which sauropod specimens belong to *Barosaurus* is a matter of debate. The holotype remains were excavated in 1889, and since that time, numerous sets of sauropod remains—ranging from the highly fragmentary to the nearly complete—have also been pulled from the Morrison Formation of the US Midwest. Whether an individual specimen represents *Diplodocus, Kaatedocus, Supersaurus,* or *Barosaurus*, or whether the bones are even distinct enough to be identifiable, is a matter that is continuously in dispute. The skeletal representation below combines the holotype YPM 429 with AMNH 6341 and AMNH 7535 (Tschopp et al., 2015). As a consequence of these uncertainties, estimates of the overall size of *Barosaurus* vary widely, with some length estimates reaching past 35 meters.

The generic name *Barosaurus* combines the Greek "barys" (meaning "heavy") and "sauros" (meaning "lizard"). The specific name *lentus* is Latin for "tough." A second species, *B. affinis* (Latin for "kindred" or "affinity") was named by Marsh in 1899 based on a few foot bones, but the species has long been considered synonymous with *B. lentus*.

CLASSIFICATION
Dinosauria
 Sauropoda
  Eusauropoda
   Neosauropoda
    Diplodocoidea
     Flagellicaudata
      Diplodocidae
       Diplodocinae

LOCATION
United States

KNOWN REMAINS
Partial skeleton

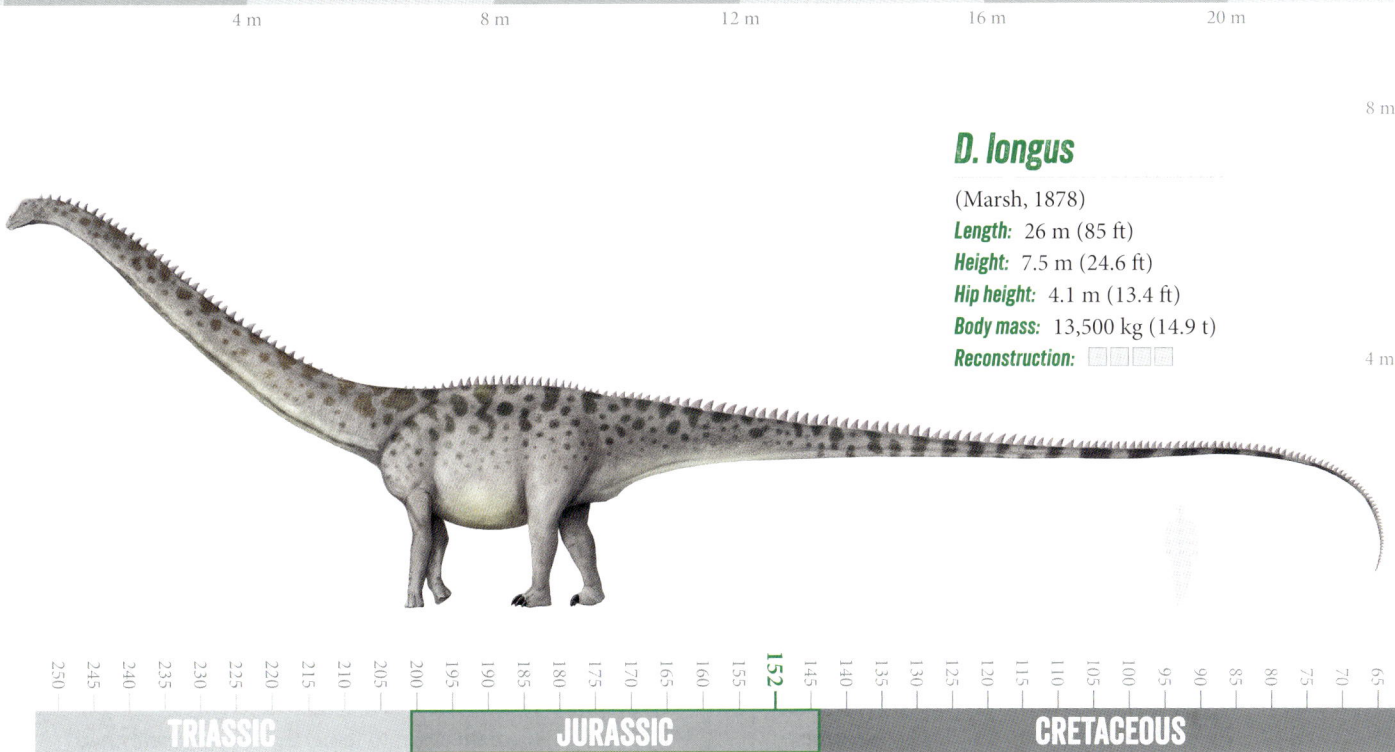

4 m   8 m   12 m   16 m   20 m

8 m

D. longus

(Marsh, 1878)
Length: 26 m (85 ft)
Height: 7.5 m (24.6 ft)
Hip height: 4.1 m (13.4 ft)
Body mass: 13,500 kg (14.9 t)
Reconstruction: ☐☐☐☐

4 m

250 245 240 235 230 225 220 215 210 205 200 195 190 185 180 175 170 165 160 155 **152** 145 140 135 130 125 120 115 110 105 100 95 90 85 80 75 70 65

TRIASSIC	JURASSIC	CRETACEOUS

Diplodocus longus (meaning "long double beam"), at one point thought to be the longest-ever land animal, is known from a plethora of specimens. It had an extremely long and thin tail and long, peg-like teeth that were used to strip vegetation from branches. At least eight skulls of *Diplodocus* are currently recognized (Woodruff et al., 2018).

The holotype specimen of *D. longus*, YPM 1920, consists only of a handful of tail vertebrae. A large study of various diplodocid specimens claimed that these bones actually bore no unique identifiable features, thus they could not form the basis for *D. longus* and therefore the species was invalid. The researchers appealed to the authority of the International Code of Zoological Nomenclature to then replace the type species (Tschopp et al., 2015). However, this notion was rejected, and other researchers have voiced dissenting opinions, pointing out potentially identifiable characteristics of the fossil material, as well as that other *D. longus* fossils

collected from the same area in the same excavation period could aid in distinguishing the species (Carpenter, 2017).

Other species considered to be valid include *D. carnegii*, named by John Bell Hatcher in 1901, and *D. hallorum*, which was originally identified as *Seismosaurus halli* in 1991 by David D. Gillette before being reidentified in 2006 (Lucas et al., 2006). Dubious species include *D. lacustris*, named in 1884 by Marsh, which is now thought to be a juvenile *Apatosaurus* or *Camarasaurus*. The species *D. hayi*, named by William Jacob Holland in 1924, has now been given its own genus, *Galeamopus*.

The generic name *Diplodocus* combines the Greek "diplo" (meaning "double") and "dokos" (meaning "beam" or "rafter"), referencing aspects of the animal's tail bones. The specific name *longus* is Latin for "long"; *carnegii* honors philanthropist Andrew Carnegie; *hallorum* honors Jim and Ruth Hall of Ghost Ranch, New Mexico.

CLASSIFICATION
Dinosauria
 Sauropoda
  Eusauropoda
   Neosauropoda
    Diplodocoidea
     Flagellicaudata
      Diplodocidae
       Diplodocinae

LOCATION
United States

KNOWN REMAINS
Nearly complete

D. polyonychius

(McIntosh et al., 1992)
Length: 16 m (52 ft)
Height: 3.1 m (10.2 ft)
Reconstruction: ☐☐☐☐

155?

| TRIASSIC | JURASSIC | CRETACEOUS |

Sometime prior to his death in 1937, geologist F. B. Loomis unearthed the set of fossils in Wyoming now known as AC 663, which would become the holotype of the diplodocid **Dyslocosaurus polyonychius** (meaning "poorly placed, many-clawed lizard"). However, there was next to no documentation produced for these fossils, so there is nothing definitive known about exactly where they were found, or even if they were found together. On this basis, some researchers have considered the genus *nomen dubium*. In

1998, Sereno and Wilson proposed that the specimen was a chimera, made of titanosaur limb bones and digits belonging to a theropod. A later examination would lend further credence to the case that at least one of the foot bones did not originate with the rest of the fossils (Tschopp et al., 2015).

The generic name refers to the uncertainty regarding the origin of the fossils. The specific name references *Dyslocosaurus*'s main claim to fame: that it supposedly had four, or even five, clawed toes, as opposed to the typical three.

A. montanus

(Marsh, 1877a)
Length: 23 m (75 ft)
Height: 4.3 m (14.1 ft)
Reconstruction: ☐☐☐☐

150

| TRIASSIC | JURASSIC | CRETACEOUS |

Atlantosaurus montanus (meaning "Atlas's lizard from Montana") has had a fraught history, nearly from the moment of its discovery. The specimen known as YPM 1835 consists of a partial sacrum (fused hip vertebrae) and at the time of its discovery was considered remarkable due to the presence of air-filled pneumatic chambers. This pneumaticity was used as a defining characteristic of the genus. However, it is now known that such features were very common among sauropods. As such, there is nothing uniquely identifiable about the specimen, rendering *Atlantosaurus* a dubious genus.

The following year, in 1878, the specimen YPM 1840 was used as the basis for a second species, *A. immanis*. However, during shipping, several elements of the skeleton were inadvertently intermingled with elements from the sauropod *Apatosaurus ajax*, and it remains uncertain to this day which bones belong to which specimen. Since *Atlantosaurus* has been invalidated, it has been suggested that a new genus be erected for these remains (Tschopp et al., 2015).

4 m   8 m   12 m   16 m

A. viator

(van der Linden et al., 2024)
Length: 19.4 m (63.3 ft)
Height: 7 m (23 ft)
Hip height: 3.5 m (11.5 ft)
Body mass: 6,000 kg (13.2 t)
Reconstruction: ☐ ☐ ☐ ☐

4 m

250 | 245 | 240 | 235 | 230 | 225 | 220 | 215 | 210 | 205 | 200 | 195 | 190 | 185 | 180 | 175 | 170 | 165 | 160 | 155 | **150** | 145 | 140 | 135 | 130 | 125 | 120 | 115 | 110 | 105 | 100 | 95 | 90 | 85 | 80 | 75 | 70 | 65

TRIASSIC   **JURASSIC**   **CRETACEOUS**

The remains of *Ardetosaurus viator* (meaning "burned traveler lizard") were discovered in 1992 at a fossil-rich bonebed in the famous Morrison Formation of the western United States. The carcasses of numerous sauropods (along with a stegosaur, an allosaur, and other creatures) seem to have been floating down a river before a log jam (which also fossilized) impeded their path.

The specimen was shipped to Europe, and some of the bones were sent to be prepared at the Dinosaurier Freilichtmuseum in Germany. In October of 2003, a fire caused by malicious arson damaged some of the leg bones and destroyed three neck vertebrae. The bones now reside at the Oertijdmuseum in the Netherlands.

The holotype of *Ardetosaurus* is one of the few sauropod specimens that preserves the first chevron—the foremost "tail rib." It has been suggested that the morphology of this bone may be one of the only ways to differentiate the gender of a sauropod fossil, owing to the bone's proximity to the animal's cloaca. Further research will need to be conducted to determine if this is indeed the case.

The generic name *Ardetosaurus* derives from "ardeto" (Latin meaning "to burn," referring to the history of some of the fossil elements). The specific name *viator* is Latin for "traveler," referring to the journey of the specimen from the United States, via Switzerland and Germany, to the Netherlands.

CLASSIFICATION
Dinosauria
 Sauropoda
  Eusauropoda
   Neosauropoda
    Diplodocoidea
     Flagellicaudata
      Diplodocidae
       Diplodocinae

LOCATION
Wyoming, United States

KNOWN REMAINS
Partial skeleton

Rebbachisauridae

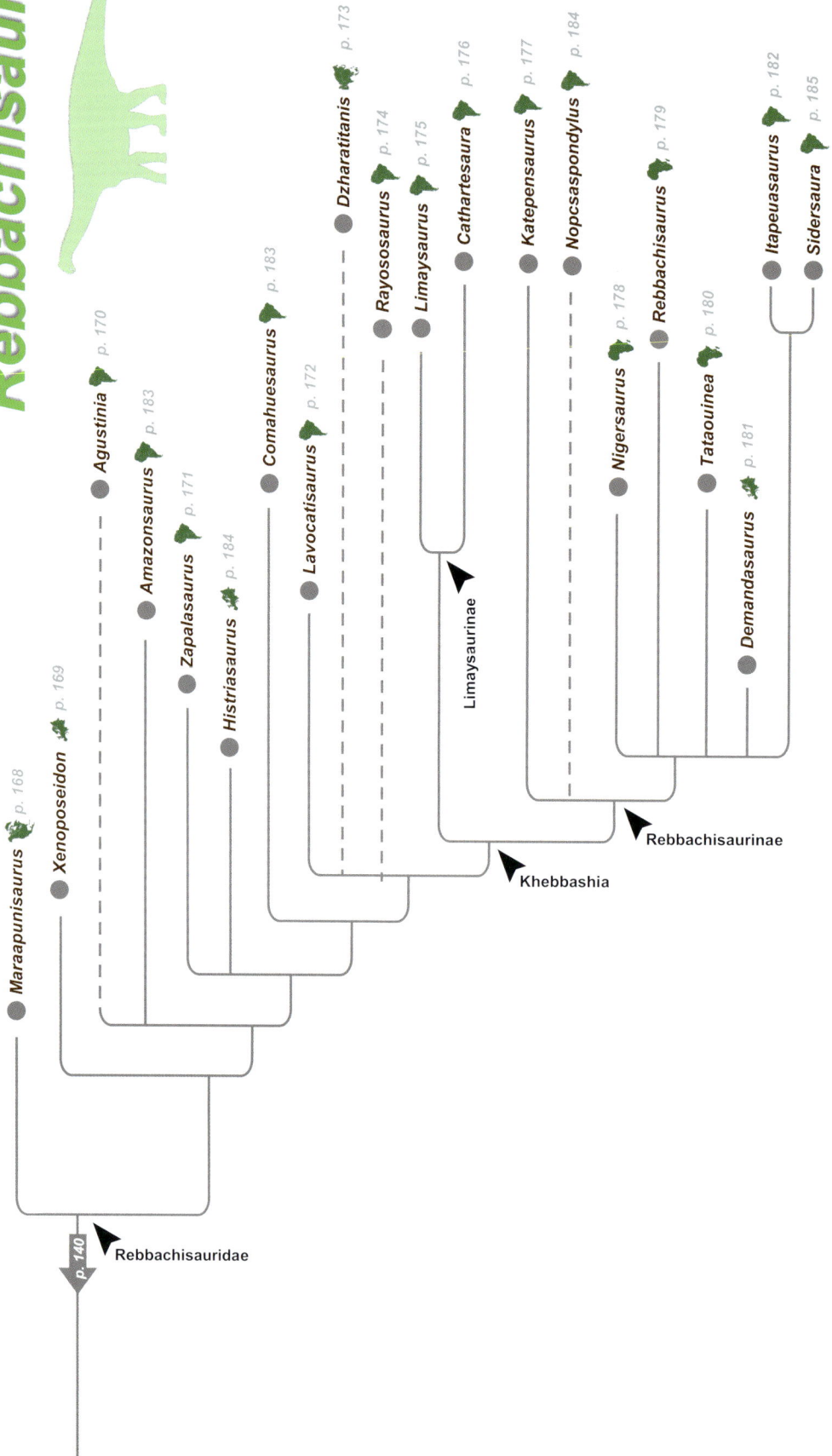

Rebbachisauridae *p. 140*

Khebbashia

Rebbachisaurinae

Limaysaurinae

Maraapunisaurus	p. 168
Xenoposeidon	p. 169
Agustinia	p. 170
Amazonsaurus	p. 183
Zapalasaurus	p. 171
Histriasaurus	p. 184
Comahuesaurus	p. 183
Lavocatisaurus	p. 172
Dzharatitanis	p. 173
Rayososaurus	p. 174
Limaysaurus	p. 175
Cathartesaura	p. 176
Katepensaurus	p. 177
Nopcsaspondylus	p. 184
Nigersaurus	p. 178
Rebbachisaurus	p. 179
Tataouinea	p. 180
Demandasaurus	p. 181
Itapeuasaurus	p. 182
Sidersaura	p. 185

Jurassic — Middle / Late
Cretaceous — Early / Late
Pal.

170 160 150 140 130 120 110 100 90 80 70 60

Unlike the other diplodocoid lineages, which did not survive past the mid–Early Cretaceous, the rebbachisaurids persisted into the Late Cretaceous. They are predominately, although not exclusively, known from South America and Africa, suggesting that the group emerged far earlier than we currently have fossils to indicate, as those two continents are thought to have split during the Middle Jurassic. Although rebbachisaurid cranial material is particularly rare, the group is perhaps best known for the highly unusual skull morphology of *Nigersaurus.*

According to Carpenter (2018), the oldest and most basal rebbachisaurid is *Maraapunisaurus*. The same study also considered *Xenoposeidon* to be a member of the group, following (Taylor, 2018), although there are cases to be made that it was actually a titanosaur. The long-misunderstood *Agustinia* has recently been placed as one of the basalmost members of the group, although its exact position is unclear (Bellardini et al., 2022).

When taking into account only genera whose positions are more firmly established, recent consensus shows *Amazonsaurus* as the most "primitive" rebbachisaurid, followed by *Zapalasaurus* and *Histriasaurus* (Canudo et al., 2018; Whitlock and Mantilla, 2020; Averianov and Sues, 2021; Ren et al., 2023). Although their exact positions shift slightly from study to study, *Comahuesaurus* and *Lavocatisaurus* reliably hold the subsequent crownward positions (Bellardini et al., 2022).

The affinities of *Rayososaurus* are slightly more uncertain, as it is often excluded from phylogenetic analyses; Royo-Torres et al. (2020) place it quite basally, Ren et al. (2023) place it within the derived Rebbachisaurinae, while Averianov and Sues (2021) take something of an uncertain middle ground. This last study, which described *Dzharatitanis*, similarly places it in an uncertain polytomy here. While Royo-Torres et al. (2020) placed *Katepensaurus* outside Khebbashia, others have favored a placement as a basal member of the group (Canudo et al., 2018; Bellardini et al., 2022).

The clade **Khebbashia** immediately forks into two subfamilies. Although other genera have been assigned to it in the past, **Limaysaurinae** is now typically shown to contain only *Limaysaurus* and *Cathartesaura* (Fanti et al., 2015; Averianov and Sues, 2021; Ren et al., 2023).

Rebbachisaurinae—typically referred to as Nigersaurinae prior to Fanti et al. (2015)—is near-universally shown to include *Nigersaurus*, *Rebbachisaurus*, *Tataouinea*, and *Demandasaurus* (Mannion et al., 2019b). The describers of *Itapeuasaurus* place it within the group (Lindoso et al., 2019). The describers of *Sidersaura* placed it as the sister taxon of *Itapeuasaurus* (although this study placed both taxa in a much more stemward position; Lerzo et al., 2024). The highly fragmentary *Nopcsaspondylus* has been speculated to belong, as well (Apesteguía, 2007).

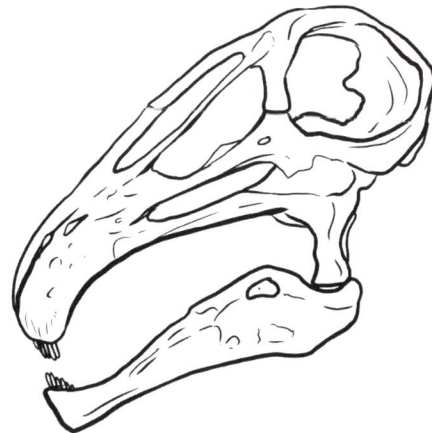

Nigersaurus taqueti
(skull in side view)

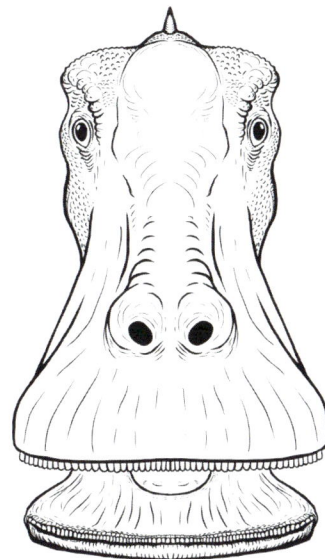

Nigersaurus taqueti
(restored cranium in front view)

MARAAPUNISAURUS

6 m   12 m   18 m   24 m   30 m   36 m

12 m

6 m

M. fragillimus

(Cope, 1878)

Length: 35 m (115 ft)
Height: 10.8 m (35.4 ft)
Hip height: 6.5 m (21.3 ft)
Body mass: 70,000 kg (77.2 t)
Reconstruction: ☐☐☐☐

250 | 245 | 240 | 235 | 230 | 225 | 220 | 215 | 210 | 205 | 200 | 195 | 190 | 185 | 180 | 175 | 170 | 165 | 160 | 155 | **150** | 145 | 140 | 135 | 130 | 125 | 120 | 115 | 110 | 105 | 100 | 95 | 90 | 85 | 80 | 75 | 70 | 65

TRIASSIC   **JURASSIC**   **CRETACEOUS**

Maraapunisaurus fragillimus (meaning "very fragile huge lizard") was originally known as *Amphicoelias fragillimus*, the second species to be named to that genus of diplodocid. However, a thorough reanalysis identified distinct **rebbachisaurid** traits in the fossil, which led to it being given its own, new genus. This reassignment also made it the oldest known rebbachisaurid, and the only one from North America (Carpenter, 2018).

Notably, Carpenter's review could be conducted only by using an original illustration of the single holotype bone, a dorsal vertebra, because the specimen itself has long been lost. Originally collected in 1877 on behalf of Edward Cope (of "Bone Wars" fame), the fossil had already gone missing by the time his collection was sold and cataloged to the American Museum of Natural History in 1897. Based on the soft mudstone present in the location in which the fossil was discovered, it has been speculated that the fragile holotype deteriorated rapidly; this possibility could also account for

why Cope only illustrated the specimen from a single viewpoint. Reportedly, a femur fragment was also uncovered in 1877 but was never studied. Subsequent attempts to find additional remains from the site revealed the area to be highly eroded.

What makes the case of *Maraapunisaurus* more than just an obscure footnote in history is the vertebra's truly colossal size. Using *Diplodocus* as a basis for comparison, some astounding size estimates have been made for *Maraapunisaurus*, ranging up to 60 meters and 120,000 kilograms (132 tons), which would make it the biggest terrestrial animal ever. However, taking into account the specimen's newer rebbachisaurid identification, and thus using *Limaysaurus* for comparison, a much more reasonable (though still huge) size estimate has been calculated.

The generic name *Maraapunisaurus* derives from the Southern Ute term "maraapuni" (meaning "huge"). The specific name *fragillimus* is Latin for "very fragile."

CLASSIFICATION

Dinosauria
  Sauropoda
   Gravisauria
    Eusauropoda
     Neosauropoda
      Diplodocoidea
       Rebbachisauridae

LOCATION

Colorado, USA

KNOWN REMAINS

Partial vertebra, (possible) partial femur

2 m   4 m   6 m   8 m   10 m   12 m   14 m

X. proneneukos

(Taylor and Naish, 2007)
Length: 14.8 m (49 ft)
Height: 4.5 m (14.7 ft)
Hip height: 3.1 m (10.2 ft)
Body mass: 7,000 kg (7.7 t)
Reconstruction: ☐☐☐☐

4 m

2 m

250 245 240 235 230 225 220 215 210 205 200 195 190 185 180 175 170 165 160 155 150 145 **140** 135 130 125 120 115 110 105 100 95 90 85 80 75 70 65

TRIASSIC | **JURASSIC** | **CRETACEOUS**

Xenoposeidon proneneukos (meaning "forward-sloping alien Poseidon") is known from a fragment of a single, peculiar vertebra that was discovered in the 1890s by Philip James Rufford, somewhere near Hastings, England. The fossil sat unnoticed in the archives of the Natural History Museum of London until its uniqueness was recognized in 2006.

One unique aspect of the vertebral bone is that the "neural arch" slopes forward instead of backward or vertically. The arch is also proportionally tall, being roughly the same height as the "centrum" portion of the bone. The vertebra was also originally thought to be asymmetric when comparing the left side to the right; however, this interpretation was later corrected after an opening "filled with matrix" was reinterpreted.

The original description of the genus was unable to pin down which kind of neosauropod that *Xenoposeidon* was, as its unique combination of traits seemed not to fit neatly into any particular category. Later analysis by the same lead author, Michael P. Taylor, would recognize a unique characteristic shared by *Rebbachisaurus*, resulting in the genus being placed within **Rebbachisauridae** (Taylor, 2018).

Although the exact location of the fossil's discovery is unrecorded, interpretations of the surviving original notes from the fossil's collector suggest that it came from strata that are 140 million years old. If accurate, this would make *Xenoposeidon* the oldest known rebbachisaurid, with the possible exception of *Maraapunisaurus* (Taylor, 2018).

The generic name *Xenoposeidon* combines the Greek "xenos" (meaning "strange" or "alien") with the name of the ancient Greek deity of earthquakes and the sea, Poseidon. The specific name *proneneukos* derives from the Latin "pronus" (meaning "forward sloping").

CLASSIFICATION
Dinosauria
 Sauropoda
  Gravisauria
   Eusauropoda
    Neosauropoda
     Diplodocoidea
      Rebbachisauridae

LOCATION
England

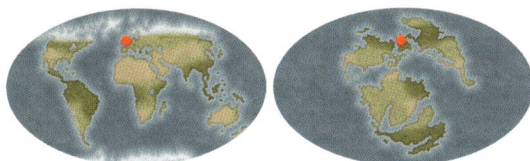

KNOWN REMAINS
Partial vertebra

4 m   8 m   12 m   16 m

4 m

A. ligabuei

(Bonaparte, 1999)
Length: 16.7 m (55 ft)
Height: 4 m (13.1 ft)
Hip height: 3.5 m (11.5 ft)
Body mass: 9,800 kg (10.8 t)
Reconstruction: ▢ ▢ ▢ ▢

| 250 | 245 | 240 | 235 | 230 | 225 | 220 | 215 | 210 | 205 | 200 | 195 | 190 | 185 | 180 | 175 | 170 | 165 | 160 | 155 | 150 | 145 | 140 | 135 | 130 | 125 | 120 | 115 | 110 | 105 | 100 | 95 | 90 | 85 | 80 | 75 | 70 | 65 |

TRIASSIC   **JURASSIC**   **CRETACEOUS**

Agustinia ligabuei (meaning "Agustin's and Ligabue's one") has the dubious distinction of having been reconstructed and illustrated very incorrectly for a long period of time.

The only known *Agustinia* specimen is very incomplete, and many of the bones are in a poor state of preservation; for instance, a partial femur found at the site was too fragile to extract adequately. These elements were recovered in 1997 and consisted of a lower hindlimb, scattered partial vertebral elements, as well as numerous elements of unclear origin.

These mysterious fragments were originally thought to be osteoderms—bony, armor-like projections that, in life, would have been embedded in the animal's skin. These elements were spike-like in appearance, and many reconstructions depicted *Agustinia* bearing these protrusions in a manner reminiscent of *Stegosaurus* or *Kentrosaurus*, with a double row of spines along the back and tail of the animal. This seemingly unique feature made *Agustinia* a sensation, standing out as distinct from any other sauropod.

Yet the interpretation of the "osteoderms" was not universally accepted, and later research called the identification of these elements into question (D'Emic et al., 2009; Mannion et al., 2013). A detailed histological study, examining the internal structure of the bones, more recently found that these skeletal elements did not match any other osteoderms known from any other variety of dinosaur. Rather than being external spikey projections, the fragments in question are most likely shards of ribs, and in one case, a sliver of pelvis (Bellardini and Cerda, 2017). With this new skeletal interpretation, a subsequent analysis placed *Agustinia* into the **Rebbachisauridae** family (Bellardini et al., 2022).

The generic name *Agustinia* honors the fossil's discoverer, Agustin Martinelli. The specific name *ligabuei* honors Ginacarlo Ligabue, who supported the expedition on which the fossils were discovered.

CLASSIFICATION
Dinosauria
 Sauropoda
  Gravisauria
   Eusauropoda
    Neosauropoda
     Diplodocoidea
      Rebbachisauridae

LOCATION
Argentina

KNOWN REMAINS
Fragments

2 m 4 m 6 m 8 m

2 m

Z. bonapartei

(Salgado et al., 2006)
Length: 10 m (33 ft)
Height: 2.4 m (7.9 ft)
Hip height: 2.2 m (7.2 ft)
Body mass: 2,300 kg (2.5 t)
Reconstruction: ☐☐☐☐

| 250 | 245 | 240 | 235 | 230 | 225 | 220 | 215 | 210 | 205 | 200 | 195 | 190 | 185 | 180 | 175 | 170 | 165 | 160 | 155 | 150 | 145 | 140 | 135 | 130 | 125 | 120 | 115 | 110 | 105 | 100 | 95 | 90 | 85 | 80 | 75 | 70 | 65 |

TRIASSIC **JURASSIC** **CRETACEOUS**

Zapalasaurus bonapartei (meaning "Bonaparte's lizard from Zapala") is known from a fragmentary skeleton, consisting primarily of a section of tail vertebrae and a poorly preserved pelvis along with a neck vertebra and hindleg fragments, which are all of poor quality. The series of tail vertebrae were not found articulated to one another; however, it is considered likely that they form a near-continuous series. The majority of the bones were excavated in 1995–1996 by teams from the Museo Argentina de Ciencias Naturales, with some remaining specimens being collected in 2004 by researchers of the Museo de Geología y Paleontología.

Zapalasaurus was unearthed from the La Amarga geological formation, specifically from a section composed of sandstone and siltstone that was deposited by a series of streams or rivers in what is known as an alluvial floodplain. A slightly older section of the formation has yielded a separate sauropod, *Amargasaurus*, showing that *Zapalasaurus* may have shared its habitat with different sauropods.

The description of *Zapalasaurus* found it to be a basal diplodocoid, being the sister clade of all of Diplodocimorpha. However, several subsequent studies have placed it in a slightly more derived position, among **Rebbachisauridae**, although it is still considered among the basalmost members of that group (Royo-Torres et al., 2020; Ren et al., 2023). Other rebbachisaurids that roamed Patagonia during the same general time span include *Limaysaurus* and *Rayososaurus*.

The generic name *Zapalasaurus* refers to the city of Zapala, which is located approximately 80 kilometers (50 miles) away from the dig site. The specific name *bonapartei* honors paleontologist José F. Bonaparte, who collected many of the fossils.

CLASSIFICATION
Dinosauria
 Sauropoda
  Gravisauria
   Eusauropoda
    Neosauropoda
     Diplodocoidea
      Rebbachisauridae

LOCATION
Argentina

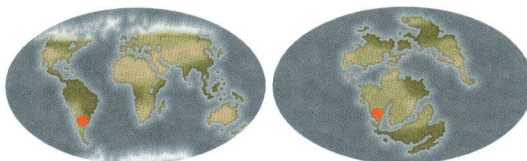

KNOWN REMAINS
Partial skeleton

LAVOCATISAURUS

L. agrioensis
(Canudo et al., 2018)
Length: 10.2 m (33.5 ft)
Height: 3 m (9.8 ft)
Hip height: 2 m (6.6 ft)
Body mass: 2,300 kg (2.5 t)
Reconstruction: ▢▢▢▢

TRIASSIC	JURASSIC	CRETACEOUS

Lavocatisaurus agrioensis (meaning "Lavocat's lizard from Agrio") is known from the partial, semi-articulated remains of one adult and two juveniles, which were discovered together at the same location. This co-occurrence potentially indicates that this was a family group—significant, as this is the first evidence of group behavior among the **rebbachisaurid** sauropods.

The fossils were recovered during a series of expeditions first launched in 2009, whose purpose was actually to discover further remains of a different rebbachisaurid, *Rayososaurus*. However, it was the bones of *Lavocatisaurus* that were discovered instead, making it the first dinosaur to be named from the Rayoso geological formation. The juvenile remains were tentatively assigned to *Zapalasaurus* (Salgado et al., 2012) before the erection of a new genus occurred in 2018.

The holotype of *Lavocatisaurus* is notable for including a nearly complete set of cranial and jaw bones. Skull material from rebbachisaurids is particularly rare, a fact that has often stymied efforts to categorize members of the group accurately. The teeth of *Lavocatisaurus* bear enamel that is highly asymmetrical, showing that this distinctive rebbachisaurid trait had appeared prior to the emergence of the group's more derived subfamilies.

It is possible that *Lavocatisaurus* possessed a keratinous beak. Although the notion of beaked sauropods is not often considered, some genera possess a combination of suggestive neurovascular openings and a horizontal edge of the maxilla, which makes the idea plausible.

The generic name *Lavocatisaurus* honors French paleontologist René Lavocat, who described *Rebbachisaurus*. The specific name *agrioensis* refers to the region of Patagonia known as Agrio del Medio.

CLASSIFICATION
Dinosauria
 Sauropoda
  Gravisauria
   Eusauropoda
    Neosauropoda
     Diplodocoidea
      Rebbachisauridae

LOCATION
Argentina

KNOWN REMAINS
Partial skull and skeleton

2 m　　4 m　　6 m　　8 m　　10 m　　12 m

4 m

D. kingi

(Averianov and Sues, 2021)

Length: 15 m (49 ft)
Height: 4 m (13.1 ft)
Hip height: 3.4 m (11.1 ft)
Body mass: 9,000 kg (9.9 t)
Reconstruction: ☐ ☐ ☐ ☐

2 m

250 245 240 235 230 225 220 215 210 205 200 195 190 185 180 175 170 165 160 155 150 145 140 135 130 125 120 115 110 105 100 95 **90** 85 80 75 70 65

TRIASSIC	JURASSIC	CRETACEOUS

Dzharatitanis kingi (meaning "King's giant from Dzharakuduk") is described on the basis of a single tail vertebra (likely the frontmost), which was unearthed in 1997. Paleontologists David J. Ward and Hans-Dieter Sues were members of the URBAC (Uzbekistan/Russian/British/American/Canadian) expedition when they found the fossil, USNM 538127, in the Bissekty Formation of Uzbekistan.

The describers of the genus conducted a phylogenetic analysis that placed *Dzharatitanis* as a member of the diplodocoid clade **Rebbachisauridae**. If accurate, this would make *Dzharatitanis* the first rebbachisaurid known from Asia and one of the latest-surviving members of the clade. The authors suggest the possibility that *Dzharatitanis* evolved from ancestral, European rebbachisaurids.

Contrastingly, a subsequent study argued that the bone was more likely to be that of a titanosaur, based both on anatomical and biogeographic evidence (Lerzo et al., 2021). A resolution of these differing conclusions has not yet been reached.

Interestingly, a second sauropod vertebra from the region has gone through similar identification interpretations: USNM 538133, the dorsal centrum of a juvenile, was first identified as being from an indeterminate titanosaur (Sues et al., 2015). But the describers of *Dzharatitanis* noted some similarities between that bone and those of *Rebbachisaurus*, suggesting the possibility that it, too, belonged to *Dzharatitanis*.

The generic name *Dzharatitanis* refers to the area Dzharakuduk in Uzbekistan; "titanis" is the feminine form of "titan," referring to the enormous mythological Greek entities. The specific name *kingi* honors geologist Christopher King for his work on the Cretaceous formations of central Asia.

CLASSIFICATION
Dinosauria
　Sauropoda
　　Gravisauria
　　　Eusauropoda
　　　　Neosauropoda
　　　　　Diplodocoidea
　　　　　　Rebbachisauridae

LOCATION
Uzbekistan

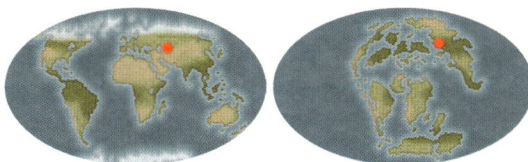

KNOWN REMAINS
Tail vertebra

R. agrioensis

(Bonaparte, 1996)

Length: 13.1 m (43 ft)
Height: 3 m (9.8 ft)
Hip height: 2.5 m (8.2 ft)
Body mass: 4,900 kg (5.4 t)
Reconstruction: ▨□□□

TRIASSIC	JURASSIC	CRETACEOUS

Rayososaurus agrioensis (meaning "lizard from the Agrio of Rayoso") is known only from a few fragments of various bones. The holotype fossils consist of the scapula and portions of the femur and tibia; these were recovered by José Bonaparte in 1991. The exact site of their discovery was not adequately recorded; it was not until 2009 that geologic studies conducted in the region were able to compare and identify the rocky matrix surrounding the bones, pinning down their location of origin (Carballido et al., 2010). Several fragmentary vertebral centra have also been assigned to the genus, although their species remains indeterminate (Medeiros and Schultz, 2004).

Despite the scant nature of the remains, it was clear to the researchers who later reexamined the fossils and redescribed the genus that *Rayososaurus* was a **rebbachisaurid**. Their analysis placed *Rayososaurus* as the sister taxon of *Cathartesaura* (Carballido et al., 2010).

Subsequent analyses have found slightly different results, placing *Rayososaurus* in a slightly more basal position within the family (Royo-Torres et al., 2020; Ren et al., 2023).

With the identification of the fossils' geographic origins, researchers were thereby able to determine details regarding the animal's habitat. It would seem that *Rayososaurus* shared its ecosystem with at least two other rebbachisaurid sauropods, *Limaysaurus* and *Nopcsaspondylus*, in addition to much larger titanosaur sauropods.

South American rebbachisaurid specimens, such as *Rayososaurus*, are important to compare against African genera, such as *Rebbachisaurus*, so that the timing of the separation of the two continents can be more accurately determined.

The animal's binomial name refers to the Rayoso geological formation, which is located near the Agrio River.

CLASSIFICATION

Dinosauria
  Sauropoda
    Eusauropoda
      Neosauropoda
        Diplodocoidea
          Rebbachisauridae
          Khebbashia

LOCATION

Argentina

KNOWN REMAINS

Fragments

2 m   4 m   6 m   8 m   10 m   12 m   14 m

4 m

2 m

L. tessonei

(Salgado et al., 2004)
Length: 15 m (49 ft)
Height: 3.8 m (12.5 ft)
Hip height: 3.2 m (10.5 ft)
Body mass: 7,500 kg (8.3 t)
Reconstruction: ▢▢▢▢▢

250 245 240 235 230 225 220 215 210 205 200 195 190 185 180 175 170 165 160 155 150 145 140 135 130 125 120 115 110 105 **98** 95 90 85 80 75 70 65

TRIASSIC	JURASSIC	CRETACEOUS

Limaysaurus tessonei (meaning "Tessone's lizard from Limay") was originally known as *Rebbachisaurus tessonei*. When the species was first described in 1995, many traits used for the specimen's identification were known to be present only in the genus *Rebbachisaurus*, and thus, this animal was placed within that genus as a separate species. This placement was despite that *Rebbachisaurus* was known from Africa, whereas this new animal was from South America, and that the two continents were likely separated by this time (Calvo and Salgado, 1995).

However, as the years passed and more closely related genera were described, it became evident that the specific features used for the identification of *R. tessonei* were actually present across an entire clade of animals. Thus, the species was reclassified into a new genus, *Limaysaurus*, after the South American rebbachisaurid *Rayososaurus* was ruled out as a possible identity.

Three specimens of the species are currently known; some were collected by J. F. Bonaparte in 1995 and 1996, while others were not unearthed until 2002. The holotype specimen (MUCPv-205) is reported to be "80% complete"; however, many of its skeletal elements have never been pictured in the scientific literature, so liberties have been taken with the skeletal diagram below.

Another series of fossils from a quarry at Cerro Aguada del León, which originated from at least three separate individuals, has been identified as the remains of *Limaysaurus*; however, it is unclear if they belong to the same exact species and, thus, have only been labeled as "*Limaysaurus* sp."

The generic name *Limaysaurus* refers to an important nearby waterway, the Rio Limay. The specific name *tessonei* honors the fossil's discoverer, Lieto Tessone.

CLASSIFICATION
Dinosauria
 Sauropoda
  Eusauropoda
   Neosauropoda
    Diplodocoidea
     Rebbachisauridae
      Khebbashia
       Limaysaurinae

LOCATION
Argentina

KNOWN REMAINS
Partial skeleton

175

2 m   4 m   6 m   8 m   10 m   12 m

4 m

2 m

C. anaerobica

(Gallina and Apesteguía, 2005)
Length: 14.8 m (49 ft)
Height: 3.2 m (10.5 ft)
Hip height: 2.6 m (8.5 ft)
Body mass: 7,000 kg (7.7 t)
Reconstruction: ☐ ☐ ☐ ☐

250 245 240 235 230 225 220 215 210 205 200 195 190 185 180 175 170 165 160 155 150 145 140 135 130 125 120 115 110 105 100 **95** 90 85 80 75 70 65

TRIASSIC   **JURASSIC**   **CRETACEOUS**

Cathartesaura anaerobica (meaning "Anaeróbicos's vulture lizard") is known from just a handful of bones that were found associated within the same quarry from 1999 through 2002; this close placement led the describers to surmise that the bones all came from the same individual. At the time of description, the fossils of the animal's ilium (pelvic bone) had yet to be completely prepared; fragments representing the ulna and dorsal vertebra were too poorly preserved to be of any diagnostic use.

Cathartesaura was unearthed from the Huincal geological formation of Argentina. *Argentinosaurus*, a titanosaur sauropod that was significantly larger than *Cathartesaura*, was discovered from the same formation, as was the abelisaurid *Skorpiovenator*, which could conceivably have preyed upon *Cathartesaura*. These dinosaurs inhabited a primarily arid environment that is thought to have hosted seasonal streams. *Cathartesaura* was among the last surviving non-titanosaur sauropods.

The original description of *Cathartesaura* placed it within **Rebbachisauridae**, an interpretation that remains the consensus. Subsequent works have divided the group into smaller clades, specifically placing *Cathartesaura* within Limaysaurinae (Mannion et al., 2019b). During excavation, *Cathartesaura* was speculated to be a titanosaur; the partially concealed vertebra and scapula seemed, at first, to be one massive vertebra.

The generic name *Cathartesaura* is derived from the scientific name of the turkey vulture, *Cathartes*, as the animals were common at the dig site; this is combined with the feminine suffix "saura." The feminine form was chosen because this makes the dinosaur's name sound just like a particular species of vulture, *Cathartes aura*. The specific name *anaerobica* acknowledges Anaeróbicos, an Argentine company that produces adhesives and provided fieldwork and lab support during the excavation.

CLASSIFICATION
Dinosauria
 Sauropoda
  Eusauropoda
   Neosauropoda
    Diplodocoidea
     Rebbachisauridae
      Khebbashia
       Limaysaurinae

LOCATION
Argentina

KNOWN REMAINS
Skeletal fragments

K. goicoecheai

(Ibiricu et al., 2013)
Length: 13 m (42 ft)
Height: 3.3 m (10.8 ft)
Hip height: 2.5 m (8.2 ft)
Body mass: 4,700 kg (5.2 t)
Reconstruction: ▦▦▢▢▢

TRIASSIC	JURASSIC	CRETACEOUS

Katepensaurus goicoecheai (meaning "Goicoecheai's hole lizard") was described on the basis of a handful of vertebral elements that were excavated in association sometime after 2005. Later, the identity of some elements that had at first been deemed "indeterminate" were deduced, and additional fossils were identified from the same location; it is believed that all of these *Katepensaurus* fossils once belonged to the same individual (Ibiricu et al., 2015).

Notably, *Katepensaurus* possessed a degree of skeletal pneumaticity never before seen in sauropods. Utilizing CT scans, researchers discovered that the transverse processes of the dorsal vertebrae (the sideways-pointing, wing-like projections on the bones) had a series of interconnected air sacs within them. While the majority of sauropods are known to have had extensive systems of air sacs throughout their bodies, this particular variety has only ever been observed in birds and some theropod dinosaurs. The reduced weight of these bones would have made the animal lighter, requiring less energy to move and generating less body heat that would need to be dissipated (Ibiricu et al., 2017).

While undoubtedly a member of **Rebbachisauridae**, the exact phylogenetic position of *Katepensaurus* is somewhat uncertain. The original description of the genus placed it within the subfamily Limaysaurinae, whereas other analyses have placed it within the subfamily Rebbachisaurinae (Fanti et al., 2015), or in a more basal position just outside Khebbashia (Royo-Torres et al., 2020).

The generic name *Katepensaurus* derives from the indigenous Tehuelche word "katepenk" (meaning "hole"), referring to one of the unique features of the genus—a distinctive opening in the transverse processes of the dorsal vertebrae. The specific name *goicoecheai* honors landowner Alejandro Goicoecheai for his long-term paleontological cooperation.

CLASSIFICATION
Dinosauria
 Sauropoda
  Eusauropoda
   Neosauropoda
    Diplodocoidea
     Rebbachisauridae
      Khebbashia

LOCATION
Argentina

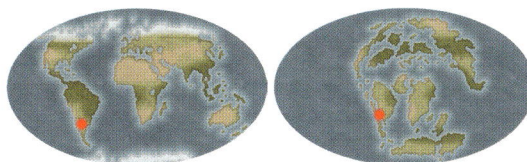

KNOWN REMAINS
Fragments

2 m   4 m   6 m   8 m

N. taqueti

(Sereno et al., 1999)
Length: 9 m (29 ft)
Height: 2.6 m (8.5 ft)
Hip height: 2.2 m (7.2 ft)
Body mass: 1,900 kg (2.1 t)
Reconstruction: ▢▢▢▢▢

2 m

250 245 240 235 230 225 220 215 210 205 200 195 190 185 180 175 170 165 160 155 150 145 140 135 130 125 120 115 110 105 100 95 90 85 80 75 70 65

TRIASSIC   **JURASSIC**   **CRETACEOUS**

Nigersaurus taqueti (meaning "Taquet's lizard from Niger") has been described by lead author Paul Sereno as "the weirdest dinosaur I've ever seen." It had a feeding apparatus unlike that of any other known sauropod, with both its upper and lower jaws forming a flat, front-facing surface, reminiscent of a wide-angle vacuum cleaner attachment. The bones of the skull were extremely thin and fragile, even more so than in most sauropods, making it something of a minor miracle that any cranial material at all survived to be fossilized (Joyce, 2007).

The front surface of these jaws was packed with a staggering number of small, slender teeth: 68 in the upper jaw, and 60 in the lower. Not only that, but numerous replacement teeth were already lined up and ready to go, in columns up to five teeth deep. All told, the animal possessed over 500 teeth at any given time. Further, the structure of the teeth was also unique among sauropods, with the enamel being 10 times thicker on the front than on the back.

(Similarly asymmetrical enamel has since been found on some of the animal's relatives, such as *Lavocatisaurus*). It seems likely that *Nigersaurus* was a low grazer, efficiently snipping short, tough vegetation with its shear-like jaws, which may have even been covered with a keratinous beak. Uniquely among tetrapods, its snout was wider than its skull.

Fragmentary fossils of *Nigersaurus* had been discovered as early as the 1960s, but the animal was poorly understood until more complete remains began to be unearthed in 1997. While the animal's skull has been heavily studied and analyzed, the rest of the known remains have by and large only received a cursory description. Reportedly, several substantial specimens are known (Sereno et al., 2007).

The generic name *Nigersaurus* refers to the African nation of Niger. The specific name *taqueti* honors Philippe Taquet, a pioneer in Nigerian paleontological expeditions during the 1960s and 1970s.

CLASSIFICATION

Dinosauria
  Sauropoda
    Eusauropoda
      Neosauropoda
        Diplodocoidea
          Rebbachisauridae
            Khebbashia
              Rebbachisaurinae

LOCATION

Niger

KNOWN REMAINS

Skull and skeleton

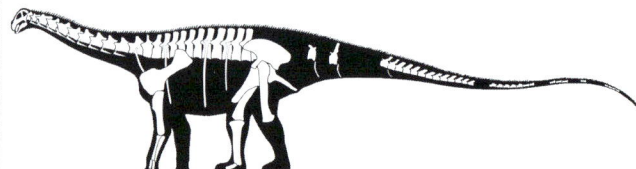

R. garasbae

(Lavocat, 1954)
Length: 26 m (85 ft)
Height: 7 m (23 ft)
Hip height: 6.2 m (20.3 ft)
Body mass: 12,000 kg (13.2 t)
Reconstruction: ☐☐☐☐

8 m

4 m

250 245 240 235 230 225 220 215 210 205 200 195 190 185 180 175 170 165 160 155 150 145 140 135 130 125 120 115 110 105 100 **97** 90 85 80 75 70 65

TRIASSIC   **JURASSIC**   **CRETACEOUS**

Rebbachisaurus garasbae (meaning "Rebbach's lizard from Gara Sbaa") is known primarily from the holotype specimen, which was unearthed from 1949 through 1952 by French paleontologist René Lavocat. He only ever described the scapula and a dorsal vertebra in any detail, although he listed a handful of other elements, and limited photographic evidence exists that was taken during the original fieldwork. However, when researchers undertook a thorough, detailed reanalysis of the *Rebbachisaurus* fossils, several could not be located in museum archives, including several tail vertebrae, ribs, and a pelvic bone (Wilson and Allain, 2015).

Many other fragmentary specimens have been ascribed to *Rebbachisaurus* at various times, but most of these remains have either been reclassified as belonging to different genera or have been deemed unidentifiable. A second species within the genus, *R. tamesnensis*, was named in 1960 by Albert-Félix de Lapparent, but it has been determined that this species is not valid as the skeletal specimen was actually composed of pieces from numerous different localities.

The distinct **rebbachisaurid** dorsal vertebrae possessed transverse processes (the sideways-jutting, wing-like portions of the bones) that would have made them "especially resistant to downwardly directed loads," presumably originating from the ribs. The purpose for such an adaptation is unknown (Wilson and Allain, 2015).

The etymology of the generic name *Rebbachisaurus* is muddled; the general locality where the holotype fossils were discovered was called, according to Lavocat, "the territory of the people of Rebbach," but whom he meant to be referencing is unclear. It is speculated that he meant to be referring to the Indigenous tribe known as the "Khebbash." The specific name *garasbae* references a more precise location within the region, Gara Sbaa, which means "Lion Hill" in Arabic.

CLASSIFICATION
Dinosauria
  Sauropoda
   Eusauropoda
    Neosauropoda
     Diplodocoidea
      Rebbachisauridae
       Khebbashia
        Rebbachisaurinae

LOCATION
Morocco

KNOWN REMAINS
Fragments

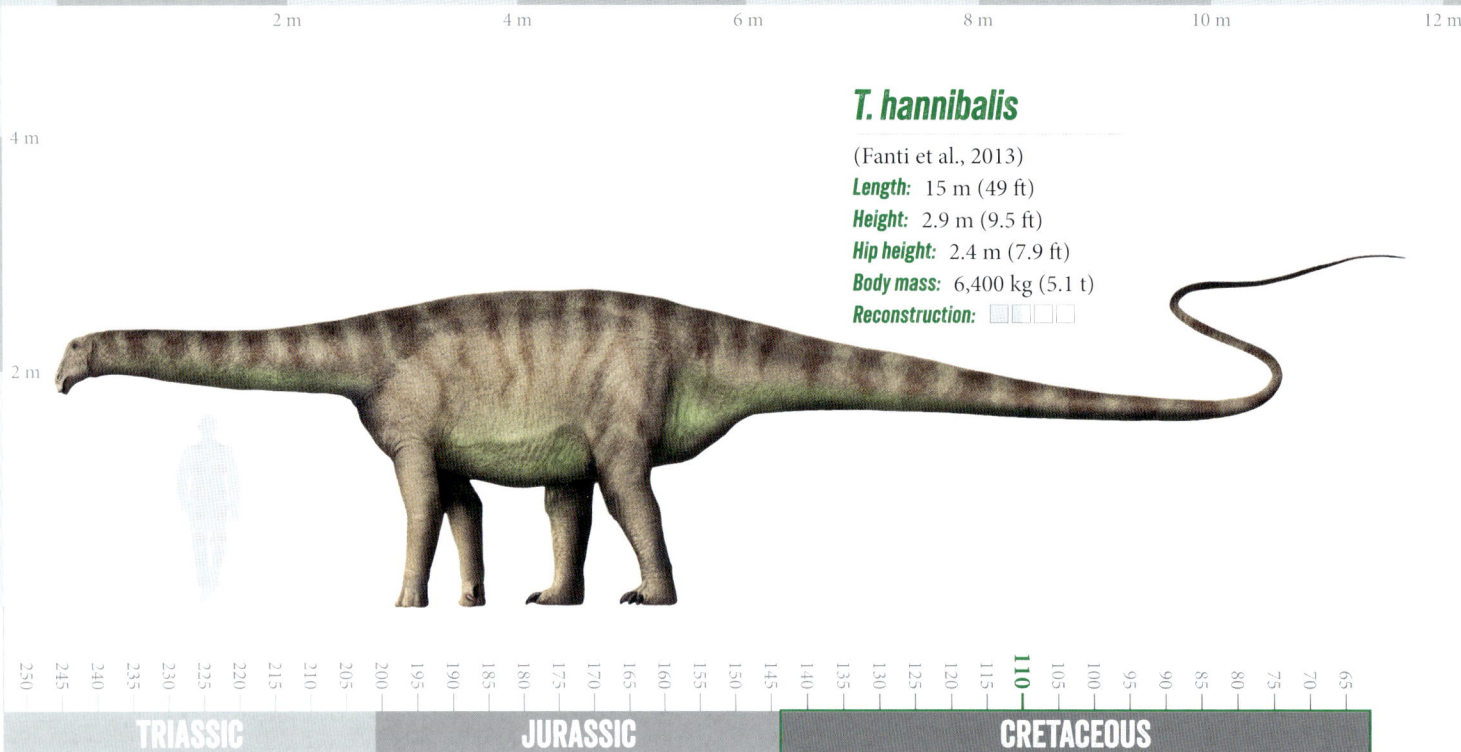

2 m   4 m   6 m   8 m   10 m   12 m

4 m

2 m

T. hannibalis

(Fanti et al., 2013)
Length: 15 m (49 ft)
Height: 2.9 m (9.5 ft)
Hip height: 2.4 m (7.9 ft)
Body mass: 6,400 kg (5.1 t)
Reconstruction: ☐☐☐☐

250 — 245 — 240 — 235 — 230 — 225 — 220 — 215 — 210 — 205 — 200 — 195 — 190 — 185 — 180 — 175 — 170 — 165 — 160 — 155 — 150 — 145 — 140 — 135 — 130 — 125 — 120 — 115 — **110** — 105 — 100 — 95 — 90 — 85 — 80 — 75 — 70 — 65

TRIASSIC	JURASSIC	CRETACEOUS

Tataouinea hannibalis (meaning "Hannibal's one from Tataouine") is notable for the high degree of pneumaticity (i.e., air-filled passageways) that is present within its skeleton, particularly within the sacral (hip) vertebrae and the ischium bone of the pelvis. In fact, *Tataouinea* is the first dinosaur to ever be discovered with an ischial pneumatic opening.

It has long been accepted that bird-like respiratory systems—complex passageways running throughout much of the body—had been present in the various dinosaurian lineages, particularly the theropods and sauropods. But details such as the degree of pneumaticity, or the evolutionary stages through which these features emerged, are often less clear. The high degree to which these features are present in *Tataouinea*, and other **rebbachisaurids**, is interesting, because an entirely different group of sauropods, the titanosaurian family Saltasauridae, seems to have

independently evolved a very similar configuration. This concurrence means that whatever advantages were presented by this sort of pneumatization were significant and likely played a key role in the longevity of these lineages.

Tataouinea is known from a single set of remains that were first found in 2011. The description of the genus focused on the pelvic region and base of the tail, but further excavations continued to unearth more tail vertebrae, leading to an even deeper analysis of the genus. *Tataouinea* was the first articulated dinosaur skeleton to be discovered from Tunisia (Fanti et al., 2015).

The generic name *Tataouinea* refers to the city of Tataouine in Tunisia. The specific name *hannibalis* references Hannibal Barca, a Carthaginian military commander (247–183 BC) who was known to utilize war elephants.

CLASSIFICATION

Dinosauria
 Sauropoda
  Eusauropoda
   Neosauropoda
    Diplodocoidea
     Rebbachisauridae
      Khebbashia
       Rebbachisaurinae

LOCATION

Tunisia

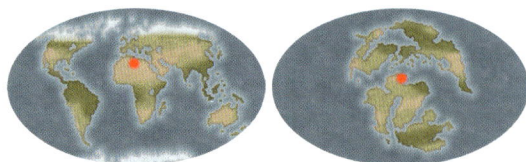

KNOWN REMAINS

Partial hip and tail

2 m   4 m   6 m   8 m   10 m

4 m

D. darwini

(Fernández-Baldor et al., 2011)
Length: 10.8 m (35.4 ft)
Height: 2.8 m (9.2 ft)
Hip height: 2.4 m (7.9 ft)
Body mass: 2,400 kg (2.6 t)
Reconstruction: ☐ ☐ ☐ ☐

250 245 240 235 230 225 220 215 210 205 200 195 190 185 180 175 170 165 160 155 150 145 140 135 130 125 120 115 110 105 100 95 90 85 80 75 70 65

TRIASSIC   **JURASSIC**   **CRETACEOUS**

Demandasaurus darwini (meaning "Darwin's lizard from Demanda") is known from a specimen unearthed from 2002 through 2004. The specimen consists primarily of vertebral and rib elements but also includes the ischia (rear hip bone), femur, and portions of the snout. The hundreds of fragments that make up the specimen are believed to have belonged to a single individual, as no repeated elements were found. As the bones were jumbled and disarticulated, it is thought that the animal's remains were buried in a heap by deposited river sediment. Evidence of damage caused by burrowing insects (likely beetles) can be seen on the specimens (Fernández-Baldor, 2012). Fusion between the vertebral bones suggests that the specimen was an adult at the time of its death.

The teeth of *Demandasaurus* are quite distinct from those of its closest relatives, being fairly pointed and slender. Unlike its cousin *Nigersaurus*, which had teeth that were routinely worn to a significant degree, *Demandasaurus* seems to have had teeth that were lacking any detectable wear patterns whatsoever. It also had significantly fewer teeth in total than *Nigersaurus*, only having about one-third as many located in the lower jaw. Additionally, its jaw was substantially rounded in shape, which contrasts with the triangular or square-shaped structures seen in other diplodocoids—and neosauropods in general.

Prior to the description of *Demandasaurus*, **rebbachisaurids** had only been definitively known from Africa and South America. The existence of a European (and therefore Laurasian) rebbachisaurid shows that there must have been a continental connection during this time capable of bearing large terrestrial vertebrates.

The generic name *Demandasaurus* refers to the Sierra de la Demanda mountain chain. The specific name *darwini* honors naturalist Charles Darwin.

CLASSIFICATION
Dinosauria
 Sauropoda
  Eusauropoda
   Neosauropoda
    Diplodocoidea
     Rebbachisauridae
      Khebbashia
       Rebbachisaurinae

LOCATION
Spain

KNOWN REMAINS
Partial skull and skeleton

1 m 2 m 3 m 4 m 5 m 6 m 7 m

2 m

1 m

I. cajapioensis

(Lindoso et al., 2019)

Length: 7.5 m (25 ft)
Height: 2 m (6.6 ft)
Hip height: 1.5 m (4.9 ft)
Body mass: 800 kg (1,700 lb)
Reconstruction: ☐☐☐☐

250 245 240 235 230 225 220 215 210 205 200 195 190 185 180 175 170 165 160 155 150 145 140 135 130 125 120 115 110 105 100 95 90 85 80 75 70 65

TRIASSIC **JURASSIC** **CRETACEOUS**

While *Itapeuasaurus cajapioensis* (meaning "lizard from Itapeua, Cajapió") is known from only a handful of fossils, it is clear from this material that the animal was a **rebbachisaurid**. It is also evident that it differs from previously described rebbachisaurids in unique ways, such as bearing a complex system of fossae on the dorsal vertebra. The phylogenetic analysis accompanying the description of the genus placed it as the sister taxon to *Demandasaurus*, a genus known from Spain.

The few bones known from *Itapeuasaurus* were particularly difficult to extract. They were first seen by a local fisherman, who spotted them protruding from a beach. More specifically, the bones were well within the intertidal area, meaning that during periods of high tide the entire area was submerged underwater. In 2015, field workers were forced to improvise, developing new techniques to extract and stabilize the bones during the

brief periods when the tide receded. A combination of waterproof foam sealant, wire, and splints were needed to finish the job.

All told, a handful of vertebral elements were assigned as the holotype of the genus, while additional vertebral elements and an incomplete ischium were designated as a paratype specimen. While the bones of the holotype have been figured, the bones of the paratype have not been shown or described in any detail. A preliminary report about the excavation mentioned the purported presence of ribs, but none were named in the final description of *Itapeuasaurus* (Medeiros et al., 2015). Many unofficial news sources mention a humerus being present among the fossil material, but no mention of any limb bones is made in the publication that described Itapeuasaurus.

The sauropod's binomial name refers to the Itapeua beach, which is located in the Cajapió municipality of Brazil.

CLASSIFICATION

Dinosauria
 Sauropoda
  Eusauropoda
   Neosauropoda
    Diplodocoidea
     Rebbachisauridae
      Khebbashia
       Rebbachisaurinae

LOCATION

Brazil

KNOWN REMAINS

Fragments

C. windhauseni

(Carballido et al., 2012)
Length: 14 m (46 ft)
Height: 2.8 m (9.2 ft)
Reconstruction: ☐☐☐☐

| TRIASSIC | JURASSIC | CRETACEOUS |

110

Comahuesaurus windhauseni (meaning "Windhausen's lizard from Comahue") hails from the Lohan Cura geological formation of Argentina, which means it is approximately 117 to 100 million years old. The fragmentary remains of at least three individuals were discovered prior to 2004 as part of a single bone bed, likely created by the flow of a river. The remains were originally attributed to *Limaysaurus* (Salgado et al., 2004) before being assigned to a new genus.

The describers placed *Comahuesaurus* as a fairly basal member of **Rebbachisauridae**, an interpretation upheld by later studies (Mannion et al., 2019b; Ren et al., 2020). The generic name *Comahuesaurus* refers to the region Comahue (a Mapuche term meaning "place of abundance"), located in northern Patagonia. The specific name *windhauseni* honors geologist Anselmo Windhausen.

A. maranhensis

(Carvalho et al., 2003)
Length: 10.5 m (34 ft)
Height: 2.5 m (8.2 ft)
Reconstruction: ☐☐☐☐

| TRIASSIC | JURASSIC | CRETACEOUS |

120

Amazonsaurus maranhensis (meaning "Amazon lizard from Maranhão") is known only from a partial set of poorly preserved, disarticulated bones composed mainly of pelvic fragments and fragmentary vertebrae of the tail and back. This scant material has been sufficient to confidently place *Amazonsaurus* as a member of Diplodocidae but not to pin down a more detailed placement within the group. One interpretation put forth by the describers is that it is a late-surviving, basally placed diplodocoid. Other papers suggest a basal **rebbachisaurid** placement (Bellardini et al., 2022).

The binomial name refers to the Amazon region of Brazil, specifically within the Brazilian state of Maranhão. The fossils were unearthed from the Itapecuru geological formation, which historically has yielded only scant dinosaurian remains, with *Amazonsaurus* being the most substantial.

H. boscarollii

(Dalla Vecchia, 1998)
Length: 11.5 m (38 ft)
Height: 3 m (9.8 ft)
Reconstruction: ☐☐☐☐

1 m

130

TRIASSIC	JURASSIC	CRETACEOUS

Histriasaurus boscarollii (meaning "Boscarolli's lizard from Istria") is known from a single dorsal (back) vertebra that was discovered in Croatia during the 1980s. Numerous other fragments of sauropod bones have been unearthed from the same region, but all have proven too fragmentary to identify with any certainty. The sites in question have been exposed to marine conditions, which in many cases has led to the damage or degradation of the remains and the limestone layers from which they originate.

Some subsequent works considered the genus *nomen dubium*, but recent analyses continue to include the genus in phylogenetic studies; these consistently find *Histriasaurus* to be a **rebbachisaurid**. *Histriasaurus* is among the oldest known rebbachisaurids.

The generic name *Histriasaurus* refers to the Istria Peninsula, as "Histria" is the Latin name for the region. The specific name *boscarollii* honors Darío Boscarolli, who first discovered the site of the fossils.

N. alarconensis

(Apesteguía, 2007)
Length: 11.5 m (38 ft)
Height: 3 m (9.8 ft)
Reconstruction: ☐☐☐☐

1 m

95

TRIASSIC	JURASSIC	CRETACEOUS

Nopcsaspondylus alarconensis (Nopcsa's spine from Alarcón) is known only from illustrations of a single vertebral bone, which was discovered in 1902 in Argentina. It had first been referred to the genus *Bothriospondylus* by paleontologist Franz Nopcsa, but this taxon is now considered to be *nomen dubium*—not to mention that the

first fossils attributed to *Bothriospondylus* were found in England, not South America. At some point, the fossil specimen was lost. Based on its surviving depictions, it seems to have possessed traits that are now known to be indicative of **Rebbachisauridae** (Royo-Torres et al., 2020).

S. marae

(Lerzo et al., 2024)
Length: 21 m (69 ft)
Height: 6 m (19.7 ft)
Reconstruction: ☐☐☐☐

95

3 m

Sidersaura marae (meaning "Mara's star lizard") is one of the largest rebbachisaurids known to date. The four partial specimens that were used to describe the genus were first found in 2012. *Sidersaura* had a proportionately long tail and cranial bones that were unexpectedly robust. It is also the only known **rebbachisaurid** to have a certain opening in its skull known as a frontoparietal foramen.

The generic name *Sidersaura* derives from the Latin "sideris" (meaning "star"), referring to the shape of some of the tail chevrons. The specific name *marae* honors fossil preparator Mara Ripoll.

Tapuiasaurus macedoi

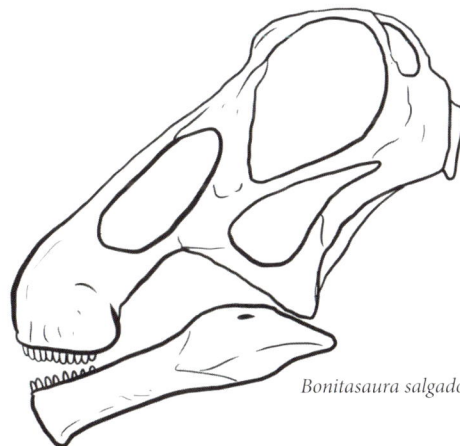

Bonitasaura salgadoi

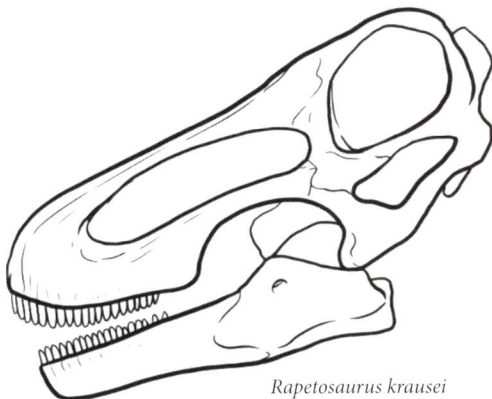

Rapetosaurus krausei

Nemegtosaurus mongoliensis

185

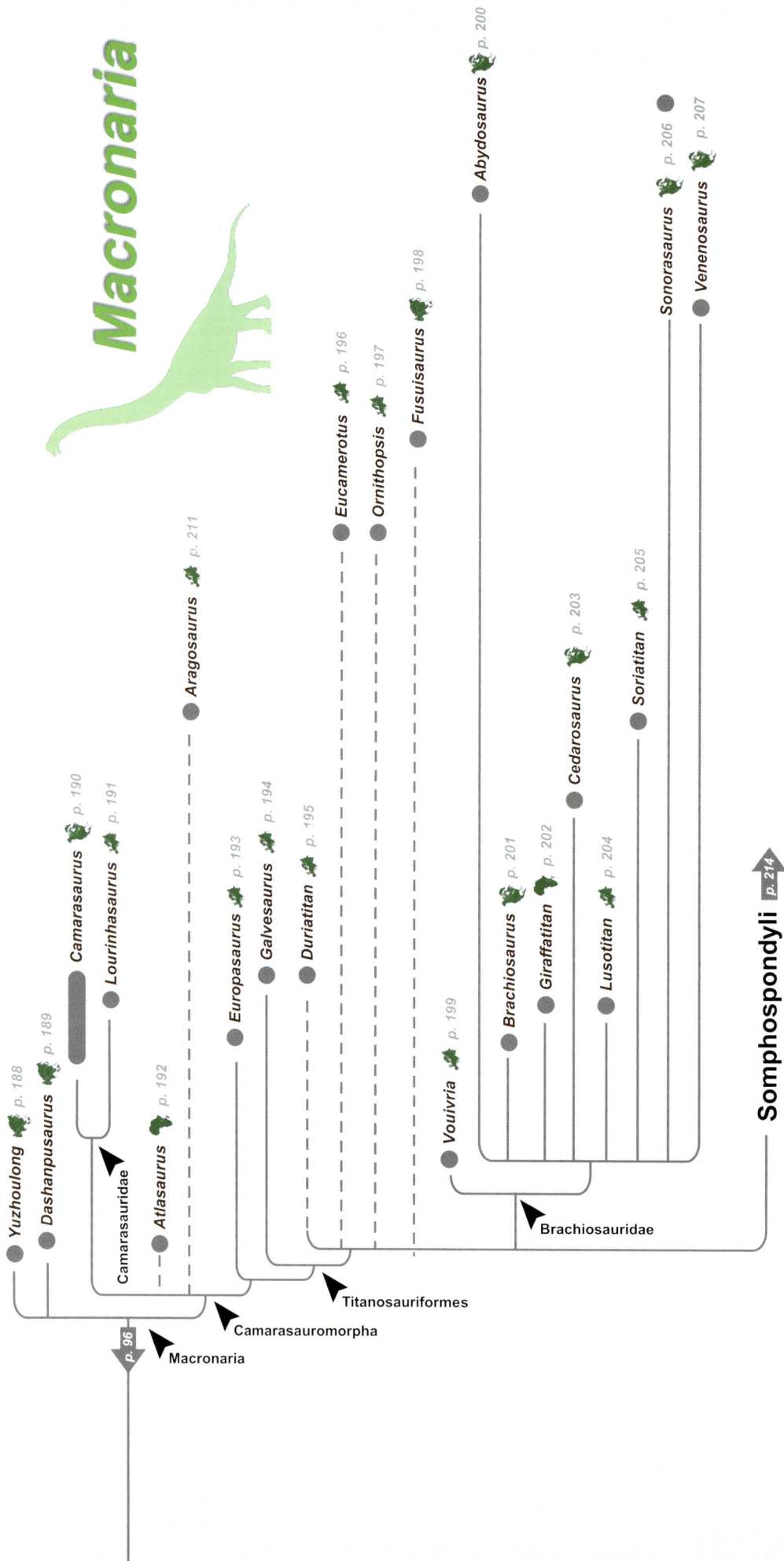

The macronarians derive their name (meaning "large nose") from the expanded nasal opening present in their skulls, sometimes being larger than the orbital opening that housed the eye. Many species of macronarians were colossal in size.

The describers of *Yuzhoulong* place it as the basalmost macronarian (Dai et al., 2022). The reexamination of *Dashanpusaurus* found a similar placement for that genus (Ren et al., 2023). Both of these analyses then find *Camarasaurus* in the next most derived position, and modern analyses that do not include the previous two genera reliably find *Camarasaurus* in this position (Moore et al., 2020; An et al., 2023), although it should be noted that some analyses place it outside Neosauropoda altogether (Mannion et al., 2019). When the often-overlooked *Lourinhasaurus* is included in datasets, it tends to be shown as the sister taxon of *Camarasaurus* (Mocho et al., 2014; Royo-Torres et al., 2020). (Note that *Bellusaurus* is also commonly shown in a similar position, but recent work [Moore et al., 2023] instead suggests it to be a juvenile form of a mamenchisaurid.)

The affinities of *Atlasaurus* are ambiguous; Mannion et al. (2017) and Ren et al. (2023) place it outside Macronaria, Moore et al. (2020) places it as a basal macronarian, and Royo-Torres et al. (2020) places it as a brachiosaur. The affinities of *Aragosaurus* are also ill defined, but it might be a basal macronarian (Mannion et al., 2019; Moore et al., 2020). Although their exact placement varies between studies, *Europasaurus* and *Galvesaurus* are often shown in the next most derived positions (Ren et al., 2023). *Tehuelchesaurus* is also often calculated to be in a similar position, but other works have instead found it to be within Turiasauria (Mannion et al., 2019).

The enigmatic taxa *Duriatitan* (Mannion et al., 2013), *Fusuisaurus* (Mannion et al., 2019), *Eucamerotus*, and *Ornithopsis* (Upchurch et al., 2011) have all been suggested to have **titanosauriform** affinities, but their limited remains render their positions uncertain.

Among **Brachiosauridae**, *Vouivria* has been placed as among the first of the clade (Royo-Torres et al., 2020).

Within the family, the relationships among the genera are often imprecise, but the group is generally considered to include *Abydosaurus*, *Brachiosaurus*, *Giraffatitan*, *Cedarosaurus*, *Lusotitan*, *Sonorasaurus*, *Soriatitan*, and *Venenosaurus* (Pérez-Pueyo et al., 2019; Carballido et al., 2020; Moore et al., 2020; Royo-Torres et al., 2020; Dai et al., 2022).

Brachiosaurus altithorax

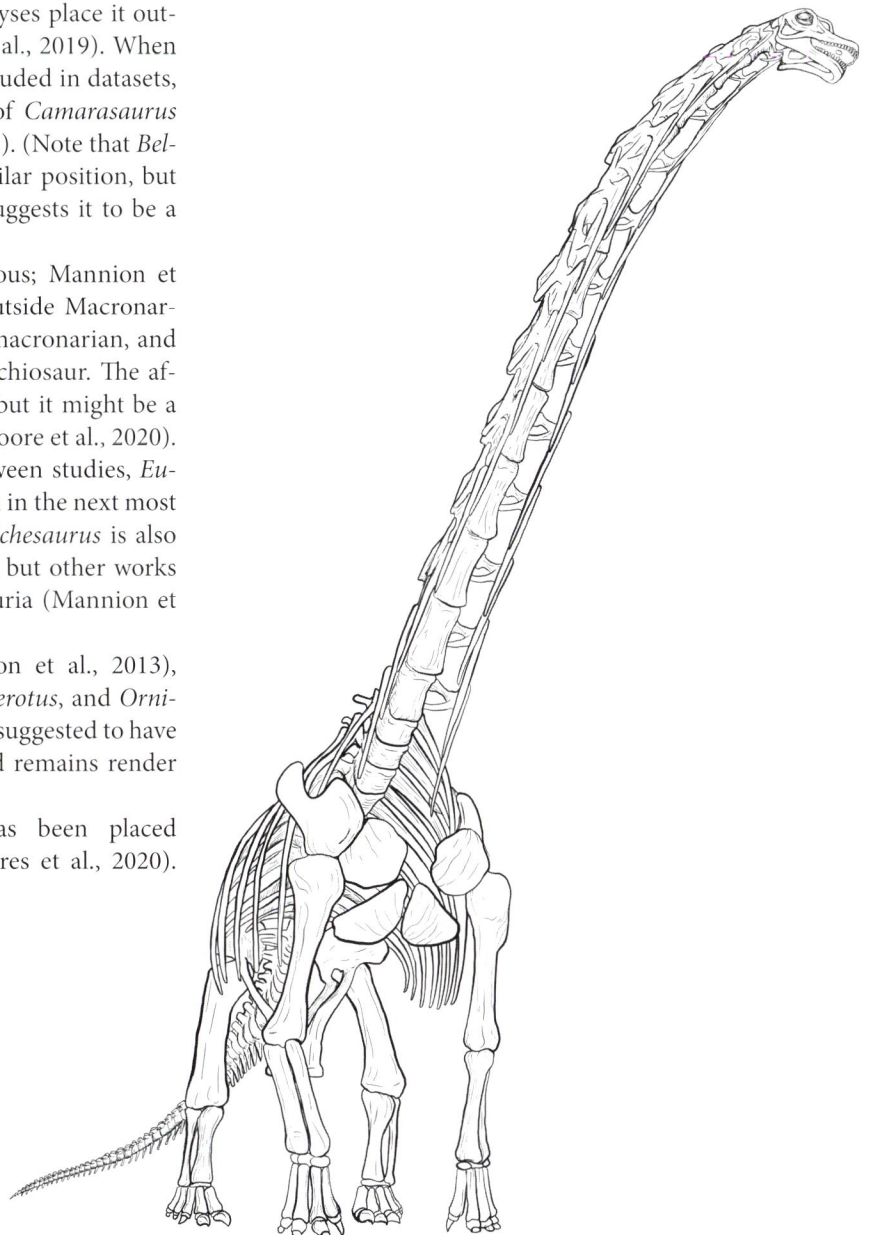

2 m   4 m   6 m   8 m   10 m

Y. qurenensis

(Dai et al., 2022)
Length: 9.5 m (31 ft)
Height: 5.6 m (18.3 ft)
Hip height: 2.2 m (7.2 ft)
Body mass: 3,000 kg (3.3 t)
Reconstruction: ☐☐☐☐

| TRIASSIC | JURASSIC | CRETACEOUS |

Yuzhoulong qurenensis (meaning "dragon from Quren, Yuzhou") is, according to its describers, perhaps the basalmost known **macronarian**. The significance of this fact relates directly to when, exactly, the macronarians dispersed and diversified. The presence of *Yuzhoulong* in China, in the Middle Jurassic, is strong evidence that the group was already well established by this point, meaning that their development and radiation would have occurred during the Early Jurassic—earlier than many popular hypotheses would suggest.

In particular, the existence of *Yuzhoulong* is detrimental to the widely held East Asia Isolation Hypothesis, which posits that the region was geographically isolated at this time due to the presence of a shallow seaway. This hypothesis would explain why, by and large, sauropods of the region belong to different families than those of, say, Europe. But discoveries

of other sauropod lineages present in China, such as macronarians, muddies this neat and tidy picture.

At the time of *Yuzhoulong*'s description, only some of its known skeletal elements had been completely prepared; many remained still buried in the quarry, intermingled with the bones of a different, as yet unidentified, sauropod specimen. This fossiliferous location, which is a component of the famous Shaximiao geological formation, was first discovered in 2016. The individual is thought to have been "submature" at the time of its death, meaning that the size of a fully grown adult is currently unknown.

The generic name *Yuzhoulong* derives from "Yuzhou," which is the ancient name of the city of Chongqing; "long" means "dragon" in Chinese Pinyin. The specific name *qurenensis* derives from "Quren," which is the ancient name of Yunyang County.

CLASSIFICATION
Dinosauria
 Sauropoda
  Gravisauria
   Eusauropoda
    Neosauropoda
     Macronaria

LOCATION
China

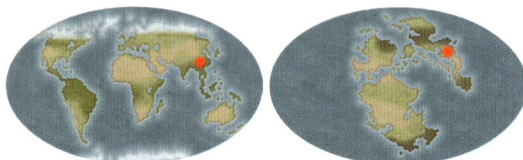

KNOWN REMAINS
Partial skull and skeleton

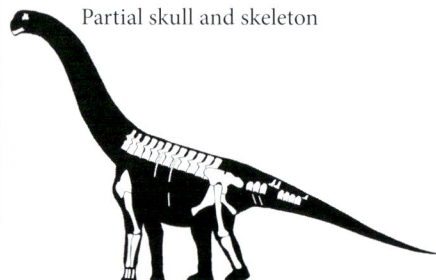

2 m   4 m   6 m   8 m   10 m   12 m

6 m

D. dongi

(Peng et al., 2005)
Length: 13.8 m (45 ft)
Height: 7.5 m (24.6 ft)
Hip height: 3.7 m (12.1 ft)
Body mass: 5,200 kg (5.7 t)
Reconstruction: ☐☐☐☐

4 m

2 m

250 245 240 235 230 225 220 215 210 205 200 195 190 185 180 175 170 **165** 160 155 150 145 140 135 130 125 120 115 110 105 100 95 90 85 80 75 70 65

TRIASSIC   **JURASSIC**   **CRETACEOUS**

Dashanpusaurus dongi (meaning "Dong's lizard from Dashanpu") is a Chinese sauropod that has only recently received detailed analysis in an English-language publication (Ren et al., 2022).

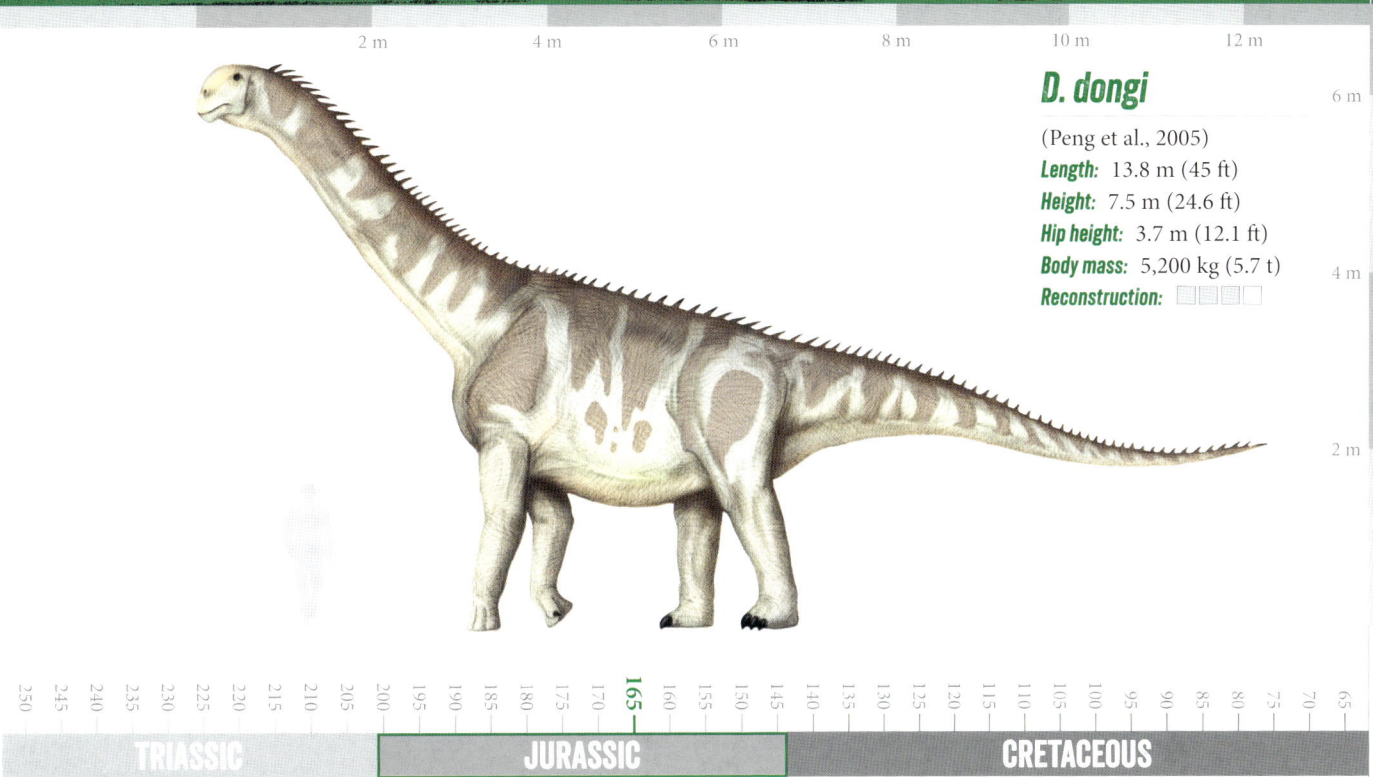

Dashanpusaurus is currently known from two partial skeletons, unearthed from the lower Shaximiao geological formation (although the possibility of additional remains has been briefly mentioned). Altogether, the known skeleton is mainly just lacking the animal's feet and skull. During the Middle Jurassic, the Shaximiao region held a lush forest, as well as a large, river-fed lake. Numerous dinosaurian discoveries of substantial nature have been made at this fossil-rich hotspot.

Many of the largest and most famous of sauropods belonged to the group Neosauropoda, which dominated the Late Jurassic and Cretaceous. Despite this, the origins of the clade have remained somewhat uncertain.

A recent study, though, has found *Dashanpusaurus* to be the earliest known neosauropod, on the sub-branch **Macronaria**. This not only shows that the two main divisions of Neosauropoda (Macronaria and Diplodocoidea) diverged earlier than was previously thought (during the early Middle Jurassic, at least) but also that the neosauropods had achieved a global dispersion before the final breakup of Pangaea (since its closest relatives would have been *Atlasaurus* of India and *Camarasaurus* of North America) (Ren et al., 2023).

The generic name *Dashanpusaurus* references the town of Dashanpu, China. The specific name *dongi* honors paleontologist Dong Zhiming, who has researched the Shaximiao formation extensively.

CLASSIFICATION
Dinosauria
  Sauropoda
    Gravisauria
      Eusauropoda
        Neosauropoda
        Macronaria

LOCATION
China

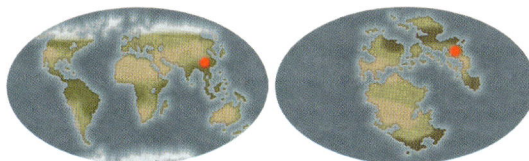

KNOWN REMAINS
Partial skeleton

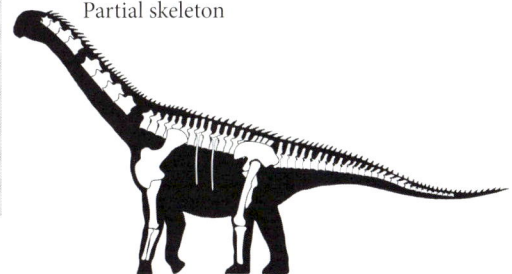

MACRONARIA

4 m   8 m   12 m   16 m   20 m

8 m

4 m

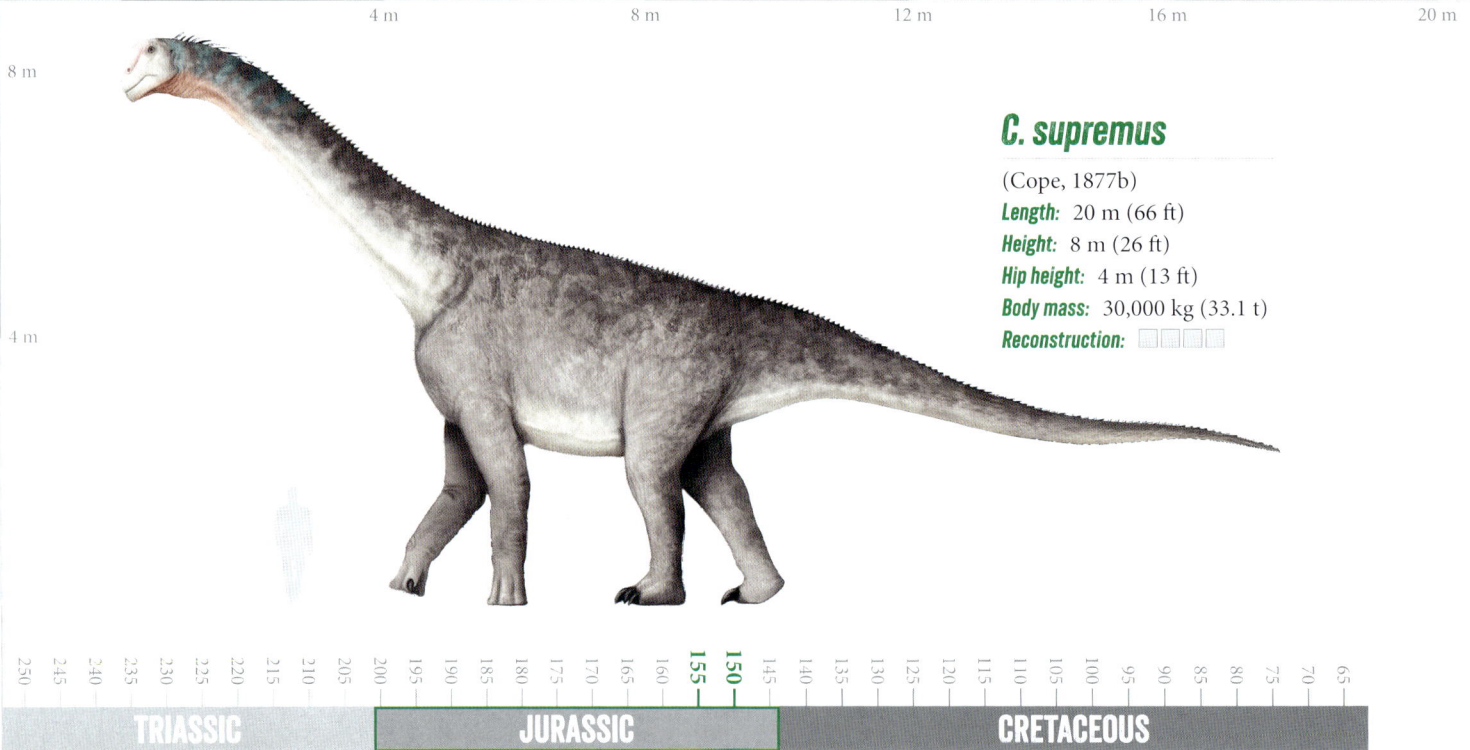

C. supremus

(Cope, 1877b)
Length: 20 m (66 ft)
Height: 8 m (26 ft)
Hip height: 4 m (13 ft)
Body mass: 30,000 kg (33.1 t)
Reconstruction: ☐☐☐☐

| 250 | 245 | 240 | 235 | 230 | 225 | 220 | 215 | 210 | 205 | 200 | 195 | 190 | 185 | 180 | 175 | 170 | 165 | 160 | **155** | **150** | 145 | 140 | 135 | 130 | 125 | 120 | 115 | 110 | 105 | 100 | 95 | 90 | 85 | 80 | 75 | 70 | 65 |

TRIASSIC   **JURASSIC**   **CRETACEOUS**

Camarasaurus supremus (meaning "supreme chambered lizard") is among the most common of sauropods in the fossil record. The famous Morrison Formation of the western United States has produced approximately 50 partial skeletons of the genus (Woodruff and Foster, 2017). Numerous skulls of *Camarasaurus* are also known, which allows for comparative analyses—a rare opportunity among sauropods. It has been found that *Camarasaurus* skull proportions do not necessarily scale with body proportions; whether this is due to sexual dimorphism or merely a high degree of variation between individuals remains an open question (Woodruff et al., 2021).

Camarasaurus was named by Edward Cope, of the infamous "Bone Wars" of the late nineteenth century. His rival, Othniel Marsh, named his own genus of sauropod, *Morosaurus*, but this later turned out to be synonymous with *Camarasaurus*. In the last century and a half, numerous

changes in specimen identification have been made. Of the four commonly recognized species within the genus, *C. grandis* and *C. lentus* are the most common, with *C. lewisi* and *C. supremus* being less well represented. *Camarasaurus supremus* was the largest of these and also the latest surviving.

Compared with the diplodocid sauropods who shared its habitat, *Camarasaurus* had teeth that were stouter and flatter, suggesting that it subsisted on sturdier vegetation. Its skull was also distinctively blunt compared with the elongated skull of *Diplodocus*.

The generic name *Camarasaurus* combines the Greek "kamara" (meaning "chamber") and "sauros" (meaning "lizard"), referring to the hollow chambers that are present in its cervical (neck) vertebrae. The specific name *supremus* is Latin for "supreme," while *grandis* is Latin for "large," *lentus* is Latin for "tough," and *lewisi* honors fossil preparator Arnold D. Lewis.

CLASSIFICATION
Dinosauria
 Sauropoda
  Gravisauria
   Eusauropoda
    Neosauropoda
     Macronaria
      Camarasauridae

LOCATION
United States

KNOWN REMAINS
Complete

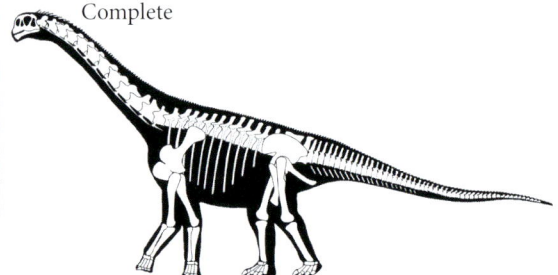

4 m   8 m   12 m   16 m   20 m

L. alenquerensis

8 m

(Dantas et al., 1998)
Length: 17.5 m (87 ft)
Height: 8.4 m (27.5 ft)
Hip height: 4 m (13.1 ft)
Body mass: 19,000 kg (20.9 t)
Reconstruction:

4 m

250 245 240 235 230 225 220 215 210 205 200 195 190 185 180 175 170 165 160 155 **152** 145 140 135 130 125 120 115 110 105 100 95 90 85 80 75 70 65

| TRIASSIC | JURASSIC | CRETACEOUS |

Lourinhasaurus alenquerensis (meaning "lizard from Alenquer and Lourinhã") has been known by many names. The first bones attributed to the animal were found in 1949 and were subsequently named to a new species of *Apatosaurus* in 1957 by Albert-Félix de Lapparent. In 1970, the species was reassigned to the genus *Atlantosaurus*; in 1978, to *Brontosaurus*; and in 1990, to *Camarasaurus*.

Originally, several partial specimens from the same strata were considered to be syntypes of the species—a holotype was never designated. In 2003, researchers chose the remains of one particular individual to represent *Lourinhasaurus*. Since many of the various bones were assigned differing catalog numbers, the collection is considered a lectotype (Antunes and Mateus, 2003).

The taxonomic affinities of *Lourinhasaurus* have proven difficult to pin down, with hypotheses ranging from a basal position within Eusauropoda all the way to Diplodocidae.

Most recently, a dedicated study found that differing computational methods still arrived at the same conclusion, namely that *Lourinhasaurus* was most closely related to *Camarasaurus*. Thus, the two were placed in the same family, **Camarasauridae**, a group that was first proposed in 1877 but whose use had fallen out of favor since it had apparently seemed to contain only *Camarasaurus* (Mocho et al., 2014).

Compared with the exclusively North American *Camarasaurus*, the Portuguese *Lourinhasaurus* had proportionally longer forelimbs.

The generic name *Lourinhasaurus* refers to the town of Lourinhã, where one of the original specimens was discovered, although those fossils (ML 414) are now considered to belong to *Dinheirosaurus*. The specific name *alenquerensis* refers to the town of Alenquer, near where the series of lectotype fossils was discovered.

CLASSIFICATION
Dinosauria
 Sauropoda
  Gravisauria
   Eusauropoda
    Neosauropoda
     Macronaria
      Camarasauridae

LOCATION
Portugal

KNOWN REMAINS
Partial skeleton

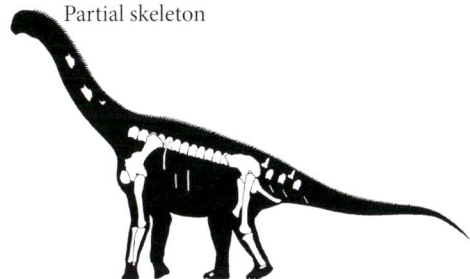

191

4 m   8 m   12 m   16 m

8 m

4 m

A. imelakei

(Monbaron et al., 1999)
Length: 15.5 m (27.9 ft)
Height: 8.5 m (28 ft)
Hip height: 3.5 m (11.5 ft)
Body mass: 21,000 kg (23.1 t)
Reconstruction: ▢▢▢▢

250 245 240 235 230 225 220 215 210 205 200 195 190 185 180 175 170 **165** 160 155 150 145 140 135 130 125 120 115 110 105 100 95 90 85 80 75 70 65

TRIASSIC　　**JURASSIC**　　**CRETACEOUS**

Atlasaurus imelakei (meaning "giant lizard from Atlas") is an understudied sauropod. The nearly complete specimen housed in the Musée des Sciences de la Terre de Rabat was unearthed from strata that, during the animal's lifetime, would have been a large, river-strewn, seaside plain. The individual seems to have been buried during a flood, judging partially by the high volume of fossilized plant debris and the completeness of the specimen. Notably, the specimen includes a partial jaw and skull. Fused vertebral elements indicate that the animal was nearly fully grown.

A second specimen, consisting of a nearly complete tail, underwent preparation in a facility in Utah, before being put on display in Mexico City, and then was eventually auctioned off to a private collector (Reuters, 2018).

The original paper that first described *Atlasaurus* was barely five pages in length and only figured a small number of the specimen's bones. More than two decades later, there has yet to be a more detailed description of the specimen, despite its tantalizing completeness.

Owing to this surprising lack of detail, scoring the species for inclusion in phylogenetic studies has been challenging, and subsequent results have been inconclusive. Its describers hypothesized that *Atlasaurus* was a brachiosaurid, but some subsequent analyses found it to be much more basally placed, outside Neosauropoda entirely (D'Emic et al., 2016; Mannion et al., 2017). Conversely, some of the most recent studies have once again found it to be included within Brachiosauridae (Mannion et al., 2019; Royo-Torres et al., 2020).

In comparison to *Brachiosaurus*, *Atlasaurus* is approximately 15 million years older. Proportionately, its limbs and tail are longer (very much so, in the case of the limbs), its neck is shorter, and its skull is larger.

The generic name *Atlasaurus* refers to the Atlas Mountains of Morocco. The specific name *imelakei* derives from the Arabic "imelake" (meaning "giant").

CLASSIFICATION
Dinosauria
　Sauropoda
　　Gravisauria
　　　Eusauropoda
　　　　Neosauropoda
　　　　　Macronaria
　　　　　　Camarasauromorpha

LOCATION
Morocco

KNOWN REMAINS
Nearly complete

E. holgeri

(Sander et al., 2006)
Length: 6 m (20 ft)
Height: 4.6 m (15.1 ft)
Hip height: 1.6 m (5.2 ft)
Body mass: 800 kg (1,700 lb)
Reconstruction: ▢▢▢▢

TRIASSIC **JURASSIC** **CRETACEOUS**

Europasaurus holgeri (meaning "Holger's lizard from Europe") is a very small sauropod thought to be the result of a phenomenon known as insular dwarfism, also termed island dwarfism. This dwarfism results when the body size of a species shrinks when the population is limited to a small habitat with limited resources. During this time period, Europe consisted primarily of a series of isolated, small landmasses.

A single sauropod tooth was discovered in 1998 in an extensively excavated chalk quarry. Although fossils of aquatic life are common in these rocks, as they were originally laid down about 30 meters beneath the surface of a shallow sea, fossils of terrestrial animals have been rare. Soon after 1998, years of more dedicated effort ensued as more sauropod remains were prospected. The task was difficult, though, as the fossil-bearing layers were nearly vertical, as a result of Early Cretaceous mountain-building events. Most of

the remains were hauled away in large chunks of stone that had already been subject to disruptive blasting efforts; thus, there is limited information about exactly where each bone originated. Based on repetitive skeletal elements, researchers determined more than 21 individuals contributed to these fossils (Carballido et al., 2020). It seems plausible that a herd of the animals had been traversing a tidal zone when they drowned; about one-third of the bones feature tooth marks from fish or crocodyliforms (Wings, 2015).

While *Europasaurus* was undoubtedly a basal **macronarian** of some description, it remains unclear whether it fell within Camarasauridae or Brachiosauridae or if it was outside both clades (Carballido et al., 2020).

The generic name *Europasaurus* refers to Europe, as the fossils represent the very first sauropod skull ever to be discovered on the continent. The specific name *holgeri* honors Holger Lüdtke, who discovered the holotype fossils.

CLASSIFICATION
Dinosauria
 Sauropoda
  Gravisauria
   Eusauropoda
    Neosauropoda
     Macronaria
      Camarasauromorpha

LOCATION
Germany

KNOWN REMAINS
Nearly complete

12 m 4 m 8 m 12 m 16 m 20 m

8 m

4 m

G. herreroi

(Barco et al., 2005)
Length: 18 m (59 ft)
Height: 11 m (36 ft)
Hip height: 5 m (16.4 ft)
Body mass: 14,000 kg (15.4 t)
Reconstruction: ☐☐☐☐

250 245 240 235 230 225 220 215 210 205 200 195 190 185 180 175 170 165 160 155 **150** 145 140 135 130 125 120 115 110 105 100 95 90 85 80 75 70 65
TRIASSIC	JURASSIC	CRETACEOUS

Galvesaurus herreroi (meaning "Herrero's lizard from Galve") is a relatively gracile sauropod known only from fragmentary remains.

The fossil fragments of *Galvesaurus* were first noticed by an amateur fossil collector in the early 1980s. Over the next decade, many individual bones were excavated and prepared by an amateur team; many had eroded away from a rock layer of higher elevation. Professional excavations to explore this layer, and others, began in 1993. Since that time, the locality has become something of a dinosaur hot spot, with dozens of fossiliferous sites being studied over the years.

In particular, the work that began with these sauropod remains helped clue scientists into the fact that not all of the nearby fossil layers were the exact same age, which had been the assumption previous to 1994. The specific stratum that bore *Galvesaurus* was deposited by a coastal climate during the Late Jurassic.

There is controversy surrounding the animal's name and its scientific description. Two separate descriptive studies were published (unbeknownst to each other) at almost the same time, and both were based on the exact same fossils. Researcher Bárbara Sánchez Hernandez named the specimen "*Galveosaurus*" (note the extra letter), and this alternate name has also been widely used.

Originally thought to be a type of turiasaur, more recent work has shown *Galvesaurus* to be among the basalmost **titanosauriforms** (Pérez-Pueyo et al., 2019).

The generic name *Galvesaurus* refers to the town of Galve, Spain. The specific name *herreroi* honors the fossil's discoverer, José María Herrero.

CLASSIFICATION
Dinosauria
 Sauropoda
  Gravisauria
   Eusauropoda
    Neosauropoda
     Macronaria
      Titanosauriformes

LOCATION
Spain

KNOWN REMAINS
Fragments

D. humerocristatus

(Barrett et al., 2010)
Length: 15 m (48 ft)
Height: 9 m (29.5 ft)
Hip height: 3 m (9.8 ft)
Body mass: 10,000 kg (11 t)
Reconstruction: ☐☐☐☐

Duriatitan humerocristatus (meaning "titan from Duria with crested humerus") is a poorly understood European neosauropod.

The holotype specimen, BMNH 44635, was found near the city of Dorset, in the United Kingdom. It consisted of a single, partially damaged left humerus. The fossil was first described in 1874 by John Hulke as belonging to a species of *Cetiosaurus* and was variously reassigned to several sauropod genera over the decades. Finally, in 2010, it received the distinction of being placed within its own, new genus, *Duriatitan*.

In 2016, a second humerus (MG 4976), this time from Portugal, was attributed to *Duriatitan*. The specimen had been a part of the archives in the Museu Geológico in the city of Lisboa. According to this interpretation, the islands that are today the Iberian Peninsula and the United Kingdom must have been connected sometime around the Late Jurassic, in order for the species to be present in both locations (Mocho et al., 2016).

A third humerus unearthed in 2018 (MDS-VPCR 214) from Spain was additionally identified as *Duriatitan*, although its age was slightly different from the other specimens (Fernández-Baldor et al., 2020).

Naturally, the limited skeletal material precludes *Duriatitan* from being categorized with much precision. It is generally treated as a basal **titanosauriform**, perhaps even a brachiosaur (Mannion et al., 2013).

The generic name *Duriatitan* derives from the Latin name for the city of Dorset, "Duria," and the Greek "titan." The specific name *humerocristatus* is Latin for "crested humerus," referencing the distinctive trait primarily used to identify the genus.

CLASSIFICATION

Dinosauria
 Sauropoda
  Gravisauria
   Eusauropoda
    Neosauropoda
     Macronaria
      Titanosauriformes

LOCATION

United Kingdom, Portugal, Spain

KNOWN REMAINS

Humerus

195

MACRONARIA

4 m   8 m   12 m   16 m

8 m

4 m

E. foxi

(Blows, 1995)
Length: 14 m (45 ft)
Height: 8.5 m (27.9 ft)
Hip height: 3.5 m (11.5 ft)
Body mass: 8,500 kg (9.4 t)
Reconstruction: ☐ ☐ ☐ ☐

250 245 240 235 230 225 220 215 210 205 200 195 190 185 180 175 170 165 160 155 150 145 140 135 130 **125** 120 115 110 105 100 95 90 85 80 75 70 65

TRIASSIC   **JURASSIC**   **CRETACEOUS**

Eucamerotus foxi (meaning "Fox's well-chambered one") is a contentious taxon.

In 1872, John Hulke created the genus *Eucamerotus* based on a single vertebral neural arch, although he failed to provide a name for the species. But in 1879, he decided that the genus was synonymous with the sauropod *Ornithopsis*, which had been named in 1870 and thus had naming priority. Later authors in the twentieth century tended to synonymize both genera with *Pelorosaurus*.

However, a review in 1995 determined that the *Eucamerotus* holotype fragment was different enough from *Ornithopsis* to be considered valid. The researchers erected a species to go along with the genus and assigned several other dorsal vertebrae as paratype specimens (shown below). Several additional dorsal vertebrae were also designated as "referred specimens" (but since their positions in the vertebral column are less certain, they mostly have not been figured in the literature and are not shown below).

Subsequent papers questioned the validity of the revitalization of *Eucamerotus* (Naish and Martill, 2001), but a later review of sauropod material from England once again validated *Eucamerotus* by providing further detail regarding the animal's distinctive traits. The same work clarified that while the animal was likely a basal **titanosauriform** of some sort, that not enough information is known about the evolution of early brachiosaurids or somphospondyls to say whether or not *Eucamerotus* was a part of either group (Upchurch et al., 2011). The infamous "Barnes High" partial skeleton (MIWG-BP001) has been suggested to represent *Eucamerotus*, but this is unclear.

The generic name *Eucamerotus* combines the Greek "eu" (meaning "good" or "well") and "kamarotos" (meaning "chambered"); this refers to the hollow chambers of the vertebrae. The specific name *foxi* honors William Fox, who collected much of the paratype.

CLASSIFICATION

Dinosauria
  Sauropoda
    Gravisauria
      Eusauropoda
        Neosauropoda
          Macronaria
            Titanosauriformes

LOCATION

England, UK

KNOWN REMAINS

Vertebrae

O. hulkei

(Seeley, 1870)

Length: 17.4 m (57 ft)
Height: 13 m (42.6 ft)
Hip height: 3.4 m (11.1 ft)
Body mass: 15,500 kg (17.1 t)
Reconstruction: ☐ ☐ ☐ ☐

TRIASSIC	JURASSIC	CRETACEOUS	

Ornithopsis hulkei (meaning "Hulke's bird-like one") has a complex, twisting history of naming and classification.

The genus was first described based on just two partial vertebrae. Since these were found at different locations, and they had no definitive traits in common, Richard Lydekker omitted one of the bones in 1888, leaving just the one that is now known as NHMUK 28632. At the same time, Lydekker decided to identify other specimens as being from *Ornithopsis*, and as the decades ensued, numerous species were erected within the genus by a host of researchers. Eventually *Ornithopsis* had become something of a wastebasket taxon.

Today, all specimens once attributed to some species of *Ornithopsis*—save for one—have either been reassigned to other genera, or have been stripped of any meaningful identification whatsoever, on the basis of being indeterminately undiagnostic. For example, the hip bones that were once the basis of "*Ornithopsis eucamerotus*" are now considered as only an "undiagnostic titanosauriform" (Upchurch et al., 2011).

The only specimen that has kept the genus "alive" is the original holotype vertebra. This bone has, at times, been considered undiagnostic itself (Naish and Martill, 2001), but other works have concluded that unique features of the bone are unlikely to be due to deformation and, thus, the genus *Ornithopsis* remains valid. It can be assigned to the basal reaches of **Titanosauriformes** with some confidence, although a more precise identification is not possible (Upchurch et al., 2011).

The generic name *Ornithopsis* combines the Greek "ornis" (meaning "bird") and "opsis" (meaning "likeness"). This refers to the bird-like air cavities within the bone. The specific name *hulkei* honors geologist John Whitaker Hulke, a colleague of Harry Seeley.

CLASSIFICATION

Dinosauria
  Sauropoda
   Gravisauria
    Eusauropoda
     Neosauropoda
      Macronaria
       Titanosauriformes

LOCATION

England, UK

KNOWN REMAINS

Vertebra

F. zhaoi

(Mo et al., 2006)
Length: 25 m (83 ft)
Height: 14.5 m (48 ft)
Hip height: 4.2 m (14 ft)
Body mass: 35,000 kg (38.6 t)
Reconstruction:

TRIASSIC   JURASSIC   CRETACEOUS

Fusuisaurus zhaoi (meaning "Zhao's lizard from Fusui") is a large sauropod, featuring an ilium (hip bone) "longer than any other known sauropod" (Mo et al., 2020), even though calculations place its overall mass as not being record breaking. It was first reported to have come from the Napai geological formation, but this was later amended to the Xinlong Formation, the age of which is somewhat uncertain but could be from the Aptian of the Early Cretaceous (Mo et al., 2016).

The holotype specimen, unearthed in 2001, comprised roughly 10% of an adult skeleton. The disarticulated remains were intermingled with those of the smaller sauropod *Liubangosaurus*, as well as other unidentified sauropod bones. Starting in 2016, the dig site was reopened, and approximately 100 additional disarticulated bones were uncovered. The majority of these fossils seem to have belonged to juvenile specimens of sauropods and

hadrosaurs. One humerus, though, has tentatively been assigned to *Fusuisaurus*; despite being found 25 meters away from the holotype remains, its size in comparison to the other specimens from the site tentatively identifies it as having come from the same individual.

With such limited skeletal material available, *Fusuisaurus* can be something of an "unstable taxon" when placed within a phylogenetic dataset—slight tweaks to the calculations involved can lead to wildly different placements for the genus (Mannion et al., 2019). Nonetheless, the general consensus among researchers is that is likely some sort of basal **titanosauriform** (Mo et al., 2020). Some versions of analyses even place it as a brachiosaur.

The generic name *Fusuisaurus* refers to Fusui County, China. The specific name *zhaoi* honors Zhao Xijin in recognition for the dinosaur research he has accomplished in Guangxi Province.

CLASSIFICATION
Dinosauria
 Sauropoda
  Gravisauria
   Eusauropoda
    Neosauropoda
     Macronaria
      Titanosauriformes

LOCATION
China

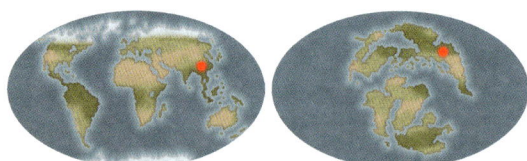

KNOWN REMAINS
Fragments

4 m   8 m   12 m   16 m   8 m

V. damparisensis

(Mannion et al., 2017)
Length: 15 m (48 ft)
Height: 8 m (26 ft)
Hip height: 3.2 m (10.5 ft)
Body mass: 9,000 kg (9.9 t)
Reconstruction: ☐☐☐☐

4 m

250 245 240 235 230 225 220 215 210 205 200 195 190 185 180 175 170 165 **160** 155 150 145 140 135 130 125 120 115 110 105 100 95 90 85 80 75 70 65

TRIASSIC   **JURASSIC**   **CRETACEOUS**

Vouivria damparisensis (meaning "viper from Damparis") is the earliest known brachiosaurid and titanosauriform.

The specimen on which *Vouivria* was described was uncovered in a French chalkstone quarry in 1934. The animal had been partially buried in a lagoon environment and, based on the presence of *Megalosaurus*-like teeth, likely had been partially scavenged by theropods at some point, although toothmarks seem to be absent from the bones themselves.

In 1943, Albert-Félix de Lapparent described the specimen as belonging to the now-dubious genus *Bothriospondylus*. That sauropod, however, was based on fragmentary remains that were found in Madagascar—far enough away that, based on geography alone, the specimens likely had little to do with one another. Indeed, as time wore on, the French specimen was widely recognized to be distinct from the Madagascar specimen, being referred to primarily as the "French *Bothriospondylus*," for lack of a more precise designation.

A new genus, *Vouivria*, was finally erected for the distinct sauropod in 2017. Careful phylogenetic study confirmed its placement within **Brachiosauridae**, a conclusion that many had long suspected. The study also clarified that Late Jurassic brachiosaurs were seemingly limited to East Africa, Western Europe, and the United States, and fossils that fall outside those boundaries do not represent brachiosaurs, despite earlier misidentifications.

The generic name *Vouivria* is derived from the French "vouivre," which itself is derived from the Latin "vipera" (meaning "viper"). Local legends speak of "la vouivre," or "the wyvern," a mythical winged reptile. The specific name *damparisensis* refers to the local region of Damparis.

CLASSIFICATION
Dinosauria
 Sauropoda
  Gravisauria
   Eusauropoda
    Neosauropoda
     Macronaria
      Titanosauriformes
       Brachiosauridae

LOCATION
France

KNOWN REMAINS
Partial skeleton

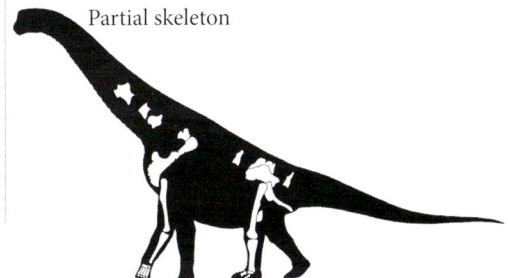

199

| 12 m | | 3 m | 6 m | 9 m | 12 m | 15 m | 18 m | 21 m | 24 m |

A. mcintoshi

(Chure et al., 2010)
Length: 20 m (66 ft)
Height: 12.3 m (40.4 ft)
Hip height: 5 m (16.4 ft)
Body mass: 26,000 kg (28.7 t)
Reconstruction: ☐☐☐☐

| 250 | 245 | 240 | 235 | 230 | 225 | 220 | 215 | 210 | 205 | 200 | 195 | 190 | 185 | 180 | 175 | 170 | 165 | 160 | 155 | 150 | 145 | 140 | 135 | 130 | 125 | 120 | 115 | 110 | **105** | 100 | 95 | 90 | 85 | 80 | 75 | 70 | 65 |

| TRIASSIC | JURASSIC | CRETACEOUS |

Abydosaurus mcintoshi (meaning "McIntosh's Abydonian lizard") was excavated from a cliff face at Dinosaur National Monument. The location is famed for its Jurassic deposits, but the fossils of *Abydosaurus* come from a Cretaceous layer situated at a higher elevation. The precarious location of the bones, combined with the extremely hard sandstone in which they were entombed, made their extraction a taxing process.

The holotype specimen, which boasts the first complete skull for a Cretaceous sauropod described in the United States, was actually a juvenile at the time of its death. The remains of at least four other individuals of differing ages have also been identified from the dig site, including significant cranial material. Taken in sum, much of the animal's appendicular skeleton is accounted for. At the time of description, much of the *Abydosaurus* fossil material was still awaiting proper preparation, and at the time of this writing, no additional publications have yet described any of these fossils.

For these reasons, the size of an adult *Abydosaurus* is fairly uncertain. Particularly hefty rib fragments hint that the animal might have grown larger than is often estimated.

Abydosaurus had teeth that were quite narrow in comparison to earlier **brachiosaurs**. It is possible that this tooth shape is an adaptation to the predominant conifer vegetation that was emerging during the middle Cretaceous.

The generic name *Abydosaurus* refers to the African city of Abydos, which legend holds as the resting place for the head and neck of the Egyptian deity Osiris; this alludes to the type specimen, which consisted of a head and neck. The specific name *mcintoshi* honors Jack McIntosh for his contributions to Dinosaur National Monument.

CLASSIFICATION
Dinosauria
 Sauropoda
  Gravisauria
   Eusauropoda
    Neosauropoda
     Macronaria
      Titanosauriformes
       Brachiosauridae

LOCATION
Utah, USA

KNOWN REMAINS
Partial skull and skeleton

| 4 m | 8 m | 12 m | 16 m | 20 m | 24 m |

12 m

B. altithorax

(Riggs, 1903)
Length: 26 m (85 ft)
Height: 12 m (39 ft)
Hip height: 5 m (16.4 ft)
Body mass: 50,000 kg (55.1 t)
Reconstruction: ☐☐☐☐

8 m

4 m

| 250 | 245 | 240 | 235 | 230 | 225 | 220 | 215 | 210 | 205 | 200 | 195 | 190 | 185 | 180 | 175 | 170 | 165 | 160 | **154** | 150 | 145 | 140 | 135 | 130 | 125 | 120 | 115 | 110 | 105 | 100 | 95 | 90 | 85 | 80 | 75 | 70 | 65 |

| TRIASSIC | JURASSIC | CRETACEOUS |

Brachiosaurus altithorax (meaning "high-chested arm lizard") is one of the rarer species of sauropod from the famous Morrison Formation. Other well-known sauropods from the formation, such as *Camarasaurus*, *Apatosaurus*, and *Diplodocus*, have far more plentiful fossil material to their name than *Brachiosaurus* does.

Brachiosaurus had forelimbs that were longer than its hindlimbs, which would have given its body and neck an upward-tilted posture. As is the case with most sauropods, there is still debate as to at what angle *Brachiosaurus* held its neck and how it would most efficiently move while feeding, but it seems likely that it specialized as a high-browser that fed on the tallest vegetation. This height would have allowed it to avoid competition with other sauropod genera with which it shared its habitat.

The odd hump-like projection on the animal's skull, a common feature among **brachiosaurids** in general, remains mysterious. Although it is commonly believed that the nostrils of *Brachiosaurus* were located high up on this protrusion, and thus not located at the tip of its snout, in truth it is impossible to determine their location from bones alone. One hypothesis is that the nostrils of brachiosaurids evolved an elevated position in order to allow the animals to breathe while drinking; *Brachiosaurus* would have been unable to maneuver its head the way a modern giraffe does and would have needed to adapt accordingly (Hallett and Wedel, 2016).

The generic name *Brachiosaurus* combines the Greek "brachon" (meaning "arm") and "sauros" (meaning "lizard"). The specific name *altithorax* combines the Latin "alti" (meaning "high") and "thorax" (meaning "chest"). Other species have previously been assigned to the genus but have now been elevated to genera of their own, such as *Lusotitan* and *Giraffatitan*.

CLASSIFICATION
Dinosauria
 Sauropoda
  Gravisauria
   Eusauropoda
    Neosauropoda
     Macronaria
      Titanosauriformes
       Brachiosauridae

LOCATION
United States

KNOWN REMAINS
Partial skull and skeleton

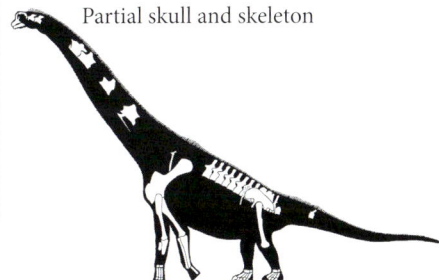

| 4 m | 8 m | 12 m | 16 m | 20 m | 24 m | 28 m |

G. brancai

(Paul, 1988)
Length: 25 m (82 ft)
Height: 12 m (39 ft)
Hip height: 4 m (13 ft)
Body mass: 48,000 kg (52.9 t)
Reconstruction: ☐☐☐☐

| 250 | 245 | 240 | 235 | 230 | 225 | 220 | 215 | 210 | 205 | 200 | 195 | 190 | 185 | 180 | 175 | 170 | 165 | 160 | 155 | 152 | 145 | 140 | 135 | 130 | 125 | 120 | 115 | 110 | 105 | 100 | 95 | 90 | 85 | 80 | 75 | 70 | 65 |

TRIASSIC **JURASSIC** **CRETACEOUS**

Giraffatitan brancai (meaning "Branca's gigantic giraffe") was originally described in 1914 as a species of *Brachiosaurus*, known as *B. brancai*. Eventually, though, researchers began to notice just how different the specimens of the African *B. brancai* were when compared against the North American species, *B. altithorax*.

Gregory S. Paul was the first (in 1988) to coin the name *Giraffatitan*, not as a genus but rather as a subgenus of *Brachiosaurus*. In 1991, George Olshevsky went one step further, suggesting that *Giraffatitan* be a distinct genus. These views did not gain much traction until a comprehensive comparison of the two animals was undertaken by Michael Taylor in 2009. This analysis showed that the two animals differed not in just a few ways but across numerous different measurable metrics. Since that time, most studies have treated *Giraffatitan* as a valid genus.

Compared with *Brachiosaurus*, *Giraffatitan* has a more slender, gracile build, with both its trunk and its tail being proportionally shorter. The "hump" on the head of *Giraffatitan* was seemingly more pronounced, as well.

The various specimens now attributed to *Giraffatitan* all came from a series of digs that were led by Werner Janensch and took place between 1909 and 1912. Dozens of different sites were explored, from which numerous specimens emerged. Additionally, scattered elements from even more nearby locations were referred to the species. Many of the original field notes from the time have been lost, leading to some uncertainties regarding various specimens' association.

The generic name *Giraffatitan* refers to the animal's huge size and long neck. The specific name *brancai* honors German paleontologist Wilhelm von Branca (1844–1929).

CLASSIFICATION
Dinosauria
  Sauropoda
    Gravisauria
      Eusauropoda
        Neosauropoda
          Macronaria
            Titanosauriformes
              Brachiosauridae

LOCATION
Tanzania

KNOWN REMAINS
Nearly complete

2 m	4 m	6 m	8 m	10 m	12 m	14 m

C. weiskopfae

(Tidwell et al., 1999)
Length: 15 m (49 ft)
Height: 6.4 m (21 ft)
Hip height: 2.3 m (7.5 ft)
Body mass: 10,500 kg (11.6 t)
Reconstruction: ☐☐☐☐

250	245	240	235	230	225	220	215	210	205	200	195	190	185	180	175	170	165	160	155	150	145	139	135	130	125	120	115	110	105	100	95	90	85	80	75	70	65

TRIASSIC · **JURASSIC** · **CRETACEOUS**

Cedarosaurus weiskopfae (meaning "Weiskopf's lizard from Cedar") is a medium-sized brachiosaur, notable for preserving a large number of gastroliths, or "gizzard stones." Among dinosaurs, gastroliths seem to have been most common in theropods; our understanding of how common and important they were among the sauropods remains incomplete. Seven kilograms of stones, of various sizes and textures, were recovered with the animal's remains (Sanders et al., 2001).

The holotype specimen, consisting of nearly the entire animal—save for the head, neck, and hips—was first discovered in 1996 by workers from the Denver Museum of Natural History. The skeleton was partially articulated and, based on fusion between shoulder and vertebral elements, is thought to have been fully grown at the time of its death. The limbs and vertebrae on the animal's left side are either missing or heavily damaged by erosion, indicating that the body was likely not buried in its entirety upon death; parts remained exposed until finally being entombed by overbank sediments.

In 2012, a lower hindlimb and foot were also referred to the genus. The bones, recovered from the Trinity Group, are younger than those of the holotype. The specimen had previously been referred to "*Pleurocoelus*" by W. Langston in 1974 (D'Emic, 2013).

The animal's membership within **Brachiosauridae** was first proposed by its describers, and that view has remained predominant, being backed up by subsequent studies (An et al., 2023).

The generic name *Cedarosaurus* refers to the Cedar Mountain geological formation. The specific name *weiskopfae* honors Carol Weiskopf "for her hard work in the field and lab."

CLASSIFICATION

Dinosauria
 Sauropoda
  Gravisauria
   Eusauropoda
    Neosauropoda
     Macronaria
      Titanosauriformes
       Brachiosauridae

LOCATION

Utah, USA

KNOWN REMAINS

Partial skeleton

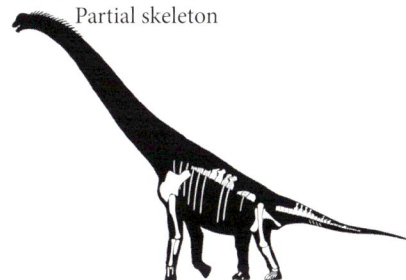

	4 m	8 m	12 m	16 m	20 m	24 m

L. atalaiensis

(Antunes and Mateus, 2003)
Length: 24 m (79 ft)
Height: 12 m (39 ft)
Hip height: 4 m (13 ft)
Body mass: 34,000 kg (37.5 t)
Reconstruction:

TRIASSIC   JURASSIC   CRETACEOUS

Lusotitan atalaiensis (meaning "Luso giant from Atalaia") was originally described in 1957 as a species of *Brachiosaurus*, known as *B. atalaiensis*. Eventually, though, researchers began to notice just how different the specimens of the European *B. atalaiensis* were when compared against the North American species, *B. altithorax*.

The original 1957 description by Lapparent and Zbyszewski based the species on remains found in several different locations but never technically designated a holotype specimen. When the species was reevaluated and raised to its own genus, only one set of remains—the most complete—was chosen as its representative. This lectotype of fossils were originally found in close association, with some still being articulated, and thus are presumed to have originated from a single individual.

When the genus was once again scrutinized in 2013, though, several of these skeletal elements "could not be located" in the museum collections. Nonetheless, this redescription of the animal once again supported the conclusion that *Lusotitan* is a distinct genus, apart from *Brachiosaurus*, by pointing out six unique characteristics (Mannion et al., 2013).

This 2013 redescription was unsure whether or not *Lusotitan* belonged within Brachiosauridae proper or whether it was on its own closely related branch of Titanosauriformes. More recent analyses tend to place it firmly within **Brachiosauridae** (Mannion et al., 2019; Carballido et al., 2020).

The generic name *Lusotitan* is derived from the term "Luso," which refers to an inhabitant of the archaic region known as Lusitania that partially encompassed Portugal at one point in time. The specific name *atalaiensis* refers to Atalaia, a town in Portugal.

CLASSIFICATION
Dinosauria
 Sauropoda
  Gravisauria
   Eusauropoda
    Neosauropoda
     Macronaria
      Titanosauriformes
       Brachiosauridae

LOCATION
Portugal

KNOWN REMAINS
Partial skeleton

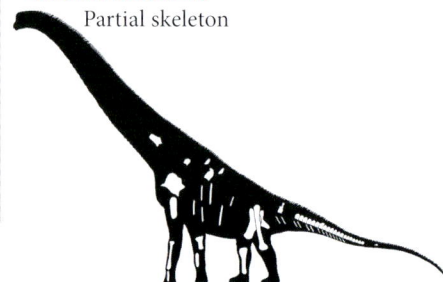

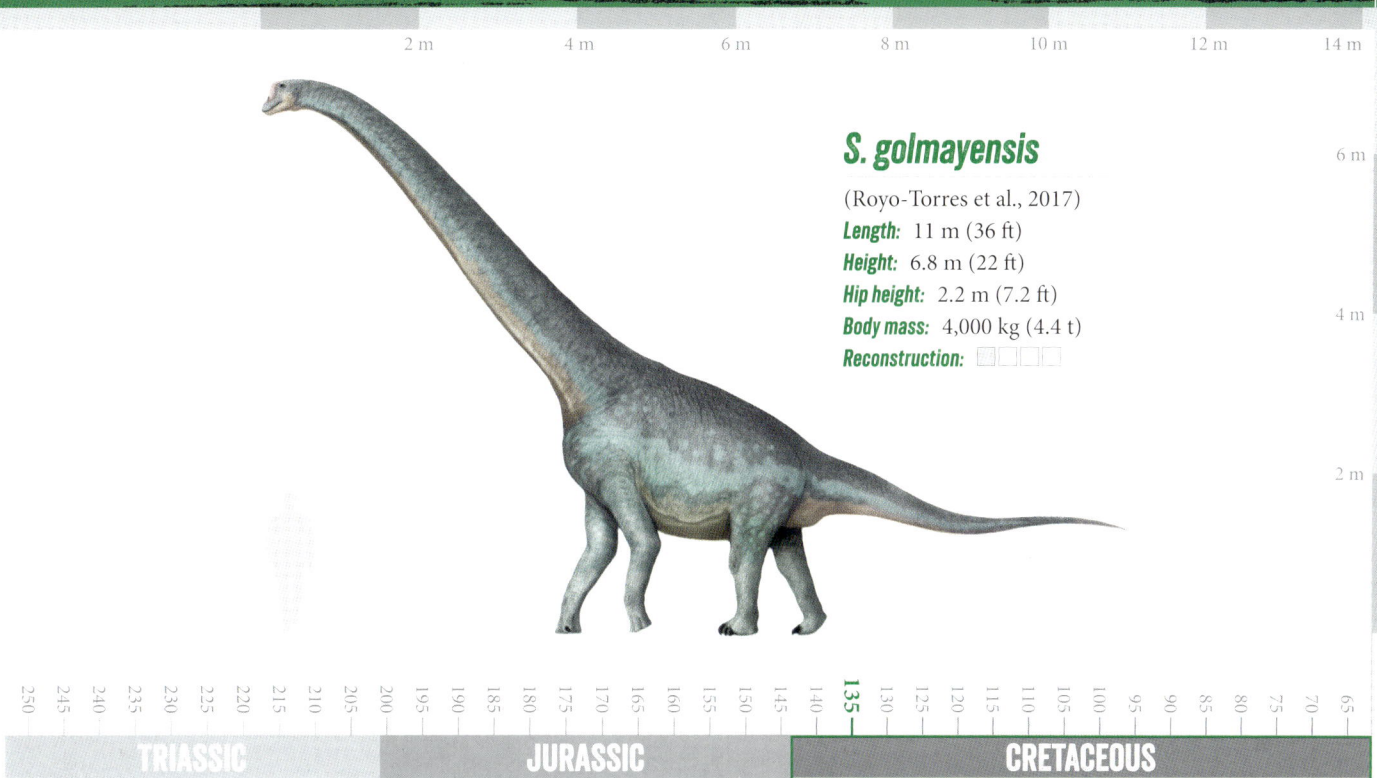

2 m   4 m   6 m   8 m   10 m   12 m   14 m

6 m

4 m

2 m

S. golmayensis

(Royo-Torres et al., 2017)

Length: 11 m (36 ft)
Height: 6.8 m (22 ft)
Hip height: 2.2 m (7.2 ft)
Body mass: 4,000 kg (4.4 t)
Reconstruction: ☐ ☐ ☐ ☐

250 245 240 235 230 225 220 215 210 205 200 195 190 185 180 175 170 165 160 155 150 145 140 **135** 130 125 120 115 110 105 100 95 90 85 80 75 70 65

TRIASSIC   **JURASSIC**   **CRETACEOUS**

Soriatitan golmayensis (meaning "giant from Golmayo, Soria") is one of the final brachiosaurs known to have inhabited Europe. Judging from the fossilized remains of plants found at the area, *Soriatitan* inhabited a subtropical climate and likely had a diet consisting of conifer trees.

The area that yielded the only known *Soriatitan* specimen also contained the remains of numerous different dinosaurs, including ankylosaurian, dromaeosaurid, and iguanodontian fossils. The animals were buried in layers of freshwater sediment, suggesting the possibility of periodic flooding. Excavations of this material began in 2000, and complete preparation of the specimens took years to complete.

The *Soriatitan* holotype consists only of scattered and disarticulated bones, primary among them being the major bones of the forelimb and the femur but also fragments of

pelvic and vertebral elements. Preliminary observations of the specimen occurred in 2005 (Vidarte et al., 2005).

The similarities between *Soriatitan* and the contemporaneous American brachiosaur *Cedarosaurus* are notable; the same can also be said of ankylosaur and iguanodontid species during the Barremian stage of the Early Cretaceous period. This occurrence could be evidence that North America and Europe were still linked, at least sporadically, during this time in history.

The describers conducted a phylogenetic study which placed *Soriatitan* within **Brachiosauridae**, a position that has been corroborated by subsequent studies (Royo-Torres et al., 2020).

The generic name *Soriatitan* refers to the province of Soria, Spain. The specific name *golmayensis* references the town of Golmayo.

CLASSIFICATION

Dinosauria
 Sauropoda
  Gravisauria
   Eusauropoda
    Neosauropoda
     Macronaria
      Titanosauriformes
       Brachiosauridae

LOCATION
Spain

KNOWN REMAINS
Fragments

4 m   8 m   12 m   16 m   20 m

8 m

4 m

S. thompsoni

(Ratkevich, 1998)
Length: 18 m (58 ft)
Height: 11 m (36 ft)
Hip height: 3.5 m (11.5 ft)
Body mass: 18,000 kg (19.8 t)
Reconstruction: ☐☐☐☐

| 250 | 245 | 240 | 235 | 230 | 225 | 220 | 215 | 210 | 205 | 200 | 195 | 190 | 185 | 180 | 175 | 170 | 165 | 160 | 155 | 150 | 145 | 140 | 135 | 130 | 125 | 120 | 115 | 110 | 105 | 100 | 95 | 90 | 85 | 80 | 75 | 70 | 65 |

TRIASSIC   **JURASSIC**   **CRETACEOUS**

Sonorasaurus thompsoni (meaning "Thompson's lizard from Sonora") is a medium-sized **brachiosaur**, and the official state dinosaur of Arizona. It lived in a semiarid environment that was subject to periodic drought, and subsisted on the tough evergreen foliage that dominated the area (D'Emic et al., 2016).

The animal's somewhat diminished size was not due to its phase of growth, as histological sampling of the inner bone material has shown that the specimen was approaching its full-grown status. However, the analysis showed that *Sonorasaurus* grew slower and more sporadically compared with other sauropods. Given knowledge of its ecosystem, it may be the case that brachiosaurs in this period were less well nourished, relying on an irregularly inhospitable environment for sustenance. The presence of digestion-aiding gastroliths in the specimen further attests to the need to heavily process a low-nutrient diet for maximum calorie intake.

These facts may be indicative of a larger trend across North American sauropods, as *Sonorasaurus* represents one of the latest-surviving sauropods on the continent. Shortly after its time, sauropods seemingly go missing in the regional fossil record for about 30 million years, a time known as the "sauropod hiatus." This absence may simply be a result of sampling bias, where conditions for fossil preservation were unfavorable or where not enough deposits from the time have been explored. But, if the hiatus represents a real, regional extinction, a change in climate and therefore a change in food availability could be a compelling reason for the absence of sauropods for such a duration.

The generic name *Sonorasaurus* refers to the Sonoran Desert in southern Arizona. The specific name *thompsoni* honors fossil discoverer Richard Thompson.

CLASSIFICATION
Dinosauria
 Sauropoda
  Gravisauria
   Eusauropoda
    Neosauropoda
     Macronaria
      Titanosauriformes
       Brachiosauridae

LOCATION
Arizona, USA

KNOWN REMAINS
Partial skeleton

4 m   8 m   12 m

4 m

V. dicrocei

(Tidwell et al., 2001)
Length: 13.5 m (44 ft)
Height: 7 m (23 ft)
Hip height: 2.2 m (7.2 ft)
Body mass: 8,000 kg (8.8 t)
Reconstruction: ☐☐☐☐

| 250 | 245 | 240 | 235 | 230 | 225 | 220 | 215 | 210 | 205 | 200 | 195 | 190 | 185 | 180 | 175 | 170 | 165 | 160 | 155 | 150 | 145 | 140 | 135 | 130 | 125 | 120 | 115 | 112 | 105 | 100 | 95 | 90 | 85 | 80 | 75 | 70 | 65 |

TRIASSIC   **JURASSIC**   **CRETACEOUS**

Venenosaurus dicrocei (meaning "DiCroce's poison lizard") is a smaller brachiosaur, with a proportionally slender forelimb.

The holotype specimen consists of various disarticulated skeletal elements. Based on similar weathering, similar size, and lack of repeated elements, the bones are all considered to have come from the same individual. The animal was likely an adult, based on the fusion between vertebral elements, although it may not have reached its maximum possible size. A second, even more fragmentary specimen was found at the same site. Only 40% as large as the first, this one was juvenile. Comparison of forelimb elements shows that the general shape of the bones did not change with age (Tidwell and Wilhite, 2005).

Additional brachiosaur remains have been reported from the vast bone bed at Dalton Wells, Utah. These 42 individual elements, loosely clustered in three regions of the bone bed, represent at least three, possibly four, individuals, at least two of which were subadults. These have tentatively been assigned to *Venenosaurus* (Britt et al., 2009), although this assignment "remains tenuous" (Britt et al., 2017).

The consensus among recent studies places *Venenosaurus* within **Brachiosauridae** (Carballido et al., 2020; An et al., 2023). This placement matches the hypothesis of the genus's describers, who considered *Venenosaurus* to be the closest known relative of *Cedarosaurus*.

The generic name *Venenosaurus* combines the Latin "venenos" (meaning "poison") with the Greek "sauros" (meaning "lizard"); this references that the fossils were found in the Poison Strip Member of the Cedar Mountain geological formation. The specific name *dicrocei* honors fossil discoverer Anthony DiCroce.

CLASSIFICATION
Dinosauria
 Sauropoda
  Gravisauria
   Eusauropoda
    Neosauropoda
     Macronaria
      Titanosauriformes
       Brachiosauridae

LOCATION
Utah, USA

KNOWN REMAINS
Fragments

| 4 m | 8 m | 12 m | 16 m | 20 m | 24 m |

8 m

4 m

A. johnstoni
(Leidy, 1865)

P. nanus
(Marsh, 1888)
Length: 22 m (71 ft)
Body mass: 23,000 kg (25.4 t)
Reconstruction: ☐ ☐ ☐ ☐

| 250 | 245 | 240 | 235 | 230 | 225 | 220 | 215 | 210 | 205 | 200 | 195 | 190 | 185 | 180 | 175 | 170 | 165 | 160 | 155 | 150 | 145 | 140 | 135 | 130 | 125 | 120 | 115 | 112 | 105 | 100 | 95 | 90 | 85 | 80 | 75 | 70 | 65 |

| TRIASSIC | JURASSIC | CRETACEOUS |

What is certain, is that a large number of individual sauropod bones have been discovered over the years in the vicinity of Muirkirk, Maryland. What is far less certain, however, is the exact identity of the animal, or animals, to which those bones belonged.

In 1865, Joseph Leidy named **Astrodon johnstoni**, one of the first dinosaurs named from the United States, based on a sauropod tooth. Later, in 1888, O. C. Marsh named two species of **Pleurocoelus** (*P. nanus* and *P. altus*) based on sauropod bones from the same general area. Over the years, numerous bones from the region would be named to one of these species, many of which were from juvenile specimens. But, are these two dinosaurs actually one and the same? Or, are *both* of these names effectively meaningless?

In 1903, John Bell Hatcher was among the first to suggest that the bones in question represented the same animal. A 2005 study accepted that *Astrodon*, as the first name to be created, had priority, and that the remains should be referred to that genus (Carpenter and Tidwell, 2005). One point in favor of using the name *Astrodon* is that it was made the official state dinosaur of Maryland in 1998.

However, other researchers have subsequently pointed out that there is no way to ensure that each of these disparate specimens did, in fact, come from the same species. Furthermore, even if they indeed had, that might not matter, since by modern standards the bones are undiagnostic—there is nothing unique about them that would differentiate them from other sauropod genera, such as *Camarasaurus*. Therefore, both genera are often considered to be *nomina dubia* in recent literature (Rose, 2007; D'Emic, 2013).

The generic name *Astrodon* combines the Greek "astron" (meaning "constellation" or "star") and "odon" (meaning "tooth")based on the shape of the cross section. The specific name *johnstoni* honors Christopher Johnston, who examined the structure of the holotype fossil tooth.

The generic name *Pleurocoelus* combines the Greek "pleuron" (meaning "side") and "koilos" (meaning "hollow"). "Nanus" is Latin for "dwarf," while "altus" is Latin for "high."

CLASSIFICATION
Dinosauria
 Sauropoda
  Gravisauria
   Eusauropoda
    Neosauropoda
     Macronaria (?)

LOCATION
Maryland, USA

KNOWN REMAINS
Uncertain

P. brevis

(Mantell, 1850)
Length: 14 m (46 ft)
Height: 6.4 m (21 ft)
Reconstruction: ☐☐☐☐

2 m

132

TRIASSIC	JURASSIC	CRETACEOUS

Pelorosaurus brevis (meaning "short monster lizard") is another genus with a convoluted naming history. A set of English fossils was originally used as the basis for one of many species of *Cetiosaurus*, *C. brevis*, by Sir Richard Owen in 1842. But, in 1849, Alexander Melville found that the group of fossils contained iguanodont bones; he renamed the remaining sauropod vertebrae *C. conybeari*. A year later, in 1850, Gideon Mantell decided that the animal was different from *Cetiosaurus*; he combined the vertebrae with a humerus from the same dig site and called the animal *Pelorosaurus*. After this point, numerous species would be erected within the genus, although these specimens have all been adequately reassigned, except for the type specimen. It remains highly ambiguous as to whether the species is valid, or even which combination of genus and species name should technically count, with one paper claiming *P. conybeari* (Upchurch et al., 2011).

D. mackesoni

(Owen, 1884)
Length: 15 m (49 ft)
Height: 7 m (23 ft)
Reconstruction: ☐☐☐☐

2 m

125

TRIASSIC	JURASSIC	CRETACEOUS

Dinodocus mackesoni (meaning "Mackeson's terrible beam") was named by Sir Richard Owen on the basis of partial forelimb elements. The poorly preserved bones were first discovered in 1840, in Kent, England, by H. B. Mackeson. At various points in time, researchers have synonymized the genus with *Pelorosaurus*. It would seem that the specimens have not undergone any kind of modern reanalysis. In 2004, it was suggested that the material, while most likely belonging to a **titanosauriform**, was undescriptive, and thus the genus should be rendered *nomen dubium* (Upchurch et al., 2004).

D. viaemalae

(Cope, 1877a)
Length: 18.7 m (61 ft)
Height: 7.8 m (26 ft)
Reconstruction: ☐☐☐☐☐

	155	
TRIASSIC	JURASSIC	CRETACEOUS

2 m

Dystrophaeus viaemalae (meaning "coarse joint of the bad road") has received little attention, despite being the first set of associated dinosaur remains ever found in western North America. The scant handful of bones were never fully prepared, which obscures key anatomical details, and the site of their origin was thought lost to history until it was finally relocated in 1987. Some surface specimens were recovered in 1989, but proper excavation did not resume until 2014, largely because of the extreme inaccessibility of the site. The bones are obscured in hard, iron-based concretions, which delays preparation. To date, a full redescription is still pending.

Long cited as a diplodocoid, that classification has now been largely ruled out, with more likely phylogenetic possibilities including basal **Macronaria** or even pre-Neosauropoda (Tschopp et al., 2015; Foster et al., 2016).

DATOUSAURUS

	165	
TRIASSIC	JURASSIC	CRETACEOUS

D. bashanensis

(Dong and Tang, 1984)
Length: 14 m (46 ft)
Height: 7.5 m (24.6 ft)
Reconstruction: ☐☐☐☐☐

2 m

Datousaurus bashanensis (meaning "big-headed lizard from Bashan") seemingly has plentiful skeletal material that could potentially be examined. One publication listed "an incomplete skeleton and incomplete skull," another "incomplete skull," and another "a complete skeleton," listing nearly the entire vertebral column, along with scapulae and at least a partial pelvis and limbs (Guangzhao, 2005). The problem is that the species has "never been properly diagnosed" (Upchurch et al., 2004), and very little information is available about the specimens. The majority of published information is from Chinese-language sources. The genus is potentially synonymous with the dubious "*Bashunosaurus*" (Molina-Perez and Larramendi, 2020).

What is known is that *Datousaurus* has only 13 cervical (neck) vertebrae, unlike some close relatives, and has teeth that are especially "spatulate" or "spoon shaped." The describers suggested mamenchisaurids affinities, while a later source listed it under **Camarasauridae** (Guangzhao, 2005).

A. dongpoi

(Ouyang, 1989)
Length: 9.5 m (31 ft)
Height: 4.9 m (16 ft)
Reconstruction: ☐☐☐☐

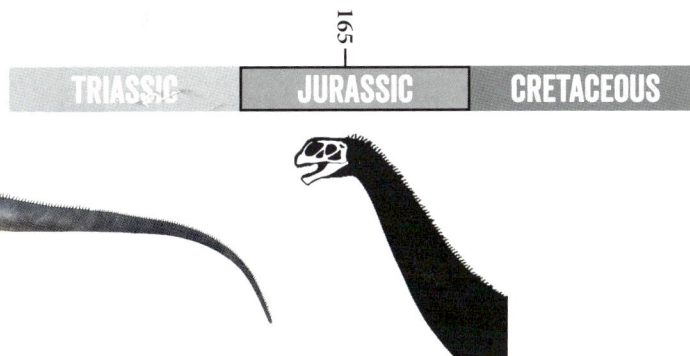

165 —

TRIASSIC | JURASSIC | CRETACEOUS

1 m

Abrosaurus dongpoi is known from two cranial specimens, one of which is nearly complete. These were unearthed in 1984 from the main dig site of the Zigong Dinosaur Museum. In the original description, "postcranial specimens" were mentioned but had "not been excavated" and have remained undescribed. It was hypothesized to be a camarasaurid, while subsequent sources tend to place it merely as **Macronaria** *incertae sedis* (Upchurch et al., 2004; Paul, 2016).

The generic name *Abrosaurus* combines the Greek "habrós" (meaning "delicate") and "sauros" (meaning "lizard"). The specific name *dongpoi* is meant to honor the eleventh century Chinese poet Su Shi, also known as Su Dongpo, as he was born in the same region where the holotype was found, the Sichuan Province of China. When originally published, the specific name was incorrectly given as "dongpoensis."

2 m

A. ischiaticus

(Sanz et al., 1987)
Length: 18 m (59 ft)
Height: 7.2 m (24 ft)
Reconstruction: ☐☐☐☐

135 —

TRIASSIC | JURASSIC | CRETACEOUS

Aragosaurus ischiaticus was first discovered in 1958 by José María Herrero Marzo, an amateur paleontologist. In 1960, paleontologist A. F. Lapparent mentioned the specimen, but it would not be until 1987 that the animal was described and named as *Aragosaurus*. It was originally placed within Camarasauridae, although modern studies tend to find it outside that clade while still being one of the basalmost **macronarians** (Mannion et al., 2019; Moore et al., 2020).

The exact location of its discovery (which had been ill documented), and thus its age, was finally determined in 2011 (Canudo et al., 2012).

The generic name *Aragosaurus* refers to the territory of Aragón, Spain. The specific name *ischiaticus* refers to the animal's ischiatic peduncle, which is the tip of the ischium bone of the pelvis; the area is quite distinctive for *Aragosaurus* and is used as a defining characteristic.

FUSHANOSAURUS

F. qitaiensis

(Wang et al., 2019)
Length: 20 m (66 ft)
Height: 8 m (26 ft)
Reconstruction:

2 m

TRIASSIC	JURASSIC	CRETACEOUS

160

Fushanosaurus qitaiensis has been studied only within the brief, Chinese-language publication that was used to first describe the genus. It has, as of this time, not been included in any broad phylogenetic analyses. The describers noted features of the holotype, consisting of a sole femur, which they viewed as clear indications of **titanosauriform** affiliation. However, a subsequent study reassessing the characteristics of **mamenchisaurid** sauropods notes that the

specific "titanosauriform" features actually have "broader distribution in eusauropods," suggesting that *Fushanosaurus* might be mamenchisaurid, which would be much more typical for a sauropod from the Shishugou Formation (Moore et al., 2023).

The generic name *Fushanosaurus* refers to the Fushan Museum, where the specimen is housed. The specific name *qitaiensis* refers to the Qitai area of Xinjiang, China.

RUGOCAUDIA

R. cooneyi

(Woodruff, 2012)
Length: 14 m (46 ft)
Height: 7.5 m (24.6 ft)
Reconstruction:

2 m

TRIASSIC	JURASSIC	CRETACEOUS

100

Rugocaudia cooneyi is known from a fragmentary set of remains that were collected in Montana in 1985 by a team from the Museum of the Rockies. The fossils had eroded out of the mudstone and were considered "float" material on the surface of the ground. Although spread apart, the bones were deemed to have come from one single individual, judging by the lack of repeated elements and the consistent size of the bones.

Although its describer found the specimen's features to be unique enough to erect a new genus, later studies cast this into doubt by asserting that the remains possessed only general, nonspecific **titanosauriform** features. The validity of the genus has often, therefore, been treated as dubious (D'Emic and Foreman, 2012; Mannion et al., 2013).

The generic name *Rugocaudia* combines the Latin "rugo" (meaning "wrinkled") and "caudia" (meaning "tail"). The specific name *cooneyi* honors landowner J. P. Cooney.

Brachiosaurus

Somphospondyli

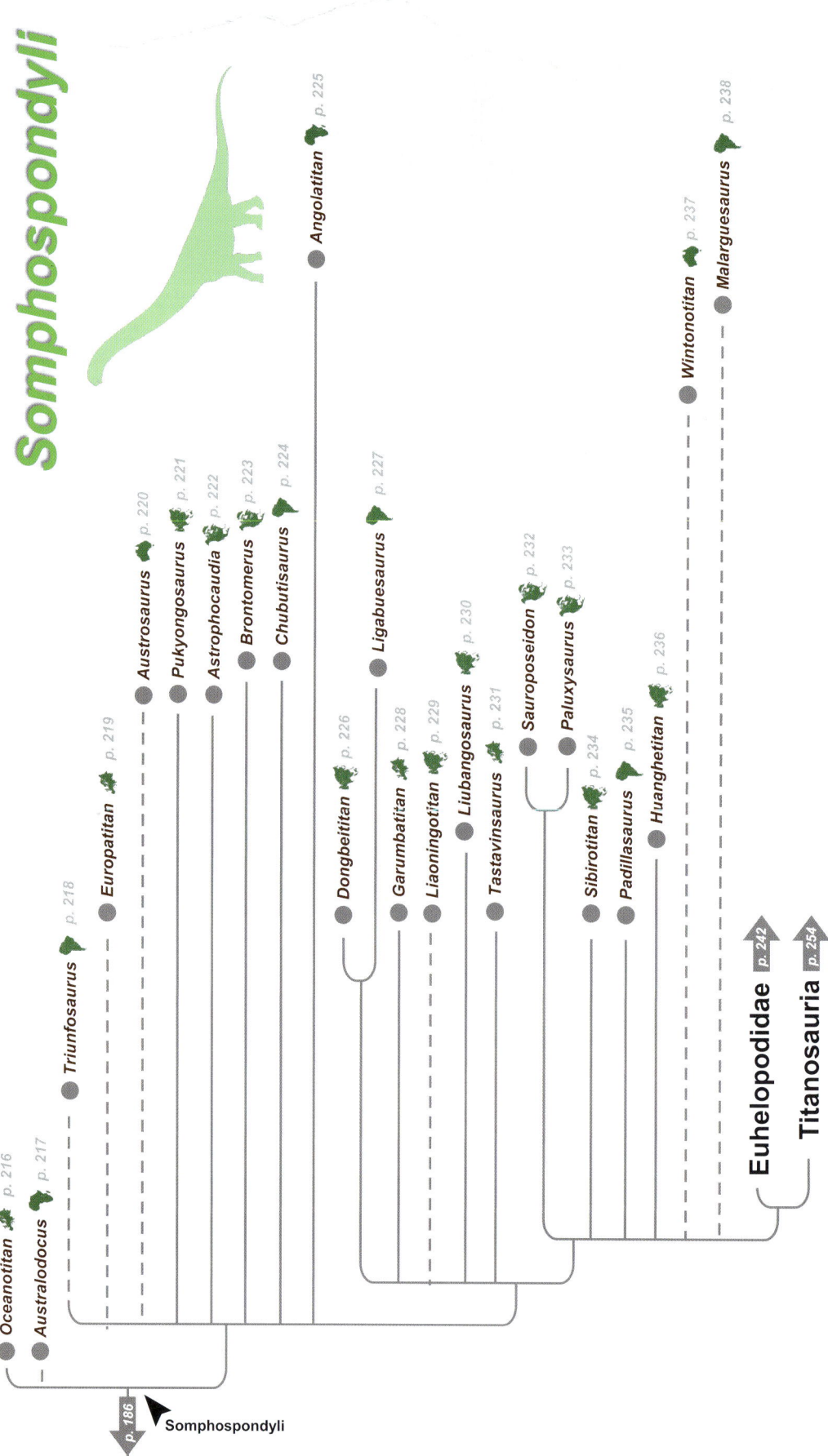

Oceanotitan *p. 216*

Australodocus *p. 217*

Triunfosaurus *p. 218*

Europatitan *p. 219*

Austrosaurus *p. 220*

Pukyongosaurus *p. 221*

Astrophocaudia *p. 222*

Brontomerus *p. 223*

Chubutisaurus *p. 224*

Angolatitan *p. 225*

Dongbeititan *p. 226*

Ligabuesaurus *p. 227*

Garumbatitan *p. 228*

Liaoningotitan *p. 229*

Liubangosaurus *p. 230*

Tastavinsaurus *p. 231*

Sauroposeidon *p. 232*

Paluxysaurus *p. 233*

Sibirotitan *p. 234*

Padillasaurus *p. 235*

Huanghetitan *p. 236*

Wintonotitan *p. 237*

Malarguesaurus *p. 238*

Euhelopodidae p. 242

Titanosauria p. 254

p. 186 ▶ Somphospondyli

It is quite common across all of paleontology for there to be multiple contemporary interpretations of phylogenetic relationships. As different authors incorporate different datasets, it is not unusual for the placement of one species or another to shift around. However, this is all doubly true for the somphospondylan sauropods. Even the most recent of analyses often differ considerably in their definitions and placements. These differences are partly due to the highly fragmentary nature of many of the known specimens. As such, the interpretation presented here of the basal reaches of Somphospondyli is particularly precarious.

The placement of *Austrosaurus* is uncertain within **Somphospondyli** (Poropat et al., 2017), as is that of *Australodocus* (Mannion et al., 2019). *Astrophocaudia* has been shown near the base of the branch (D'Emic, 2013; Averianov et al., 2018), although it has also been placed in Euhelopodidae (Mo et al., 2022). Similar placements seem plausible for *Angolatitan* (Wang et al., 2021), *Brontomerus* (Averianov et al., 2018), *Chubutisaurus* (Mo et al., 2022), *Oceanotitan* (Mocho et al., 2019; Apesteguía et al., 2021), and *Pukyongosaurus* (Poropat et al., 2022). The placement of the possibly chimeric *Triunfosaurus* is uncertain (Poropat et al., 2017).

Dongbeititan and *Ligabuesaurus* have sometimes been recovered within the same smaller clade (Mannion et al., 2019; Mo et al., 2022). The affinities of *Europatitan* are especially uncertain, with it having been placed as a non-somphospondylan (Mo et al., 2022), an intermediate somphospondylan (Wang et al., 2021), or a titanosaur (Mannion et al., 2019). *Liaoningotitan* has not been studied extensively; it has been suggested as an intermediate somphospondylan (Zhou et al., 2018). *Liubangosaurus* could possibly be similarly positioned (Mo et al., 2022).

Tastavinsaurus is another example of highly uncertain placement. While Mannion et al. (2019), Pérez-Pueyo et al. (2019), Moore et al. (2020), Apesteguía et al. (2021), Dai et al. (2022), and Mo et al. (2022) show it in a more derived position within Somphospondyli, research by Carballido et al. (2020), An et al. (2023), and Ren et al. (2023) show it as actually a basal macronarian. *Huanghetitan* is likely among the most derived of the non-titanosaur somphospondylans (Moore et al., 2020).

Sauroposeidon and *Paluxysaurus*, if they are not synonymous, are at least sister taxa whose placement is approaching the Titanosauria (Averianov et al., 2018; Wang et al., 2021). The fragmentary *Sibirotitan* has been suggested to be still more derived (Averianov et al., 2018). *Padillasaurus* is also likely among the more derived non-titanosaurs (Moore et al., 2020; Royo-Torres et al., 2020). *Wintonotitan* is often shown as one of the most derived non-titanosaurs (Bellardini et al., 2022), but this depiction is not universal (Wang et al., 2021). It is possible, but not certain, that *Malarguesaurus* occupies a similar position (Carballido et al., 2022).

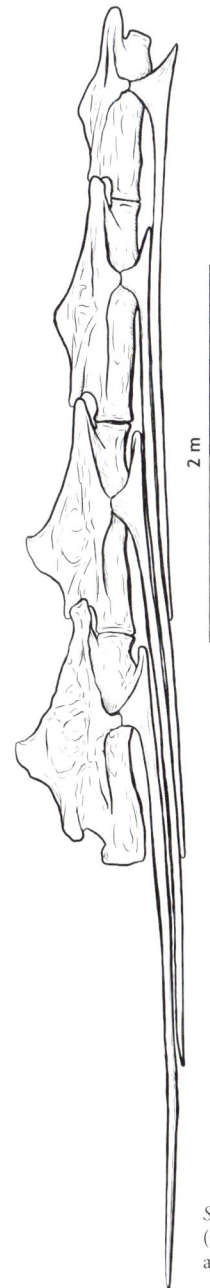

2 m

Sauroposeidon proteles (cervical vertebrae and ribs)

8 m

4 m

4 m   8 m   12 m   16 m

O. dantasi

(Mocho et al., 2019)
Length: 16 m (52 ft)
Height: 7.8 m (25.5 ft)
Hip height: 3.1 m (10.2 ft)
Body mass: 10,000 kg (11 t)
Reconstruction: ☐☐☐☐

250 | 245 | 240 | 235 | 230 | 225 | 220 | 215 | 210 | 205 | 200 | 195 | 190 | 185 | 180 | 175 | 170 | 165 | 160 | 155 | 152 | 145 | 140 | 135 | 130 | 125 | 120 | 115 | 110 | 105 | 100 | 95 | 90 | 85 | 80 | 75 | 70 | 65

TRIASSIC   **JURASSIC**   **CRETACEOUS**

Oceanotitan dantasi (meaning "Dantas's ocean giant") is a medium-sized sauropod known only from incomplete remains.

The holotype specimen of *Oceanotitan* was found by a private fossil collector near the Atlantic Ocean, at the bottom of a coastal cliff. The individual donated the specimen to the Sociedade de História Natural in 2008. The various bones are likely those of a single individual, given that they were found very close together.

Taxonomically, the describers of *Oceanotitan* found it to be a basal member of **Somphospondyli**, stemward of the major branches Euhelopodidae and Titanosauria. One defining characteristic of somphospondylans possessed by *Oceanotitan* is that its pelvic ischium bone is shorter than its pubis bone. A subsequent study also came to a very similar phylogenetic conclusion (Apesteguía et al., 2021). If accurate, this would make *Oceanotitan* one of the earliest

known somphospondylans. As the evolutionary origins of many sauropod lineages are still poorly understood, this evidence provides a key piece of information that can help untangle the sauropod family tree.

With the discovery of *Oceanotitan*, it can now be said that during the Late Jurassic, the Iberian Peninsula was the home of titanosauriform sauropods, in addition to turiasaurs, diplodocids, and brachiosaurids. This high level of sauropod diversity could suggest that the region was integral to the evolution and dispersal of multiple lineages of sauropods between the continents of Europe, North America, and Africa.

The generic name *Oceanotitan* derives from the Latin word "oceanus" (meaning "ocean"). The specific name *dantasi* honors Portuguese paleontologist Pedro Dantas for his dedicated work within the country.

CLASSIFICATION
Dinosauria
  Sauropoda
    Gravisauria
      Eusauropoda
        Neosauropoda
          Macronaria
            Titanosauriformes
              Somphospondyli

LOCATION
Portugal

KNOWN REMAINS
Fragments

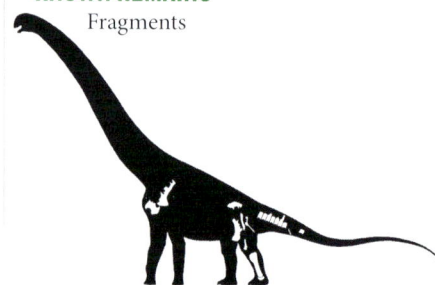

4 m   8 m   12 m   16 m   20 m

8 m

4 m

A. bohetii

(Remes, 2007)
Length: 20 m (66 ft)
Height: 9.5 m (31 ft)
Hip height: 3.5 m (11.5 ft)
Body mass: 18,000 kg (19.8 t)
Reconstruction: ▢▢▢▢

| 250 | 245 | 240 | 235 | 230 | 225 | 220 | 215 | 210 | 205 | 200 | 195 | 190 | 185 | 180 | 175 | 170 | 165 | 160 | 155 | **150** | 145 | 140 | 135 | 130 | 125 | 120 | 115 | 110 | 105 | 100 | 95 | 90 | 85 | 80 | 75 | 70 | 65 |

TRIASSIC | **JURASSIC** | **CRETACEOUS**

Australodocus bohetii (meaning "Boheti's southern beam") is known from two neck vertebrae that were collected in 1909 during an expedition headed by Werner Janensch. There were originally four vertebrae collected, but two of them (along with much of the other fossil material collected from the expedition) were destroyed during World War II.

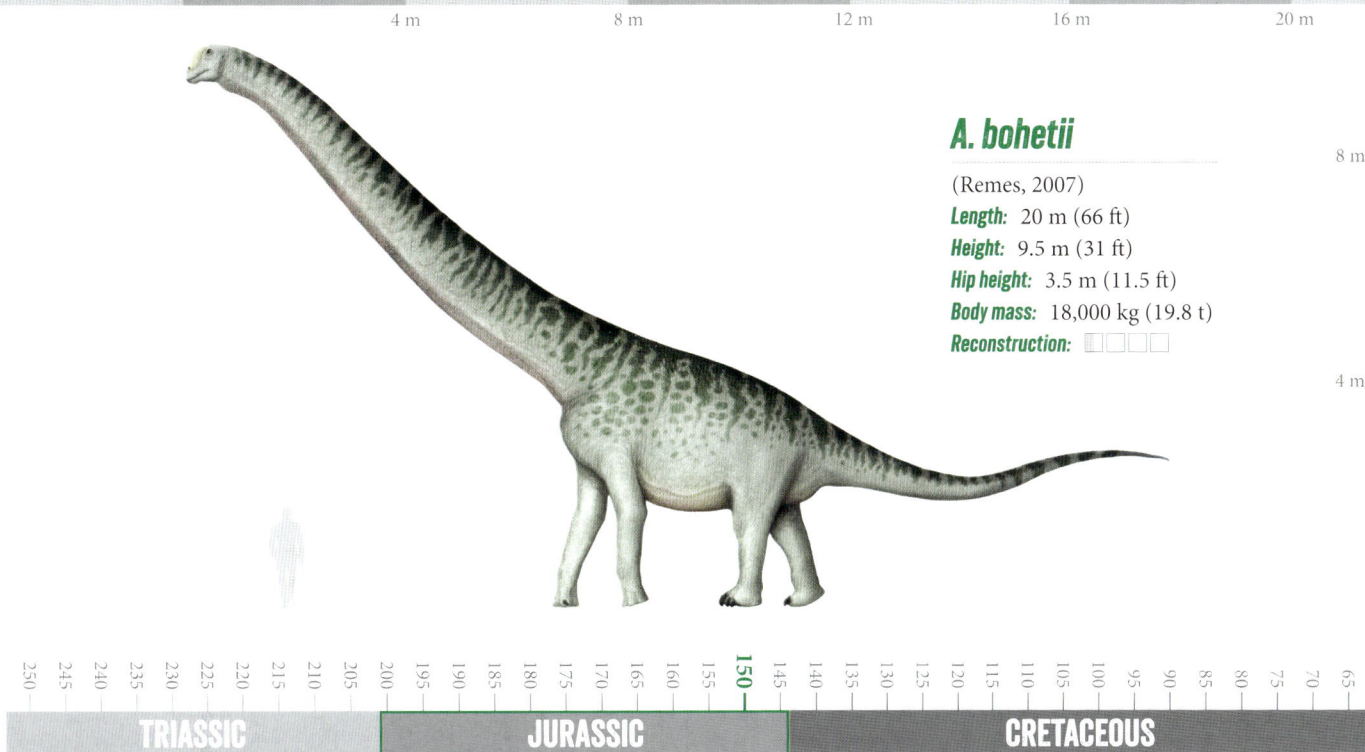

The two surviving bones were labeled as "*Barosaurus africanus*" in the archives of the Museum für Naturkunde in Berlin, and were never described in literature until they were eventually reexamined by Kristian Remes.

Taxonomically placing *Australodocus* has been challenging. The genus was originally thought to be a diplodocoid, based on the distinctive bifurcated shape of the vertebra's neural spine. But subsequent studies found *Australodocus* was more likely to be a titanosauriform, possibly a **somphospondylan** (Whitlock, 2011; Poropat et al., 2015). This placement was further corroborated (albeit tentatively) by CT scans that examined the internal pneumatic characteristics of the bones (Mannion et al., 2019).

It is possible that the split neural spines instead represent a trait of the ill-defined Euhelopodidae, which some studies place as a basal somphospondylan clade. However, since the trait has therefore evolved independently more than once, this alone is not reason enough to definitively place *Australodocus* within Euhelopodidae. Furthermore, the unambiguous members of Euhelopodidae have all occurred in East Asian strata.

The generic name *Australodocus* combines the Latin "australis" (meaning "southern") and the Greek "dokus" (meaning "beam"), alluding to the assumed close relationship to *Diplodocus*. The specific name *bohetii* honors Boheti bin Amrani, African crew supervisor and chief preparator of the German Tendaguru Expedition.

CLASSIFICATION

Dinosauria
 Sauropoda
  Gravisauria
   Eusauropoda
    Neosauropoda
     Macronaria
      Titanosauriformes
       Somphospondyli

LOCATION

Tanzania

KNOWN REMAINS

Neck vertebrae

217

4 m 8 m 12 m 16 m 20 m 24 m

8 m

4 m

T. leonardii

(Carvalho et al., 2017)
Length: 21 m (68 ft)
Height: 11 m (36 ft)
Hip height: 4.5 m (14.5 ft)
Body mass: 17,000 kg (18.7 t)
Reconstruction: ☐☐☐☐

250 245 240 235 230 225 220 215 210 205 200 195 190 185 180 175 170 165 160 155 150 145 140 **135** 130 125 120 115 110 105 100 95 90 85 80 75 70 65

TRIASSIC	JURASSIC	CRETACEOUS

Triunfosaurus leonardii (meaning "Leonardi's lizard from Triunfo") is a potentially problematic taxon, as there are disagreements about which bones actually belong to the genus.

Triunfosaurus is known only from a small segment of the tail, as well as the ischium bone of the pelvis. The proportions of the ischium are quite distinctive, with the surface that would touch the forward-facing pubis bone being half as high as the entire bone's length.

Traits of the tail vertebrae and chevrons—originally interpreted as being from the middle portion of the tail—were used to make the determination that *Triunfosaurus* belonged to the derived clade Titanosauria (although this determination was admittedly made with a low level of confidence). This finding would be significant, as this would make *Triunfosaurus* potentially the oldest known titanosaur; if this were the case, the evolutionary origins of titanosaurs would need to be reevaluated, as would the timing of the breakup of Gondwana.

However, a subsequent study questioned the validity of several traits that were used to assign *Triunfosaurus* to Titanosauria. Most notably, it was noticed that the tail vertebrae were proportioned in such a way that they were unlikely to have come from the middle portion of the tail; instead, it would be more likely that they came from the front portion of the tail. With that in mind, the bones seem to be too small to have come from the same individual as the ischium bone. Thus, it is possible that the *Triunfosaurus* specimen is actually a chimera. This analysis placed the genus as an indeterminate **somphospondylan** (Poropat et al., 2017).

The generic name *Triunfosaurus* refers to Triunfo County, Brazil. The specific name *leonardii* honors paleontologist Giuseppe Leonardi for his work dedicated to Brazilian paleontology.

CLASSIFICATION
Dinosauria
 Sauropoda
  Gravisauria
   Eusauropoda
    Neosauropoda
     Macronaria
      Titanosauriformes
       Somphospondyli

LOCATION
Brazil

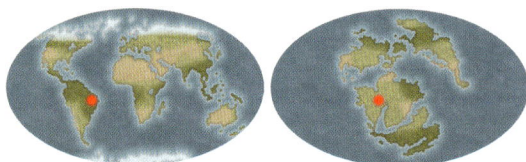

KNOWN REMAINS
Fragments

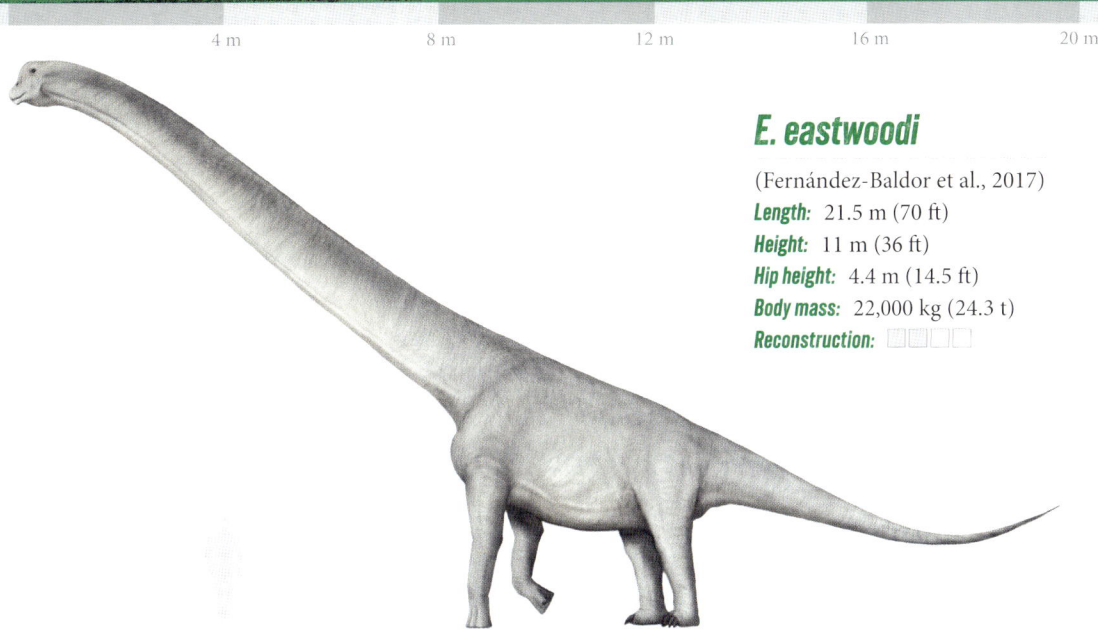

4 m   8 m   12 m   16 m   20 m

8 m

4 m

E. eastwoodi

(Fernández-Baldor et al., 2017)
Length: 21.5 m (70 ft)
Height: 11 m (36 ft)
Hip height: 4.4 m (14.5 ft)
Body mass: 22,000 kg (24.3 t)
Reconstruction: ☐☐☐☐

250 245 240 235 230 225 220 215 210 205 200 195 190 185 180 175 170 165 160 155 150 145 140 135 130 125 120 115 110 105 100 95 90 85 80 75 70 65

TRIASSIC   **JURASSIC**   **CRETACEOUS**

Europatitan eastwoodi (meaning "Eastwood's European giant") is a stout sauropod unearthed from the Castrillo de la Reina geological formation of Spain.

The only known specimen of *Europatitan* was buried in river deposits, in what is thought to have once been a floodplain environment. Some of the bones show toothmarks from scavengers, and indeed, more than one kind of theropod tooth was discovered among the remains (which raises entirely different questions regarding theropod diversity and behavior). Only the bones of the front portion of the animal's tail were articulated in their natural position, whereas the other remains were slightly displaced. The bones were excavated between 2004 and 2006.

The describers concluded that *Europatitan* was one of the basalmost **somphospondylans**. A subsequent study produced several versions of somphospondylan phylogenetic relationships, some of which placed *Europatitan* in a slightly more derived position as a basal titanosaur within the superfamily Andesauroidea (Mannion et al., 2019). Both studies have confirmed that *Europatitan* is separate and distinct from another sauropod that lived in the same time and place, *Tastavinsaurus*. The rebbachisaurid sauropod *Demandasaurus* is also found in the same geological formation.

Europatitan lived in a time of titanosauriform dispersion among the continents. Its presence in Europe at this point in time provides researchers with one more data point with which to nail down how and when this dispersion occurred.

The generic name *Europatitan* refers to the continent of Europe. The specific name *eastwoodi* honors actor Clint Eastwood, because part of his famous film *The Good, the Bad and the Ugly* was filmed at a location nearby the dig site.

CLASSIFICATION
Dinosauria
 Sauropoda
  Gravisauria
   Eusauropoda
    Neosauropoda
     Macronaria
      Titanosauriformes
       Somphospondyli

LOCATION
Spain

KNOWN REMAINS
Partial skeleton

219

4 m   8 m   12 m   16 m   20 m

8 m

4 m

A. mckillopi

(Longman, 1933)
Length: 18 m (59 ft)
Height: 8.7 m (28.5 ft)
Hip height: 3.5 m (11.5 ft)
Body mass: 13,000 kg (14.3 t)
Reconstruction:

250 245 240 235 230 225 220 215 210 205 200 195 190 185 180 175 170 165 160 155 150 145 140 135 130 125 120 115 **112** 105 100 95 90 85 80 75 70 65

TRIASSIC   **JURASSIC**   **CRETACEOUS**

Austrosaurus mckillopi (meaning "southern lizard") was the first Cretaceous dinosaur to be named from Australia.

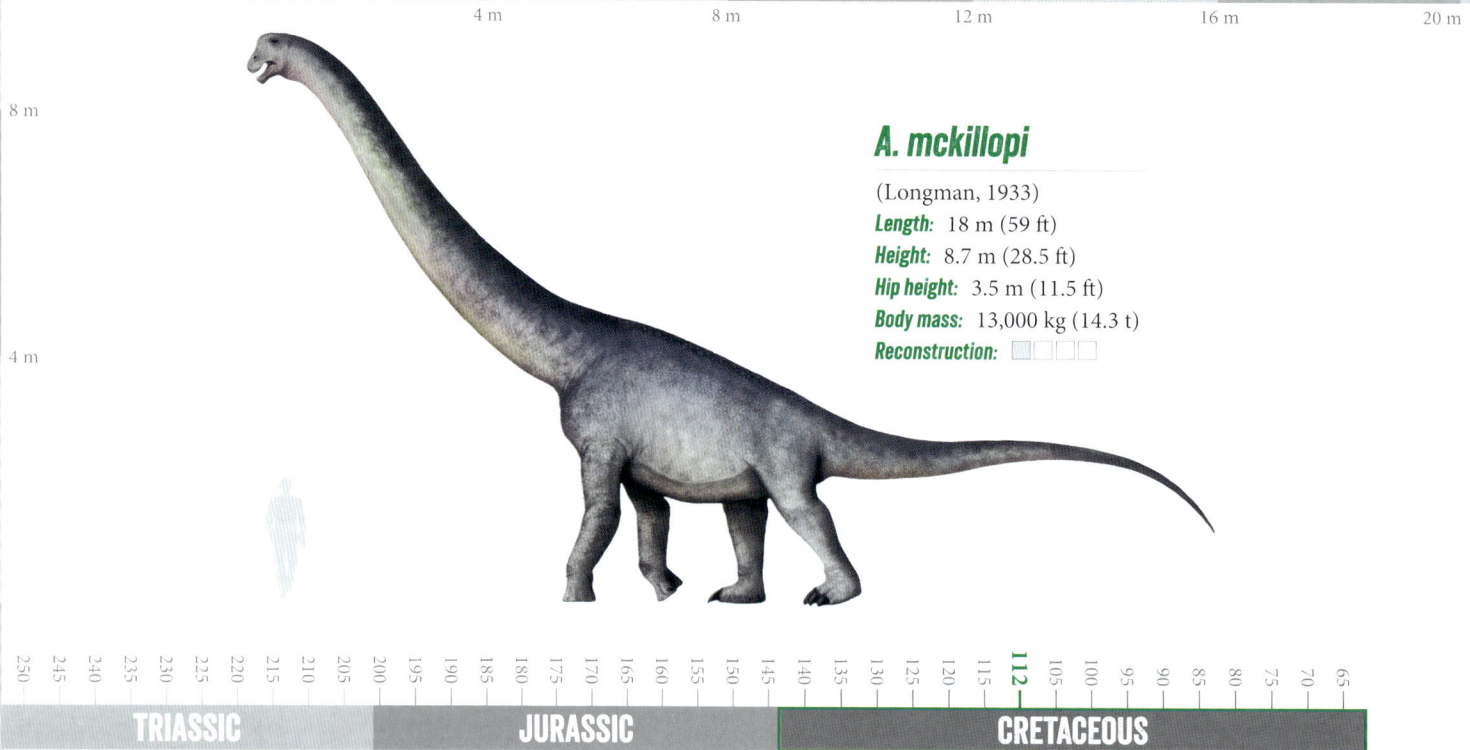

The holotype material (QM F2316) originally consisted of the damaged, central portions of three vertebrae from the base of the neck and the upper back. However, surviving correspondence from the time suggests that additional elements might have been collected from the same specimen. In a 2017 reappraisal of the specimen, not only were additional adjacent elements prepared and described from the original collection, but new elements were described that had been recovered during a 2014 expedition that rediscovered the original dig site. The newly unearthed material (KK F1020) consists of partial ribs and vertebral fragments that align with the holotype. The remains were embedded in marine sediment, suggesting that a partially scavenged carcass of *Austrosaurus* had washed out to sea (Poropat et al., 2017).

Additional sauropod remains from the general vicinity were attributed to *Austrosaurus* in 1981 (Coombs and Molnar, 1981). However, with little to no skeletal overlap present on which to build a comparison, these fossils cannot be confidently attributed to the genus. One such set of remains is now the basis for the genus *Wintonotitan*. One additional neck vertebra (QM F6142, the "Hughenden sauropod") could potentially be referable to *Austrosaurus* (Poropat et al., 2017).

The 2017 reassessment was able to confidently classify *Austrosaurus* as a **titanosauriform**, most likely a somphospondylan.

The generic name *Austrosaurus* combines the Latin "austri" (meaning "southern"), and the Greek "sauros" (meaning "lizard"). The specific name *mckillopi* honors H. J. McKillop, then manager of the sheep station where the fossils were found.

CLASSIFICATION
Dinosauria
  Sauropoda
    Gravisauria
      Eusauropoda
        Neosauropoda
          Macronaria
            Titanosauriformes
              Somphospondyli

LOCATION
Australia

KNOWN REMAINS
Vertebrae and ribs

4 m   8 m   12 m

4 m

P. millenniumi

(Dong et al., 2001)

Length: 14 m (45 ft)
Height: 7.5 m (24.5 ft)
Hip height: 3 m (10 ft)
Body mass: 8,500 kg (9.4 t)
Reconstruction: ▢▢▢▢▢

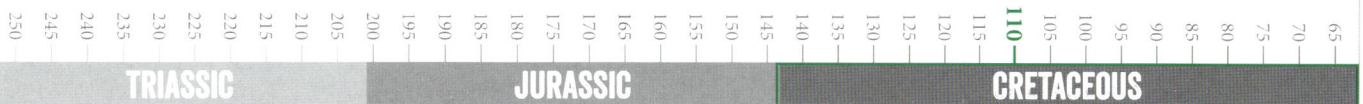

| 250 | 245 | 240 | 235 | 230 | 225 | 220 | 215 | 210 | 205 | 200 | 195 | 190 | 185 | 180 | 175 | 170 | 165 | 160 | 155 | 150 | 145 | 140 | 135 | 130 | 125 | 120 | 115 | 110 | 105 | 100 | 95 | 90 | 85 | 80 | 75 | 70 | 65 |

TRIASSIC   **JURASSIC**   **CRETACEOUS**

Pukyongosaurus millenniumi (meaning "millennial lizard from Pukyong") is a possibly dubious genus that is based on very scant remains.

Isolated sauropod fragments pulled from the same strata of the Hasandong geological formation in 1998 form the constituents of the *Pukyongosaurus* holotype. The bones were assumed to have come from the same animal based on their proximity. The paper that introduced *Pukyongosaurus* was a self-described "preliminary report" that was presented at a conference, the Eighth Annual Meeting of the Chinese Society of Vertebrate Paleontology. The description mentioned seven cervical (neck) vertebrae, but only four were actually shown.

The ensuing years failed to produce a more thorough description of the fossils. Other authors generally referred to the genus as *nomen dubium* in their own analyses. In 2016, researcher Jin-Young Park provided a breakdown of the

rationale behind considering the taxon as dubious, noting that numerous characteristics used to identify the genus are, in fact, present throughout **Titanosauriformes**. This analysis also reidentified one of the bones, which was first interpreted as a portion of the clavicle but in actuality represented a chevron of the tail (Park, 2016).

An analysis of theropod feeding behavior, focusing on the tooth marks left on a single sauropod tail vertebra, considered the sauropod in question to be *Pukyongosaurus*. Besides the tail bone, the work also referred to other fragments not mentioned elsewhere, such as a cervical rib and a tooth (Paik et al., 2011).

The generic name *Pukyongosaurus* refers to Pukyong National University, in South Korea. The specific name *millenniumi* commemorates the new millennium, beginning in the year 2000.

CLASSIFICATION

Dinosauria
  Sauropoda
    Gravisauria
      Eusauropoda
        Neosauropoda
          Macronaria
            Titanosauriformes
              Somphospondyli

LOCATION

South Korea

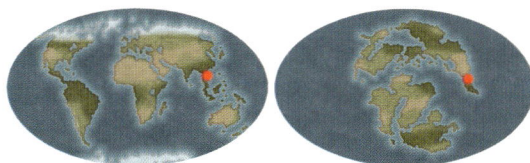

KNOWN REMAINS

Fragments

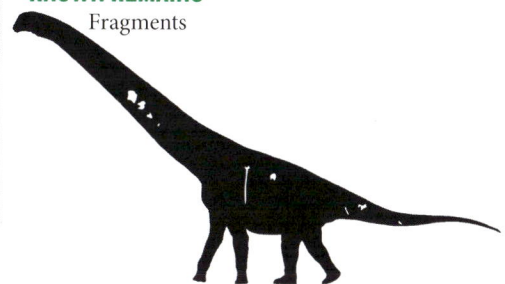

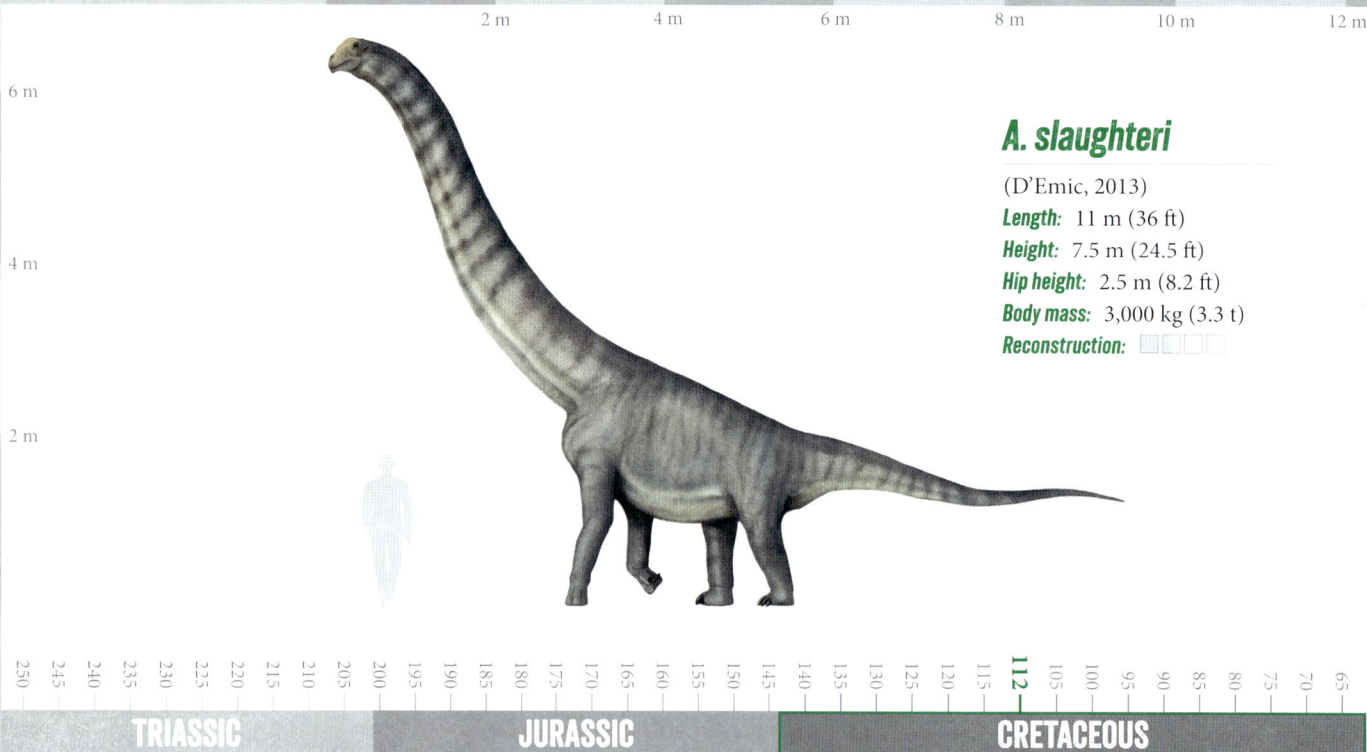

2 m   4 m   6 m   8 m   10 m   12 m

6 m

4 m

2 m

A. slaughteri

(D'Emic, 2013)
Length: 11 m (36 ft)
Height: 7.5 m (24.5 ft)
Hip height: 2.5 m (8.2 ft)
Body mass: 3,000 kg (3.3 t)
Reconstruction:

250 245 240 235 230 225 220 215 210 205 200 195 190 185 180 175 170 165 160 155 150 145 140 135 130 125 120 115 **112** 105 100 95 90 85 80 75 70 65

| TRIASSIC | JURASSIC | CRETACEOUS |

Astrophocaudia slaughteri (meaning "Slaughter's nontwisting tail") is known from a set of fossils that were unearthed in the 1960s and were first mentioned by Maria A. Marques-Bilelo in 1969. In 1974, they were attributed to an indeterminate species of *Pleurocoelus* by Wann Langston Jr. Later researchers, though, began to realize that these fossils were noticeably distinct from others that had been labeled *Pleurocoelus* (Tidwell et al., 1999; Wedel et al., 2000).

In 2012, the remains were reexamined and reclassified as a new genus by Michael D'Emic. In addition to the material described by Langston in 1974, several additional elements were identified from museum collections. For instance, three tail vertebrae had originally been misidentified as "turtle vertebrae," and a second tail chevron was hidden among the material.

Altogether, the portion of the specimen in the best preservational state is the tail. Except for the base, vertebrae from all segments of the tail are well represented. Numerous fragments, most of which are too damaged to be useful, are also present from other parts of the vertebral column and the ribcage. Significant portions of the shoulder blade and hip (ilium) are also preserved.

The describer's phylogenetic analysis paced *Astrophocaudia* as a basal **somphospondylan** sauropod; some ensuing studies have supported this (Averianov et al., 2018).

The generic name *Astrophocaudia* combines the Greek "a-" (meaning "non-"), "stropho" (meaning "twisting"), and "caud" (meaning "tail"). The name also plays off of the Greek "astro" (meaning "star"), as the midtail vertebrae resemble a star shape when viewed from behind; it is also a reference to the sauropod *Astrodon*, one of the first to be named from North America. The specific name *slaughteri* honors Robert H. Slaughter, the excavator of the specimen.

CLASSIFICATION
Dinosauria
 Sauropoda
  Gravisauria
   Eusauropoda
    Neosauropoda
     Macronaria
      Titanosauriformes
       Somphospondyli

LOCATION
Texas, USA

KNOWN REMAINS
Fragments

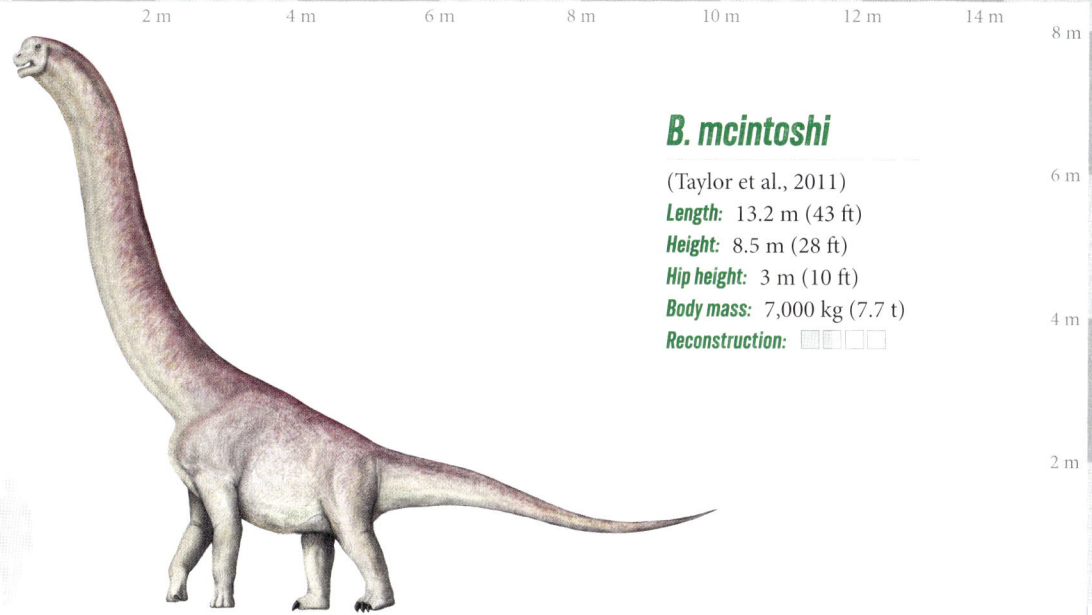

B. mcintoshi

(Taylor et al., 2011)
Length: 13.2 m (43 ft)
Height: 8.5 m (28 ft)
Hip height: 3 m (10 ft)
Body mass: 7,000 kg (7.7 t)
Reconstruction:

TRIASSIC   JURASSIC   CRETACEOUS

Brontomerus mcintoshi (meaning "McIntosh's thunder thigh") is notable for having extremely hefty thigh muscles, proportionally speaking. This is known because of the structure of the animal's hip bone (ilium), where the associated musculature would be attached.

Specifically, the muscles used to move the hindlimb in the forward direction are enlarged, but the same cannot be said for the muscles that would move the leg backward. This configuration suggests that the purpose of the enhanced musculature was not to provide the animal with greater speed; rather, the describers consider it most plausible that *Brontomerus* boasted a powerful "kick." This could have been useful for intraspecific fights against rival males or for defense against predators.

The known specimens of *Brontomerus* come from at least two individuals—a young juvenile and an adult. One possibility is that the remains came from a mother and her young calf. The locality from which the fossils were recovered had previously been extensively excavated by private fossil hunters, and paleontologists believe a significant amount of valuable material has been lost as a result of this unsanctioned activity.

The original phylogenetic comparison made by the describers placed *Brontomerus* as a camarasauromorph. A subsequent study shortly thereafter considered the genus to be *nomen dubium* and judged the remains to be titanosauriform (D'Emic, 2013). A more recent set of work validated *Brontomerus* and placed it as a basal **somphospondylan** (Mannion et al., 2013; Averianov et al., 2018).

The generic name *Brontomerus* combines the Greek "bronto" (meaning "thunder") and "meros" (meaning "thigh"). The specific name *mcintoshi* honors physicist and avocational sauropod paleontologist, John S. McIntosh.

CLASSIFICATION
Dinosauria
 Sauropoda
  Gravisauria
   Eusauropoda
    Neosauropoda
     Macronaria
      Titanosauriformes
       Somphospondyli

LOCATION
Utah, USA

KNOWN REMAINS
Fragments

CHUBUTISAURUS

	4 m	8 m	12 m	16 m	20 m

C. insignis

(del Corro, 1975)
Length: 18.5 m (61 ft)
Height: 11 m (36 ft)
Hip height: 3.5 m (11.5 ft)
Body mass: 14,000 kg (15.4 t)
Reconstruction: ☐☐☐☐

8 m

4 m

| 250 | 245 | 240 | 235 | 230 | 225 | 220 | 215 | 210 | 205 | 200 | 195 | 190 | 185 | 180 | 175 | 170 | 165 | 160 | 155 | 150 | 145 | 140 | 135 | 130 | 125 | 120 | 115 | **110** | 105 | 100 | 95 | 90 | 85 | 80 | 75 | 70 | 65 |

TRIASSIC	JURASSIC	CRETACEOUS

Chubutisaurus insignis (meaning "remarkable lizard from Chubut") is an Argentine sauropod that has thankfully received its fair share of examinations over the years.

The lone specimen of *Chubutisaurus* was found by a farmer, Mr. Martínez, in 1961 near the village of El Escorial and was excavated in 1965 by means of dynamite. A decade later, Guillermo del Corro preliminarily described the specimen, but this work was far from exhaustive. Several authors have subsequently reexamined the remains.

The most comprehensive redescription of *Chubutisaurus* (by José Carballido in 2011) included revisions that corrected previously misidentified elements as well as new elements that had been collected from the original site between 1991 and 2007. As no skeletal elements are duplicated, it is assumed that all of the remains came from a single individual (Carballido et al., 2011).

A histological analysis of the specimen's inner bone tissue revealed that the animal was an adult at the time of its death. Whereas the fossils were originally thought to have been Late Cretaceous in origin, it is now known that they originated from an earlier time.

Recent consensus places *Chubutisaurus* as a non-titanosaur **somphospondylan** (Averianov et al., 2018; Mo et al., 2022). With the titanosaurs dominating the South American sauropod ecospace, studying how *Chubutisaurus* differs from true titanosaurs can help pin down the slippery definition of what traits truly define the group.

The generic name *Chubutisaurus* refers to the province of Chubut, Argentina. The specific name *insignis* is Latin for "remarkable" or "notable."

CLASSIFICATION
Dinosauria
  Sauropoda
   Gravisauria
    Eusauropoda
     Neosauropoda
      Macronaria
       Titanosauriformes
        Somphospondyli

LOCATION
Argentina

KNOWN REMAINS
Partial skeleton

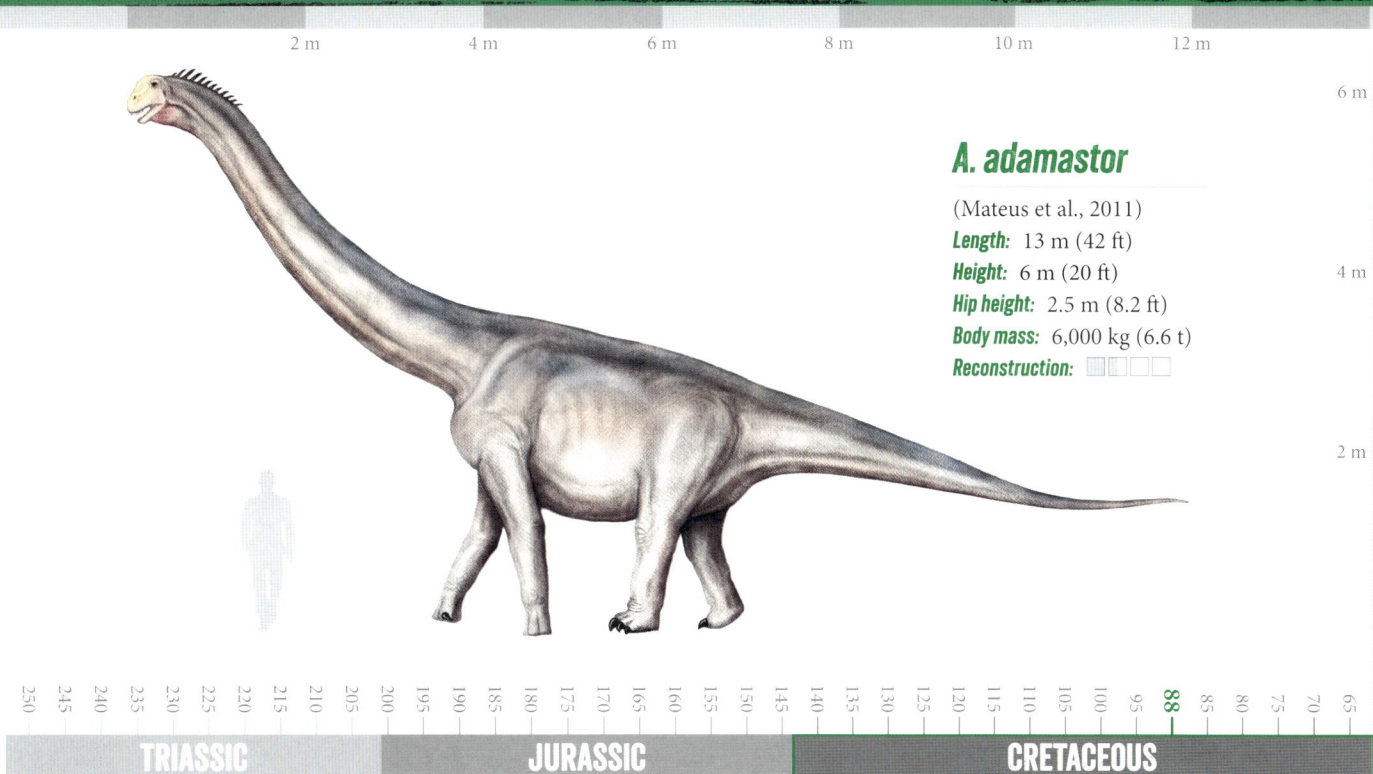

2 m   4 m   6 m   8 m   10 m   12 m

A. adamastor

(Mateus et al., 2011)
Length: 13 m (42 ft)
Height: 6 m (20 ft)
Hip height: 2.5 m (8.2 ft)
Body mass: 6,000 kg (6.6 t)
Reconstruction: ▣▢▢▢

6 m

4 m

2 m

250 245 240 235 230 225 220 215 210 205 200 195 190 185 180 175 170 165 160 155 150 145 140 135 130 125 120 115 110 105 100 95 **88** 85 80 75 70 65

TRIASSIC   **JURASSIC**   **CRETACEOUS**

Angolatitan adamastor (meaning "Adamastor giant from Angola") is the first dinosaur from the nation of Angola to be named. Although it has long been known that fossiliferous Cretaceous deposits were present there, political instability prevented any sort of fossil hunting expeditions until the early 2000s.

Researchers from the subsequent PaleoAngola Project discovered the remains of *Angolatitan* in 2005 in the predominately marine "Tadi Beds," which are dominated by fossils of ammonites, fish, and sharks as well as some turtles, mosasaurs, and plesiosaurs. It is thought that the sauropod's remains were washed out to sea, where they were scavenged by fish and sharks—a number of shark teeth were found among the remains.

Although only the forelimb of the animal was preserved, it displayed some rather unexpected characteristics. Despite being a Late Cretaceous animal, *Angolatitan* did not bear many of the skeletal traits associated with Titanosauria, the group of sauropods that dominated the world at the time. For instance, its radius and ulna were less robust, the elbow projection of the ulna was far less pronounced, and its first metacarpal was not bowed in its shape.

Consequently, *Angolatitan* has been considered a late-surviving, basal **somphospondylan**—a holdover from a lineage originating tens of millions of years earlier, in the Early Cretaceous. This hints that African sauropod fauna were perhaps more diverse than once thought (Gorscak et al., 2017).

The generic name *Angolatitan* refers to the African nation of Angola. The specific name *adamastor* refers to a mythological sea giant from the South Atlantic that would endanger Portuguese sailors; Angola was formerly a Portuguese colony.

CLASSIFICATION
Dinosauria
 Sauropoda
  Gravisauria
   Eusauropoda
    Neosauropoda
     Macronaria
      Titanosauriformes
       Somphospondyli

LOCATION
Angola

KNOWN REMAINS
Forelimb

225

3 m 6 m 9 m 12 m 15 m

6 m

3 m

D. dongi

(Wang et al., 2007)
Length: 14 m (45 ft)
Height: 6 m (20 ft)
Hip height: 2 m (6.5 ft)
Body mass: 6,000 kg (6.6 t)
Reconstruction: ☐☐☐☐

250 245 240 235 230 225 220 215 210 205 200 195 190 185 180 175 170 165 160 155 150 145 140 135 130 **125** 120 115 110 105 100 95 90 85 80 75 70 65

TRIASSIC **JURASSIC** **CRETACEOUS**

Dongbeititan dongi (meaning "Dong's giant from Dongbei") is known from the Yixian geological formation, which itself is part of the famous Jehol group. The climate of the Yixian appears to have been temperate, enduring periodic rainy and dry seasons as well as moderately fluctuating temperatures between the winter and summer months.

At the time of *Dongbeititan*'s discovery, the Yixian Formation had provided dozens of new genera of both avian and nonavian dinosaurs (particularly, a number of important feather-bearing fossils) but had never produced any substantial sauropod remains, only some indeterminate fragments. Therefore, the description of the first significant sauropod from the formation filled in some crucial missing information regarding the animals that inhabited this Early Cretaceous ecosystem.

The describers of *Dongbeititan* placed the genus within **Somphospondyli**. More recent analyses have continued to support this position, with one suggesting it to be the basalmost member of the clade (Mannion et al., 2019).

The majority of the specimen's neck bones are present—a rarity among sauropod remains; however, all are crushed and/or broken into segments. Likewise, the dorsal vertebrae that are present are also compressed, as are the bones of the lower leg. The best-preserved bones are those of the pelvis, tail, and thigh. Features that distinguish *Dongbeititan* from other genera are present in the coracoid bone of the shoulder girdle and the pubis bone of the pelvic girdle.

The generic name *Dongbeititan* refers to Dongbei, a name for the northeastern region of China. The specific name *dongi* honors Chinese paleontologist Dong Zhiming, known for his prolific rate of naming dinosaur genera.

CLASSIFICATION
Dinosauria
 Sauropoda
  Gravisauria
   Eusauropoda
    Neosauropoda
     Macronaria
      Titanosauriformes
       Somphospondyli

LOCATION
China

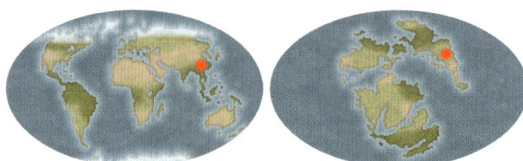

KNOWN REMAINS
Partial skeleton

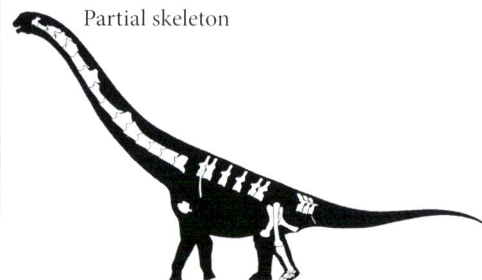

L. leanzai

(Bonaparte et al., 2006)
Length: 23 m (75 ft)
Height: 13 m (42.6 ft)
Hip height: 4.2 m (14 ft)
Body mass: 27,000 kg (29.8 t)
Reconstruction: ▢▢▢▢

TRIASSIC   **JURASSIC**   **CRETACEOUS**

Ligabuesaurus leanzai (meaning "Ligabue's and Leanza's lizard") is among the earliest known somphospondylans of South America. It inhabited what is now the Lohan Cura geological formation; during the Early Cretaceous, the environment would have been a nonarid coastal woodland.

Several partial specimens of *Ligabuesaurus* were unearthed between 1998 and 2000. At first, only one of these sets of remains was described, forming the basis for the genus. After further preparation and study, more of the abundant sauropod remains from the same location were confidently referred to the genus, as well as additional elements of the holotype specimen itself (Bellardini et al., 2022).

A portion of *Ligabuesaurus*'s upper jaw is known, containing partial teeth that were evenly spaced. The animal's forelimbs were proportionally rather long, showing that this trait is not exclusive to Brachiosauridae. In addition, its tibia and fibula were both proportionally slender. The dorsal (back) vertebrae of *Ligabuesaurus* were partitioned with camellate air cavities—meaning that instead of having fewer, bigger pneumatic chambers, numerous smaller openings were present.

The describers of *Ligabuesaurus* placed it among Titanosauria. However, subsequent studies tend to find its position to be slightly more basal, among the non-titanosaurian **somphospondylans** (Ren et al., 2018; Mannion et al., 2019).

The generic name *Ligabuesaurus* honors Italian paleontologist and philanthropist Giancarlo Ligabue (1931–2015), who was responsible for numerous dinosaurian discoveries. The specific name *leanzai* honors geologist Héctor A. Leanza, whose work led to the discovery of the holotype specimen.

CLASSIFICATION

Dinosauria
 Sauropoda
  Gravisauria
   Eusauropoda
    Neosauropoda
     Macronaria
      Titanosauriformes
       Somphospondyli

LOCATION

Argentina

KNOWN REMAINS

Partial skeleton

3 m   6 m   9 m   12 m   15 m   18 m

6 m

3 m

G. morellensis

(Mocho et al., 2024)
Length: 18 m (59 ft)
Height: 8.6 m (28 ft)
Hip height: 3.4 m (11.1 ft)
Body mass: 14,000 kg (15.4 t)
Reconstruction: ☐☐☐☐

250 245 240 235 230 225 220 215 210 205 200 195 190 185 180 175 170 165 160 155 150 145 140 135 130 **123** 120 115 110 105 100 95 90 85 80 75 70 65

| TRIASSIC | JURASSIC | CRETACEOUS |

Garumbatitan morellensis (meaning "giant from Garumba, Morella") is known from remains that preserve two nearly complete hindfeet, a rarity among titanosauriforms. *Garumbatitan* is notable for lacking the calcaneum bone and for having elongated inner metatarsal bones while simultaneously having reduced outer metatarsals.

Garumbatitan is known from one of the most fruitful titanosauriform fossil sites in Europe, the "Sant Antoni de la Vespa" locality, which was discovered in 1998 by Miquel G. Fígols. The known fossils of *Garumbatitan* were unearthed during expeditions that took place in 2005 and 2008.

The remains of four individual sauropods were recovered, three of which have been identified as *Garumbatitan*. Two of the individuals were noticeably smaller than the third and were likely subadults. As of the genus's initial description, only some of the fossil elements had been fully prepared and

examined, with many more still being entombed in stone and plaster jackets, including what is believed to be a nearly complete dorsal vertebral series as well as elements of the tail.

Garumbatitan has been identified as being one of the earliest-branching **somphospondylans**. It lacks some features that are typically seen as being somphospondylan identifiers, such as having elongated, forward-jutting, interlocking facets of the vertebrae; this lack suggests that this feature might actually have been absent in the most basal somphospondylans and only became present later in the group's evolution.

The generic name *Garumbatitan* refers to one of the highest peaks in the region, Mola de la Garumba. The specific name *morellensis* refers to the nearby town of Morella and the Arcillas de Morella geological formation.

CLASSIFICATION
Dinosauria
  Sauropoda
    Gravisauria
      Eusauropoda
        Neosauropoda
          Macronaria
            Titanosauriformes
              Somphospondyli

LOCATION
Spain

KNOWN REMAINS
Partial skeleton

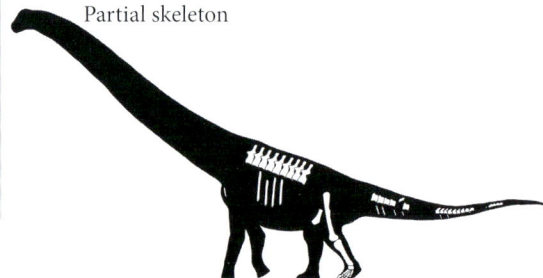

3 m   6 m   9 m   12 m

6 m

L. sinensis

(Zhou et al., 2018)
Length: 13.3 m (44 ft)
Height: 6 m (20 ft)
Hip height: 2 m (6.5 ft)
Body mass: 4,500 kg (5 t)
Reconstruction: ▣▢▢▢

3 m

| 250 | 245 | 240 | 235 | 230 | 225 | 220 | 215 | 210 | 205 | 200 | 195 | 190 | 185 | 180 | 175 | 170 | 165 | 160 | 155 | 150 | 145 | 140 | 135 | 130 | **125** | 120 | 115 | 110 | 105 | 100 | 95 | 90 | 85 | 80 | 75 | 70 | 65 |

| TRIASSIC | JURASSIC | CRETACEOUS |

Liaoningotitan sinensis (meaning "giant from Liaoning, China") is, thus far, an understudied sauropod. Its remains were unearthed in 2007, and it was informally named and displayed as early as 2012.

At the time of *Liaoningotitan's* description, the Yixian Formation had provided dozens of new genera of both avian and nonavian dinosaurs (particularly, a number of important feather-bearing fossils) but had only bore one substantial sauropod genus, *Dongbeititan*. Thus, *Liaoningotitan* offers an important data point in understanding the prevalence and role of sauropods in this ecosystem, which is known as the Jehol Biota.

The describers repeatedly refer to the specimen as "nearly complete," but the descriptive study (published exclusively in a Chinese-language source) provides only a "brief" overview of the material. Most notable among the remains are the frontmost portions of the animal's skull and jaw, each of which still contains teeth. From this distorted material, the describers conclude that the full skull would have been "short and tall" in shape, with a maxilla that was "slightly elongated."

Liaoningotitan was preliminarily calculated to be a basal **somphospondylan**, although several traits of its skull were compared to more derived titanosaurs, such as *Nemegtosaurus*. Without a fully detailed description of the remains, though, placing the available data into large phylogenetic studies would likely lead to dubious results. For instance, a more recent study found it to be an "unstable taxon" in its analysis, prone to shifting position considerably after only small changes in calculations (Mo et al., 2022).

The animal's binomial name refers to Liaoning Province, China; "sinae" is the Latin word for the Chinese.

CLASSIFICATION
Dinosauria
 Sauropoda
  Gravisauria
   Eusauropoda
    Neosauropoda
     Macronaria
      Titanosauriformes
       Somphospondyli

LOCATION
China

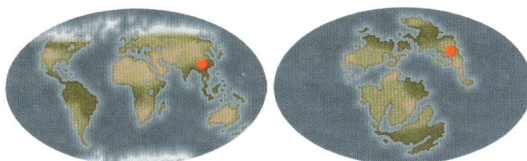

KNOWN REMAINS
Partial skull and skeleton

L. hei

(Mo et al., 2010)
Length: 21 m (69 ft)
Height: 11 m (36 ft)
Hip height: 4.5 m (14.5 ft)
Body mass: 26,000 kg (28.7 t)
Reconstruction: ☐☐☐☐

TRIASSIC | JURASSIC | CRETACEOUS

Liubangosaurus hei (meaning "He's lizard from Liubang") is known from five articulated dorsal vertebrae. These bones were discovered in 2001 among the remains of several individual sauropods. One set was distinguishable based on its size and was described as *Fusuisaurus*. The other two sets of remains were mostly disarticulated and intermingled, making it difficult or impossible to discern which bones belonged to which creature. These five dorsal vertebrae, though, clearly displayed unique characteristics and thus were used to describe this new genus. It is possible, or even likely, that some of the other skeletal remains at the site belonged to the same individual, but unless more complete, comparable specimens are unearthed in the future, this cannot be known for certain. The *Liubangosaurus* fossils likely came from an adult individual.

As for where *Liubangosaurus* fits in the sauropod family tree, its describers came to a rather bold conclusion,

namely that it was a very late-surviving basal eusauropod. This would be extremely unexpected, as not only are Chinese sauropods of this time dominated by the Titanosauriformes, but further, there are no definitive instances of Cretaceous non-neosauropods existing in any part of the globe.

Phylogenetic results that were more reasonable were subsequently calculated by later studies. In 2013, one result placed it among the **somphospondylans**, although a more precise placement remained elusive (Mannion et al., 2013). A 2022 study found it to be a basal member of the group, stemward of the major branches Euhelopodidae and Titanosauria (Mo et al., 2022).

The generic name *Liubangosaurus* refers to the name of the fossil site, the Liubang quarry, in Guangxi Province, China. The specific name *hei* honors He Wenjian, who brought the fossil site to the attention of paleontologists.

CLASSIFICATION
Dinosauria
 Sauropoda
  Gravisauria
   Eusauropoda
    Neosauropoda
     Macronaria
      Titanosauriformes
      Somphospondyli

LOCATION
China

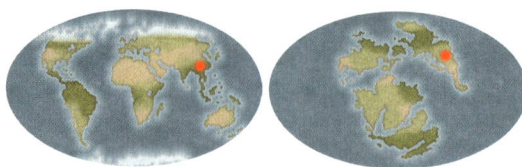

KNOWN REMAINS
Vertebrae

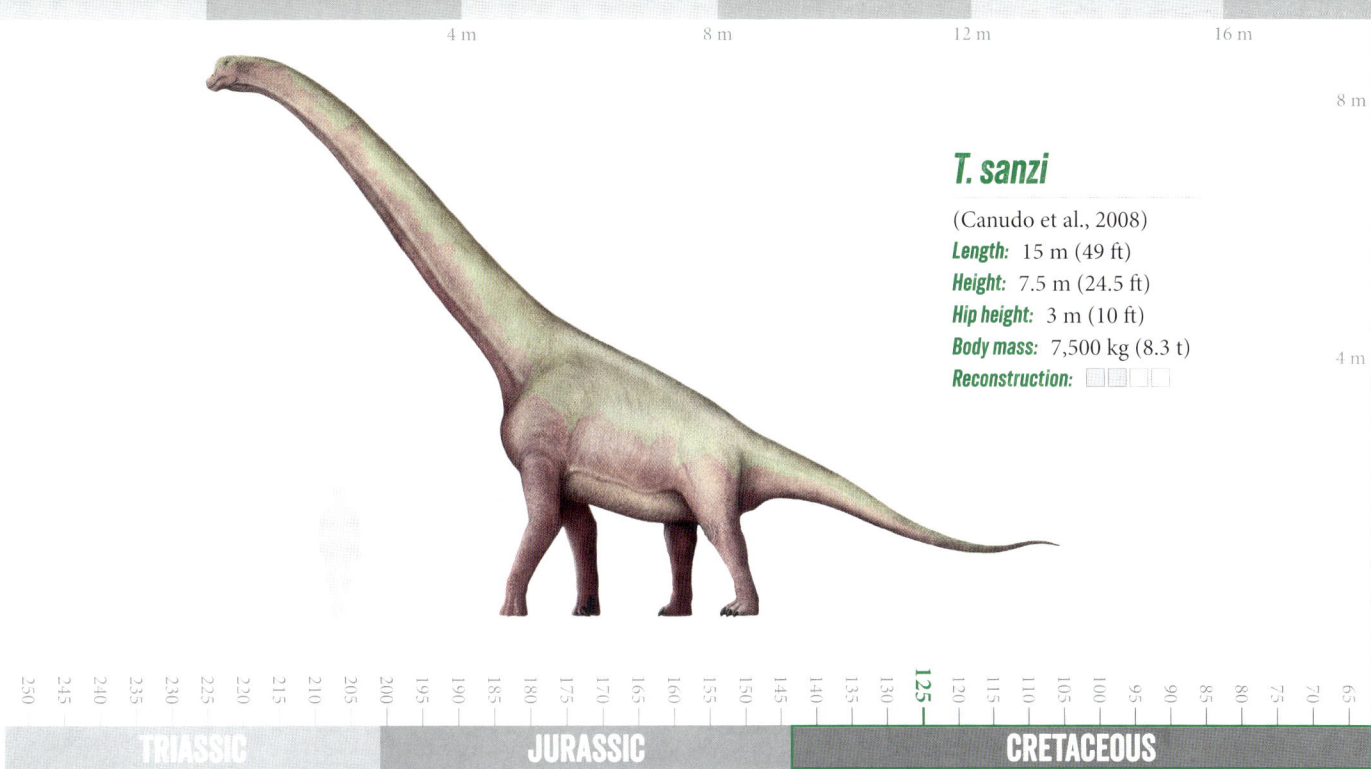

4 m 8 m 12 m 16 m

8 m

4 m

T. sanzi

(Canudo et al., 2008)
Length: 15 m (49 ft)
Height: 7.5 m (24.5 ft)
Hip height: 3 m (10 ft)
Body mass: 7,500 kg (8.3 t)
Reconstruction: ☐☐☐☐

250 245 240 235 230 225 220 215 210 205 200 195 190 185 180 175 170 165 160 155 150 145 140 135 130 **125** 120 115 110 105 100 95 90 85 80 75 70 65

TRIASSIC   **JURASSIC**   **CRETACEOUS**

Tastavinsaurus sanzi (meaning "Sanz's lizard from Tastavins") is known from two partial specimens. The holotype preserves essentially the entire rear half of the animal.

Opinions on the taxonomic status of *Tastavinsaurus* have varied considerably between different studies. Researcher Rafael Royo-Torres erected the clade Laurasiformes, placed as a basal branch of Titanosauriformes, to contain *Tastavinsaurus* and what were then thought to be its closest relatives, *Cedarosaurus* and *Venenosaurus* (Royo-Torres et al., 2012). As time moved forward, though, more studies began to show *Cedarosaurus* and *Venenosaurus* as being brachiosaurs; consequently, some studies moved the entirety of Laurasiformes within Brachiosauridae (Royo-Torres et al., 2017).

Rather contrastingly, numerous subsequent analyses have found *Tastavinsaurus* to be quite far removed from Brachiosauridae, casting the validity of Laurasiformes into doubt. Rather, the current majority tend to place the genus as a basal **somphospondylan** (Mannion et al., 2019 Pérez-Pueyo et al., 2019; Moore et al., 2020; Apesteguía et al., 2021; Dai et al., 2022). On the other hand, there are those that place it much more stemward, as a basal macronarian (Carballido et al., 2020; An et al., 2023; Ren et al., 2023).

Within Somphospondyli, some works have shown *Tastavinsaurus* to be the sister taxon to *Europatitan* (Wang et al., 2021; Mo et al., 2022), while others have shown them to not be that closely related to one another (Fernández-Baldor et al., 2017; Mannion et al., 2019).

The generic name *Tastavinsaurus* refers to the village of Peñarroya de Tastavins, near where the fossils were found; "tastavin" is also Catalan for "wine taster." The specific name *sanzi* honors José Luis Sanz for his work studying the dinosaurs of Spain.

CLASSIFICATION
Dinosauria
 Sauropoda
  Gravisauria
   Eusauropoda
    Neosauropoda
     Macronaria
      Titanosauriformes
       Somphospondyli

LOCATION
Spain

KNOWN REMAINS
Partial skeleton

8 m   16 m   24 m   32 m

16 m

8 m

S. proteles

(Wedel et al., 2000)
Length: 29 m (95 ft)
Height: 16 m (52.5 ft)
Hip height: 4.5 m (14.7 ft)
Body mass: 40,000 kg (44.1 t)
Reconstruction: ☐☐☐☐

250 245 240 235 230 225 220 215 210 205 200 195 190 185 180 175 170 165 160 155 150 145 140 135 130 125 120 **115** 110 105 100 95 90 85 80 75 70 65

TRIASSIC   **JURASSIC**   **CRETACEOUS**

Sauroposeidon proteles (meaning "Poseidon's lizard at the end") lived in North America during a time when most sauropod species on the continent were shrinking to smaller sizes. This is why, when the series of enormous sauropod neck bones were discovered in Oklahoma in 1994, they were just assumed to be pieces of petrified wood—surely, it was thought, no bones could be that big.

In 1999, the excavator, Richard Cifelli, directed graduate student Matt Wedel to analyze the specimens, and their true nature was revealed. Having been partially exposed, the left side of the bones was in a worse state of preservation than the right, and the fossils were generally quite fragile. Media coverage at the time often claimed *Sauroposeidon* to be the largest sauropod of all time, but this is not likely to be the case.

Additional remains from a locality in Wyoming have also been referred to *Sauroposeidon*. At least three individuals are represented: a juvenile, a subadult, and an adult. A popular case has also been made that *Paluxysaurus* is synonymous with *Sauroposeidon* (D'Emic and Foreman, 2012).

Early interpretations of the holotype remains placed *Sauroposeidon* as a brachiosaur. However, the current consensus places it among the non-titanosaurian **somphospondylans** (Fernández-Baldor et al., 2017; Bellardini et al., 2022). This subsequently alters the original reconstruction of the animal, which had been viewed as something like a "thin-necked" version of *Brachiosaurus*.

The generic name *Sauroposeidon* combines the Greek "sauros" (meaning "lizard") with the name of the ancient Greek deity of earthquakes and the sea, Poseidon. The specific name *proteles* is Greek for "result" or "the end." This is meant to refer to that at the time of its description, *Sauroposeidon* was the final sauropod known in North America prior to the "sauropod hiatus."

CLASSIFICATION

Dinosauria
 Sauropoda
  Gravisauria
   Eusauropoda
    Neosauropoda
     Macronaria
      Titanosauriformes
       Somphospondyli

LOCATION

United States

KNOWN REMAINS

Partial skeleton

4 m   8 m   12 m   16 m   20 m

8 m

P. jonesi

(Rose, 2007)
Length: 20 m (64 ft)
Height: 11 m (36 ft)
Hip height: 4.5 m (14.5 ft)
Body mass: 12,500 kg (13.8 t)
Reconstruction: ☐☐☐☐☐

4 m

250 245 240 235 230 225 220 215 210 205 200 195 190 185 180 175 170 165 160 155 150 145 140 135 130 125 120 **115** 110 105 100 95 90 85 80 75 70 65

| TRIASSIC | JURASSIC | CRETACEOUS |

Paluxysaurus jonesi (meaning "Jones's lizard from Paluxy") is potentially the same animal as *Sauroposeidon*. According to work done by Michael D'Emic, it is likely that the two genera are synonymous. Although the remains of the two sauropods were not found at the same location, the strata that they were embedded in are "laterally equivalent," being of the same age. Additionally, D'Emic claims that traits originally proposed to differentiate *Paluxysaurus* from *Sauroposeidon* are not actually distinguishable (D'Emic, 2013).

Still, even the largest size estimates of *Paluxysaurus* indicate that the animal was significantly smaller than *Sauroposeidon*. D'Emic explains this difference by showing that the known *Paluxysaurus* specimens were not fully grown, as was originally thought, by pointing to some vertebral features that are not fully fused, as they would be if the animal had reached its maximum size. A histological

examination of the inner bone tissue also came to the same conclusion (D'Emic, 2013).

While the synonymy of the two sauropods has been largely accepted, some researchers allow for the possibility that they may yet be distinguishable; if not separate genera, then potentially at least separate species (Averianov et al., 2018).

The remains attributed to *Paluxysaurus* come from the same location and originally belonged to at least four different individuals of approximately the same age. The partially disarticulated remains were collected from 1985 through 1987, and then from 1993 onward. The animals were buried in fluvial (river-based) deposits, along with portions of petrified wood.

The generic name *Paluxysaurus* refers to the town of Paluxy, Texas. The specific name *jonesi* honors landowner William R. (Bill) Jones for his cooperation.

CLASSIFICATION
Dinosauria
 Sauropoda
  Gravisauria
   Eusauropoda
    Neosauropoda
     Macronaria
      Titanosauriformes
       Somphospondyli

LOCATION
Texas, USA

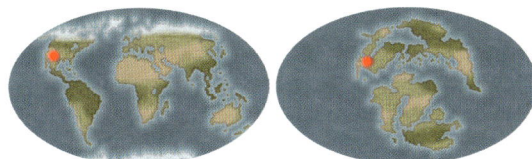

KNOWN REMAINS
Partial skull and skeleton

233

SOMPHOSPONDYLI

8 m

4 m

S. astrosacralis

(Averianov et al., 2018)
Length: 21.5 m (70 ft)
Height: 8.3 m (27.2 ft)
Hip height: 4.2 m (13.8 ft)
Body mass: 21,500 kg (23.7 t)
Reconstruction: ☐☐☐☐

| 250 | 245 | 240 | 235 | 230 | 225 | 220 | 215 | 210 | 205 | 200 | 195 | 190 | 185 | 180 | 175 | 170 | 165 | 160 | 155 | 150 | 145 | 140 | 135 | 130 | 125 | 120 | 115 | 110 | 105 | 100 | 95 | 90 | 85 | 80 | 75 | 70 | 65 |

| TRIASSIC | JURASSIC | CRETACEOUS |

Sibirotitan astrosacralis (meaning "star-boned giant from Siberia") is among the earliest known Asian titanosauriforms.

In 2002, a partial sauropod foot was described from the Ilek Formation of Siberia, at one of the richest fossil-bearing sites in all of Russia. Although it was identifiable as belonging to a titanosauriform, it did not have the unique traits necessary to attribute it to a particular genus. Over the years, more sauropod remains were uncovered from the same location, including vertebrae found in 2008 and 2011. Taken altogether, the material preserves enough distinctive characteristics to justify the erection of a new genus, *Sibirotitan*.

Most of the material is believed to have come from a single individual, an adult, but the presence of a partial juvenile vertebra indicates that the remains of more than one individual were present. Reportedly, large dinosaur bones were found at the location in the early 1960s, which could hypothetically have belonged to the adult, but the current whereabouts of these fossils are unknown.

Another bone, the second neck vertebrae (the axis) was identified in 2022. It had been part of a collection of fossils gathered by a local teacher, G. A. Chudovoy (Averianov and Lopatin, 2022).

The describers taxonomically placed *Sibirotitan* as a titanosauriform, with at least one trait indicating that it was a comparatively advanced non-titanosaurian **somphospondylan**.

The generic name *Sibirotitan* refers to the region of Siberia, Russia. The specific name *astrosacralis* combines the Greek "astro" (meaning "star") and the Latin "os sacrum" (meaning "sacred bone"), referring to the configuration of the animal's sacral ribs, as seen from above.

CLASSIFICATION
Dinosauria
 Sauropoda
  Gravisauria
   Eusauropoda
    Neosauropoda
     Macronaria
      Titanosauriformes
       Somphospondyli

LOCATION
Siberia, Russia

KNOWN REMAINS
Fragments

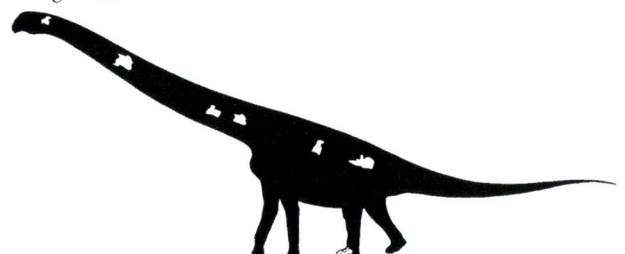

4 m   8 m   12 m   16 m

P. leivaensis

(Carballido et al., 2015)
Length: 16 m (53 ft)
Height: 8 m (26 ft)
Hip height: 3.1 m (10.2 ft)
Body mass: 12,000 kg (13.2 t)
Reconstruction: ☐☐☐☐

4 m

250 245 240 235 230 225 220 215 210 205 200 195 190 185 180 175 170 165 160 155 150 145 140 135 130 125 120 115 110 105 100 95 90 85 80 75 70 65

TRIASSIC   **JURASSIC**   **CRETACEOUS**

Padillasaurus leivaensis (meaning "Padilla's lizard from Leiva") is known only from a series of incomplete vertebrae. These fossils were discovered and unearthed by local farmers in the 1990s. As such, the exact location of the specimen's discovery is unknown. Ammonoids that were included with the remains show that the animal was buried in marine sediment, having presumably been washed out to sea after its death.

Initially, *Padillasaurus* was classified within Brachiosauridae. This apparent finding was significant, as at that point no brachiosaur species had ever appeared from South America. As such, this was not a determination that the describers made lightly. One key attribute in this determination was the presence of a certain opening in some of the tail vertebrae, a trait only seen previously in brachiosaurs. Still, the authors made clear that this

determination was "weak" given the overall lack of skeletal remains with which to work.

However, a subsequent study that was specifically focused on the evolution of Brachiosauridae found *Padillasaurus* to be a basal **somphospondylan** instead. This finding was partly due to the interim discovery of *Savannasaurus*, a non-brachiosaur that nonetheless possessed that particular vertebral opening (Mannion et al., 2017). Ensuing studies have agreed with this somphospondylan placement, although *Padillasaurus*'s exact place within the group is still debatable (Moore et al., 2020; Wang et al., 2021).

The generic name *Padillasaurus* honors paleontological enthusiast Carlos Bernardo Padilla Bernal (1957–2013), who helped create the Centro de Investigaciones Paleontologicas in Villa de Leiva, Colombia; the specific name *leivaensis* refers to this location.

CLASSIFICATION
Dinosauria
　Sauropoda
　　Gravisauria
　　　Eusauropoda
　　　　Neosauropoda
　　　　　Macronaria
　　　　　　Titanosauriformes
　　　　　　　Somphospondyli

LOCATION
Colombia

KNOWN REMAINS
Vertebrae

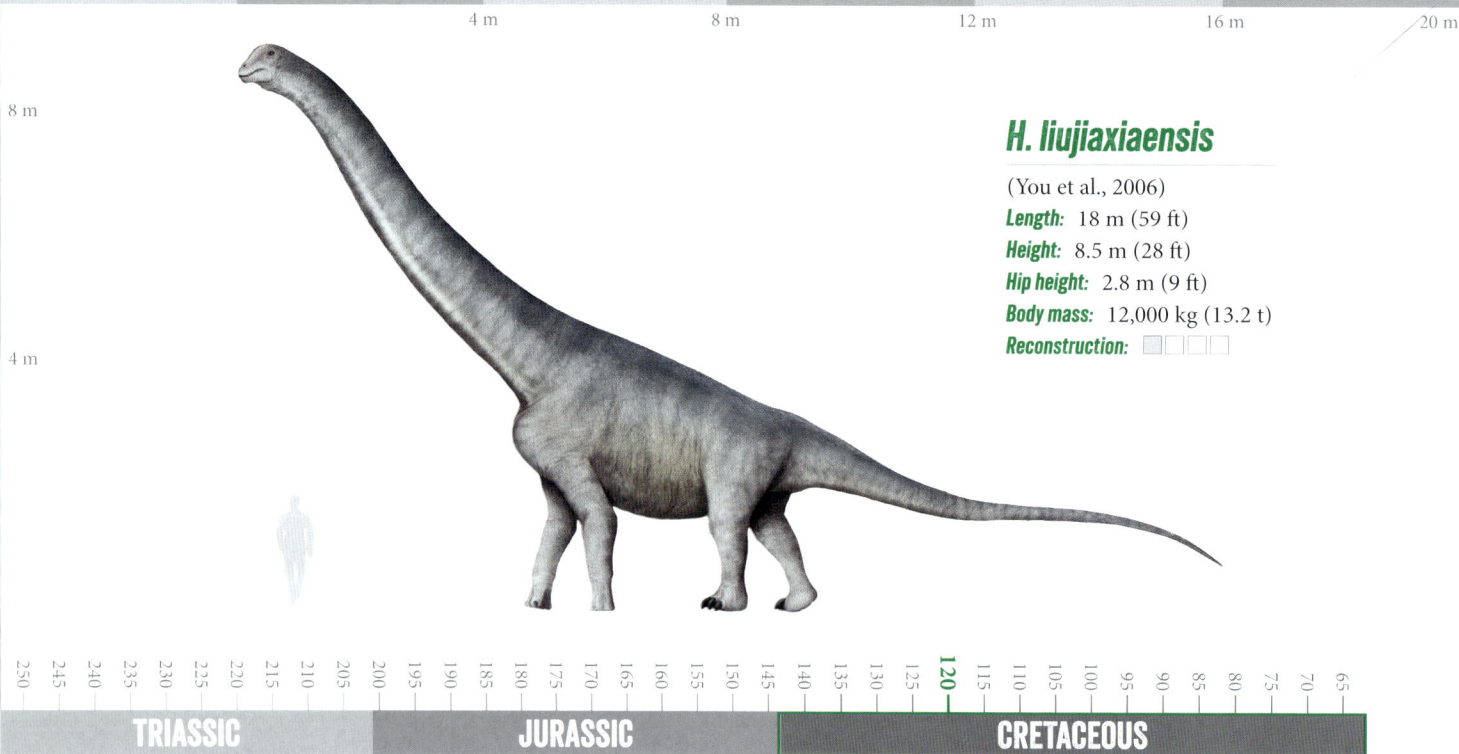

H. liujiaxiaensis

(You et al., 2006)
Length: 18 m (59 ft)
Height: 8.5 m (28 ft)
Hip height: 2.8 m (9 ft)
Body mass: 12,000 kg (13.2 t)
Reconstruction: �

| | | | | | |

TRIASSIC · **JURASSIC** · **CRETACEOUS**

Huanghetitan liujiaxiaensis (meaning "giant from the Liujia Gorge of Yellow River") is an enigmatic sauropod. It is currently known only from a sacrum, two tail vertebrae, a few ribs, and shoulder elements that were unearthed in 2004 and has thus far only been described in a "preliminary" review.

Phylogenetic studies tend to place *H. liujiaxiaensis* among the most derived of the non-titanosaur **somphospondylans** (Moore et al., 2020; Wang et al., 2021; Mo et al., 2022), although it has also been recovered as a basal titanosaur (Mannion et al., 2019).

In 2007, a second species within the genus was described, *H. ruyangensis*. The specimen included the sacrum and the foremost portion of the tail. Numerous other sauropod remains from various sites in the vicinity may or may not be attributable to the same species, including numerous limb elements (Lü et al., 2007, 2009). The animal also dwarfed the type specimen, weighing in at an estimated 40,800–49,900 kilograms (45–55 tons; Paul, 2019).

However, several subsequent studies have found that the two species are not actually very closely related. This means that "*Huanghetitan*" *ruyangensis* should not be included in the same genus and should be assigned a new generic name (Mannion et al., 2013). (Only the fossils of *H. liujiaxiaensis* are depicted below.) A detailed analysis of all possible "*H.*" *ruyangensis* material has yet to be published. With what data are available, the species could potentially belong to Euhelopodidae (Wang et al., 2021).

The binomial name derives from the Chinese "huanghe," meaning "Yellow River," which flows near the fossil sites, and "liujiaxia," meaning "Liujia Gorge," which is part of the Yellow River. The name *ruyangensis* refers to Ruyang County of Henan Province, China.

CLASSIFICATION
Dinosauria
 Sauropoda
  Gravisauria
   Eusauropoda
    Neosauropoda
     Macronaria
      Titanosauriformes
       Somphospondyli

LOCATION
China

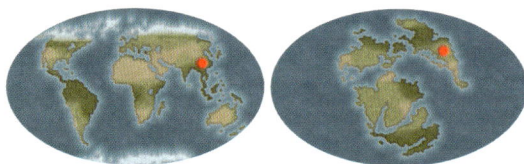

KNOWN REMAINS
Fragments

4 m   8 m   12 m   16 m

4 m

W. wattsi

(Hocknull et al., 2009)
Length: 15 m (49 ft)
Height: 7.5 m (24.5 ft)
Hip height: 3.5 m (11.5 ft)
Body mass: 7,000 kg (7.7 t)
Reconstruction: ☐☐☐☐

250 245 240 235 230 225 220 215 210 205 200 195 190 185 180 175 170 165 160 155 150 145 140 135 130 125 120 115 110 105 100 **95** 90 85 80 75 70 65

| TRIASSIC | JURASSIC | CRETACEOUS |

Wintonotitan wattsi (meaning "Watts's giant from Winton") is primarily known from a set of remains that had long been considered to represent the sauropod *Austrosaurus*, which at the time was the only known Cretaceous sauropod from the Australian continent. (Some additional tail vertebrae from a separate location have also been attributed.)

A reexamination of the fossils, which was summarized in 2009, concluded that the remains actually represented a distinct genus, and thus *Wintonotitan* was erected. This analysis concluded that *Wintonotitan* was likely a basal titanosauriform, although the describers also expressed the possibility that it may belong within Titanosauria.

A comprehensive and dedicated revaluation of the fossils in 2014 found evidence that *Wintonotitan* was a non-titanosaurian **somphospondylan**. The redescription also corrected several errors present in the original manuscript, including the misidentification of several skeletal elements. For instance, elements previously hypothesized to be armor-like osteoderms were instead identified as vertebral neural spines from *Wintonotitan* (Poropat et al., 2014).

Wintonotitan shared its environment with the titanosaur *Diamantinasaurus*, which shows that multiple varieties of sauropods were able to coexist in Australia at the time. As Australia's dinosaurian fossil record is notoriously poor, any information regarding species diversity is valuable.

The generic name *Wintonotitan* refers to the town of Winton, Australia. The specific name *wattsi* honors fossil discoverer Keith Watts, who donated the fossils to the Queensland Museum in 1974 after their discovery at the Elderslie Sheep Station, which he owned.

CLASSIFICATION
Dinosauria
 Sauropoda
  Gravisauria
   Eusauropoda
    Neosauropoda
     Macronaria
      Titanosauriformes
      Somphospondyli

LOCATION
Australia

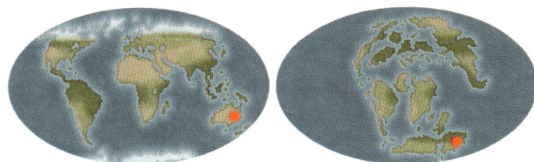

KNOWN REMAINS
Partial skeleton

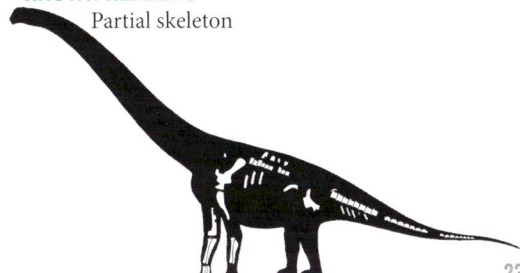

237

| 4 m | 8 m | 12 m | 16 m | 20 m | 24m |

M. florenciae

(González Riga et al., 2009)
Length: 20 m (66 ft)
Height: 10 m (33 ft)
Hip height: 3.5 m (11.5 ft)
Body mass: 16,000 kg (17.6 t)
Reconstruction: ☐☐☐☐

| | | TRIASSIC | | | JURASSIC | | | CRETACEOUS | | |

Malarguesaurus florenciae (meaning "Florencia's lizard from Malargüe") is one of the many Late Cretaceous sauropods known from the strata of Argentina.

Two fragmentary sets of remains are known of *Malarguesaurus*. In 2005, the fossils of the holotype were found jumbled in a small area of approximately 16 square meters; additionally, two tail bones and a partial fibula, found nearby, were designated as paratype specimens. The remains suffered some weathering before their burial; no other kinds of skeletal remains were found at the site. The bones were first thought to have come from an adult individual, but later evidence suggested that this was not the case (Previtera, 2017).

The describers of *Malarguesaurus* categorized it as a non-titanosaurian **somphospondylan**, a position that has also been found by at least one subsequent study (Carballido

et al., 2022). If accurate, this would make *Malarguesaurus* one of the last surviving non-titanosaurs.

This phylogenetic designation is not completely clearcut, though, as other results have indeed found it to be a basal titanosaur (Carballido et al., 2017), while still others have not even been able to hazard a guess, as the genus behaves as an "unstable taxon" when included in analyses (Mannion et al., 2019). This uncertainty is likely due to the partial nature of the remains; in addition, the animal's known tail vertebrae feature an "unusual combination of articulations" unlike those of any other comparable genera (Carballido et al., 2022).

The generic name *Malarguesaurus* refers to the Malargüe region in Mendoza Province, Argentina, combined with the Greek "sauros" (meaning "lizard"). The specific name *florenciae* honors fossil discoverer Florencia Fernández Favarón for her fieldwork collaborations.

CLASSIFICATION
Dinosauria
  Sauropoda
    Gravisauria
      Eusauropoda
        Neosauropoda
          Macronaria
            Titanosauriformes
              Somphospondyli

LOCATION
Argentina

KNOWN REMAINS
Fragments

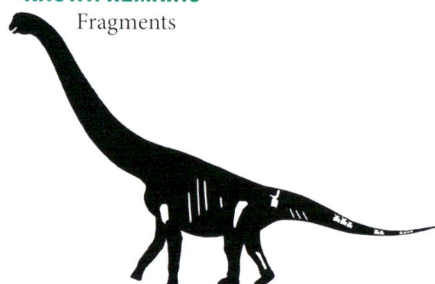

ARKHARAVIA

TRIASSIC	JURASSIC	CRETACEOUS

A. heterocoelica

(Alifanov and Bolotsky, 2010)
Length: 6.3 m (21 ft)
Height: 3.9 m (12.8 ft)
Reconstruction: ☐☐☐☐

Arkharavia heterocoelica is a dubious species. It was described based on one partial tail vertebra (the holotype) as well as several other tail vertebrae and a tooth (referred specimens). Shortly thereafter, though, a study in which one of the original coauthors participated came to the conclusion that these referred vertebrae actually belonged to a hadrosaur (Godefroit et al., 2011). The remaining holotype does belong to some sort of **somphospondylan** sauropod, but does not have any unique diagnostic characteristics (Mannion et al., 2013).

The generic name *Arkharavia* refers to the village of Arkhara, in Russia; this is combined with the Latin "via" (meaning "road"). The specific name *heterocoelica* refers to the heterocoelous shape of the vertebral centra, meaning that the surfaces where the bone joins its neighboring vertebrae are saddle shaped.

ALGOASAURUS

TRIASSIC	JURASSIC	CRETACEOUS

A. bauri

(Broom, 1904)
Length: 7 m (23 ft)
Height: 3 m (10 ft)
Reconstruction: ☐☐☐☐

Algoasaurus bauri is a dubious genus described from fossils that were belatedly recovered from a clay quarry—several bones were reportedly destroyed before they could be salvaged. What survived were "very imperfect fragments" of vertebrae and ribs, a foot bone, a partial femur, and a partial scapula. Unfortunately, most of these remains have subsequently disappeared, leaving only one tail bone and a foot bone. At various points, the genus has been considered a titanosaurian or rebbachisaurid, but most researchers now consider it *nomen dubium*. The most that can be said for its taxonomy is that is an indeterminate **eusauropod** (McPhee et al., 2016).

The generic name *Algoasaurus* refers to Algoa Bay, South Africa. The specific name *bauri* honors German paleontologist Georg Baur (1859–1898).

J. xidiensis

(Wu et al., 2006)
Length: 16.3 m (54 ft)
Height: 9.5 m (31 ft)
Reconstruction:

2 m

Jiutaisaurus xidiensis is known from 18 articulated caudal (tail) vertebrae and thirteen chevrons (tail ribs) that were unearthed in 2003. Its describers placed it within Titanosauria, in a very brief Chinese-language publication.

However, numerous subsequent studies have dismissed the validity of the genus, citing that the "unique" features used to differentiate it are actually present throughout numerous genera in **Titanosauriformes** (Wilson and Upchurch, 2009; Mannion et al., 2013).

The binomial name refers to the district of Jiutai in the city of Changchun, as well as the village of Xidi in Jilin Province, China.

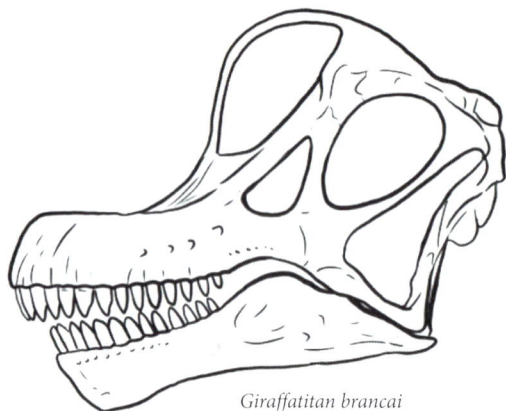

Giraffatitan brancai

Abydosaurus mcintoshi

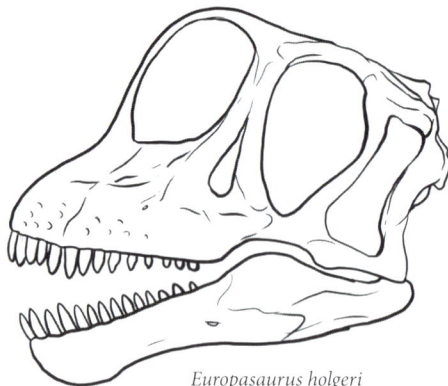

Europasaurus holgeri

Sarmientosaurus musacchioi

4 m   8 m   12 m   16 m

8 m

4 m

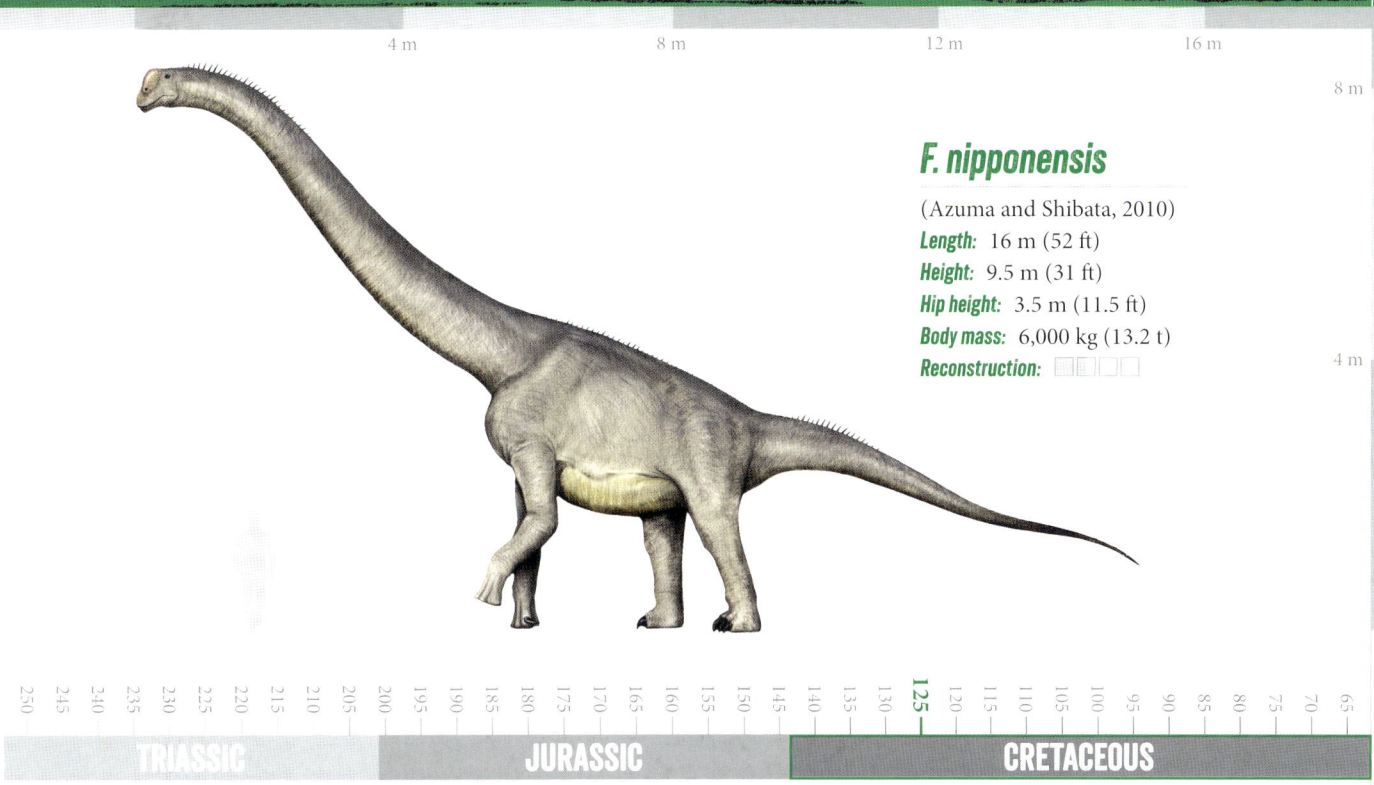

F. nipponensis

(Azuma and Shibata, 2010)

Length: 16 m (52 ft)
Height: 9.5 m (31 ft)
Hip height: 3.5 m (11.5 ft)
Body mass: 6,000 kg (13.2 t)
Reconstruction: ☐☐☐☐

| 250 | 245 | 240 | 235 | 230 | 225 | 220 | 215 | 210 | 205 | 200 | 195 | 190 | 185 | 180 | 175 | 170 | 165 | 160 | 155 | 150 | 145 | 140 | 135 | 130 | 125 | 120 | 115 | 110 | 105 | 100 | 95 | 90 | 85 | 80 | 75 | 70 | 65 |

TRIASSIC   **JURASSIC**   **CRETACEOUS**

Fukuititan nipponensis (meaning "giant from Fukui, Nippon") is the first titanosauriform to be named from Japan; prior to this, only partially identifiable fragments from this clade had been found in the country. Its remains were uncovered in 2007 at the fossiliferous Kitadani Dinosaur Quarry, which has also produced the theropod dinosaurs *Fukuiraptor* and *Fukuisaurus*, along with numerous aquatic vertebrates, such as fish and turtles.

It is possible that the individual was not fully grown at the time of its death (Paul, 2024). The describing authors nonetheless estimated the holotype individual to be 16 meters in length.

Some features that distinguish *Fukuititan* as a unique animal include the shape of its teeth, its relatively wide humerus, and metacarpals that are proportionally long in comparison to the radius.

While the describers placed *Fukuititan* as a titanosauriform, its remains are so fragmentary that attempting to place it within a phylogenetic dataset results in it being an "unstable" taxon, prone to shifting its taxonomic position widely. Thus, its exact affinities remain unclear (Mannion et al., 2013). The most recent analysis to include the animal in its calculations places *Fukuititan* as being very closely related to *Dongbeititan* and *Ligabuesaurus* (Han et al., 2024).

The generic name *Fukuititan* refers to the Japanese region of Fukui. The specific name *nipponensis* means "from Japan," as "Nippon" is one way of pronouncing Japan's native name.

CLASSIFICATION

Dinosauria
 Sauropoda
  Gravisauria
   Eusauropoda
    Neosauropoda
     Macronaria
      Titanosauriformes
       Somphospondyli

LOCATION

Japan

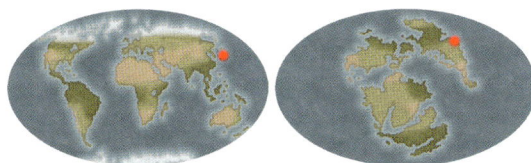

KNOWN REMAINS

Partial limbs and fragments

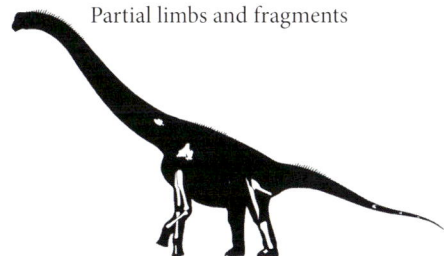

Euhelopodidae

Time scale (across top)

Jurassic — Late — Kimm. | Tithonian
Cretaceous — Early — Berria. | Valanginian | Haut. | Barrem. | Aptian | Albian
Cretaceous — Late — Cenomanian | Turon. | Coniac. | Sant. | Campanian | Maastrichtian

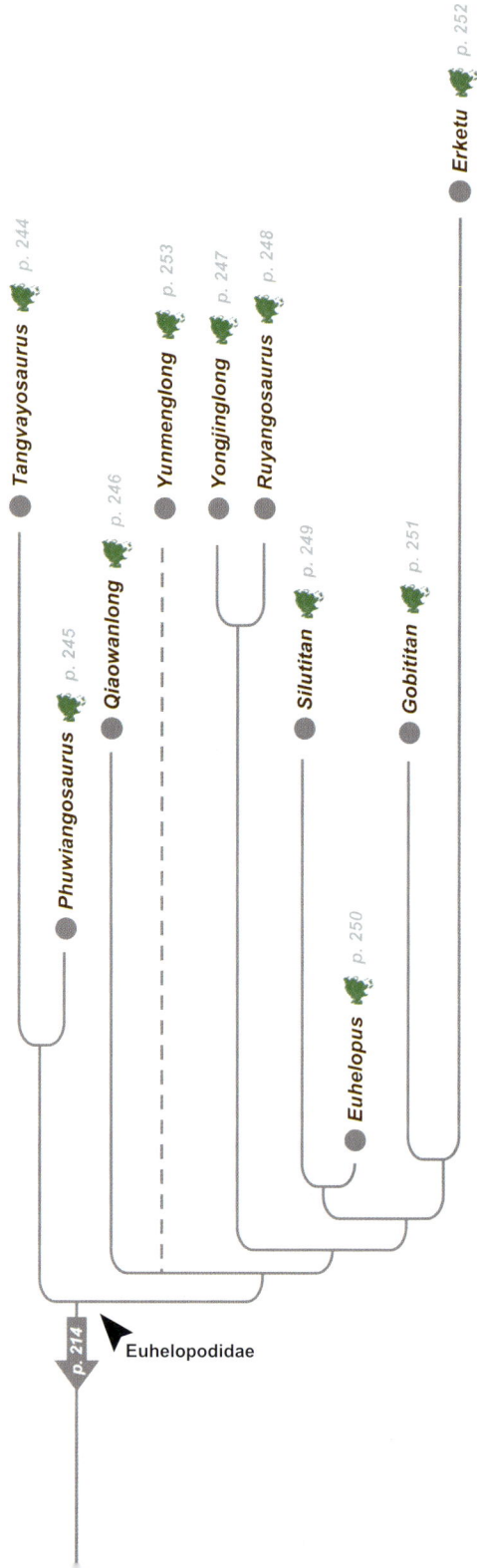

155 | 150 | 145 | 140 | 135 | 130 | 125 | 120 | 115 | 110 | 105 | 100 | 95 | 90 | 85 | 80 | 75 | 70 | 65

Taxa

Tangvayosaurus — p. 244
Phuwiangosaurus — p. 245
Qiaowanlong — p. 246
Yunmenglong — p. 253
Yongjinglong — p. 247
Ruyangosaurus — p. 248
Silutitan — p. 249
Euhelopus — p. 250
Gobititan — p. 251
Erketu — p. 252

p. 214

▶ Euhelopodidae

The family Euhelopodidae has a long history, with its first iteration appearing in 1929. The notion of the group was largely abandoned for decades, and to this day, not all studies agree on the group's validity—and fewer agree on where it should be placed within the broader context of the sauropod family tree. Still, the recent majority of studies find Euhelopodidae to be a somphospondylan clade, roughly sister to Titanosauria; this is the arrangement depicted in this volume.

The analyses conducted by Mannion et al. (2017, 2019b), Averianov et al. (2018), Mocho et al. (2019), and Wang et al. (2021) each place *Euhelopus, Erketu, Gobititan, Qiaowanlong, Phuwiangosaurus,* and *Tangvayosaurus* as being members of the **Euhelopodidae** clade. The exact relationship among these genera differs from study to study, although all find *Phuwiangosaurus* and *Tangvayosaurus* to be sister taxa.

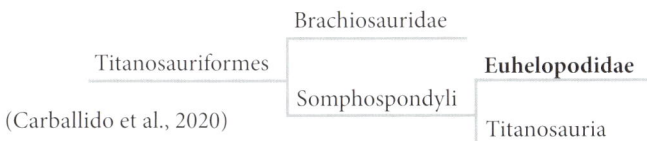

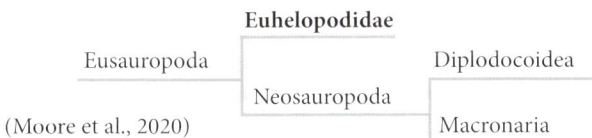

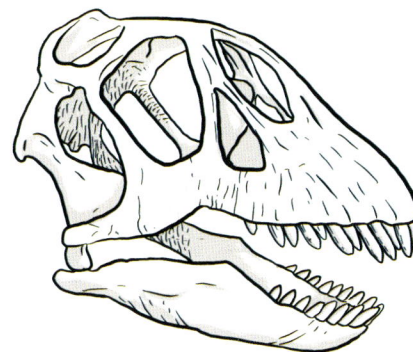

Titanosauriformes — Brachiosauridae / Somphospondyli — **Euhelopodidae** / Titanosauria
(Carballido et al., 2020)

Macronaria — *Euhelopus* / Titanosauriformes — Brachiosauridae / Somphospondyli
(Ren et al., 2023)

Eusauropoda — Euhelopodidae / Neosauropoda — Diplodocoidea / Macronaria
(Moore et al., 2020)

The describers of *Silutitan* identify it as an euhelopid (Wang et al., 2021). *Yongjinglong* was considered a member of the clade by Mannion et al. (2019a), Wang et al. (2021), and Mo et al. (2023), with Averianov and Sues (2017) noting its "*Euhelopus*-like vertebral column."

Although a number of analyses find *Ruyangosaurus* to be a basal titanosaur (Moore et al., 2020; Mo et al., 2023), there is also a growing sentiment that *Ruyangosaurus* is a euhelopodid (Mannion et al., 2019a; Gallina et al., 2021; Wang et al., 2021; Bellardini et al., 2022).

To date, *Yunmenglong* has not been included in any phylogenetic studies beyond that which was conducted by its describers (Lü et al., 2013). They found *Yunmenglong* to be most closely related to *Erketu* and *Qiaowanlong,* which are now generally considered to be members of Euhelopodidae.

Interestingly, each of these genera is East Asian in origin.

Euhelopus zdanskyi

4 m	8 m	12 m	16 m	20 m

8 m

4 m

T. hoffeti

(Allain et al., 1999)
Length: 19 m (62 ft)
Height: 9 m (30 ft)
Hip height: 3.5 m (11.5 ft)
Body mass: 17,000 kg (18.7 t)
Reconstruction:

TRIASSIC	JURASSIC	CRETACEOUS

Tangvayosaurus hoffeti (meaning "Hoffet's lizard from Tang Vay") is a little-studied dinosaur as its remains have received little direct attention since their first brief description.

Tangvayosaurus is primarily known from two separate partial skeletons found at locations known as Tang Vay 2 and Tang Vay 4. Both sets of remains were disarticulated but mostly associated within a small area. The best-represented portions of *Tangvayosaurus* are the hindlimb and a large portion of the tail.

Tangvayosaurus is possibly the same animal as *Titanosaurus falloti*, a species described by Josué-Heilmann Hoffet in 1942 and based primarily on a femur that was collected from the same general area.

In general, *Tangvayosaurus* has consistently been placed somewhere near the base of Titanosauria by various studies, although within this region of the family tree, its exact position has historically been less clear. Out of the various recent analyses that have recovered **Euhelopodidae** as a distinct somphospondylan clade, the majority have included *Tangvayosaurus* in their number (Averianov et al., 2018; Mannion et al., 2019a, b; Wang et al., 2021; Mo et al., 2023).

The Grès supérieurs geological formation that has borne *Tangvayosaurus* is not generally known for being particularly fossil rich. One of the only other notable dinosaurian discoveries from the formation is the spinosaurid *Ichthyovenator*. The formation has yet to be precisely dated and could range anywhere from 100 to 120 million years of age.

The generic name *Tangvayosaurus* refers to the region Tang Vay in Khon Kaen Province, Thailand. The specific name *hoffeti* honors French paleontologist Josué-Heilmann Hoffet.

CLASSIFICATION
Sauropoda
　Gravisauria
　　Eusauropoda
　　　Neosauropoda
　　　　Macronaria
　　　　　Titanosauriformes
　　　　　　Somphospondyli
　　　　　　　Euhelopodidae

LOCATION
Laos

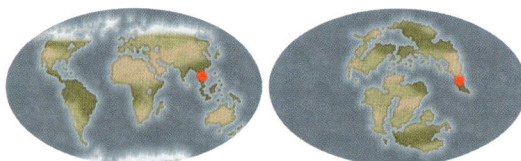

KNOWN REMAINS
Partial skeleton

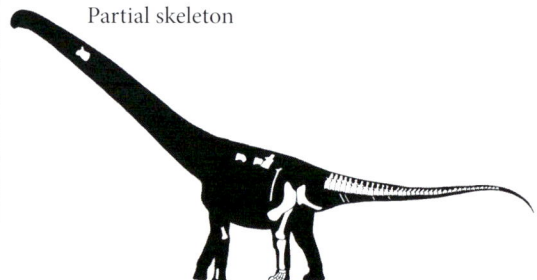

4 m   8 m   12 m   16 m   20 m

P. sirindhornae

8 m

(Martin et al., 1994)
Length: 19 m (61 ft)
Height: 9.5 m (31.1 ft)
Hip height: 4 m (13.1 ft)
Body mass: 11,000 kg (12.1 t)
Reconstruction: ☐☐☐☐

4 m

| 250 | 245 | 240 | 235 | 230 | 225 | 220 | 215 | 210 | 205 | 200 | 195 | 190 | 185 | 180 | 175 | 170 | 165 | 160 | 155 | 150 | 145 | 140 | **133** | 130 | 125 | 120 | 115 | 110 | 105 | 100 | 95 | 90 | 85 | 80 | 75 | 70 | 65 |

TRIASSIC   **JURASSIC**   **CRETACEOUS**

Phuwiangosaurus sirindhornae (meaning "Sirindhorn's lizard from Phu Wiang") is the first sauropod to have been described and named from Thailand.

The holotype specimen, discovered in 1982, comprised only about 10% of the animal's skeleton. However, 2009 saw the description of a much more complete specimen, preserving approximately 60% of the skeleton, including fragments of cranial material (Suteethorn et al., 2009). Shortly thereafter, a few more bones that belonged to the original specimen were revealed (Suteethorn et al., 2010). Neither of these two individuals was fully grown, so determining the potential size of a fully grown adult is problematic (Klein et al., 2009).

Additionally, it seems likely that numerous fossils found in the vicinity belonging to juvenile and baby sauropods also represent *Phuwiangosaurus*, but much of this material remains largely undescribed (Martin et al., 1999). Various

samples have undergone histological examination, though, and it is likely that evidence of sexual dimorphism has been observed (Klein et al., 2009).

Out of the various recent analyses that have recovered **Euhelopodidae** as a distinct somphospondylan clade, the majority have included *Phuwiangosaurus* in their number (Averianov et al., 2018; Mannion et al., 2019a; Wang et al., 2021; Mo et al., 2023). Specific analysis of the brain endocast, obtained from the preserved portions of the skull, led researchers to conclude that *Phuwiangosaurus* represented a "transitional form between the somphospondylan and titanosaurian" sauropods (Kaikaew et al., 2022).

The generic name *Phuwiangosaurus* refers to the region of Phu Wiang in northeastern Thailand. The specific name *sirindhornae* honors Princess Maha Chakri Sirindhorn of Thailand for her interest in Thai paleontology.

CLASSIFICATION
Sauropoda
  Gravisauria
   Eusauropoda
    Neosauropoda
     Macronaria
      Titanosauriformes
       Somphospondyli
        Euhelopodidae

LOCATION
Thailand

KNOWN REMAINS
Partial skull and skeleton

245

4 m   8 m   12 m   16 m   20 m

8 m

4 m

Q. kangxii

(You and Li, 2009)
Length: 19 m (61 ft)
Height: 9.4 m (31 ft)
Hip height: 2.9 m (9.5 ft)
Body mass: 13,000 kg (14.3 t)
Reconstruction:

| 250 | 245 | 240 | 235 | 230 | 225 | 220 | 215 | 210 | 205 | 200 | 195 | 190 | 185 | 180 | 175 | 170 | 165 | 160 | 155 | 150 | 145 | 140 | 135 | 130 | 125 | 120 | 115 | 110 | 105 | 100 | 95 | 90 | 85 | 80 | 75 | 70 | 65 |

TRIASSIC   **JURASSIC**   **CRETACEOUS**

Qiaowanlong kangxii (meaning "Kangxi's dragon bridge over the riverbend") is known from a series of eight articulated cervical (neck) vertebrae as well as the right half of the pelvic girdle, along with some other unidentified fragments. These fossils were excavated in 2007 from the Yujingzi Basin in Gansu Province, China.

The pubis bone of the pelvis is particularly massive, proportionally speaking. The pelvis and the neck are considered to have come from the same animal because the pelvic remains were directly covering the neck bones as they were unearthed.

Qiaowanlong was originally reported as being the first brachiosaurid genus to have been discovered in China, but this view has not been supported by subsequent works as several of the traits used to make this determination have been shown as being widely distributed outside of just that clade (Ksepka and Norel, 2010).

Out of the various recent analyses that have recovered **Euhelopodidae** as a distinct somphospondylan clade, the majority have included *Qiaowanlong* in their number (Averianov et al., 2018; Mannion et al., 2019b; Wang et al., 2021; Mo et al., 2023). Euhelopodidae is, thus far, considered to be an Asian-centric clade, so this assignment is fitting in a biogeographical sense.

The bifurcated shape of its vertebral neural spines would have been a unique feature among Brachiosauridae, but this anatomical characteristic is in line with other euhelopodids.

The generic name *Qiaowanlong* combines the Chinese "qiao" (meaning "bridge"), "wan" (meaning "bend in a stream"), and "long" (meaning "dragon"). This is also the name of a "cultural relic" near the fossil site. The specific name *kangxii* refers to Kangxi, an emperor of the Qing Dynasty, who was known to appreciate and even dream of the natural beauty of the region.

CLASSIFICATION

Sauropoda
 Gravisauria
  Eusauropoda
   Neosauropoda
    Macronaria
     Titanosauriformes
      Somphospondyli
       Euhelopodidae

LOCATION

China

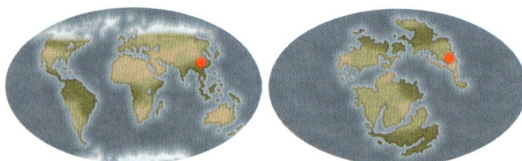

KNOWN REMAINS

Partial neck and pelvis

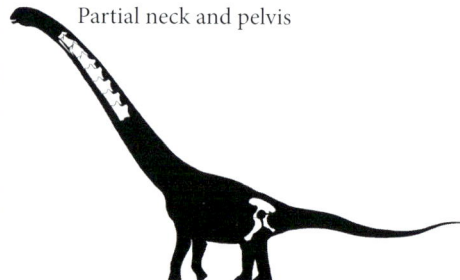

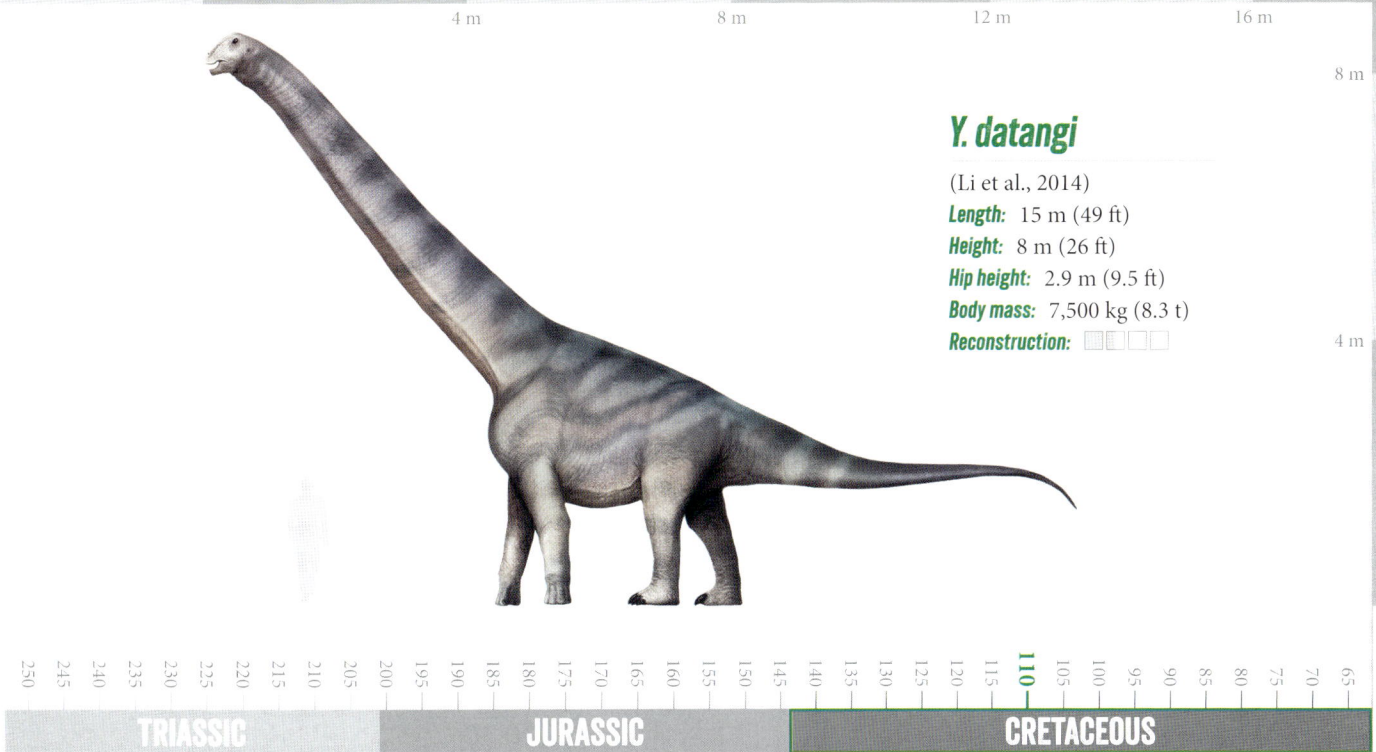

4 m | 8 m | 12 m | 16 m

8 m

4 m

Y. datangi

(Li et al., 2014)
Length: 15 m (49 ft)
Height: 8 m (26 ft)
Hip height: 2.9 m (9.5 ft)
Body mass: 7,500 kg (8.3 t)
Reconstruction: ▢▢▢▢

| 250 | 245 | 240 | 235 | 230 | 225 | 220 | 215 | 210 | 205 | 200 | 195 | 190 | 185 | 180 | 175 | 170 | 165 | 160 | 155 | 150 | 145 | 140 | 135 | 130 | 125 | 120 | 115 | 110 | 105 | 100 | 95 | 90 | 85 | 80 | 75 | 70 | 65 |

TRIASSIC | **JURASSIC** | **CRETACEOUS**

Yongjinglong datangi (meaning "Tang's dragon from Yongjing") is a medium-sized sauropod featuring very stout forelimbs. The only known remains, collected in 2008, are from an adult individual, judging from fusion among the preserved shoulder elements.

Yongjinglong was originally interpreted as a saltasaurid, which would be quite unusual considering when the animal was alive: the Early Cretaceous. It would also be odd given that, proportionately, the shoulder blade of *Yongjinglong* would be record-breakingly large for a saltasaurid.

While the describers of *Yongjinglong* placed it as a fairly derived titanosaurian saltasaurid, this has not borne out. Out of the various recent analyses that have recovered **Euhelopodidae** as a distinct somphospondylan clade, the majority have included *Yongjinglong* in their number (Mannion et al., 2019a; Wang et al., 2021; Mo et al., 2022).

Not only does a euhelopodid placement for *Yongjinglong* make more sense on the timeline of the titanosauriform family tree, but the animal's large shoulders are also much less strange in this case as *Euhelopus* itself is known for its massive shoulders, which were proportionally even larger than those of *Yongjinglong*.

Although only a single neck vertebra of *Yongjinglong* is known, when comparing its proportions against those of *Euhelopus*, it would seem that *Yongjinglong's* neck was actually rather short, proportionally speaking, for a euhelopodid.

The generic name *Yongjinglong* refers to Yongjing County, China. The specific name *datangi* honors the Tang dynasty and more specifically Zhi-Lu Tang from the Institute of Vertebrate Paleontology and Paleoanthropology.

CLASSIFICATION
Sauropoda
 Gravisauria
  Eusauropoda
   Neosauropoda
    Macronaria
     Titanosauriformes
      Somphospondyli
       Euhelopodidae

LOCATION
China

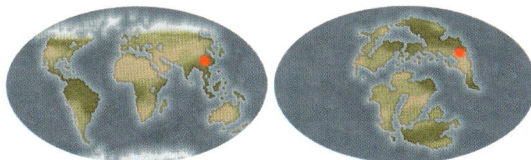

KNOWN REMAINS
Partial skeleton

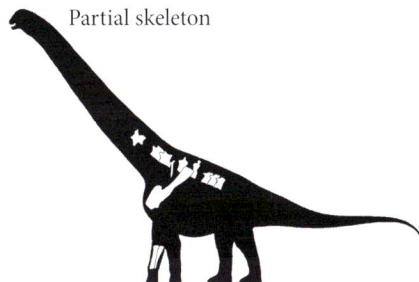

247

| 5 m | 10 m | 15 m | 20 m | 25 m | 30 m |

R. giganteus

(Lü et al., 2009)
Length: 27 m (89 ft)
Height: 15 m (49 ft)
Hip height: 4 m (13 ft)
Body mass: 46,000 kg (50.7 t)
Reconstruction: ▢▢▢▢

| 250 | 245 | 240 | 235 | 230 | 225 | 220 | 215 | 210 | 205 | 200 | 195 | 190 | 185 | 180 | 175 | 170 | 165 | 160 | 155 | 150 | 145 | 140 | 135 | 130 | 125 | 120 | 115 | 110 | 105 | 100 | 95 | 90 | 85 | 80 | 75 | 70 | 65 |

TRIASSIC — **JURASSIC** — **CRETACEOUS**

Ruyangosaurus giganteus (meaning "gigantic lizard from Ruyang") is one of the most massive sauropods known from Asia. It had a proportionately short torso and long neck, featuring some neck vertebrae that are on the verge of being record breaking in length.

The original holotype specimen included only six bones, two of which were ribs. Naturally, this severely limited the information that was obtainable regarding *Ruyangosaurus*. Five years later, though, information was released regarding several new specimens, greatly increasing the number of fossils attributable to the animal (Lü et al., 2014).

Most of the new bones that were excavated from the original quarry may be from the same exact animal as the holotype specimen; these remains represent the largest known *Ruyangosaurus* individual. The quarry also yielded some bones from at least two individuals that were of smaller size. The authors also referred some additional sauropod bones that were found at different locations.

The phylogenetic position of *Ruyangosaurus* is uncertain. Its describers considered the possibility that it represented a non-titanosauriform macronarian. A number of later analyses instead found it to be a basal titanosaur (Moore et al., 2020; Mo et al., 2022). There is also a growing sentiment that *Ruyangosaurus* is a **euhelopodid** (Mannion et al., 2019a; Gallina et al., 2021; Wang et al., 2021; Bellardini et al., 2022).

The Haoling geological formation that yielded the holotype has yet to be precisely dated and could range anywhere from 100 to 120 million years of age.

The generic name *Ruyangosaurus* refers to Ruyang County in Henan Province, China. The specific name *giganteus* is Greek for "gigantic."

CLASSIFICATION
Sauropoda
 Gravisauria
  Eusauropoda
   Neosauropoda
    Macronaria
     Titanosauriformes
      Somphospondyli
       Euhelopodidae

LOCATION
China

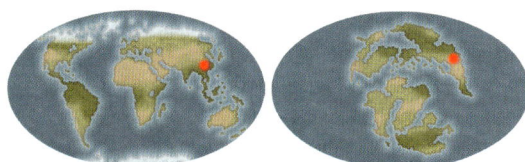

KNOWN REMAINS
Partial skeleton

S. sinensis
(Wang et al., 2021)
Length: 20 m (66 ft)
Height: 11 m (36 ft)
Hip height: 3.9 m (13 ft)
Body mass: 22,000 kg (24.3 t)
Reconstruction: ☐☐☐☐☐

TRIASSIC	JURASSIC	CRETACEOUS

Silutitan sinensis (meaning "Chinese giant of the silk road") is one of the few dinosaurs known from the Hami Pterosaur Fauna.

The year 2006 marked the discovery of a lagerstätte—an area of exceptional fossil preservation—in the Turpan-Hami Basin of western China. The area is one of the few known pterosaur bone beds, containing numerous specimens of very well-preserved adults, eggs, and embryos of the genus *Hamipterus*. Only a few non-pterosaur vertebrate fossils have been found in the area, one of which consists of an articulated set of six sauropod neck vertebrae, unearthed in 2016. These are now known as the holotype specimen of *Silutitan*.

The describers of the genus were able to determine that *Silutitan* is a **euhelopodid**, based on several traits shared with *Euhelopus* and other genera such as *Qiaowanlong*. Specifically, various versions of their analysis found *Silutitan* to be the sister taxon of *Euhelopus*. The majority of euhelopodids have been described from eastern China, so the discovery of *Silutitan* in the western part of the country expands the known geographical range of this particular clade.

A few kilometers away, the tail vertebrae of another sauropod, *Hamititan*, were discovered. With no overlapping material between the two specimens, it cannot be conclusively ruled out that they actually represent the same genus. However, the phylogenetic study that compared the two came to the conclusion that they are likely to be different animals.

The generic name *Silutitan* derives from the Chinese term "silu," meaning "silk road" and commemorating the connection between the East and the West. The specific name *sinensis* refers to China; "sinae" is the Latin word for the Chinese.

CLASSIFICATION
Sauropoda
 Gravisauria
  Eusauropoda
   Neosauropoda
    Macronaria
     Titanosauriformes
      Somphospondyli
       Euhelopodidae

LOCATION
China

KNOWN REMAINS
Partial neck

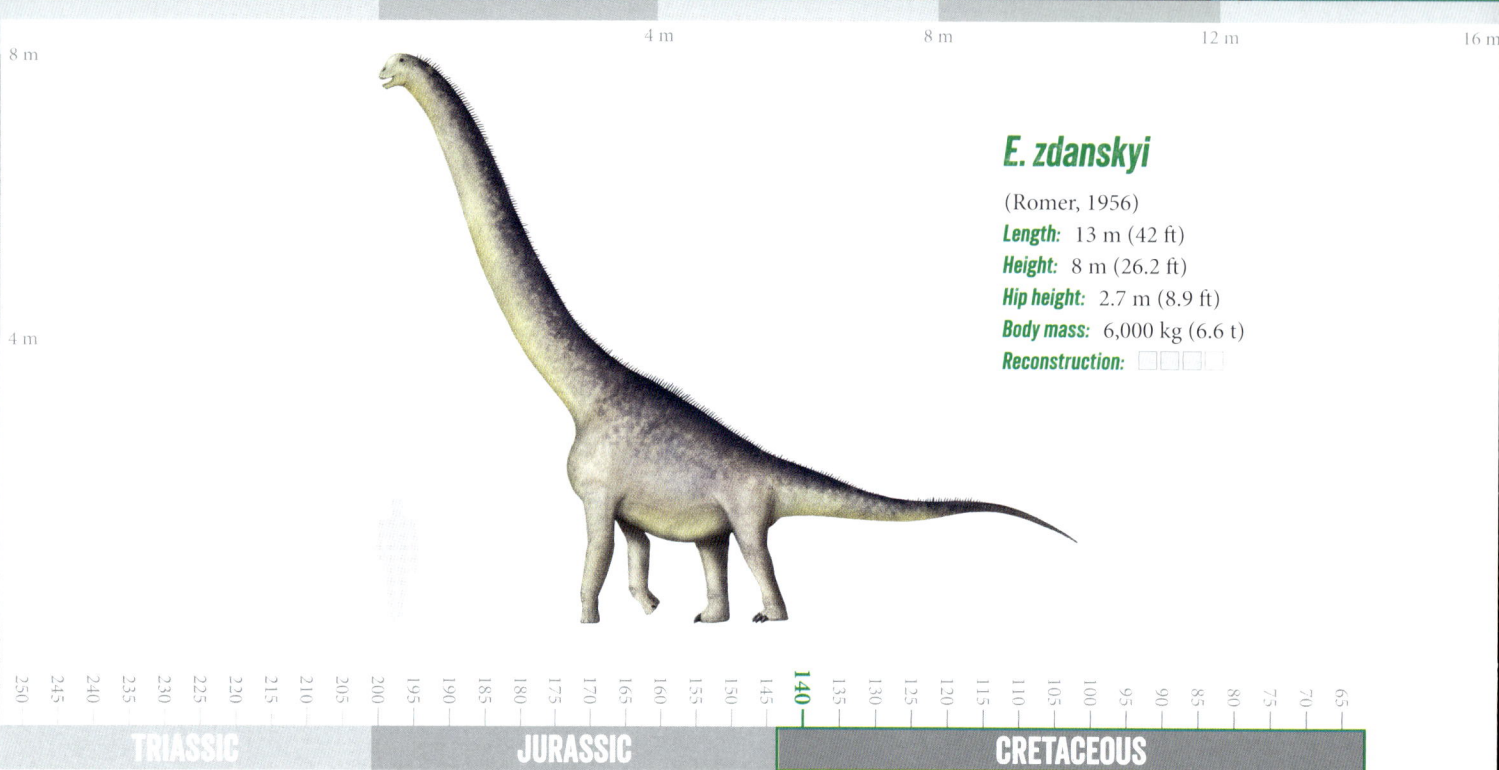

8 m

4 m

4 m   8 m   12 m   16 m

E. zdanskyi

(Romer, 1956)
Length: 13 m (42 ft)
Height: 8 m (26.2 ft)
Hip height: 2.7 m (8.9 ft)
Body mass: 6,000 kg (6.6 t)
Reconstruction: ☐☐☐☐

250 245 240 235 230 225 220 215 210 205 200 195 190 185 180 175 170 165 160 155 150 145 **140** 135 130 125 120 115 110 105 100 95 90 85 80 75 70 65

| TRIASSIC | JURASSIC | CRETACEOUS |

Euhelopus zdanskyi (meaning "Zdansky's true *Helopus*") was the first dinosaur from China to be scientifically examined. The fossils were first noticed in 1913 by Father R. Mertens, a priest who excavated some of the vertebrae. These bones made their way into the hands of the Chinese Geological Survey in 1916. The fossil-bearing site was revisited in 1923, when Otto Zdansky unearthed two sets of remains from locations that were a few kilometers apart. A decade later, more bones were discovered by C. C. Young that likely belonged to the same specimen as the holotype, but the current whereabouts of these later fossils are unknown (Wilson and Upchurch, 2009).

Unlike the titanosaurs and diplodocoids, which had elongated skulls and narrow teeth that were restricted to the front part of the mouth, *Euhelopus* had a boxy skull and robust teeth throughout its jaws. This arrangement might indicate that euhelopodids focused on hardier vegetation, avoiding competition with the contemporary, leaf-stripping titanosaurs (Poropat and Kear, 2013).

To this day, the taxonomic affinities of *Euhelopus* remain highly contentious. Some have interpreted the genus as possibly being a basal titanosauriform, lying stemward of Brachiosauridae (Pérez-Pueyo et al., 2019; Dai et al., 2022; An et al., 2023). Having said that, the interpretation of *Euhelopus* being among the most derived non-titanosaurian **somphospondylans** is gaining an increasing amount of favor among many studies (Mocho et al., 2019; Wang et al., 2021; Mo et al., 2022).

The generic name *Euhelopus* combines the Greek "eu" (meaning "true") with *Helopus*, which was the originally intended name (given by Carl Wiman in 1929) but was preoccupied by a living bird species, leading to the 1956 renaming by Alfred Sherwood Romer. *Helopus* itself combines the Greek "helos" (meaning "swamp") and "pous" (meaning "foot"). The specific name *zdanskyi* honors paleontologist Otto Karl Josef Zdansky.

CLASSIFICATION
Sauropoda
 Gravisauria
  Eusauropoda
   Neosauropoda
    Macronaria
     Titanosauriformes
      Somphospondyli
      Euhelopodidae

LOCATION
China

KNOWN REMAINS
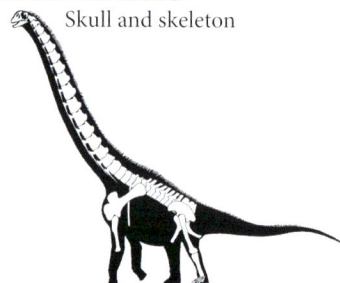
Skull and skeleton

4 m   8 m   12 m

8 m

G. shenzhouensis

(You et al., 2003)
Length: 15 m (49 ft)
Height: 8.5 m (27.9 ft)
Hip height: 2.8 m (9.2 ft)
Body mass: 8,000 kg (8.8 t)
Reconstruction: ☐☐☐☐

4 m

250 245 240 235 230 225 220 215 210 205 200 195 190 185 180 175 170 165 160 155 150 145 140 135 130 125 **120** 115 110 105 100 95 90 85 80 75 70 65

TRIASSIC   **JURASSIC**   **CRETACEOUS**

Gobititan shenzhouensis (meaning "giant from Gobi, China") is known from a single set of remains, consisting of a mostly complete tail and a complete lower hindlimb. These fossils were recovered in 1999 from the middle unit of the Xiagou geological formation; excavations in the general area had begun in 1997.

Unlike the more derived titanosaurs, which tended to have short tails that were composed of fewer than 35 vertebrae, *Gobititan* had a tail that was significantly longer. Forty-one tail vertebrae were recovered, the foremost of which is thought to have been the animal's 13th, meaning that it possessed at least 53, possibly having even more near the tip, which were not recovered as the very last recovered bone was found in a damaged state.

Additionally, the foot of *Gobititan* still retained a fifth digit, a trait that is absent in typical titanosaurs.

Still, the describers considered *Gobititan* to be a basal titanosaur, one that was not particularly closely related to *Euhelopus*, as determined by differences in the anatomy of their feet. *Euhelopus* had fewer "toe" bones and had its largest claw on the second toe, as opposed to *Gobititan*, which had its largest claw on the first toe. *Gobititan* had sizable claws on the first three of its digits.

However, out of the various recent analyses that have recovered **Euhelopodidae** as a distinct somphospondylan clade, the majority have included *Gobititan* in their number (Averianov et al., 2018; Mannion et al., 2019b; Wang et al., 2021; Mo et al., 2022).

The generic name *Gobititan* refers to the Gobi Desert, which stretches across northern China, where the holotype was discovered. The specific name *shenzhouensis* is named for "Shenzhou," which is an ancient name for China.

CLASSIFICATION

Sauropoda
  Gravisauria
   Eusauropoda
    Neosauropoda
     Macronaria
      Titanosauriformes
       Somphospondyli
        Euhelopodidae

LOCATION

China

KNOWN REMAINS

Tail and hindlimb

3 m	6 m	9 m	12 m	15 m

E. ellisoni

(Ksepka and Norell, 2006)
Length: 16 m (52 ft)
Height: 8.2 m (27 ft)
Hip height: 2.8 m (9.1 ft)
Body mass: 5,500 kg (6.1 t)
Reconstruction: ☐☐☐☐☐

250	245	240	235	230	225	220	215	210	205	200	195	190	185	180	175	170	165	160	155	150	145	140	135	130	125	120	115	110	105	100	95	90	85	80	75	70	65

TRIASSIC   **JURASSIC**   **CRETACEOUS**

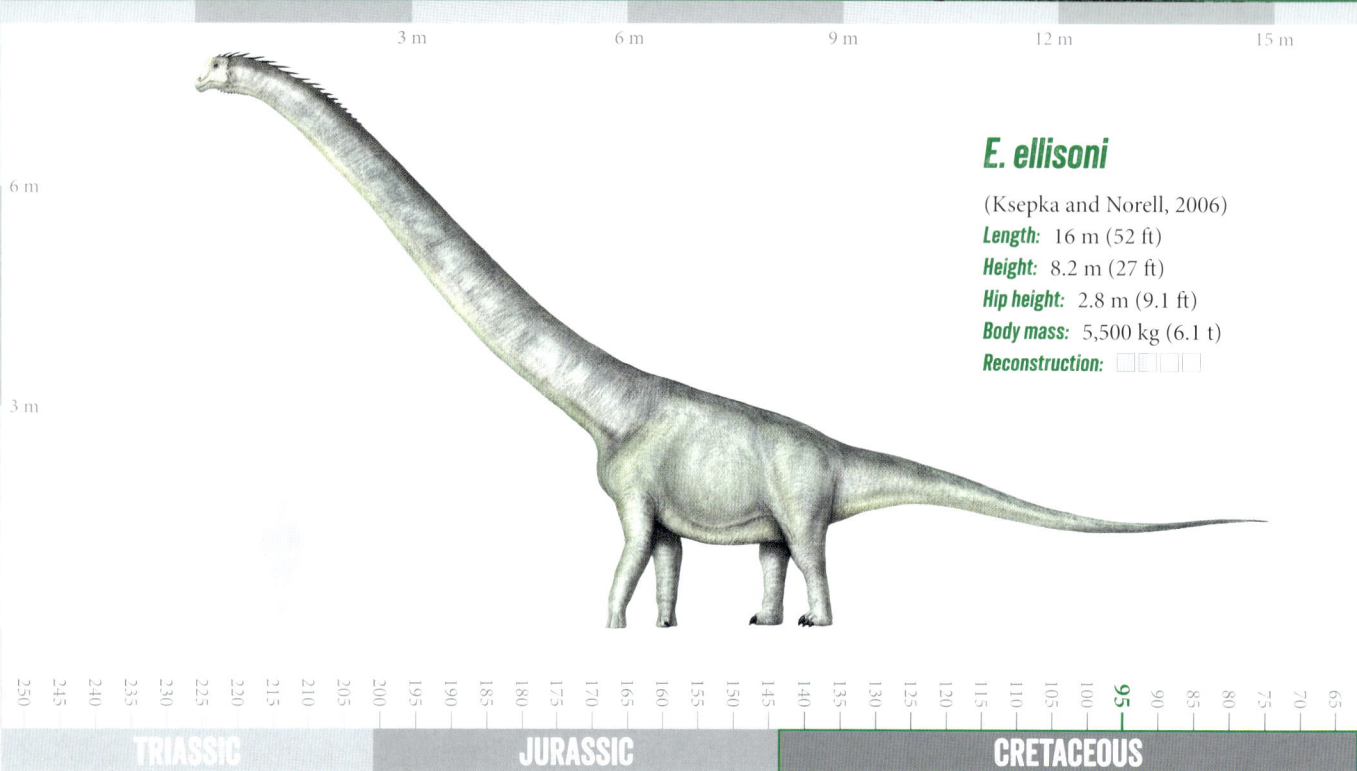

Erketu ellisoni (meaning "Ellison's mighty creator") likely had one of the most elongated necks, in proportion to its body size, of any sauropod. In particular, the third through fifth bones of the neck were extremely long in proportion. The morphology of the bones suggests that they were well equipped to deal with the stress forces that would necessarily be endured by having such a lengthy neck. The bifurcated nature of the neural spines, in particular, would be beneficial in this regard. Nonetheless, *Erketu* was only of moderate overall mass for a sauropod.

In 2002, the first five cervical (neck) vertebrae of *Erketu* were recovered in articulation, along with portions of the sixth. The three successive vertebrae were then recovered during the following year's expedition (Ksepka and Norell, 2010). An exhaustive search for skull material was conducted nearby, but none was recovered. Also unearthed were the lower leg bones and a portion of the sternal plate.

The animal appears to have been buried in a floodplain environment, along with the remains of turtles and some kind of fruit. Today, the region is a part of the Gobi Desert. While this part of the world is famous for its theropod fossils, comparatively few sauropod remains have been described from the region.

Erketu is most often recovered as a close relative of *Euhelopus* within the family **Euhelopodidae** (Mannion et al., 2019b; Wang et al., 2021), although there have been some analyses that have placed it in a significantly more derived position within Titanosauria (Royo-Torres et al., 2020).

The generic name *Erketu* refers to the deity Erketü Tengri of Mongolian shamanistic tradition, the "mighty creator" and chief of the pantheon. The specific name *ellisoni* honors paleoartist Mick Ellison of the American Museum of Natural History.

CLASSIFICATION

Sauropoda
 Gravisauria
  Eusauropoda
   Neosauropoda
    Macronaria
     Titanosauriformes
      Somphospondyli
       Euhelopodidae

LOCATION

Mongolia

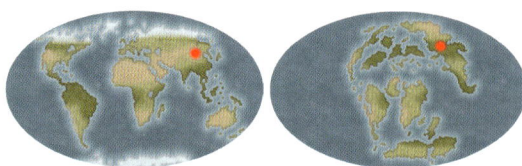

KNOWN REMAINS

Partial neck and leg

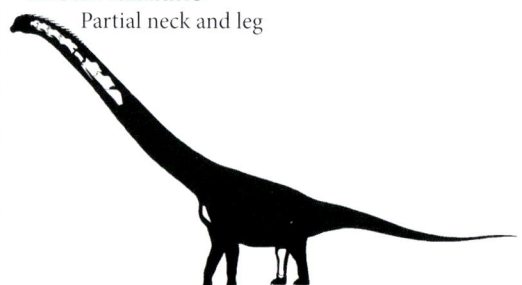

M. haplodont

(Gilmore, 1933)
Length: 16 m (52 ft)
Height: 6 m (20 ft)
Reconstruction:

2 m

110

TRIASSIC | JURASSIC | CRETACEOUS

Mongolosaurus haplodont is known from only a few teeth and fragmentary bones and thus has often been overlooked or neglected. It has occasionally been considered *nomen dubium* (Barrett et al., 2002), but many works consider it to be valid.

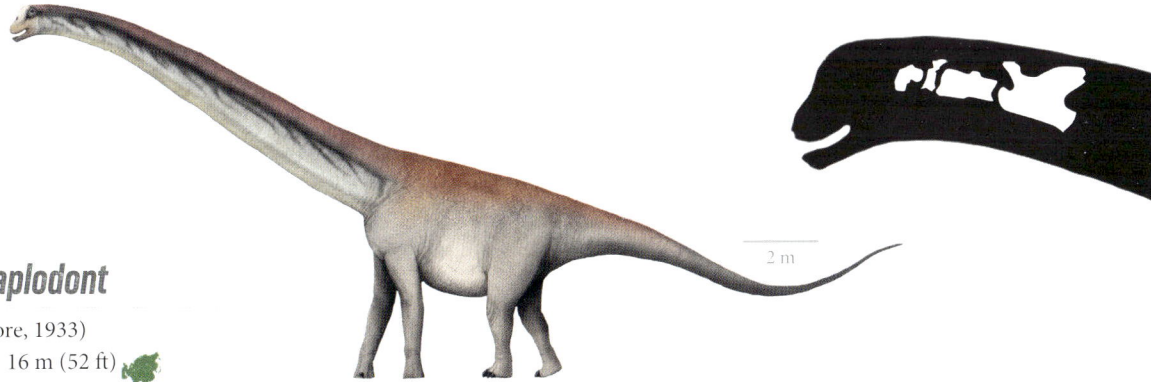

In particular, a thorough redescription of the remains provided an in-depth analysis of each bone, comparing them with specimens of similar taxa in order to justify the continued existence of the genus (Mannion, 2011). This work found *Mongolosaurus* to be a titanosaur, which some subsequent studies have also found (Mo et al., 2022), although others have considered it more likely to be a non-titanosaurian somphospondylan (D'Emic, 2012; Averianov and Sues, 2017).

The generic name *Mongolosaurus* refers to the Inner Mongolia region of China. The specific name *haplodont* combines the Greek "haploos" (meaning "single") and "odontos" (meaning "tooth").

Y. ruyangensis

(Lü et al., 2013)
Length: 27 m (89 ft)
Height: 13 m (43 ft)
Reconstruction:

3 m

110

TRIASSIC | JURASSIC | CRETACEOUS

Yunmenglong ruyangensis (meaning "dragon from Yunmengshan, Ruyang") is a seldom-mentioned sauropod from central China. The specimen preserved the second through the forward portion of the eighth neck vertebrae in their natural position. Disarticulated remains included the presumed ninth and tenth neck bones, one dorsal vertebra, four tail vertebrae, two ribs, and the femur.

To date, *Yunmenglong* has not been included in any phylogenetic studies beyond that which was conducted by its describers. They found *Yunmenglong* to be most closely related to *Erketu* and *Qiaowanlong*, which are now generally considered to be members of **Euhelopodidae**.

The animal's binomial name refers to the Yunmengshan area in Ruyang County, in Henan Province, China.

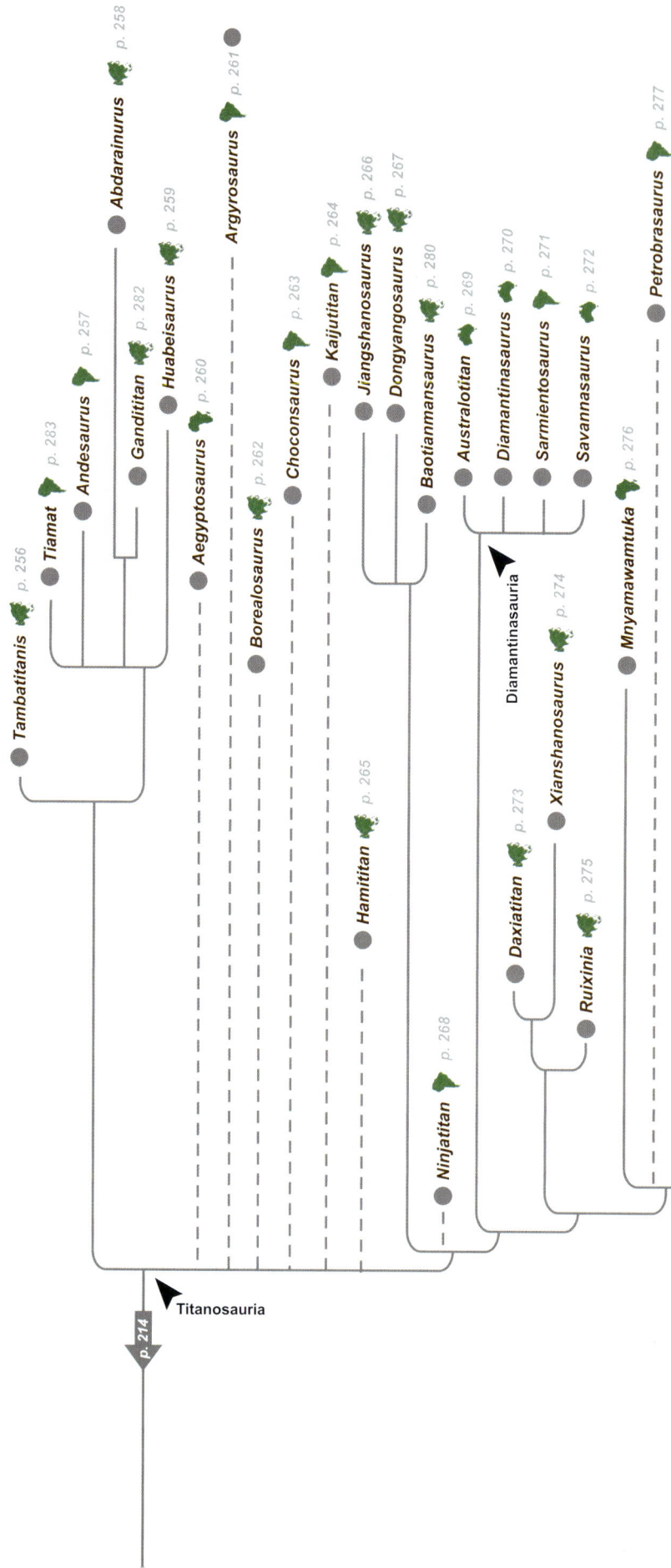

Titanosauria

Cretaceous

Early						Late						
Berria.	Valanginian	Haut.	Barrem.	Aptian	Albian	Cenomanian	Turon.	Coniac.	Sant.	Campanian	Maastrichtian	

145 140 135 130 125 120 115 110 105 100 95 90 85 80 75 70 65

p. 214 ◄ Titanosauria

Tambatitanis p. 256

Tiamat p. 283

Andesaurus p. 257

Ganditaitan p. 282

Abdarainurus p. 258

Huabeisaurus p. 259

Aegyptosaurus p. 260

Argyrosaurus p. 261

Borealosaurus p. 262

Choconsaurus p. 263

Kaijutitan p. 264

Jiangshanosaurus p. 266

Dongyangosaurus p. 267

Hamititan p. 265

Ninjatitan p. 268

Baotiannmansaurus p. 280

Australotitan p. 269

Diamantinasauria ◄

Diamantinasaurus p. 270

Sarmientosaurus p. 271

Savannasaurus p. 272

Daxiatitan p. 273

Xianshanosaurus p. 274

Ruixinia p. 275

Mnyamawamtuka p. 276

Petrobrasaurus p. 277

Lithostrotia p. 284 ►

By some accounts, there are currently more than 100 named titanosaur genera. The majority of these fall within one of the two primary branches of the family tree, Colossosauria and Saltasauroidea, and include a diverse array of sauropods. The base of the titanosaur family tree, though, is a murky area, composed of a tangle of uncertain sub-branches that have an ever-changing membership. Most of these ambiguously placed animals are known from only fragmentary sets of remains. Although numerous titanosaurs have been described in recent years, the taxonomic relationships within the group are sure to be the subject of debate for years to come.

The most commonly used definition of **Titanosauria** itself essentially begins the clade wherever *Andesaurus* is positioned (Carballido et al., 2022). *Abdarainurus* has been recovered as being in a small clade with *Andesaurus*, along with *Huabeisaurus* (Averianov and Lopatin, 2020; Wang et al., 2021; Mo et al., 2023). *Tambatitanis* is quite understudied; euhelopodid (Saegusa and Ikeda, 2014) and/or somphospondylan affinities have been put forth (Mannion et al., 2019a), but the most recent analysis in which it has been included suggests basal titanosaur status (Mo et al., 2023). The describers of *Tiamat* placed it sister to *Andesaurus* (Pereira et al., 2024); likewise with *Ganditian* to *Abdarainurus* (Han et al., 2024).

Aegyptosaurus, based on lost specimens, is thought to be a titanosaur (Gorscak et al., 2014). The understudied *Argyrosaurus* has traits that are very suggestive of basal status (Mannion and Otero, 2012), even though its late occurrence could be indicative of a derived nature (Gorscak and O'Connor, 2019). The describers of *Ninjatitan* put forth two hypotheses, one for a basal titanosaur position and one for a lognkosaurian (Gallina et al., 2021).

Borealosaurus has also been suggested to have basal titanosaurian qualities (Averianov and Sues, 2017). The describers of *Choconsaurus* note that calculations for two positionings have near-equal probability: non-lithostrotian titanosaur or non-titanosaur somphospondylan (Simón et al., 2017).

Calculations by the describers of *Hamititan* proved inconclusive with regard to its placement (Wang et al., 2021). The describers of *Kaijutitan* place it stemward of both Lithostrotia and Eutitanosauria (Filippi et al., 2019).

Sarmientosaurus musacchioi

Jiangshanosaurus has been given various placements, but one common outcome makes it the sister taxon of *Dongyangosaurus* (Mannion et al., 2019a; Wang et al., 2021), the latter of which may be closely related to the fragmentary *Baotianmansaurus* (Poropat et al., 2021).

The position of *Daxiatitan* is uncertain, having been variously recovered as a non-titanosaur somphospondylan (Mannion et al., 2019b; Wang et al., 2021; Bellardini et al., 2022), a basal titanosaur (Averianov and Sues, 2017; Poropat et al., 2021; Mo et al., 2023), or even a euhelopodid (Moore et al. 2018; Gallina et al., 2021). The majority (but not all) of these works find it to be sister to *Xianshanosaurus*. The describers of *Ruixinia* tentatively place the genus in a similar position (Mo et al., 2023).

Mnyamawamtuka has been considered by its describers as being placed near the transition to Lithostrotia (Gorscak and O'Connor, 2019). The describers of *Petrobrasaurus* place it near the base of Eutitanosauria (Filippi et al., 2011).

The clade **Diamantinasauria** has varied in its placement (Carballido et al., 2022), with (Navarro et al., 2022) putting it within Saltasauroidea, while others make it a basal titanosaurian group (Poropat et al., 2021) or potentially a non-titanosaurian somphospondylan group (Poropat et al., 2023). Members are considered to be *Diamantinasaurus*, *Sarmientosaurus*, *Savannasaurus*, and *Australotitan* (Hocknull et al., 2021).

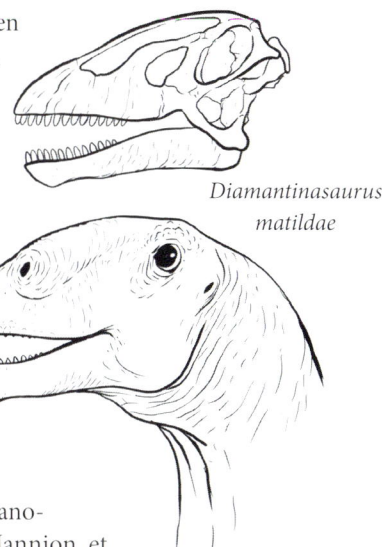

Diamantinasaurus matildae

TAMBATITANIS

T. amicitiae

(Saegusa and Ikeda, 2014)
Length: 18 m (58 ft)
Height: 8 m (26 ft)
Hip height: 2.7 m (9 ft)
Body mass: 8,000 kg (8.8 t)
Reconstruction:

8 m · 4 m · 8 m · 12 m · 16 m

250 245 240 235 230 225 220 215 210 205 200 195 190 185 180 175 170 165 160 155 150 145 140 135 130 125 120 115 110 105 100 95 90 85 80 75 70 65

TRIASSIC · **JURASSIC** · **CRETACEOUS**

Tambatitanis amicitiae (meaning "giant of friendship from Tamba") is a Japanese sauropod featuring highly distinctive tail vertebrae. Rather than having spines that point straight up or angle slightly backward, these bones of *Tambatitanis* have a highly curved shape that orients them forward. It can only be presumed that the tail of *Tambatitanis* featured a unique degree of associated musculature, which could have altered the appendage's range of motion.

The holotype specimen was found in a semi-articulated state, situated in a vaguely lifelike arrangement. The vertebrae of the tail were mostly well preserved, but the majority of the other elements were crushed. The ilium of the pelvis had fractured into 19 fragments that were too delicate to be physically glued together and thus were reconstructed digitally. Thousands of bone shards, too splintered to be reconstructed, are presumed to be all that remains of the missing sacral and dorsal vertebrae. Other

known elements include a partial braincase, a partial jaw, and teeth. At the time the genus was described, though, "much of the material" was still awaiting preparation, so further bones could still be identified in the future. These specimens were all unearthed between 2006 and 2010.

The preliminary investigation of the taxonomic position of *Tambatitanis* placed the genus within Euhelopodidae. Since that time, for some reason it has not been included in many phylogenetic studies. One recent paper found it to be in a slightly different position, as the basalmost **titanosaur** (Mo et al., 2023).

The generic name *Tambatitanis* references the southwest region of Japan, Tamba. The specific name *amicitiae* is Latin for "friendship," signifying the relationship between the two discoverers of the specimen, Murakami Shigeru and Adachi Kiyoshi.

CLASSIFICATION
Sauropoda
 Gravisauria
  Eusauropoda
   Neosauropoda
    Macronaria
     Titanosauriformes
      Somphospondyli
       Titanosauria

LOCATION
Japan

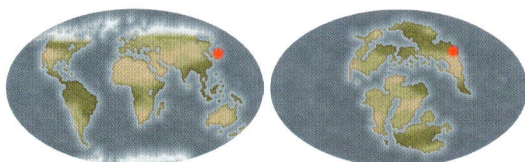

KNOWN REMAINS
Partial skull and skeleton

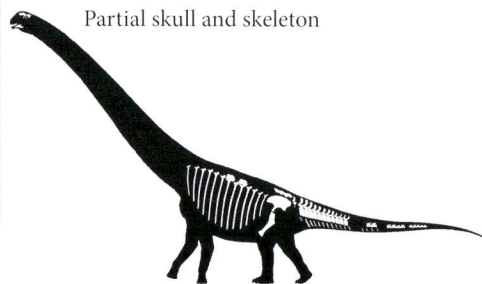

256
TITANOSAURIA

4 m   8 m   12 m   16 m   8 m

4 m

A. delgadoi

(Calvo and Bonaparte, 1991)
Length: 18 m (60 ft)
Height: 9 m (30 ft)
Hip height: 3.5 m (11.5 ft)
Body mass: 20,000 kg (22 t)
Reconstruction: ▢▢▢▢

250 245 240 235 230 225 220 215 210 205 200 195 190 185 180 175 170 165 160 155 150 145 140 135 130 125 120 115 110 105 100 97 90 85 80 75 70 65

| TRIASSIC | JURASSIC | CRETACEOUS |

Andesaurus delgadoi (meaning "Delgado's lizard from the Andes") was originally estimated to be at least 30 meters in length; modern calculations, though, have pared this value down to a much more modest 18 meters. This value relies heavily on comparisons between *Andesaurus* and other closely related genera since no neck bones are known from the animal. Still, it is unclear what stage of growth the animal was in when it died, so the true size of an adult *Andesaurus* technically remains unknown.

Andesaurus possesses some unusual features, such as having exceedingly tall neural spines along its preserved dorsal (back) vertebrae. For this and other reasons, it was originally considered by its describers to represent not only a new genus but an entirely new family, dubbed Andesauridae. However, most modern phylogenetic studies do not show any particular grouping of "andesaurids" on the sauropod family tree, so the existence of this family is often regarded

as doubtful, though not impossible, as some versions of analyses still occasionally return suggestive results (Mannion et al., 2019a).

Still, *Andesaurus* is generally regarded by many studies as one of the most "primitive" **titanosaurs**. In fact, some formal definitions of Titanosauria define the entire clade based on the phylogenetic position of *Andesaurus*.

Currently, the only recognized remains of *Andesaurus* are those of the original holotype specimen. Claims have been made assigning additional fossils to the genus; however, a thorough redescription of *Andesaurus* in 2011 concluded that these other purported specimens were not referrable (Mannion and Calvo, 2011).

The generic name *Andesaurus* refers to the Andes Mountains. The specific name *delgadoi* honors fossil discoverer Alejandro Delgado.

CLASSIFICATION
Sauropoda
 Gravisauria
  Eusauropoda
   Neosauropoda
    Macronaria
     Titanosauriformes
      Somphospondyli
       Titanosauria

LOCATION
Argentina

KNOWN REMAINS
Partial skeleton

A. barsboldi

(Averianov and Lopatin, 2020)
Length: 15 m (49 ft)
Height: 6 m (20 ft)
Hip height: 2.6 m (8.5 ft)
Body mass: 10,000 kg (11 t)
Reconstruction: ☐☐☐☐

TRIASSIC   JURASSIC   CRETACEOUS

Abdarainurus barsboldi (meaning "Barsbold's tail from Abdarain") is known from a set of tail bones that exhibit an unusual suite of features. Some aspects of their anatomy are shared with only two other titanosaurs, *Andesaurus* and *Huabeisaurus*, but other traits mark *Abdarainurus* as being distinct from these two genera. Perhaps most distinctively, the neural spines of the anterior (frontmost) vertebrae sprout from the middle of the top surface of the bone as opposed to the front, which is the condition in every other known titanosauriform.

The bones representing *Abdarainurus* have actually been known since 1970, when a Soviet-Mongolian Paleontological expedition explored the area known locally as Abdrant Nuru. V. P. Tverdokhlebov of Saratov State University excavated the remains, which then remained in an unprepared state until the 2000s, when Andrei Podlesnov recognized their potential value during preparation.

The describers concluded that a basal **titanosaurian** designation was most likely for *Abdarainurus* but stressed that this part of the sauropod family tree contains so many uncertainties that this designation is somewhat ambiguous. At least one subsequent study has agreed with the placement of *Abdarainurus* as one of the basalmost titanosaurs (Wang et al., 2021). The describers speculate that *Abdarainurus* might represent a heretofore unknown "highly specialized" lineage of macronarians that was present in Asia.

The age of the Alagteeg geological formation is somewhat uncertain and could range from 85 to 72 million years of age.

The generic name *Abdarainurus* combines the Russian name for the dig site locality, Abdarain, with the Greek "urus" (meaning "tail"). The specific name *barsboldi* honors Mongolian paleontologist, Rinchen Barsbold.

CLASSIFICATION
Sauropoda
  Gravisauria
   Eusauropoda
    Neosauropoda
     Macronaria
      Titanosauriformes
       Somphospondyli
        Titanosauria

LOCATION
Mongolia

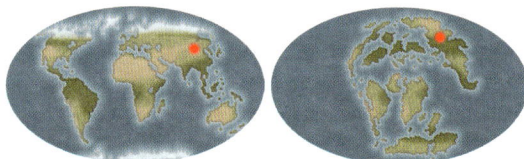

KNOWN REMAINS
Tail vertebrae

3 m   6 m   9 m   12 m   15 m   18 m   21 m

H. allocotus

(Pang and Cheng, 2000)
Length: 20 m (66 ft)
Height: 10 m (33 ft)
Hip height: 3.4 m (11 ft)
Body mass: 15,500 kg (17.1 t)
Reconstruction: ☐☐☐☐

9 m

6 m

3 m

250 – 245 – 240 – 235 – 230 – 225 – 220 – 215 – 210 – 205 – 200 – 195 – 190 – 185 – 180 – 175 – 170 – 165 – 160 – 155 – 150 – 145 – 140 – 135 – 130 – 125 – 120 – 115 – 110 – 105 – 100 – 95 – 90 – 85 – 80 – 75 – 70 – 65

TRIASSIC   **JURASSIC**   **CRETACEOUS**

Huabeisaurus allocotus (meaning "unusual north-Chinese lizard") lived during a time for which sauropod evolution is still poorly understood. As such, its discovery was hailed as an important step forward in comprehending the evolution of titanosaurs and for interpreting other sets of remains that are more fragmentary in nature.

The site that bore *Huabeisaurus* was first studied by paleontologists in 1983, when several dinosaurian tail bones were discovered. A series of digs over the next decade revealed thousands of individual bones from several dinosaurs, including those of this medium-to-large-sized sauropod. (A humerus found 200 meters from the site was also tentatively assigned to the genus, although later examinations pointed out that there was little justification for this.)

The original description of *Huabeisaurus* was, in actuality, only a brief overview. A much more comprehensive investigation of the specimen was later conducted, which greatly refined the defining characteristics of the genus. The individual animal was confirmed to have been an adult at the time of its death but, based on the presence and/or absence of various skeletal sutures, not quite fully grown (D'Emic et al., 2013).

The in-depth analysis of the remains raised the possibility that *Huabeisaurus* was a late-surviving member of Euhelopodidae, based on a number of skeletal similarities (D'Emic et al., 2013). However, many of the most recent taxonomic calculations have placed *Huabeisaurus* in a slightly more derived position, among the basalmost **titanosaurs** (Mannion et al., 2019a; Wang et al., 2021; Bellardini et al., 2022; Mo et al., 2023).

The generic name *Huabeisaurus* derives from the term "Huabei," a name for the northern region of China. The specific name derives from the Greek "allocot" (meaning "unusual").

CLASSIFICATION

Sauropoda
 Gravisauria
  Eusauropoda
   Neosauropoda
    Macronaria
     Titanosauriformes
      Somphospondyli
       Titanosauria

LOCATION

China

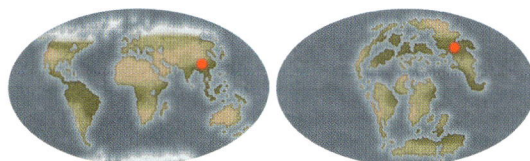

KNOWN REMAINS

Partial skeleton

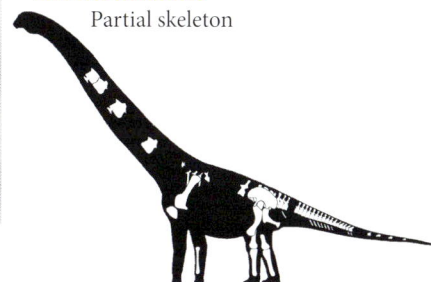

| 3 m | 6 m | 9 m | 12 m | 15 m | 18 m |

A. baharijensis

(Stromer, 1932)
Length: 16 m (52 ft)
Height: 8 m (26.2 t)
Hip height: 3 m (9.8 ft)
Body mass: 7,500 kg (8.3 t)
Reconstruction: ☐ ☐ ☐ ☐

| 250 | 245 | 240 | 235 | 230 | 225 | 220 | 215 | 210 | 205 | 200 | 195 | 190 | 185 | 180 | 175 | 170 | 165 | 160 | 155 | 150 | 145 | 140 | 135 | 130 | 125 | 120 | 115 | 110 | 105 | 100 | 95 | 90 | 85 | 80 | 75 | 70 | 65 |

TRIASSIC | **JURASSIC** | **CRETACEOUS**

Aegyptosaurus baharijensis (meaning "lizard from Baharija, Egypt") is something of a phantom because its fossils can no longer be studied.

Unearthed between 1911 and 1913, the bones of *Aegyptosaurus* were sent to Munich, Germany, in 1915 (along with the holotype of *Spinosaurus*). Sadly, these specimens were all destroyed during an Allied bombing raid that took place in 1944. As such, all that remains to work with are a handful of illustrations of these original fossils. In 1960, French paleontologist Albert-Félix de Lapparent assigned some tail bones to the genus, but this decision has been regarded as dubious.

Based on the limited data available, it is generally thought that *Aegyptosaurus* was a **titanosaur** (Gorscak et al., 2014). But determining a more precise placement within that rather large clade is just not possible.

Despite the loss of the fossils themselves, a few key insights regarding titanosaur anatomy can still be gleaned from their historical record. For one, having essentially complete long bones from both sets of limbs is something of a rarity among titanosaurian specimens. Having the measurements of these elements provided useful data regarding limb proportions, which can potentially be applied to other possible titanosaur specimens, such as a jumbled collection of fossils from a collective bone bed.

Another titanosaur from the same geological formation was described in 2001, *Paralititan*. The describers showed that the animal was distinct from *Aegyptosaurus*. Thus, even after their destruction, the remains of *Aegyptosaurus* show us that titanosaurs were diverse in this habitat (Smith et al., 2001).

The animal's binomial name refers to the Baharija geological formation of Egypt.

CLASSIFICATION

Sauropoda
 Gravisauria
  Eusauropoda
   Neosauropoda
    Macronaria
     Titanosauriformes
      Somphospondyli
       Titanosauria

LOCATION

Egypt

KNOWN REMAINS

Fragments (destroyed)

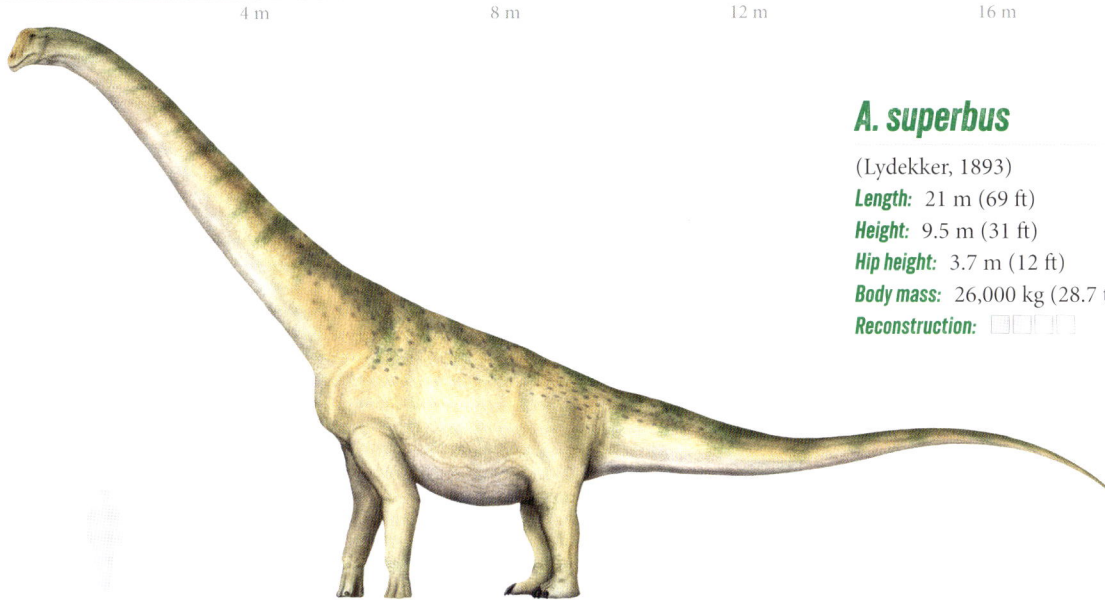

A. superbus

(Lydekker, 1893)
Length: 21 m (69 ft)
Height: 9.5 m (31 ft)
Hip height: 3.7 m (12 ft)
Body mass: 26,000 kg (28.7 t)
Reconstruction: ☐ ☐ ☐ ☐

TRIASSIC	JURASSIC	CRETACEOUS	

Argyrosaurus superbus (meaning "superb silver lizard") was among the first titanosaurs to be described, and for a long span of time it was one of the only South American sauropods known to science.

The only currently accepted specimen of *Argyrosaurus* is the holotype, which consists of the majority of a single forelimb. The bones were found in 1888 by Carlos Ameghino. Reportedly, the limb was only a part of a larger specimen, but excavation efforts destroyed any additional fossils that were present. The limb specimen originally contained two carpals (wrist bones), but as of 2012, these could not be located in museum archives. The presence of these bones is itself very distinctive, as they are almost universally absent in known titanosaur genera.

A number of specimens, mostly solitary bones, were at various times assigned to *Argyrosaurus*, but these assignments were typically based only on superficial similarities of size and shape in relation to the holotype; none are currently linked to the genus. One more substantial specimen has since been erected to its own unique genus, *Elaltitan* (Mannion and Otero, 2012).

Argyrosaurus is generally overlooked by phylogenetic studies. According to a comprehensive redescription of the genus, the traits of *Argyrosaurus* are mostly indicative of a placement among the basal **titanosaurs**, although it does bear a few traits in common with more derived titanosaurs.

The generic name *Argyrosaurus* combines the Greek "argyros" (meaning "silver") and "sauros" (meaning "lizard"); this alludes to one of the monikers for Argentina, "the silver land." The specific name *superbus* is Latin for "excellent" or "superb."

CLASSIFICATION

Sauropoda
 Gravisauria
  Eusauropoda
   Neosauropoda
    Macronaria
     Titanosauriformes
      Somphospondyli
       Titanosauria

LOCATION

Argentina

KNOWN REMAINS

Forelimb

4 m   8 m   12 m   16 m

4 m

B. wimani

(You et al., 2004)
Length: 15 m (49 ft)
Height: 6 m (20 ft)
Hip height: 2.4 m (8 ft)
Body mass: 5,500 kg (6.1 t)
Reconstruction: ▢ ▢ ▢ ▢

| 250 | 245 | 240 | 235 | 230 | 225 | 220 | 215 | 210 | 205 | 200 | 195 | 190 | 185 | 180 | 175 | 170 | 165 | 160 | 155 | 150 | 145 | 140 | 135 | 130 | 125 | 120 | 115 | 110 | 105 | 100 | 95 | 90 | 85 | 80 | 75 | 70 | 65 |

| TRIASSIC | JURASSIC | CRETACEOUS |

Borealosaurus wimani (meaning "Wiman's northern lizard") is a somewhat problematic taxon known from just a few bones that were unearthed sometime around 2000.

The holotype fossil of *Borealosaurus* is a single tail vertebra from the middle-rear portion of the tail. Three other bones were provisionally assigned to the genus as well, on the rationale that they were found nearby, but there is no way to definitively conclude that they actually came from the same animal, or even from the same species. These include a vertebra from the midtail, a humerus, and a single tooth.

Borealosaurus was distinguished based on the opisthocoelous shape of the holotype vertebra (being convex in the front and concave in the back), a rare arrangement for titanosaurs. It was thusly considered by the describers to be a member of the derived titanosaur subfamily, Opisthocoelicaudiinae.

However, it was later pointed out that the opisthocoelous condition is more common among titanosauriform middle tail vertebrae than was once thought. This condition also occurs in different regions of the tail between *Borealosaurus* and *Opisthocoelicaudia*. Based on this and other traits, *Borealosaurus* is likely a basal **titanosaur** (Averianov and Sues, 2017).

The Sunjiawan geological formation that bore the fossils was considered by the describers to be 95–90 million years of age, but later work has suggested that it is likely closer to 110–100 million years old (Wang et al., 2015).

The generic name *Borealosaurus* is derived from the Greek "borealis" (meaning "northern"), referring to the specimen's discovery from northern China. The specific name *wimani* honors Swedish paleontologist Carl Wiman (1867–1944), who named the first dinosaur from China.

CLASSIFICATION
Sauropoda
  Gravisauria
    Eusauropoda
      Neosauropoda
        Macronaria
          Titanosauriformes
          Somphospondyli
          Titanosauria

LOCATION
China

KNOWN REMAINS
Isolated bones

4 m   8 m   12 m   16 m   20 m

8 m

C. baileywillisi

(Simón et al., 2017)
Length: 21 m (70 ft)
Height: 8 m (26 ft)
Hip height: 3.1 m (10 ft)
Body mass: 18,000 kg (19.8 t)
Reconstruction: ☐ ☐ ☐ ☐

4 m

250 245 240 235 230 225 220 215 210 205 200 195 190 185 180 175 170 165 160 155 150 145 140 135 130 125 120 115 110 105 100 95 90 85 80 75 70 65

| TRIASSIC | JURASSIC | CRETACEOUS |

Choconsaurus baileywillisi (meaning "Bailey Willis's lizard from Chocon") is represented by one of the most complete sets of remains for any basal titanosaur. It includes a good portion of the vertebral column, a shoulder blade, a collection of hand bones, and a portion of the upper and lower jaws.

The bones closest to the tip of the tail are unusual, having a shape that is platycoelous—being flat in front and concave in the back. This condition is seen in only one other somphospondylan, *Andesaurus*. In all other titanosaurs, these bones are procoelous—being concave in front and convex in the back.

Some of these fossils were found in the same small area, within a radius of less than 4 meters; these were designated as the holotype specimen. Other macronarian bones found nearby were then assigned to the genus as paratypes.

The few cranial fragments available suggest that the jaws of *Choconsaurus* were U-shaped, rather than being squared off, as they would be in more derived species of titanosaur. Its teeth were also of an intermediate shape, being somewhere between the broader teeth of basal macronarians and the narrow peg-like teeth of derived titanosaurs.

These traits are consistent with the placement of *Choconsaurus* as a basal **titanosaur**, which is the preferred interpretation put forth by the authors. However, calculations show that it is almost just as likely to be a non-titanosaurian derived somphospondylan.

The generic name *Choconsaurus* refers to the locality of Villa El Chocón, in Neuquen Province, Argentina. The specific name *baileywillisi* honors American geologist Bailey Willis (1857–1949) for his works done in Argentina.

CLASSIFICATION
Sauropoda
  Gravisauria
   Eusauropoda
    Neosauropoda
     Macronaria
      Titanosauriformes
       Somphospondyli
        Titanosauria

LOCATION
Argentina

KNOWN REMAINS
Partial skeleton and cranial fragments

| | 4 m | 8 m | 12 m | 16 m | 20 m |

K. maui

(Filippi et al., 2019)
Length: 19 m (62 ft)
Height: 10 m (32.8 ft)
Hip height: 4 m (13.1 ft)
Body mass: 15,000 kg (16.5 t)
Reconstruction:

| TRIASSIC | JURASSIC | CRETACEOUS |

Kaijutitan maui (meaning "MAU's giant monster") is a sauropod that features exceptionally stout limbs, similar in proportion to the more famous titanosaur *Dreadnoughtus* (González Riga et al., 2019).

The holotype specimen's disarticulated bones were found within a relatively small area (20 square meters) buried in ancient floodplain deposits. Many of the bones were in an incomplete or damaged state. The specimen consists primarily of limb elements but also includes several ribs, three vertebrae, and the rear portion of the skull. Regarding the braincase region of the cranium, no clear sutures are visible between the elements, which suggests that the animal was an adult at the time of its death.

The shape of *Kaijutitan*'s frontmost preserved neck vertebra is notable because its neural spine features a "bifid" split arrangement. Among sauropods as a group, this feature is not uncommon, but it is quite rare specifically among the titanosaurs. Other skeletal characteristics of *Kaijutitan* feature an unusual mixture of plesiomorphic (i.e., ancestral) and derived traits.

The describers' taxonomic study found it to be a basal **titanosaur**, most likely falling outside the more derived clade of Eutitanosauria. One subsequent study also arrived at a similar conclusion, placing it within Lithostrotia (Navarro et al., 2022). This makes *Kaijutitan* among the latest surviving members of the "ancestral" Titanosauria.

The generic name *Kaijutitan* combines the Japanese "kaiju" (meaning "strange beast" or "monster") and the Greek "titan" (meaning "giant"). The specific name *maui* refers to the acronym for the Museo Municipal Argentino Urquiza.

CLASSIFICATION
Sauropoda
 Gravisauria
  Eusauropoda
   Neosauropoda
    Macronaria
     Titanosauriformes
      Somphospondyli
       Titanosauria

LOCATION
Argentina

KNOWN REMAINS
Partial skull and skeleton

3 m   6 m   9 m   12 m   15 m

6 m

H. xinjiangensis

(Wang et al., 2021)
Length: 17 m (56 ft)
Height: 6.5 m (21 ft)
Hip height: 3.4 m (11 ft)
Body mass: 22,000 kg (24.3 t)
Reconstruction: ☐☐☐☐

3 m

| 250 | 245 | 240 | 235 | 230 | 225 | 220 | 215 | 210 | 205 | 200 | 195 | 190 | 185 | 180 | 175 | 170 | 165 | 160 | 155 | 150 | 145 | 140 | 135 | 130 | 125 | **120** | 115 | 110 | 105 | 100 | 95 | 90 | 85 | 80 | 75 | 70 | 65 |

TRIASSIC   **JURASSIC**   **CRETACEOUS**

Hamititan xinjiangensis (meaning "giant from Hami, Xinjiang") is one of the few dinosaurs known from the Hami Pterosaur Fauna.

The year 2006 marked the discovery of a lagerstätte—an area of exceptional fossil preservation—in the Turpan-Hami Basin of western China. The area is one of the few known pterosaur bone beds, containing numerous specimens of very well-preserved adults, eggs, and embryos of the genus *Hamipterus*. Only a few non-pterosaur vertebrate fossils have been found in the area, one of which consists of an articulated set of seven sauropod tail vertebrae, unearthed in 2013. These are now considered the holotype specimen of *Hamititan*.

A few kilometers away, the neck vertebrae of another sauropod, *Silutitan*, were discovered. With no overlapping material between the two specimens, it cannot be conclusively ruled out that they actually represent the same genus.

However, the phylogenetic study that compared the two came to the conclusion that they are likely to be different animals.

The aforementioned study ran several versions of phylogenetic comparisons, utilizing two different sauropod datasets. These different tests produced several distinctly different results. While it is clear that *Hamititan* represents some sort of **titanosaur**, a more precise placement within that group remains elusive. *Hamititan* could represent a very basal member of the clade or, conversely, could be a derived saltasauroid.

The bones of *Hamititan* were found adjacent to a curved tooth from an unknown theropod, but none of the bones show any evidence of tooth marks.

The animal's binomial name refers to Hami City in the Xinjiang region of China.

CLASSIFICATION

Sauropoda
 Gravisauria
  Eusauropoda
   Neosauropoda
    Macronaria
     Titanosauriformes
      Somphospondyli
      Titanosauria

LOCATION

China

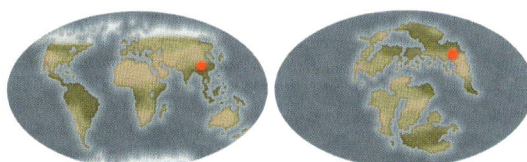

KNOWN REMAINS

Partial tail

265

JIANGSHANOSAURUS

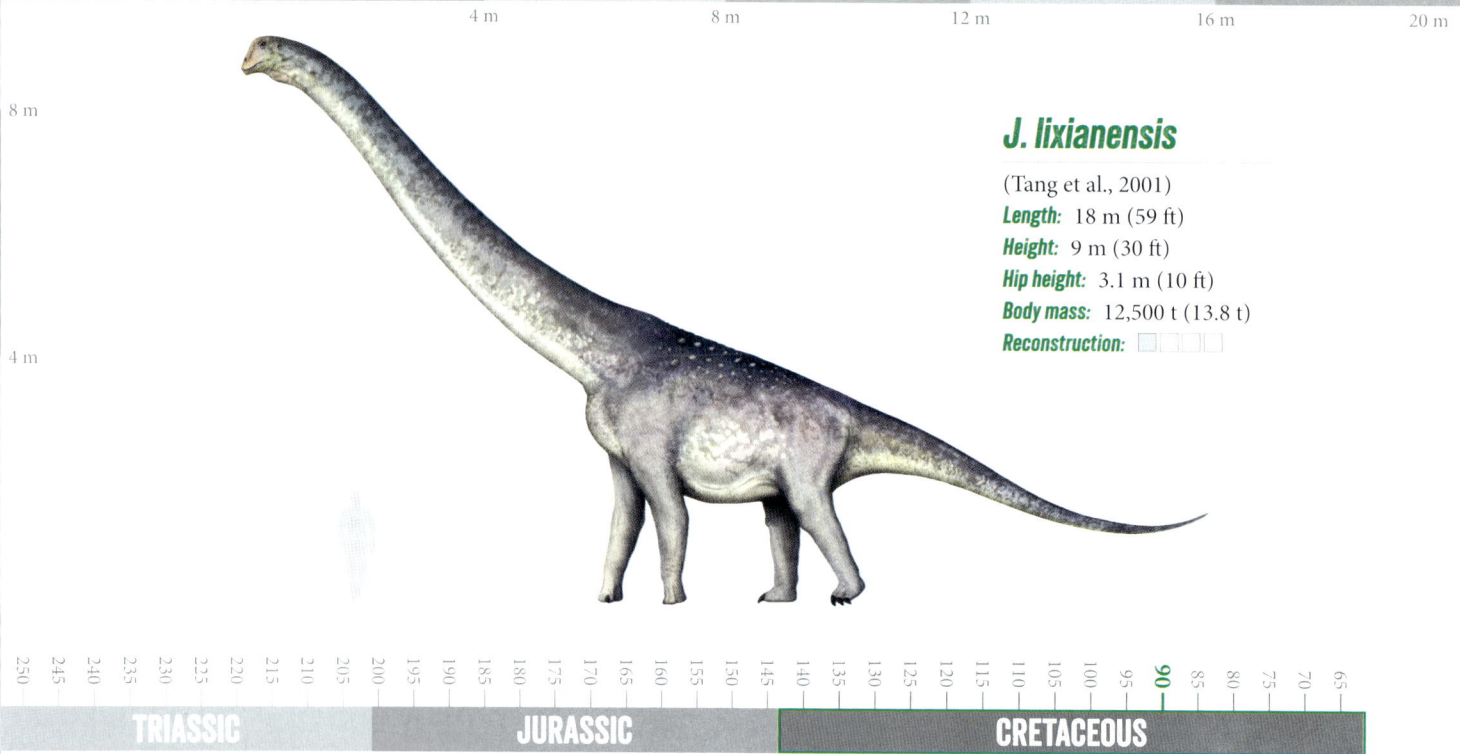

J. lixianensis

(Tang et al., 2001)
Length: 18 m (59 ft)
Height: 9 m (30 ft)
Hip height: 3.1 m (10 ft)
Body mass: 12,500 t (13.8 t)
Reconstruction: ☐ ☐ ☐ ☐

TRIASSIC JURASSIC CRETACEOUS

Jiangshanosaurus lixianensis (meaning "lizard from Lixian, Jiangshan") is a poorly understood sauropod, known only from a handful of vertebrae and portions of the shoulder and pelvic girdles.

The fragmentary remains of *Jiangshanosaurus* were discovered and unearthed in 1977 and 1978 and subsequently rested in the archives of the Zhejiang Natural Museum before finally receiving more attention after the turn of the century. They came from an area of the Jinhua geological formation once considered to be approximately 105 million years of age but which is now thought to be closer to 90 million years of age (Xi et al., 2018).

The original description of *Jiangshanosaurus* (which was brief and published in an exclusively Chinese-language source) assigned the genus to **Titanosauria**, although without much information regarding a more precise placement. The describers only noted some supposed similarities between *Jiangshanosaurus* and *Alamosaurus*, a genus from the end of the Cretaceous. Contrarily, some studies that followed found it to be among the most derived of all titanosaurs, among Saltasauridae (Mannion et al., 2013; Averianov et al., 2018).

Conversely, the most recent spate of papers has recognized that *Jiangshanosaurus* actually has very few traits in common with derived titanosaurs. Some have recovered it as being among the basalmost of titanosaurs (Wang et al., 2021), among Euhelopodidae (Mannion et al., 2019a), or even slightly stemward of Titanosauria entirely (Mo et al., 2023). Without the discovery of further remains to diagnose, it seems unlikely that these moderately differing results will be confidently settled.

The animal's binomial name refers to the village of Lixian in Jiangshan County, which is in the Zhejiang Province of China.

CLASSIFICATION
Sauropoda
 Gravisauria
  Eusauropoda
   Neosauropoda
    Macronaria
     Titanosauriformes
      Somphospondyli
       Titanosauria

LOCATION
China

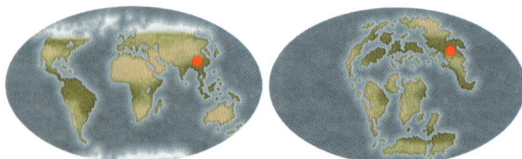

KNOWN REMAINS
Fragments

4 m   8 m   12 m   16 m

D. sinensis

8 m

(Lü et al., 2008)
Length: 15 m (49 ft)
Height: 9.5 m (31.1 ft)
Hip height: 3.1 m (10.2 ft)
Body mass: 7,000 kg (7.7 t)
Reconstruction: ▢▢▢▢▢

4 m

| 250 | 245 | 240 | 235 | 230 | 225 | 220 | 215 | 210 | 205 | 200 | 195 | 190 | 185 | 180 | 175 | 170 | 165 | 160 | 155 | 150 | 145 | 140 | 135 | 130 | 125 | 120 | 115 | 110 | 105 | 100 | 95 | 90 | 85 | 80 | 75 | 70 | 65 |

TRIASSIC   **JURASSIC**   **CRETACEOUS**

Dongyangosaurus sinensis (meaning "lizard from Dongyang, China") is known from a single specimen unearthed in 2007 that was preserved in articulation, such that it was posed nearly as it would have been positioned in life (with only minor hip displacement).

The dorsal vertebrae of *Dongyangosaurus* are pneumatic (featuring hollowed air-sacs) to a very high degree. Pneumatic vertebrae are to be expected in sauropods, but those of *Dongyangosaurus* are particularly so. They also possess very peculiar hollow cavities in their exterior surface, which feature thin sheets of bone that could have been used to house and separate additional air sacs.

Dongyangosaurus is generally placed as a basal **titanosaur** (Averianov and Lopatin, 2020; Mo et al., 2023), although some versions of analyses have assigned it to positions either much more derived (Mannion et al., 2013) or, contrastingly, more basal (Mannion et al., 2019b).

The *Dongyangosaurus* holotype specimen is from the Jinhua (formerly Fangyan) geological formation. Although the formation is well known for its preserved dinosaur eggs, dinosaur skeletal remains are much rarer here. Only one other sauropod is known from contemporaneous strata: *Jiangshanosaurus*. An analysis from 2019 confirmed that the two animals are separate and distinct genera, although one version of Mannion et al.'s (2019a) taxonomic study did place the two sauropods as sister taxa. This study was also able to partially reexamine the remains and to provide an updated description of some aspects of the animal's anatomy.

The generic name *Dongyangosaurus* refers to Dongyang City in Zhejiang Province, China. The specific name *sinensis* refers to China; "sinae" is the Latin word for the Chinese.

CLASSIFICATION
Sauropoda
 Gravisauria
  Eusauropoda
   Neosauropoda
    Macronaria
     Titanosauriformes
      Somphospondyli
       Titanosauria

LOCATION
China

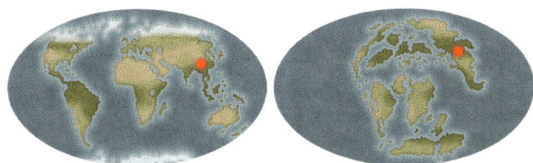

KNOWN REMAINS
Partial skeleton

4 m   8 m   12 m   16 m

4 m

N. zapatai

(Gallina et al., 2021)
Length: 15 m (49 ft)
Height: 7.8 m (25.6 ft)
Hip height: 2.6 m (8.5 ft)
Body mass: 6,000 kg (6.6 t)
Reconstruction: ☐☐☐

| 250 | 245 | 240 | 235 | 230 | 225 | 220 | 215 | 210 | 205 | 200 | 195 | 190 | 185 | 180 | 175 | 170 | 165 | 160 | 155 | 150 | 145 | 140 | 135 | 130 | 125 | 120 | 115 | 110 | 105 | 100 | 95 | 90 | 85 | 80 | 75 | 70 | 65 |

TRIASSIC   **JURASSIC**   **CRETACEOUS**

Ninjatitan zapatai (meaning "Ninja and Zapata's giant") is potentially the oldest known titanosaur.

Analytics of titanosaur phylogeny had previously hypothesized that the clade likely originated in South America during the earliest Cretaceous (Gorscak and O'Connor, 2016), but definitive titanosaurian remains from this interval proved elusive, with many of the "earliest" known remains at the time being roughly 125 million years of age.

The discovery of *Ninjatitan* from the Bajada Colorada geological formation (aged somewhere between 140 and 134 million years) lends support to this notion of an early emerging, Gondwanan-based Titanosauria.

Owing to the fragmentary state of the remains, getting a more precise taxonomic placement of *Ninjatitan* within **Titanosauria** is challenging and has led to differing hypotheses. The analyses conducted by the describers

calculated two potential outcomes: one in which *Ninjatitan* is a very basal titanosaur and one in which the genus is in the more derived subfamily Lognkosauria. A later analysis found a result similar to the second hypothesis, placing *Ninjatitan* in the closely related subfamily Rinconsauria (Moreno et al., 2023). However, such a derived position would require a certain degree of temporal contortion, as the other members of those subfamilies tend to be Late Cretaceous in origin. Thus, in the absence of more complete remains that can provide a clearer picture, a basal assignment within Titanosauria seems to currently be the best-fitting scenario.

The generic name *Ninjatitan* honors Argentine paleontologist Sebastián "Ninja" Apesteguía, mentor to some of the describers. The specific name *zapatai* honors museum technician Rogelio "Mupi" Zapata.

CLASSIFICATION
Sauropoda
  Gravisauria
   Eusauropoda
    Neosauropoda
     Macronaria
      Titanosauriformes
      Somphospondyli
      Titanosauria

LOCATION
Argentina

KNOWN REMAINS
Fragments

A. cooperensis

(Hocknull et al., 2021)
Length: 24 m (80 ft)
Height: 10.5 m (35 ft)
Hip height: 4 m (13 ft)
Body mass: 35,000 kg (38.6 t)
Reconstruction:

8 m

4 m

| TRIASSIC | JURASSIC | CRETACEOUS |

Australotitan cooperensis (meaning "southern giant from Cooper") is currently the largest described sauropod from Australia (although footprints suggest there might have been even larger examples). Proportionally, its limbs were slightly elongated and gracile, with a particularly "thin" ulna.

Numerous scattered sauropod remains, originating from several distinct individuals, have been uncovered from the Winton Formation near Cooper Creek since 2005. Several of these specimens, consisting mainly of limb and pelvic elements, contribute to the description of *Australotitan*. Most elements have suffered from erosion and are less than whole.

Several other nearby sauropod specimens of varying levels of preservation and completeness may or may not also represent *Australotitan*. At least one specimen (EMF103) is quite deformed, having suffered what is likely to have been trampling by other sauropods. Another specimen (EMF106) appears to be the most complete in the area, likely preserving most of the rear half of the animal, but much of the animal still remained enshrouded in stone at the time of *Australotitan*'s description, so its identity is not yet known.

Preliminary results show that *Australotitan* is likely a member of the primarily Australian clade **Diamantinasauria**. Certain similarities with Asian and South American taxa suggest some amount of faunal interchange, perhaps utilizing Antarctic passageways or island-based "stepping stones" over time.

The generic name *Australotitan* combines the Greek "austral" (meaning "southern") and "titan" (meaning "giant"). The specific name *cooperensis* refers to the Cooper-Eromanga Basin, aka "Cooper Country."

CLASSIFICATION
Sauropoda
 Gravisauria
  Eusauropoda
   Neosauropoda
    Macronaria
     Titanosauriformes
      Somphospondyli
       Titanosauria
        Diamantinasauria

LOCATION
Australia

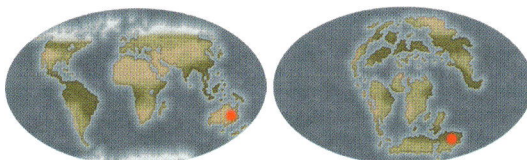

KNOWN REMAINS
Partial skeleton

4 m · 8 m · 12 m · 16 m

4 m

D. matildae

(Hocknull et al., 2009)
Length: 16 m (52 ft)
Height: 7 m (23 ft)
Hip height: 3.1 m (10 ft)
Body mass: 20,000 kg (22 t)
Reconstruction:

250 245 240 235 230 225 220 215 210 205 200 195 190 185 180 175 170 165 160 155 150 145 140 135 130 125 120 115 110 105 100 95 90 85 80 75 70 65

TRIASSIC · **JURASSIC** · **CRETACEOUS**

Diamantinasaurus matildae (meaning "Matilda lizard from Diamantina") is one of the few titanosaurs for which relatively complete cranial material is accounted for. The snout is very rounded in shape, which is interpreted as a feature of generalist browsers; sauropods that specialize in low browsing tend to have snouts that are squared-off.

The skull of *Diamantinasaurus* has been noted for the uncanny similarities it shares with that of the South American *Sarmientosaurus*. It is hypothesized that during a particularly warm interval of the Cretaceous, spanning approximately 100 to 95 MYA, faunal exchange between Australia and South America could have been possible by way of Antarctica (Poropat et al., 2023).

The relative completeness of *Diamantinasaurus*'s known osteology is due, in part, to the discovery and description of four separate specimens. One of these sets of remains is from a juvenile, which is the first juvenile sauropod material

ever found in Australia. Between these specimens, a few fragmentary bones that are rarely preserved in sauropods have been recovered, including partial gastralia ("belly ribs") and a portion of the hyoid bone (Poropat et al., 2016; Rigby et al., 2022).

Diamantinasaurus has formed the basis of the clade **Diamantinasauria**, which has been placed by different studies in various parts of the family tree but which is most likely an early branching titanosaurian group (Poropat et al., 2023).

The generic name *Diamantinasaurus* refers to the Diamantina River. The specific name *matildae* refers to one of Australia's national songs, "Waltzing Matilda," as the fossil region is known informally as "Matilda Country" and the holotype was nicknamed "Matilda." In this context, "Matilda" is not a name of a person but a slang term for a bedroll.

CLASSIFICATION
Sauropoda
 Gravisauria
  Eusauropoda
   Neosauropoda
    Macronaria
     Titanosauriformes
      Somphospondyli
       Titanosauria
        Diamantinasauria

LOCATION
Australia

KNOWN REMAINS
Partial skull and skeleton

2 m   4 m   6 m   8 m   10 m

S. musacchioi

(Martínez et al., 2016)
Length: 12 m (39 ft)
Height: 5.5 m (18 ft)
Hip height: 2.2 m (7 ft)
Body mass: 10,000 kg (11 t)
Reconstruction: ▨☐☐☐

4 m

2 m

| TRIASSIC | JURASSIC | CRETACEOUS |

Sarmientosaurus musacchioi (meaning "Musacchio's lizard from Sarmiento") is known primarily from one of the most well-preserved and complete skulls of any South American sauropod. At the time of its description, it was only the fourth titanosaur genus to incorporate complete cranial remains.

Unearthed in 1997, the skull (which came from an adult animal) seems to be a near-perfect blend of ancestral and derived characteristics, having an appearance halfway in between basal titanosauriforms (such as *Brachiosaurus*) and derived titanosaurs (such as *Nemegtosaurus*). Like *Brachiosaurus*, its skull is wide, and it has a tooth row that extends far back into the mouth. But, like *Nemegtosaurus*, it has a long snout and smaller nostril openings.

The teeth of *Sarmientosaurus* are also somewhat intermediate, being neither particularly robust nor slender. The teeth of the upper jaw are angled forward, while the teeth of the lower jaw are angled backward, a unique arrangement that left peculiar wear marks on the teeth.

The eye socket of *Sarmientosaurus* is proportionally large, leading some to hypothesize that it might have had superior vision of some sort. Computed tomography scans of the inner ear structure show that the default head posture for the animal would most likely have been with its snout facing at a downward angle. Also found among the remains was an ossified neck tendon, believed to be the first dinosaurian example of such a fossilization.

Owing to numerous similarities with the Australian sauropod *Diamantinasaurus*, recent works have found *Sarmientosaurus* to be in the clade **Diamantinasauria** (Poropat et al., 2023).

The generic name *Sarmientosaurus* refers to the town of Sarmiento. The specific name *musacchioi* honors Eduardo Musacchio.

CLASSIFICATION
Sauropoda
 Gravisauria
  Eusauropoda
   Neosauropoda
    Macronaria
     Titanosauriformes
      Somphospondyli
       Titanosauria
        Diamantinasauria

LOCATION
Argentina

KNOWN REMAINS
Skull and partial neck

271

| 2 m | 4 m | 6 m | 8 m | 10 m | 12 m | 14 m |

S. elliottorum

(Poropat et al., 2016)
Length: 15 m (49 ft)
Height: 6 m (19.5 ft)
Hip height: 2.9 m (10 ft)
Body mass: 20,000 kg (22 t)
Reconstruction: ☐☐☐ ☐ ☐

| 250 | 245 | 240 | 235 | 230 | 225 | 220 | 215 | 210 | 205 | 200 | 195 | 190 | 185 | 180 | 175 | 170 | 165 | 160 | 155 | 150 | 145 | 140 | 135 | 130 | 125 | 120 | 115 | 110 | 105 | 100 | 95 | 90 | 85 | 80 | 75 | 70 | 65 |

| TRIASSIC | JURASSIC | CRETACEOUS |

Although **Savannasaurus elliottorum** (meaning "Elliotts' savanna lizard") is known only from a partial skeleton, several unique attributes of the animal's anatomy can be determined and together paint an intriguing picture of the creature's connection to its environment.

For starters, *Savannasaurus* had very wide hips, proportionally speaking. Although the animal was only of moderate size, for a sauropod, the narrowest part of its sacrum is still more than a meter in width. Combining this measurement with measurements of its ribs and dorsal vertebrae, it has been calculated that *Savannasaurus* was overall quite barrel shaped, with a thorax circumference exceeding 5 meters.

Savannasaurus would also have had a "wide-gauge" stance and gait, placing its feet further apart than other sauropods would. This stance has been deduced based on the unusual shape of the sternum, the curiously broad base of the first metacarpal, and features of the humerus and ankle that could have been adapted for the purpose of bearing weight in new ways.

Additionally, the spinal column of *Savannasaurus* would have been particularly flexible, as the vertebrae lacked certain interconnecting features that are shared by most sauropods. Taken all together, one possible explanation for these features is that they allowed *Savannasaurus* to be uniquely suited to its floodplain environment. With a "hippo-shaped" body, it could possibly have been suited for life in the shallows: a wider stance would help prevent it from becoming mired, and a flexible body would better allow it to free itself from muddy entrapment (Poropat et al., 2020).

The generic name *Savannasaurus* refers to the "countryside in which the specimen was found." The specific name *elliottorum* honors the Elliott family for their "ongoing contributions to Australian paleontology."

CLASSIFICATION
Sauropoda
　Gravisauria
　　Eusauropoda
　　　Neosauropoda
　　　　Macronaria
　　　　　Titanosauriformes
　　　　　　Somphospondyli
　　　　　　　Titanosauria
　　　　　　　　Diamantinasauria

LOCATION
Australia

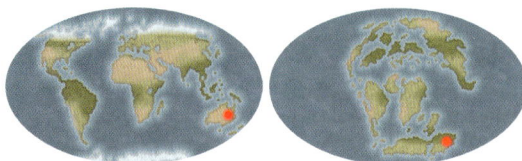

KNOWN REMAINS
Partial skeleton

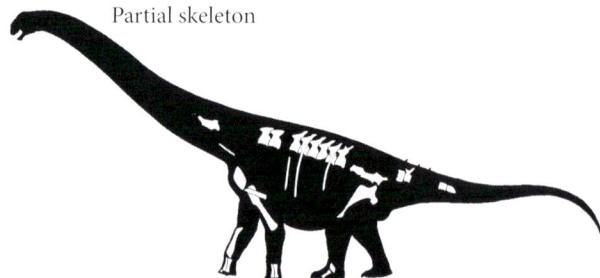

| | 4 m | 8 m | 12 m | 16 m | 20 m | 24 m | 12 m |

D. binglingi

(You et al., 2008)
Length: 25 m (82 ft)
Height: 11 m (36 ft)
Hip height: 3.7 m (12 ft)
Body mass: 23,000 kg (25.4 t)
Reconstruction: ☐ ☐ ☐ ☐ ☐

| 250 | 245 | 240 | 235 | 230 | 225 | 220 | 215 | 210 | 205 | 200 | 195 | 190 | 185 | 180 | 175 | 170 | 165 | 160 | 155 | 150 | 145 | 140 | 135 | 130 | 125 | 122 | 115 | 110 | 105 | 100 | 95 | 90 | 85 | 80 | 75 | 70 | 65 |

TRIASSIC **JURASSIC** **CRETACEOUS**

Daxiatitan binglingi (meaning "Bingling's giant from Daxia") was regarded as one of the largest of all Chinese sauropods, at the time of its description. Proportionately speaking, its limbs were of average dimensions (based on the single limb bone known, the femur), whereas its neck bones were quite robust, especially in regard to their width; this would have given *Daxiatitan* a neck that was nearly round in cross section.

Daxiatitan's femur might not have been particularly massive, but it was uniquely shaped. According to the describers, the animal's hindlimbs would have been oriented outward, resulting in a "wide-gauge" stance. This has led to speculation that *Daxiatitan* is responsible for certain "strongly outwardly rotated" sauropod trackways that have been discovered in rocks of the same age as the remains, about 50 kilometers (30 miles) away from the fossil site.

Daxiatitan has a somewhat uncertain phylogenetic placement. The describers considered it to be a **titanosaur**, and regarded its similarities to *Euhelopus*. In recent works, it has been variously recovered as a non-titanosaur somphospondylan (Mannion et al., 2019a; Wang et al., 2021; Bellardini et al., 2022), a basal titanosaur (Averianov and Sues, 2017; Poropat et al., 2021; Mo et al., 2023), or even a euhelopodid (Moore et al., 2018; Gallina et al., 2021).

Daxiatitan shared a habitat with the sauropod *Huanghetitan*. Differences in the animal's scapulae show that the two are distinct and separate genera.

The generic name *Daxiatitan* refers to the Daxia River, a branch of the Yellow River that runs through Gansu Province. The specific name *binglingi* refers to Bingling Temple, a nearby collection of Buddhist sculptures. Originally, the genus was going to be named "*Gansutitan*."

CLASSIFICATION
Sauropoda
 Gravisauria
  Eusauropoda
   Neosauropoda
    Macronaria
     Titanosauriformes
      Somphospondyli
       Titanosauria

LOCATION
China

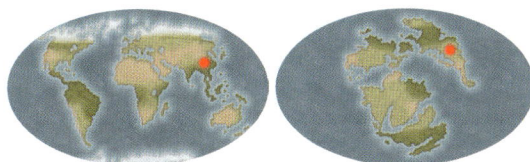

KNOWN REMAINS
Partial skeleton

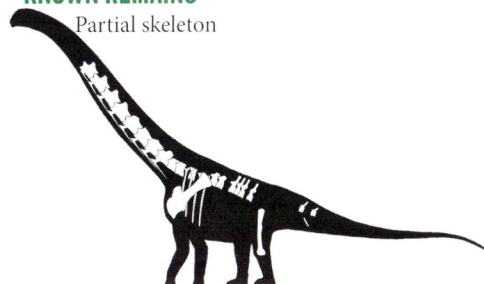

8 m

4 m

4 m 8 m 12 m 16 m

X. shijiagouensis

(Lü et al., 2009)
Length: 15 m (50 ft)
Height: 8 m (26 ft)
Hip height: 2.7 m (9 ft)
Body mass: 7,000 kg (7.7 t)
Reconstruction:

250 245 240 235 230 225 220 215 210 205 200 195 190 185 180 175 170 165 160 155 150 145 140 135 130 125 120 115 110 105 100 95 90 85 80 75 70 65

TRIASSIC **JURASSIC** **CRETACEOUS**

Xianshanosaurus shijiagouensis (meaning "lizard from Shijiagou, Xianshan") is known from only a handful of bones, but these elements have some intriguing features. For instance, the vertebrae of the animal's tail are decidedly short and wide. Additionally, the single recovered chevron (tail rib) is deflected forward to a degree that is seen in only a few other sauropods, and the arch near the top of the bone is closed, which is not the norm among titanosaurs (Mannion et al., 2013; Averianov and Sues, 2017). The individual was likely not fully grown.

In addition to these tail bones, the specimen includes only a femur, a portion of the shoulder blade, and some ribs. The tail bones were articulated, while the others were closely associated. (Originally, a tooth was also included among the finds but was later excluded on the basis of having been found too far away from the rest of the fossils.) The remains were first marked as being from the Mangchuan geological formation, but the area has since been revised as the Haoling Formation (Xu et al., 2012).

Disagreements among studies have resulted in an uncertain taxonomic placement for *Xianshanosaurus*. It has been variously recovered as a non-titanosaur somphospondylan (Bellardini et al., 2022), a euhelopodid (Gallina et al., 2021), a basal titanosaur (Averianov et al., 2018; Poropat et al., 2021; Mo et al., 2023), or a lithostrotian (Mannion et al., 2019a; Wang et al., 2021).

Many analyses find *Xianshanosaurus* to be the sister taxon of *Daxiatitan* (Averianov et al., 2018; Poropat et al., 2021; Bellardini et al., 2022; Mo et al., 2023), although this is not universally held (Royo-Torres et al., 2020; Wang et al., 2021).

The animal's binomial name refers to the Shijiagou quarry of Haoling village near Xian Mountain, China. This quarry has yielded numerous ornithomimosaur remains.

CLASSIFICATION
Sauropoda
 Gravisauria
  Eusauropoda
   Neosauropoda
    Macronaria
     Titanosauriformes
      Somphospondyli
       Titanosauria

LOCATION
China

KNOWN REMAINS
Fragments

2 m	4 m	6 m	8 m	10 m	12 m

R. zhangi

(Mo et al., 2023)
Length: 12 m (39 ft)
Height: 6 m (19.7 ft)
Hip height: 2.5 m (8 ft)
Body mass: 4,000 kg (4.4 t)
Reconstruction: ☐☐☐☐

250	245	240	235	230	225	220	215	210	205	200	195	190	185	180	175	170	165	160	155	150	145	140	135	130	125	120	115	110	105	100	95	90	85	80	75	70	65

TRIASSIC	JURASSIC	CRETACEOUS

Ruixinia zhangi (meaning "Ruixin Zhang's one") is the third sauropod genus to be named from the Jehol Biota of China. It is also one of the only titanosauriforms that preserves essentially the entire tail of the animal, providing a crucial insight into the anatomy of these dinosaurs.

Several features of *Ruixinia*'s tail are unique among sauropods. There is a clear separation between the front and rear portions of the tail, whereby the size and shape of the bones markedly change, with some individual elements being distinct from either of their neighbors. The bones of the tip of the tail are fused, a feature seen in only a few sauropod taxa, and none that are closely related.

Beyond just the tail, nearly the entire vertebral column of the individual was preserved. However, the neck and back vertebrae have been notably compressed, and so the level of detail that they can reveal is limited. The bones of the leg are also valuable, as they allow for another point of comparison between *Ruixinia* and various other titanosauriform taxa.

The specimen is believed to have been an adult, based on fusion between vertebral elements; however, there has been no histological examination of the interior bone structure to confirm this. This is partially due to that, at the time of description, many of the fossils were still halfway embedded within a rocky matrix, with only one side being exposed.

Despite the relative completeness of the holotype specimen, phylogenetically placing any titanosauriform can be challenging, as so many taxa from the group are known only from fragmentary remains. The describers confidently placed *Ruixinia* as a somphospondylan; their favored interpretation went one step further by placing it in the more derived clade **Titanosauria**.

The binomial name honors Ruixin Zhang for his work at the Erlianhaote Dinosaur Museum.

CLASSIFICATION
Sauropoda
 Gravisauria
  Eusauropoda
   Neosauropoda
    Macronaria
     Titanosauriformes
      Somphospondyli
       Titanosauria

LOCATION
China

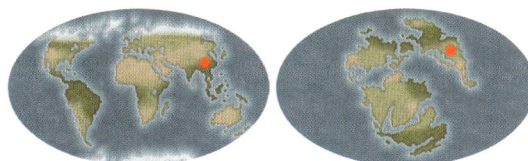

KNOWN REMAINS
Partial skeleton

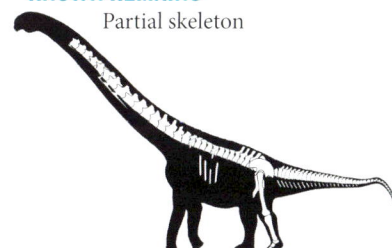

4 m | 2 m | 4 m | 6 m | 8 m

2 m

M. moyowamkia

(Gorscak and O'Connor, 2019)
Length: 9 m (30 ft)
Height: 4 m (13.1 ft)
Hip height: 1.8 m (5.9 ft)
Body mass: 1,700 kg (1.9 t)
Reconstruction: ☐☐☐☐

250 245 240 235 230 225 220 215 210 205 200 195 190 185 180 175 170 165 160 155 150 145 140 135 130 125 120 115 110 **105** 100 95 90 85 80 75 70 65

TRIASSIC | **JURASSIC** | **CRETACEOUS**

Mnyamawamtuka moyowamkia (meaning "beastly tail's heart of the Mtuka") is known from one of the most complete sets of early titanosaur remains ever discovered. First spotted in 2004, the specimen was excavated during yearly digs through 2008, providing a key glimpse into early titanosaur evolution.

The bones of *Mnyamawamtuka* were disarticulated and spread over an area of several square meters. As the fossils were embedded in fluvial (river-based) deposits, it is possible that this arrangement of the remains is due to the motion of flowing water. The individual was not fully grown at the time of its death, as evidenced by numerous unfused vertebral elements. No osteoderms (armored skin elements) were found among the remains.

Four individual teeth were recovered at the scene, and since there is no evidence of any different sauropods among the remains, it is assumed that these teeth all belong to *Mnyamawamtuka*. This is notable, because the teeth have differing cross-sectional shapes, ranging from D-shaped to round (the latter of which being the norm among titanosaurs).

Malawisaurus displays a mix of derived lithostrotian traits as well as ancestral titanosaurian traits. Various phylogenetic calculations conducted by the describers recover *Mnyamawamtuka* as either being the most derived of the non-lithostrotian **titanosaurs** or as being the sister taxon of *Malawisaurus*, thereby tying it for being the basalmost of the lithostrotians.

The generic name *Mnyamawamtuka* combines the Kiswahili terms "mnyama" (meaning "animal" or "beast") and "wa Mtuka" (meaning "of the Mtuka", referring to the Mtuka River). The specific name *moyowamkia* combines the Kiswahili terms "moyo" (meaning "heart") and "wa mkia" (meaning "of the tail"), referencing the shape of one of the tail vertebrae.

CLASSIFICATION
Sauropoda
 Gravisauria
  Eusauropoda
   Neosauropoda
    Macronaria
     Titanosauriformes
      Somphospondyli
       Titanosauria

LOCATION
Tanzania

KNOWN REMAINS
Partial skeleton

4 m   8 m   12 m   16 m

P. puestohernandezi

(Filippi et al., 2011)
Length: 18 m (59 ft)
Height: 7.5 m (25 ft)
Hip height: 3.3 m (11 ft)
Body mass: 13,000 kg (14.3 t)
Reconstruction: ☐☐☐☐

4 m

| 250 | 245 | 240 | 235 | 230 | 225 | 220 | 215 | 210 | 205 | 200 | 195 | 190 | 185 | 180 | 175 | 170 | 165 | 160 | 155 | 150 | 145 | 140 | 135 | 130 | 125 | 120 | 115 | 110 | 105 | 100 | 95 | 90 | 85 | 80 | 75 | 70 | 65 |

TRIASSIC | **JURASSIC** | **CRETACEOUS**

Petrobrasaurus puestohernandezi (meaning "Petrobras's lizard from Puesto Hernández") is an understudied South American titanosaur. It was medium sized and had limb and pelvic bones that were relatively slender. Its teeth were typically titanosaurian, having a peg-like shape.

In 2006, land-leveling work was being done by the Petrobras oil company in preparation for installing the PH 1597 oil well within the Puesto Hernández oil production field. This activity was responsible for uncovering the fossil-bearing location, and the remains of a single sauropod individual were excavated through the spring of 2007.

The bones were disarticulated but still in an approximate anatomical layout, indicating that they underwent little transportation before their final burial. Five partial theropod teeth were also found at the site, possibly an indication of scavenging or predation.

As far as titanosaur specimens go, the holotype of *Petrobrasaurus* is relatively complete, and yet it has received unusually little attention across various studies, especially regarding its taxonomic affinities. Its describers were unable to firmly pin down its exact position within **Titanosauria**, recovering it "in a polytomy at the base of Eutitanosauria." They did note certain similarities between *Petrobrasaurus* and *Mendozasaurus*, a member of Lognkosauria, indicating that *Petrobrasaurus* could possibly have a similarly derived position among Titanosauria. Future analyses will be needed to pinpoint its classification.

The generic name *Petrobrasaurus* refers to the Petrobras oil company for their cooperation. The specific name *puestohernandezi* specifically references the Puesto Hernández oil field.

CLASSIFICATION
Sauropoda
 Gravisauria
  Eusauropoda
   Neosauropoda
    Macronaria
     Titanosauriformes
      Somphospondyli
       Titanosauria

LOCATION
Argentina

KNOWN REMAINS
Partial skeleton

TITANOSAURIA

T. indicus

(Lydekker, 1877)
Length: 15 m (50 ft)
Height: 6 m (20 ft)
Hip height: 3 m (9.8 ft)
Body mass: 8,000 kg (8.8 t)
Reconstruction: ☐☐☐☐

| TRIASSIC | JURASSIC | CRETACEOUS |

Titanosaurus indicus (meaning "giant lizard of India"), the original titanosaur, was the first dinosaur to be described from India. The holotype, consisting of two tail vertebrae, was discovered in 1828 by Major-General Sir William Henry Sleeman. Eventually, geologist Hugh Falconer would briefly describe them in 1862, and paleontologist Richard Lydekker would finally name them in 1877. (At some point, the remains were considered to be "lost" but were later rediscovered in misplaced museum archives [Mohabey et al., 2013].)

At the time, there were obviously very few other specimens with which compare these bones, and as such, many of their anatomical features distinctly stood out as being noteworthy. However, in the modern day, these features are known to be broadly distributed among many titanosaurian taxa, which leaves nothing "unique" to be observed in the bones of *T. indicus*. Consequently, *T. indicus* is often treated as a dubious genus.

It would take an entire chapter to elucidate the numerous now-invalid species that have been assigned to *Titanosaurus*

in the last century and a half, let alone the fossil specimens that have been shuffled about between them. In a nutshell, *Titanosaurus* became a substantial wastebasket taxon that engulfed specimens from Madagascar, Europe, Asia, and South America.

Several new genera have been erected from fossils that were once considered *Titanosaurus*: *Isisaurus*, *Laplatasaurus*, *Magyarosaurus*, and *Neuquensaurus* are among them.

Notable now-dubious species that are now considered to be too fragmentary or indistinct to properly name include *T. blanfordi*, known from a few tail vertebrae from India; *T. falloti*, known from femoral fragments from Laos; *T. lydekkeri*, known from a tail vertebra from England; *T. madagascariensis*, known from two tail vertebrae, an osteoderm, and a humerus; *T. nanus*, known from a few "small" vertebrae from Argentina; *T. rahioliensis*, known from some teeth from India; and *T. valdensis*, known from two tail vertebrae from England.

CLASSIFICATION

Sauropoda
 Gravisauria
  Eusauropoda
   Neosauropoda
    Macronaria
     Titanosauriformes
      Somphospondyli
       Titanosauria

LOCATION

India

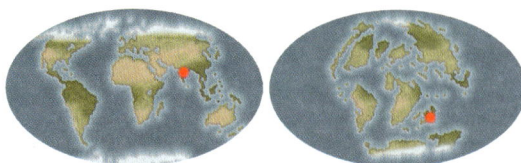

KNOWN REMAINS

Tail vertebrae

G. sinensis

(Lü et al., 2013)
Length: 26 m (85 ft)
Height: 12 m (39 ft)
Reconstruction:

3 m

66

| TRIASSIC | JURASSIC | CRETACEOUS |

Gannansaurus sinensis is known from two fragmentary vertebrae, one from the back and one from the tail. These specimens were saved (although in a damaged state) from a construction site, whereas other bones were completely destroyed by explosives.

The describers confidently placed *Gannansaurus* within Somphospondyli but then went on to note several features of the bones that were previously known only from *Euhelopus*

of the Early Cretaceous. However, the subsequent description of titanosaur remains that featured similar anatomy (Currie et al., 2018) would seem to suggest that *Gannansaurus* is most logically a **titanosaur** of some description.

The binomial name refers to the Gannan District of the city of Ganzhou in Jiangxi Province, China; "sino" is a prefix commonly used to refer to China.

K. gittelmani

(Gomani, 2005)
Length: 9.5 m (31 ft)
Height: 4 m (13 ft)
Reconstruction:

1 m

115

| TRIASSIC | JURASSIC | CRETACEOUS |

Karongasaurus gittelmani is known from a single dentary bone and a handful of isolated teeth. The jaw bone is moderately U-shaped, and the peg-like teeth are limited to roughly the front half of the mouth. These fossils come from the same bone beds that have yielded much more plentiful remains of a different titanosaur, *Malawisaurus*. Some neck and tail vertebrae from the area, which do not belong to *Malawisaurus*, may or may not pertain to *Karongasaurus*.

Several taxonomic studies have verified *Karongasaurus* as being a basal **lithostrotian** (Mannion and Calvo, 2011; Gorscak and O'Connor, 2019).

The generic name *Karongasaurus* refers to the northern Karonga District of the African nation of Malawi. The specific name *gittelmani* honors Steve Gittleman, one-time president of the Dinosaur Society.

TITANOSAURIA

NORMANNIASAURUS

N. genceyi

(Le Loeuff et al., 2013)
Length: 12 m (40 ft)
Height: 6 m (20 ft)
Reconstruction:

2 m

| TRIASSIC | JURASSIC | CRETACEOUS |

110

Normanniasaurus genceyi is known from fossil fragments that were recovered in 1990 from blocks of stone that had naturally fallen from seaside cliffs.

The shape of the articulation surfaces seen in the animal's dorsal (back) vertebrae suggests that it is a fairly basal **titanosaur**, as this particular arrangement has been seen in only a few such genera. Some phylogenetic results have supported this idea (Díez Díaz et al., 2018; Wang et al.,

2021). Contrastingly, the "spongy" texture of these bones could indicate a more "advanced" titanosaurian condition, and such a derived position has also been recovered in various taxonomies (Gorscak and O'Connor, 2019; Mannion et al., 2019a).

The generic name derives from "Normannia," the Latin name for Normandy, France. The specific name honors fossil discoverer Pierre Gencey.

BAOTIANMANSAURUS

B. henanensis

(Zhang et al., 2009)
Length: 22 m (74 ft)
Height: 12 m (39 ft)
Reconstruction:

3 m

| TRIASSIC | JURASSIC | CRETACEOUS |

95

Baotianmansaurus henanensis is one of only a handful of Chinese titanosaurs. It is known from a fragmentary set of remains that were unearthed from what used to be known as the Gaogou geological formation, but is now known as the Xiaguan Formation. The exact age of the formation is not well constrained.

Baotianmansaurus is of very uncertain taxonomic placement. It has been recovered as a non-titanosaurian somphospondylan (Mannion et al., 2019a), a basal titanosaur (Moore et al., 2020; Poropat et al., 2021), and as a derived saltasauroid (Averianov et al., 2018; Royo-Torres et al., 2020).

The animal's binomial name refers to the Baotianman National Nature Reserve, located in Henan Province, China.

A. chilensis

(Kellner et al., 2011)
Length: 13 m (42 ft)
Height: 6.5 m (21 ft)
Reconstruction: ☐☐☐☐

2 m

105—

| TRIASSIC | JURASSIC | CRETACEOUS |

Atacamatitan chilensis was the first nonavian dinosaur to be named from Chile. It is known from a fragmentary set of remains that were unearthed from the Tolar geological formation, the age of which is poorly constrained. *Atacamatitan* seems to have been a particularly gracile sauropod, especially with regard to its femur.

Atacamatitan is usually excluded from phylogenetic studies. The describers placed it somewhere within **Lithostrotia**, which was loosely corroborated by one later study (Rubilar-Rogers et al., 2012). The animal's binomial name refers to the Atacama Desert of Chile.

BARROSASAURUS

B. casamiquelai

(Salgado and Coria, 2009)
Length: 18 m (59 ft)
Height: 6 m (20 ft)
Reconstruction: ☐☐☐☐

2 m

80—

| TRIASSIC | JURASSIC | CRETACEOUS |

Barrosasaurus casamiquelai is known only from three vertebrae. The bones were found directly adjacent to one another. Although incomplete, the bones were detailed in their preservation, undamaged by any erosion. While it is clear that they belonged to a **titanosaur**, it is not currently possible to pinpoint a more precise location on the family tree for *Barrosasaurus*.

It has been suggested that *Barrosasaurus* may only represent a juvenile individual of one of the other numerous titanosaur genera that have been named from the Anacleto geological formation (Paul, 2016).

The generic name *Barrosasaurus* refers to the location Sierra Barrosa, Argentina. The specific name *casamiquelai* honors Rodolfo Casamiquela (1932–2008) for his contributions to Argentinean earth science.

TITANOSAURIA

P. hungaricus

(Díez Díaz et al., 2025)
Length: 11.3 m (37.1 ft)
Height: 7.5 m (25 ft)
Reconstruction:

1 m

70 —

| TRIASSIC | JURASSIC | CRETACEOUS |

Petrustitan hungaricus (meaning "stone titan from Hungary") is actually a reclassification of an old species into a new genus. The problematic "Magyarosaurus" *hungaricus* never really fit in with other *Magyarosaurus* species, and new analyses show it to be a genus all its own. It is known from a tibia and fibula of the lower rear leg.

G. cavocaudatus

(Han et al., 2024)
Length: 14 m (46 ft)
Height: 6.7 m (22 ft)
Reconstruction:

2 m

94 —

| TRIASSIC | JURASSIC | CRETACEOUS |

Gandititan cavocaudatus (meaning "cavity-tailed geologic giant from Ganzhou") is, as per usual, differentiated by nitty-gritty details regarding the anatomy of its vertebrae. It was considered by its describers to be a basal titanosaurian, although it could also possibly be a titanosauriform.

The generic name *Gandititan* derives from a reference to Ganzhou City, and the Chinese term "di" (meaning "earth"). The specific name *cavocaudatus* combines the Latin "cavum" (meaning "cavity") and "cauda" (meaning "tail") based on the "complicated pattern of pleurocoels and neural arch laminae seen in the anterior caudal vertebrae."

1 m 2 m 3 m 4 m 5 m 6 m 7 m 8 m

4 m

3 m

2 m

1 m

T. valdecii

(Pereira et al., 2024)
Length: 9 m (29.5 ft)
Height: 4.8 m (15.8 ft)
Hip height: 2 m (6.6 ft)
Body mass: 2,200 kg (4,850 lb)
Reconstruction: ☐☐☐☐

250 245 240 235 230 225 220 215 210 205 200 195 190 185 180 175 170 165 160 155 150 145 140 135 130 125 120 115 110 105 **100** 95 90 85 80 75 70 65

TRIASSIC	JURASSIC	CRETACEOUS

Tiamat valdecii is known only from nine fragmentary caudal (i.e., tail) vertebrae, found in Brazil. Whereas the vertebrae near the base of the tail seem to have been optimized for flexibility, those near the middle were more stiff in their connections.

Tiamat is distinguished from other sauropods by the presence of a certain kind of joint on at least some of its tail bones, known as a hyposphene-hypantrum articulation. This connection features an extension on the rear surface of the vertebra that fits into a depression on the front surface of the subsequent vertebra, granting a measure of added rigidity. The vertebrae of most tetrapods have some sort of interlocking interconnectivity between them, but this particular feature is an additional supplement. Most non-titanosaurian sauropods do have this kind of articulation among their dorsal (back) vertebrae, but *Tiamat* is one of

very few titanosaurs to have this feature present on the vertebrae of the tail.

The describers of *Tiamat* recovered it as being a basal titanosaur. Further, it was recovered as being quite closely related to *Andesaurus*. This is significant because there has been a great deal of debate as to whether or not the titanosaur family tree contains a distinct "Andesauridae" branch, and any data on these relationships can go a long way toward settling the argument.

In Mesopotamian mythology, the deity Tiamat is the "mother of dragons." This choice of name alludes to the genus's proposed place at the very base of the Titanosauria family tree, which would poetically make it the "mother of all titanosaurs." The specific name *valdecii* honors site discoverer Valdeci dos Santos Júnior "for his fundamental support during the fieldworks."

CLASSIFICATION
Sauropoda
 Gravisauria
  Eusauropoda
   Neosauropoda
    Macronaria
     Titanosauriformes
      Somphospondyli
       Titanosauria

LOCATION
Brazil

KNOWN REMAINS
Tail vertebrae

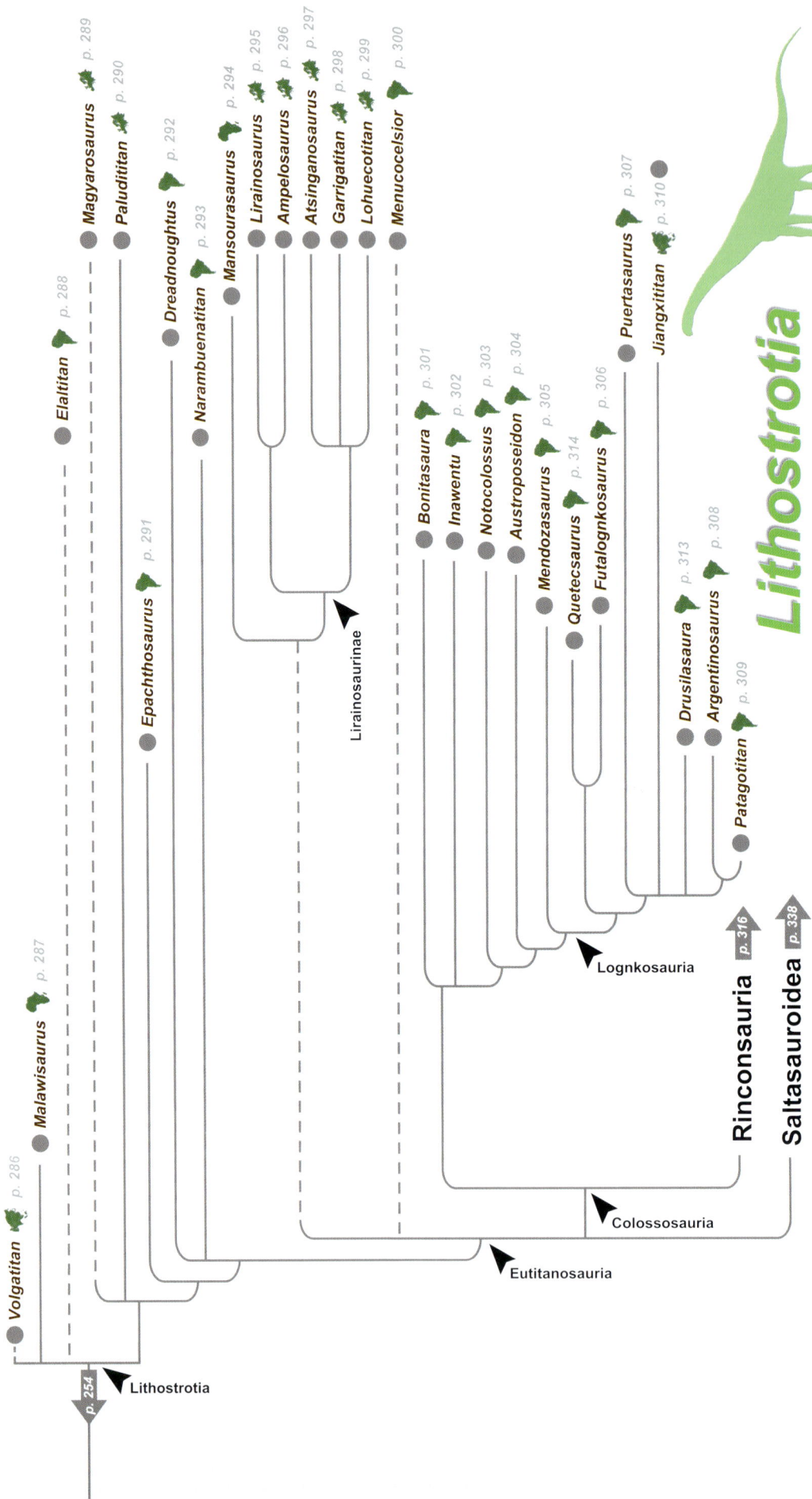

Lithostrotia

Time scale (top)

Cretaceous

Late

Early

Maastrichtian | Campanian | Sant. | Coniac. | Turon. | Cenomanian | Albian | Aptian

60 | 65 | 70 | 75 | 80 | 85 | 90 | 95 | 100 | 105 | 110 | 115 | 120 | 125

Taxa

Volgatitan p. 286

Malawisaurus p. 287

Elaltitan p. 288

Magyarosaurus p. 289

Paludititan p. 290

Epachthosaurus p. 291

Dreadnoughtus p. 292

Narambuenatitan p. 293

Mansourasaurus p. 294

Lirainosaurus p. 295

Ampelosaurus p. 296

Atsinganosaurus p. 297

Garrigatitan p. 298

Lohuecotitan p. 299

Menucocelsior p. 300

Bonitasaura p. 301

Inawentu p. 302

Notocolossus p. 303

Austroposeidon p. 304

Mendozasaurus p. 305

Quetecsaurus p. 314

Futalognkosaurus p. 306

Puertasaurus p. 307

Jiangxititan p. 310

Drusilasaura p. 313

Argentinosaurus p. 308

Patagotitan p. 309

Clade labels

Lirainosaurinae

Lognkosauria

Colossosauria

Eutitanosauria

Rinconsauria p. 316

Saltasauroidea p. 338

Lithostrotia p. 254

LITHOSTROTIA

The name Lithostrotia comes from the Greek "lithostros," meaning "inlaid with stones," which is inspired by the numerous species within the group that seem to have possessed osteoderms—armor-like boney growths embedded in the skin, a feature found much more famously in the ankylosaur dinosaurs. While these might have had a limited protection functionality for lithostrotians, it is possible that they had a different primary use, such as storing minerals for later absorption.

Eutitanosauria is sometimes considered to be more or less synonymous with Lithostrotia because the placement of the exact genus that defines each group can sometimes vary from study to study; many works from recent years have forgone the use of Eutitanosauria altogether. However, a recent redefinition proposed by Carballido et al. (2022) essentially shifts Eutitanosauria to identify the node where Colossosauria and Saltasauroidea diverge.

By common definition, **Lithostrotia** begins where *Malawisaurus* is placed (Upchurch et al., 2004). The majority of recent analyses place *Epachthosaurus* as a basal lithostrotian, outside Colossosauria (Navarro et al., 2022; Santucci and Filippi, 2022)).

Comparatively speaking, *Paludititan* has been little studied but has been recovered sandwiched between the two

Lithostrotia

Eutitanosauria

Colossosauria

(Gallina et al., 2022)
(Pérez Moreno et al., 2023)

Epachthosaurus

Lithostrotia

Eutitanosauria

(Hechenleitner et al., 2020)
(Navarro et al., 2022)

Argentinosaurus huinculensis

aforementioned genera (Díez Díaz et al., 2018; Navarro et al., 2022).

The uncertainties currently regarding the identity of *Magyarosaurus* render any precise taxonomic positioning untenable (Upchurch et al., 2004). A lack of corroborative study regarding *Volgatitan* renders its placement likewise uncertain. *Elaltitan* is woefully understudied; its describers considered it lithostrotian (Mannion and Otero, 2012).

Dreadnoughtus is often found in a basal lithostrotian position (Hechenleitner et al., 2020), although it is sometimes placed in a slightly more derived position in or near Saltasauroidea (Pérez Moreno et al., 2023). *Narambuenatitan* was placed in a similar position by Navarro et al. (2022).

The European taxa *Lirainosaurus*, *Ampelosaurus*, *Atsinganosaurus*, and *Garrigatitan* have been found to unite under **Lirainosaurinae** (Díez Díaz et al., 2021). One analysis included *Lohuecotitan* in this group (Navarro et al., 2022), although this is less certain, as it has been found in other taxonomic locations as well, such as Saltasauridae (Gorscak and O'Connor, 2019). *Mansourasaurus* has be found as being closely related to the group (Sallam et al., 2018). Lirainosaurinae, as a group, is itself lacking a definitive placement; one scenario by Carballido et al. (2022) places it roughly sister to Lognkosauria, while the large analysis performed by Navarro et al. (2022) places it as a basal clade within Saltasauroidea.

The recently described *Menucocelsior* has gained some mystique by apparently not fitting in with any eutitanosaurian clade (Rolando et al., 2022).

Lognkosauria, first proposed by Calvo et al. (2007b), has greatly expanded its ranks since that time. Just stemward of the clade are *Bonitasaura* and *Notocolossus*, according to Hechenleitner et al. (2020), Gallina et al. (2022), and Santucci and Filippi (2022). These works, plus Silva Junior et al. (2022a) and Pérez Moreno et al. (2023) agree upon the clade's members, as seen on the previous page; however, the exact positions of these members within the clade tend to vary between studies. Additionally, the describers of *Austroposeidon* considered it to fall just outside Lognkosauria (Bandeira et al., 2016). The describers of *Inawentu* placed it closely related to *Bonitasaura* (Filippi et al., 2024).

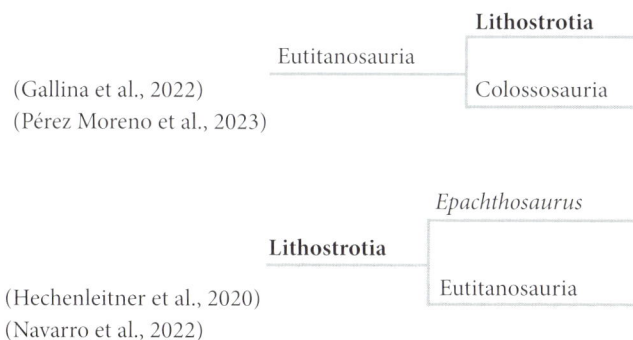

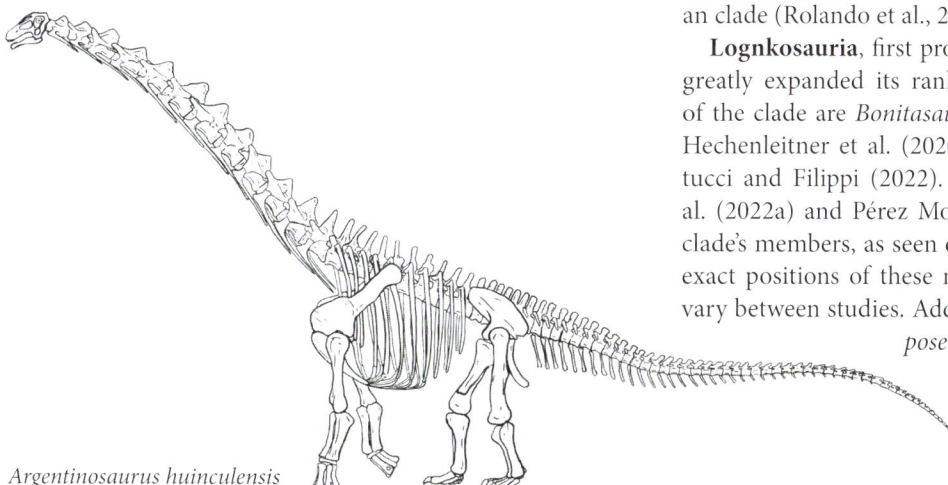

VOLGATITAN

8 m | 4 m | 8 m | 12 m | 16 m | 20 m

V. simbirskiensis

(Averianov and Efimov, 2018)
Length: 20 m (66 ft)
Height: 9 m (30 ft)
Hip height: 3.1 m (10 ft)
Body mass: 15,000 kg (16.5 t)
Reconstruction:

250 245 240 235 230 225 220 215 210 205 200 195 190 185 180 175 170 165 160 155 150 145 140 135 128 125 120 115 110 105 100 95 90 85 80 75 70 65

| TRIASSIC | JURASSIC | CRETACEOUS |

Volgatitan simbirskiensis (meaning "giant from Volga, Simbirsk") is one of the oldest known titanosaurs. At the time of its discovery, it was the oldest titanosaur known from the Northern Hemisphere.

The seven tail vertebrae that constitute the only known remains of *Volgatitan* were discovered by Vladimir Efimov in 1982, when he noticed the presence of fossils in the area. From 1984 through 1987, limestone nodules originating from a river outcrop were recovered that eventually yielded the remains. These fossils were briefly mentioned in literature throughout the years, but it would prove to be decades before they received a detailed description.

Volgatitan is distinguished from other kinds of sauropods by the presence of a certain kind of joint on at least some of its tail bones, known as a hyposphene-hypantrum articulation. This connection features an extension on the rear surface of the vertebra that fits into a depression on the front surface of the subsequent vertebra, granting a measure of added rigidity. The vertebrae of most tetrapods have some sort of interlocking interconnectivity between them, but this particular kind of feature is an additional supplement. Most non-titanosaurian sauropods do have this kind of articulation among their dorsal (back) vertebrae, but *Volgatitan* is only the third known sauropod to have this feature present on the vertebrae of the tail (after *Astrophocaudia* and *Epachthosaurus*).

Volgatitan has thus far not been included in any phylogenetic study besides that which its describers conducted. They found it to be a basal member of the lineage leading to Lognkosauria, which would most likely put it within **Colossosauria**.

The animal's binomial name refers to the Volga River and the city formerly known as Simbirsk (currently Ulyanovsk).

CLASSIFICATION

Sauropoda
  Macronaria
    Somphospondyli
      Titanosauria
        Lithostrotia
          Eutitanosauria

LOCATION

Russia

KNOWN REMAINS

Tail vertebrae

2 m   4 m   6 m   8 m   10 m

4 m

2 m

M. dixeyi

(Jacobs et al., 1993)
Length: 11 m (36 ft)
Height: 5 m (16 ft)
Hip height: 2 m (6.5 ft)
Body mass: 3,000 kg (3.3 t)
Reconstruction: ☐☐☐☐

250 245 240 235 230 225 220 215 210 205 200 195 190 185 180 175 170 165 160 155 150 145 140 135 130 125 120 **115** 110 105 100 95 90 85 80 75 70 65

| TRIASSIC | JURASSIC | CRETACEOUS |

Malawisaurus dixeyi (meaning "Dixey's lizard from Malawi") is a relatively small sauropod and is generally regarded as having one of the most complete skeletons of any known titanosaur, which includes rare material from the skull. *Malawisaurus* had an unusually blunt face, and unlike many titanosaurs, its teeth were not limited to the front of its jaw.

The fossils were unearthed in 1924 and were originally described in 1928 by Sidney H. Haughton. He named them as a new species of the now-dubious genus *Gigantosaurus*, a name that already had a complex story by that period of time. The species would go on to languish in the literature until new material discovered at the original site spurred a redescription of the material and the erection of *Malawisaurus*.

Unfortunately, questions have now arisen as to whether all of the fossil material found across the various sites actually constitutes the same genus (Gomani, 2005) and regarding just how truly diagnostic the holotype actually is (Gorscak et al., 2016).

The taxonomic clade **Lithostrotia** was first erected in 2004 and was defined as containing "the most recent common ancestor of *Malawisaurus* and *Saltasaurus* and all the descendants of that ancestor." So, by definition, *Malawisaurus* is a member of this moderately derived titanosaurian group. The clade's name comes from the Greek "lithostrotos" (meaning "inlaid with stones"), referring to the osteoderms of various sizes that have been found embedded in the skin of some (but not all) advanced titanosaurians, including *Malawisaurus* (Upchurch et al., 2004).

The generic name *Malawisaurus* refers to the African nation of Malawi. The specific name *dixeyi* honors the fossil's discoverer, Frederick Augustus Dixey.

CLASSIFICATION
Sauropoda
 Eusauropoda
  Neosauropoda
   Macronaria
    Titanosauriformes
     Somphospondyli
      Titanosauria
       Lithostrotia

LOCATION
Malawi

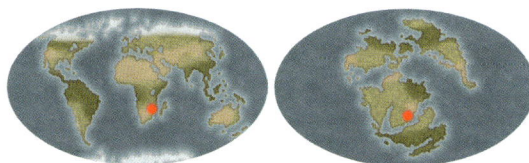

KNOWN REMAINS
Nearly complete

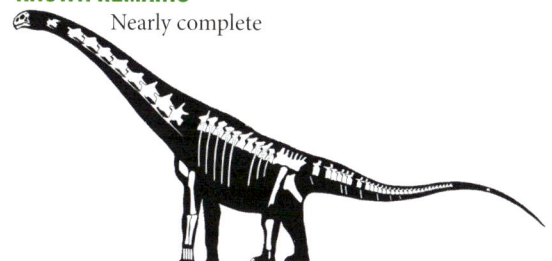

287

LITHOSTROTIA

E. lilloi

(Mannion and Otero, 2012)
Length: 20 m (66 ft)
Height: 7.5 m (25 ft)
Hip height: 3.5 m (11.5 ft)
Body mass: 23,000 kg (25.4 t)
Reconstruction: ☐☐☐☐

TRIASSIC	JURASSIC	CRETACEOUS

The only known remains of **Elaltitan lilloi** (meaning "Lillo's giant of Elal") were originally interpreted as belonging to different kinds of sauropods, first *Antarctosaurus* and then *Argyrosaurus*.

When this set of fossils (PVL 4628) was first discovered, it was originally attributed to an indeterminate species of the sauropod *Antarctosaurus* (Bonaparte and Gasparini, 1978). Later, they were reclassified as representing *Argyrosaurus* (Powell, 1986), a genus described in 1893 on the basis of a nearly complete forelimb.

When *Argyrosaurus* was thoroughly redescribed in modern literature, however, it was determined that PVL 4628 actually represented a distinct genus, and so it was erected to its own new taxon, *Elaltitan*. (The describers attributed the remains to the Bajo Barreal Formation, but this has since been revised as the Lago Colhue Huapi Formation [Casal et al., 2016]).

One notable aspect of the remains is the presence of the calcaneum ankle bone. At the time of the specimen's description, no other titanosaur calcaneum had ever been found in association with other remains. The shape of the bone is consistent with those found in other kinds of sauropod (González Riga et al., 2019).

In comparison to many titanosaurs, *Elaltitan* is represented by a relative wealth of skeletal material, and yet it has not to date been a part of any phylogenetic study. The describers considered it to be a likely **lithostrotian**, citing the need for further study.

The generic name *Elaltitan* refers to Elal, the creator god of the Indigenous Tehuelche people. The specific name *lilloi* honors naturalist Miguel Lillo (1862–1931).

CLASSIFICATION

Sauropoda
 Eusauropoda
  Neosauropoda
   Macronaria
    Titanosauriformes
     Somphospondyli
      Titanosauria
       Lithostrotia

LOCATION

Argentina

KNOWN REMAINS

Partial skeleton

M. dacus

(von Huene, 1932)
Length: 6 m (20 ft)
Height: 5.5 m (18 ft)
Hip height: 1.5 m (5 ft)
Body mass: 520 kg (1,150 lb)
Reconstruction: ☐☐☐☐

1 m · 2 m · 3 m · 4 m · 5 m · 6 m

3 m · 2 m · 1 m

250 245 240 235 230 225 220 215 210 205 200 195 190 185 180 175 170 165 160 155 150 145 140 135 130 125 120 115 110 105 100 95 90 85 80 75 70 65

TRIASSIC · **JURASSIC** · **CRETACEOUS**

Magyarosaurus dacus (meaning "Magyar's lizard from Dacia") is borderline impossible to anatomically describe because no one has yet determined which bones, exactly, belong to the genus; it is a taxon that requires "major revision" (Upchurch et al., 2004).

A great many individual sauropod bones, including osteoderms, from roughly the same locality (the Transylvania region that, at the time, was considered to be part of Hungary but is now considered Romania) were first described by Baron Nopcsa in 1915, who named them to yet another species of *Titanosaurus*. In 1932, Friedrich Richard von Hoinigen reclassified the remains as a new genus, *Magyarosaurus*, into which he erected three species: *M. dacus*, *M. transsylvanicus* (now often considered synonymous with *M. dacus*), and *M. hungaricus* (distinguished by its larger size).

What distinguishes most of the bones is their abnormally small overall size. Some researchers had hypothesized that most of the remains were from juveniles and that the rarer *M. hungaricus* fossils were from adults (Loeuff, 2005). However, a histological examination of the internal microstructures of numerous different bone samples determined that each of the small bones actually came from individuals that were fully grown adults; "*M.*" *hungaricus* was likely a different genus altogether (Stein et al., 2010).

It thus seems most likely that *Magyarosaurus* represents a case of insular dwarfism, whereby island-bound species shrink in size to compensate for the fewer available resources. The animal's Cretaceous habitat, Haţeg Island, is known to have been home to other dwarf species.

The taxonomic position of *Magyarosaurus* is, understandably, highly uncertain. Taking geography into account, it is likely part of **Lithostrotia**.

The generic name *Magyarosaurus* refers to the Magyar, the primary ethnic group of Hungary. The specific name *dacus* refers to Dacia, which is the old Roman name for the region.

CLASSIFICATION

Sauropoda
 Eusauropoda
  Neosauropoda
   Macronaria
    Titanosauriformes
     Somphospondyli
      Titanosauria
       Lithostrotia

LOCATION

Romania

KNOWN REMAINS

Uncertain

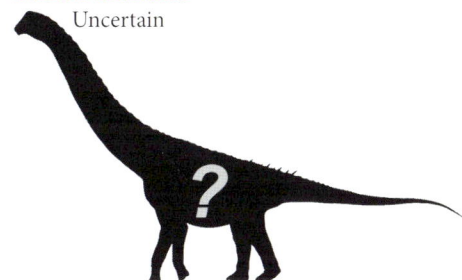

2 m   4 m   6 m   8 m

2 m

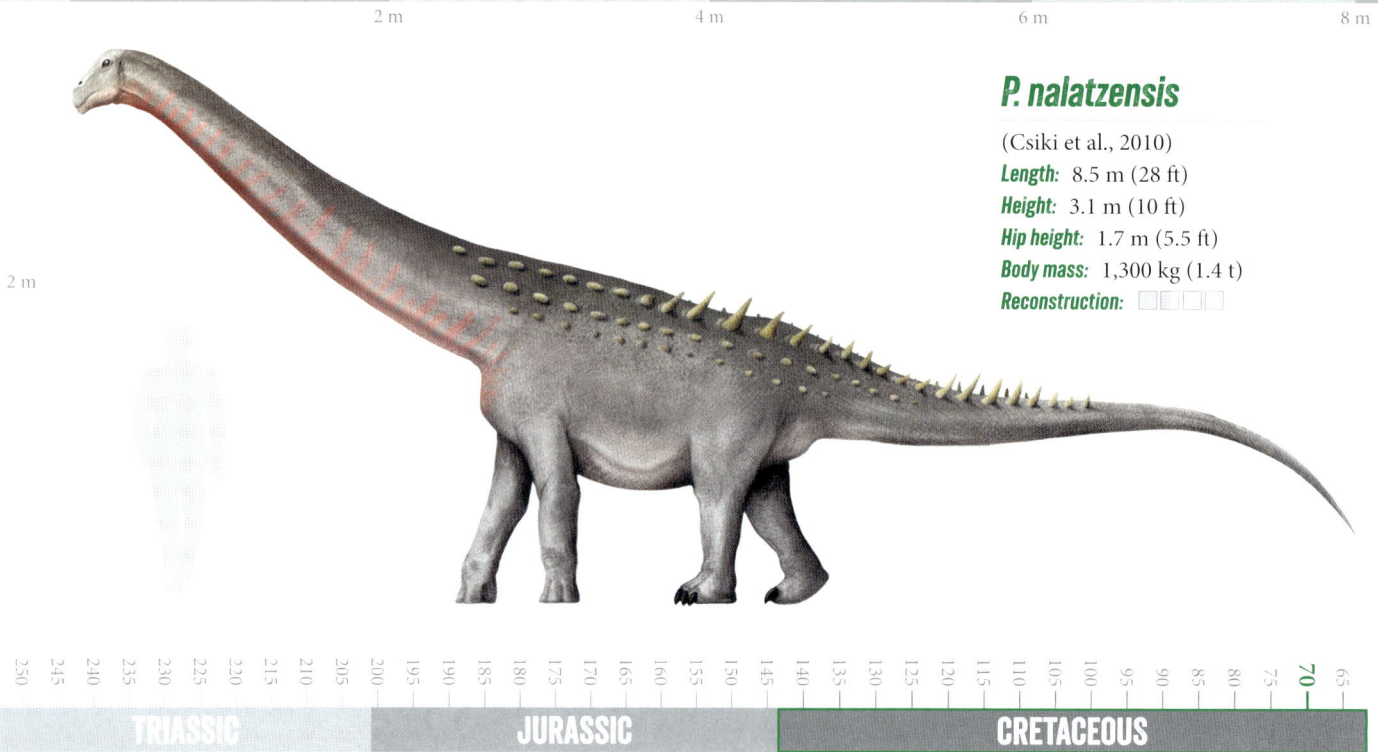

P. nalatzensis

(Csiki et al., 2010)
Length: 8.5 m (28 ft)
Height: 3.1 m (10 ft)
Hip height: 1.7 m (5.5 ft)
Body mass: 1,300 kg (1.4 t)
Reconstruction: ☐☐☐☐

250 245 240 235 230 225 220 215 210 205 200 195 190 185 180 175 170 165 160 155 150 145 140 135 130 125 120 115 110 105 100 95 90 85 80 75 **70** 65

TRIASSIC   **JURASSIC**   **CRETACEOUS**

Paludititan nalatzensis (meaning "marshy giant from Nălaţ-Vad") is known from the same general strata that have produced the vague scattershot of separate bones that have variously been attributed to the titanosaur *Magyarosaurus*. But since the question, "What really **is** *Magyarosaurus*?" currently has no good answer, the specimen UBB NVM1 was erected to its own genus, *Paludititan*. This differentiation has been validated by subsequent studies (Mannion et al., 2019).

As the bones of *Paludititan* were discovered in close association with one another, they can be surmised to have come from the same animal. This association is very useful for classifying and comparing the genus to other titanosaurs—as opposed to genera such as *Magyarosaurus*, which are based on an accumulation of bones that may or may not actually have come from the same kind of animal.

Paludititan inhabited a small landmass known as Haţeg Island. The small area available on the island is likely what led to *Paludititan*'s diminutive size.

The describers of *Paludititan* placed it within Titanosauria, but failed to calculate a more detailed classification, citing that different versions of analyses came to differing conclusions. One such conclusion, that of being a basal **lithostrotian**, has been found by some ensuing studies (Díez Díaz et al., 2018; Navarro et al., 2022).

A differing interpretation places the genus within Saltasauridae, perhaps among its most derived members (Gorscak and O'Connor, 2019).

The generic name *Paludititan* derives from the Greek "paludos" (meaning "marshy"), as the animal's habitat has been interpreted as being a wetland. The specific name *nalatzensis* refers to the name of the locality where the fossils were found, Nălaţ-Vad.

CLASSIFICATION
Sauropoda
 Eusauropoda
  Neosauropoda
   Macronaria
    Titanosauriformes
     Somphospondyli
      Titanosauria
       Lithostrotia

LOCATION
Romania

KNOWN REMAINS
Partial skeleton

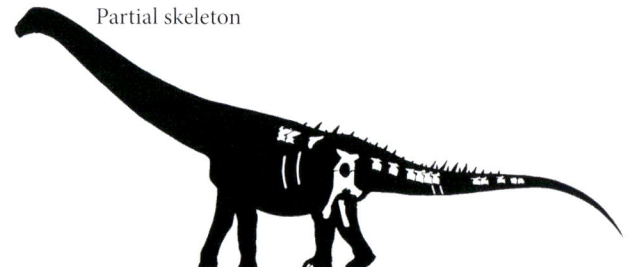

2 m   4 m   6 m   8 m   10 m   12 m

6 m

4 m

2 m

E. sciuttoi

(Powell, 1990)
Length: 13 m (45 ft)
Height: 6 m (20 ft)
Hip height: 2.3 m (7.5 ft)
Body mass: 5,000 kg (5.5 t)
Reconstruction: ☐☐☐☐

250 245 240 235 230 225 220 215 210 205 200 195 190 185 180 175 170 165 160 155 150 145 140 135 130 125 120 115 110 105 100 **95** 90 85 80 75 70 65

TRIASSIC   **JURASSIC**   **CRETACEOUS**

Epachthosaurus sciuttoi (meaning "Sciutto's heavy lizard") was first mentioned in J. E. Powell's dissertation in 1986, even though it was not formally described in a published paper until 1990. As such, there are several mentions of the genus in other works prior to 1990.

The holotype specimen used as a basis for *Epachthosaurus* is a single damaged dorsal (back) vertebra. Luckily, much more complete remains were soon found nearby—in fact, this referred specimen is one of the most complete titanosaur skeletons ever found. Furthermore, it was preserved in a completely articulated state, with nearly every bone located exactly where they would have been when the animal died.

The *Epachthosaurus* in question was, for some reason, lying directly on its belly when it died, with its rear legs folded and its front limbs splayed to the side. Nearly every bone of the rear portion of the body was preserved, past the shoulders. The foremost bones that were unearthed were

slightly eroded, suggesting that at one time in the past even more of the animal's remains had been present but had already eroded away (Martínez et al., 2004).

The high degree of fossilization even preserved a series of ossified tendons atop the pelvis. Since no ankle or wrist fossils were recovered, it seems likely that these bones did not exist in the first place; this fits with our knowledge of other titanosaurs, suggesting these structures might have been purely cartilaginous.

The majority of recent analyses place *Epachthosaurus* as a basal **lithostrotian**, outside Colossosauria (Hechenleitner et al., 2020; Wang et al., 2021; Gallina et al., 2022; Navarro et al., 2022; Santucci and Filippi, 2022).

The generic name *Epachthosaurus* combines the Greek "epachthos" (meaning "heavy") and "sauros" (meaning "lizard"). The specific name *sciuttoi* honors paleontologist J. C. Sciutto, who discovered the site.

CLASSIFICATION

Sauropoda
 Eusauropoda
  Neosauropoda
   Macronaria
    Titanosauriformes
     Somphospondyli
      Titanosauria
       Lithostrotia

LOCATION

Argentina

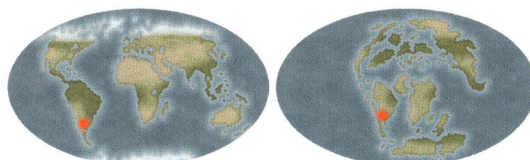

KNOWN REMAINS

Partial skeleton

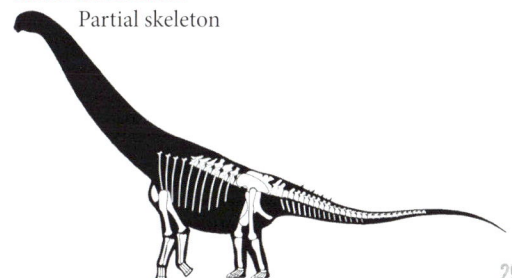

| 4 m | 8 m | 12 m | 16 m | 20 m | 24 m | 28 m |

12 m

8 m

4 m

D. schrani

(Lacovara et al., 2014)
Length: 28 m (92 ft)
Height: 12 m (39 ft)
Hip height: 4 m (13 ft)
Body mass: 48,000 kg (52.9 t)
Reconstruction: ☐☐☐☐

| 250 | 245 | 240 | 235 | 230 | 225 | 220 | 215 | 210 | 205 | 200 | 195 | 190 | 185 | 180 | 175 | 170 | 165 | 160 | 155 | 150 | 145 | 140 | 135 | 130 | 125 | 120 | 115 | 110 | 105 | 100 | 95 | 90 | 85 | 80 | 75 | 70 | 65 |

| TRIASSIC | JURASSIC | CRETACEOUS |

Dreadnoughtus schrani (meaning "Schran's dreadnought") has been the subject of some controversy regarding its size and mass. It is definitely not unheard of for the supposed mass of a sauropod to vary widely based on differing methodologies and calculation strategies, but it seems as if *Dreadnoughtus* in particular has been plagued with uncertainties and variations. The animal's estimated mass has run the gambit from 31,000 (Campione and Evans, 2020), to 48,000 (Paul, 2019), or 59,000 kilograms (original description; or, 34 t to 53 to 65 tons).

Matters are complicated even further when the animal's growth stage is taken into account. According to histological analysis of the holotype's inner bone tissue, the specimen was not even fully grown at the time of its death, having an ontogenetic score of only 9 out of 14 (Ullmann and Lacovara, 2016). Naturally, this means that the true size potential for *Dreadnoughtus* is unknown.

In any case, *Dreadnoughtus* is still in the upper echelon of massive titanosaurs. While the majority of the "giant" titanosaurs seem to have lived in the "early" Late Cretaceous, *Dreadnoughtus* is one of only two that are currently known from the "late" portion of that epoch. The other giant genus, *Puertasaurus*, is a lognkosaur, while *Dreadnoughtus* has a more basal position on the family tree. This difference indicates that supergigantism is not a unique trait and evolved multiple times throughout Titanosauria.

Dreadnoughtus is known from two impressively complete specimens that were first found in 2005. The remote location of the dig site necessitated four summers worth of excavations.

The generic name *Dreadnoughtus* derives from the Old English "dreadnought" (meaning "fears nothing"), as an adult specimen would presumably be impervious to any potential foe. The specific name *schrani* honors financial donor Adam Schran.

CLASSIFICATION
Sauropoda
 Eusauropoda
  Neosauropoda
   Macronaria
    Titanosauriformes
     Somphospondyli
      Titanosauria
       Lithostrotia

LOCATION
Argentina

KNOWN REMAINS
Partial skeleton

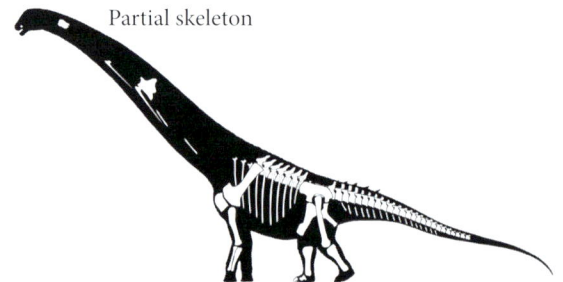

2 m   4 m   6 m   8 m   10 m   12 m

N. palomoi

(Filippi et al., 2011)
Length: 13 m (42 ft)
Height: 4.9 m (16 ft)
Hip height: 2.3 m (7.5 ft)
Body mass: 4,000 kg (4.4 t)
Reconstruction: ▢▢▢▢

4 m

2 m

250 245 240 235 230 225 220 215 210 205 200 195 190 185 180 175 170 165 160 155 150 145 140 135 130 125 120 115 110 105 100 95 90 85 **80** 75 70 65

TRIASSIC   **JURASSIC**   **CRETACEOUS**

Narambuenatitan palomoi (meaning "Palomo's giant from Narambuena") is one of a great many South American titanosaurs that are now cataloged.

Found within a 15 square meter area, the holotype remains were unearthed during several trips in 2005 and 2006. Among the mudstone and fine-grained sandstone that encased the bones, the remains of logs, plants, and gastropods were also found, suggesting that the sauropod was buried in a pond or swampy area.

The only known *Narambuenatitan* specimen consists primarily of tail vertebrae, partial pelvic elements, and long bones of the limbs (some of which have been described as "gracile"). No armored osteoderms were located. Incomplete fusion among the vertebral elements suggests that the individual was only a subadult at the time of its death. Perhaps most importantly, portions of the skull are also preserved, including the braincase region and a segment of the upper jaw.

The describers' somewhat limited phylogenetic analysis placed *Narambuenatitan* within **Lithostrotia**. For whatever reason, more than a decade would pass before any other study included the genus in an analysis, so confidence in the placement was somewhat lacking. One clue came in 2020, when CT scans of the brain and inner ear chambers showed traits indicative of derived (rather than basal) titanosaurs (Paulina Carabajal et al., 2020). Later, a review specifically targeting South American titanosaurs also found *Narambuenatitan* to be a lithostrotian, strengthening this hypothesis (Santucci and Filippi, 2022).

The generic name *Narambuenatitan* refers to the Argentine region Puesto Narambuena. The specific name *palomoi* honors fossil discoverer and museum technician Salvador Palomo.

CLASSIFICATION
Sauropoda
  Eusauropoda
    Neosauropoda
     Macronaria
       Titanosauriformes
        Somphospondyli
         Titanosauria
          Lithostrotia

LOCATION
Argentina

KNOWN REMAINS
Partial skull and skeleton

2 m 4 m 6 m 8 m

2 m

M. shahinae

(Sallam et al., 2018)
Length: 8.5 m (28 ft)
Height: 3 m (9.8 ft)
Hip height: 1.9 m (6.2 ft)
Body mass: 1,300 kg (1.4 t)
Reconstruction:

250 245 240 235 230 225 220 215 210 205 200 195 190 185 180 175 170 165 160 155 150 145 140 135 130 125 120 115 110 105 100 95 90 85 80 **73** 70 65

| TRIASSIC | JURASSIC | CRETACEOUS |

Mansourasaurus shahinae (meaning "Shahin's lizard from Mansoura") is a small sauropod from central Egypt.

At the time of its discovery, *Mansourasaurus* was the most complete Late Cretaceous dinosaur known from all of Africa. The associated skeleton was excavated from a relatively small area of approximately 25 square meters; given the propensity of titanosaurs to be found among jumbled, multispecies bone beds, this gives scientists a welcome opportunity to unambiguously study the bones confirmed to have come from one single organism.

The bones of the shoulder region (including dorsal vertebrae) and limbs were clustered most closely together. Separated from the main group by several meters were two neck vertebrae as well as two skull fragments (a partial braincase and skull roof) and a mostly complete lower jaw (devoid of exposed teeth). The jaw is of the more common U-shape, as opposed to the squared shape found in some titanosaurs. The remains come from an individual thought to have not been fully grown, judging from various bones of the shoulder girdle remaining unfused. A long ulna was also recovered 20 meters from the rest of the remains; whether or not this bone came from the same individual or the same species is unclear.

The describers concluded that *Mansourasaurus* is most closely related to Eurasian titanosaurs, thereby placing it in or close to **Lirainosaurinae**. It has been interpreted that after the continental split between Africa and South America, evolution began to take the regional population of titanosaurs in a different direction. This Eurasian relationship also strengthens other evidence that Africa and Europe maintained some sort of land connection during this time.

The generic name *Mansourasaurus* refers to Mansoura University in Mansoura, Egypt. The specific name *shahinae* honors university contributor Mona Shahin.

CLASSIFICATION
Sauropoda
 Macronaria
  Somphospondyli
   Titanosauria
    Lithostrotia
     Eutitanosauria

LOCATION
Egypt

KNOWN REMAINS
Partial skeleton

1 m   2 m   3 m   4 m   5 m   6 m

L. astibiae

(Sanz et al., 1999)
Length: 6 m (20 ft)
Height: 2.9 m (9.5 ft)
Hip height: 1.2 m (4 ft)
Body mass: 4,000 kg (4.4 t)
Reconstruction: ☐☐☐☐☐

2 m

1 m

250 245 240 235 230 225 220 215 210 205 200 195 190 185 180 175 170 165 160 155 150 145 140 135 130 125 120 115 110 105 100 95 90 85 80 75 **70** 65

TRIASSIC   **JURASSIC**   **CRETACEOUS**

Lirainosaurus astibiae (meaning "Astibia's slender lizard") is one of the smallest sauropods known. It shared its general habitat, the Ibero-Armorican Island, with several other species of small- and medium-sized titanosaurs. With a smaller body mass, less food is required, and this trend of "insular dwarfism" has shown up numerous times in disparate vertebrate lineages whenever species become isolated on a landmass of reduced area.

Most of the fossils attributed to *Lirainosaurus* come from the Laño sand quarry in northern Spain. After the initial description of the genus, comprehensive follow-up studies have attributed more fossils from the site to the genus (Díez Díaz et al., 2013b), as well as a few fossils from neighboring sites (Díez Díaz et al., 2015).

Lirainosaurus is primarily known from the larger bones of the skeleton, such as vertebrae, long limb bones, shoulder blades, and pelvic elements. Smaller bones, like those of the hands, limbs, neck, and skull are mostly lacking. As the remains from the Laño site were deposited in a river system, it seems probable that the smaller bones were more likely to wash away and thus not be preserved. Remains of at least five individuals are present at the site, including teeth from a juvenile that indicate that it likely had a different diet than the adults.

Taxonomically, *Lirainosaurus* was previously placed merely somewhere within Lithostrotia. As more Ibero-Armorican titanosaur genera have been discovered, the clade **Lirainosaurinae** has been introduced to house these closely related species (Díez Díaz et al., 2018).

The generic name *Lirainosaurus* derives from the Basque term "lirain" (meaning "slender"), as the describer considered the animal's femur to be among the most gracile of any known sauropod. The specific name *astibiae* honors paleontologist Humberto Astibia, who led initial work at the dig site.

CLASSIFICATION
Sauropoda
 Macronaria
  Somphospondyli
   Titanosauria
    Lithostrotia
     Eutitanosauria
      Colossosauria
       Lirainosaurinae

LOCATION
Spain

KNOWN REMAINS
Partial skeleton

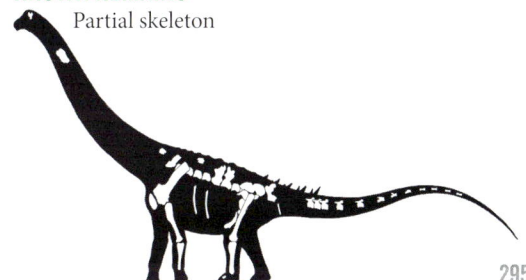

LITHOSTROTIA

3 m   6 m   9 m   12 m   15 m

6 m

3 m

A. atacis

(Le Loeuff, 1995)
Length: 16 m (53 ft)
Height: 6 m (20 ft)
Hip height: 3.2 m (10.5 ft)
Body mass: 15,000 kg (16.5 t)
Reconstruction: ☐ ☐ ☐ ☐

250 245 240 235 230 225 220 215 210 205 200 195 190 185 180 175 170 165 160 155 150 145 140 135 130 125 120 115 110 105 100 95 90 85 80 75 71 65

TRIASSIC   **JURASSIC**   **CRETACEOUS**

Ampelosaurus atacis (meaning "vine lizard from Atax") is known (most likely) from abundant skeletal material, but studies regarding the genus have thus far been quite scattershot in their interpretations.

Sauropod remains from Bellevue farm were first excavated in 1989, revealing an extensive bone bed consisting of predominately disarticulated remains. *Ampelosaurus* was very briefly described for the first time in a French journal, using three articulated dorsal (back) vertebrae as the holotype. It would be a decade before a more thorough description would be published (upon which the skeletal diagram below is based), although many aspects of the description were still quite brief. At this time, "more than 500 titanosaur bones" had been excavated at the site (Le Loeuff, 2005). Informal news bulletins over the next few years alluded to the discovery of a much more complete skull and other remains, which have yet to be published.

Despite this lack of information, one aspect of *Ampelosaurus* that seems clear is that the animal possessed a large number of armor-like osteoderms, ranging in shape from flat to spike-like. These accessories have had a large impact on the depiction of various titanosaurs in works of paleoart.

Although several studies have been conducted using individual elements of *Ampelosaurus* (Klein et al., 2012; Knoll et al., 2013), further analysis of the genus as a whole has been lacking. Hopefully, this might soon change, as Bernat Vila has announced an impending comprehensive study of the now more than "1,400 dinosaur remains" from the site. Preliminary results suggest that there are, in fact, at least two different titanosaur species present at the site, meaning that perhaps not all of the bones used in various studies and calculations actually belong to *Ampelosaurus* (Vila et al., 2023).

The generic name *Ampelosaurus* derives from the Greek "ampelos" (meaning "vine"), as the first specimens were found at a vineyard. The specific name *atacis* derives from the Latin "Atax," referring to the Aude River.

CLASSIFICATION
Sauropoda
 Macronaria
  Somphospondyli
   Titanosauria
    Lithostrotia
     Eutitanosauria
      Colossosauria
       Lirainosaurinae

LOCATION
France

KNOWN REMAINS
Partial skull and skeleton

2 m   4 m   6 m   8 m

A. velauciensis

(Garcia et al., 2010)
Length: 8 m (27 ft)
Height: 3.5 m (11.5 ft)
Hip height: 1.7 m (5.5 ft)
Body mass: 2,000 kg (2.2 t)
Reconstruction: ☐☐☐☐

2 m

| 250 | 245 | 240 | 235 | 230 | 225 | 220 | 215 | 210 | 205 | 200 | 195 | 190 | 185 | 180 | 175 | 170 | 165 | 160 | 155 | 150 | 145 | 140 | 135 | 130 | 125 | 120 | 115 | 110 | 105 | 100 | 95 | 90 | 85 | 80 | 75 | 70 | 65 |

| TRIASSIC | JURASSIC | CRETACEOUS |

Atsinganosaurus velauciensis (meaning "gypsy lizard from Velaucio") is a small European titanosaur. As the continent was divided into multiple smaller landmasses during this time, the habitat available to *Atsinganosaurus* (the Ibero-Armorican Island) was limited in size. As a means to cope with the limited available resources, various species oftentimes shrunk in size.

The first fossilized bones at the locality were observed in 1992, and the first of the specimens that would eventually be known as *Atsinganosaurus* were unearthed the following year. These remains belonged to at least two mature individuals. In 2009 and 2012, more remains were uncovered from the same site. Aside from the four dorsal (back) vertebrae that were designated as the holotype specimen, none of the remains were articulated or associated. In addition to these sauropod remains, fossils from other dinosaurs (including a different titanosaur,

Garrigatitan), pterosaurs, crocodyliforms, turtles, and fish were also found. It is thought that these remains accumulated in a river channel (Díez Díaz et al., 2018).

Along with other small-sized European titanosaurs, it has been suggested that *Atsinganosaurus* is a member of the subfamily **Lirainosaurinae**, although it remains unclear where exactly this branch belongs on the larger titanosaurian family tree (Carballido et al., 2022). It is plausible that lirainosaurs could be a "relict" lineage which broke off from other Laurasian clades, or their origin could lie with African fauna, instead (Díez Díaz et al., 2018).

The generic name *Atsinganosaurus* derives from the Greek term "atsingános" (meaning "gypsy"), alluding to evidence of sauropod migrations between eastern and western Europe suggested by the remains. The specific name *velauciensis* derives from "Velaucio," the Latin name of the city Velaux.

CLASSIFICATION

Sauropoda
  Macronaria
    Somphospondyli
      Titanosauria
        Lithostrotia
          Eutitanosauria
            Colossosauria
              Lirainosaurinae

LOCATION

France

KNOWN REMAINS

Partial skull and skeleton

G. meridionalis

(Díez Díaz et al., 2021)
Length: 14 m (46 ft)
Height: 7 m (23 ft)
Hip height: 2.7 m (9 ft)
Body mass: 8,000 kg (8.8 t)
Reconstruction: ☐☐☐☐

TRIASSIC · JURASSIC · CRETACEOUS

Garrigatitan meridionalis (meaning "southern dry thicket giant") is known from fossil material that was discovered in close proximity to the remains of the closely related *Atsinganosaurus*. The remains of *Garrigatitan* were not articulated but could be differentiated from those of *Atsinganosaurus* based on several traits: differently shaped vertebral neural spines, differently sized neural canals, and different shapes of the ischium, ilium, and femur.

Estimates of the size of *Garrigatitan* are difficult to determine. Several bones (one femur and three humeri) had their internal structure analyzed in order to determine the growth stage of the individual from which they came. All of the bones seem to have come from adults; however, they do not indicate similarly sized individuals. Three of the bones indicate a very small sauropod, approximately 5 meters in length. However, a very large ulna—the largest found in Europe, at the time of description—suggests an animal

roughly 14 meters in length instead. This could possibly mean that the aforementioned bones were only from subadults instead of adults, or it might be an indication of sexual dimorphism.

Alongside several other European titanosaurs, the describers placed *Garrigatitan* in the subfamily **Lirainosaurinae**. Specifically, members of this clade seem to be limited to the area that was, during the Late Cretaceous, the isolated landmass known as Ibero-Armorican Island. The presence of multiple titanosaur genera in this location indicates a high level of biodiversity and, most likely, niche partitioning.

The generic name *Garrigatitan* derives from the Occitan term "garriga" (meaning "dry thicket"). The specific name *meridionalis* is Latin for "southern," as the specimen was unearthed in the south of France.

CLASSIFICATION
Sauropoda
 Macronaria
  Somphospondyli
   Titanosauria
    Lithostrotia
     Eutitanosauria
      Colossosauria
       Lirainosaurinae

LOCATION
France

KNOWN REMAINS
Partial skeleton

2 m 4 m 6 m 8 m 10 m

L. pandafilandi

(Díez Díaz et al., 2016)
Length: 11 m (36 ft)
Height: 5 m (16.5 ft)
Hip height: 2 m (6.5 ft)
Body mass: 3,000 kg (3.3 t)
Reconstruction: ☐☐☐☐

4 m

2 m

250 245 240 235 230 225 220 215 210 205 200 195 190 185 180 175 170 165 160 155 150 145 140 135 130 125 120 115 110 105 100 95 90 85 80 75 72 65

TRIASSIC   **JURASSIC**   **CRETACEOUS**

Lohuecotitan pandafilandi (meaning "Pandafilando's giant from Lo Hueco") is known from partial remains, consisting primarily of the pelvis and front of the tail.

In 2007, an expansive fossil site was discovered during the excavation of a railway near Cuenca, Spain. It has since yielded thousands of individual fossils, a large percentage of which are titanosaurian in origin (Ortega et al., 2015).

Numerous partial skeletons are present, identifiable as individual entities thanks to the "low dispersion" of their various bones. At least two distinct genera are believed to be represented by these remains. One such set served as the basis for describing *Lohuecotitan*. Later, on the basis that it differed from the only other braincase thus far found, a separate braincase was potentially attributed to the genus (Knoll et al., 2019).

Presently, there's not much agreement when it comes to the phylogenetic position of *Lohuecotitan*. The describers found it to be a basal lithostrotian, a position that was also found in a subsequent work by the same lead author (Díez Díaz et al., 2018). In contrast, other works have recovered it in quite a derived position (Sallam et al., 2018), possibly within Saltasauridae (Gorscak and O'Connor, 2019). Each of the aforementioned studies found *Lohuecotitan* to be closely related to *Paludititan*, which differs considerably from the findings of the most recent analysis, which finds *Paludititan* to be a basal lithostrotian and *Lohuecotitan* to be within **Lirainosaurinae** (Navarro et al., 2022).

The generic name *Lohuecotitan* refers to the name of the locality where the fossils were found, Lo Hueco. The specific name *pandafilandi* refers to a fictional character created by Miguel de Cervantes, named Pandafilando de la Fosca Vista, who was a "giant" adversary of Don Quixote.

CLASSIFICATION
Sauropoda
 Macronaria
  Somphospondyli
   Titanosauria
    Lithostrotia
     Eutitanosauria
      Colossosauria
       Lirainosaurinae

LOCATION
Spain

KNOWN REMAINS
Partial skeleton

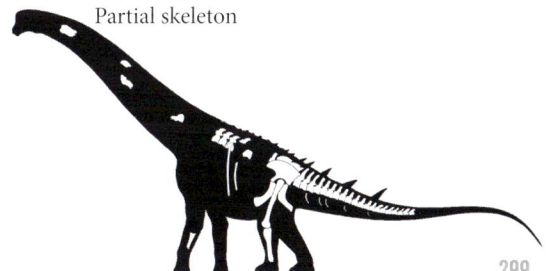

4 m 2 m 4 m 6 m 8 m 10 m

2 m

M. arriagadai

(Rolando et al., 2022)
Length: 9 m (29 ft)
Height: 3.8 m (12.5 ft)
Hip height: 2.1 m (6.9 ft)
Body mass: 6,000 kg (6.6 t)
Reconstruction: ☐☐☐☐

250 245 240 235 230 225 220 215 210 205 200 195 190 185 180 175 170 165 160 155 150 145 140 135 130 125 120 115 110 105 100 95 90 85 80 75 70 65

TRIASSIC **JURASSIC** **CRETACEOUS**

Menucocelsior arriagadai (meaning "Arriagada's major watering hole") is perhaps most notable for its seeming inability to fit in with any other titanosaurs. According to its describers, *Menucocelsior* does not belong to any of the major eutitanosaurian clades: Colossosauria, Saltasaurinae, or Aeolosaurini.

This situation would be less surprising if *Menucocelsior* lived when those branches of the family tree were just beginning to bloom, but instead, *Menucocelsior* lived much later than that, close to the very end of the Cretaceous. There are several possible explanations for this occurrence: (1) The describers have simply misinterpreted the anatomy of the fossils, and future works will reclassify *Menucocelsior*. (2) A mysterious "ghost lineage" of titanosaurs exists for which we have never discovered any trace and which stretched nearly across the entire Cretaceous. (3) *Menucocelsior* represents a new, distinct sub-branch of

Colossosauria, Saltasaurinae, or Aeolosaurini, which would require only a much shorter ghost lineage to be missing from the fossil record.

Until more complete remains are discovered for comparison, these questions will likely remain unanswered. This prospect is not quite hopeless, though, as the area where the fossils were found (the Allen geological formation) is actually quite plentiful with a multitude of different titanosaur remains, in fact being the most diverse assemblage of titanosaurs currently known.

The generic name *Menucocelsior* combines the indigenous Mapudungun word "menuco" (meaning "watering hole") with the Latin "celsior" (meaning "higher" or "major"). This was inspired by the name of the local region, Salitral Ojo de Agua, which partially translates as "water hole" in Spanish. The specific name *arriagadai* honors landowner "Beto" Arriagada and his family.

CLASSIFICATION
Sauropoda
  Macronaria
    Titanosauriformes
     Somphospondyli
      Titanosauria
       Lithostrotia
        Eutitanosauria

LOCATION
Argentina

KNOWN REMAINS
Partial skeleton

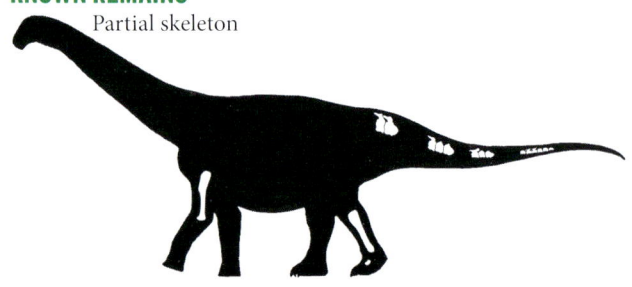

2 m   4 m   6 m   8 m   10 m

B. salgadoi

(Apesteguía, 2004)
Length: 10 m (33 ft)
Height: 4 m (13.1 ft)
Hip height: 2.3 m (7.5 ft)
Body mass: 5,500 kg (6.1 t)
Reconstruction: ☐☐☐☐

4 m

2 m

250 245 240 235 230 225 220 215 210 205 200 195 190 185 180 175 170 165 160 155 150 145 140 135 130 125 120 115 110 105 100 95 90 **85** 80 75 70 65

TRIASSIC   **JURASSIC**   **CRETACEOUS**

Bonitasaura salgadoi (meaning "Salgado's lizard from La Bonita") is known from a few sets of remains that were found in the same area. The most complete set, the holotype, was disarticulated but associated within a fairly tight area; other specimens were found roughly 20 meters away from these. Each of these individuals was mature but likely not fully grown, as indicated by histological examination of bone tissues and incomplete fusion of vertebral elements (Gallina, 2012).

Interestingly, the holotype individual had several pathologies, or injuries, that have been preserved, including an infection of a tail bone, a tumor on the femur, and a bone spur on the foot (González et al., 2017).

Additional sauropod remains found nearby in the early 1920s were originally assigned to *Laplatasaurus*, but it now seems likely that they belong to some species of *Bonitasaura* (Gallina and Otero, 2015).

The jaw of *Bonitasaura* seems to be uniquely intermediate between the shapes of other known titanosaurian jaws. Rather than a squared-off L-shape or a rounded U-shape, the jaws of *Bonitasaura* fall somewhere in between (Gallina and Apesteguía, 2011). The animal's limbs were slender in comparison to the saltasaurid sauropods that also inhabited the region, and the base of its neck was quite broad.

Recent phylogenetic studies largely agree on the classification of *Bonitasaura* as a basal **colossosaurian**, stemward of Lognkosauria (Díez Díaz et al., 2018; Santucci and Filippi, 2022; Agnolin et al., 2023). One large analysis shifted this position slightly stemward, out of Colossosauria and into Eutitanosauria (Navarro et al., 2022).

The generic name *Bonitasaura* refers to the name of the quarry where the fossils were located, "La Bonita." The specific name *salgadoi* honors Argentine paleontologist Leonardo Salgado.

CLASSIFICATION
Sauropoda
 Macronaria
  Somphospondyli
   Titanosauria
    Lithostrotia
     Eutitanosauria
      Colossosauria

LOCATION
Argentina

KNOWN REMAINS
Partial skull and skeleton

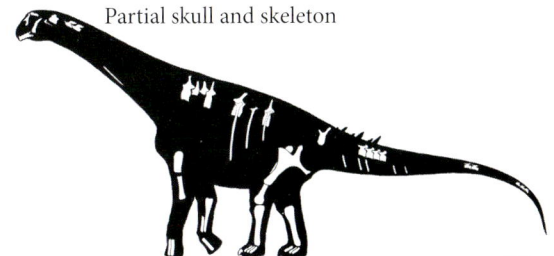

301

4 m

2 m

I. oslatus

(Filippi et al., 2024)
Length: 10 m (32.8 ft)
Height: 3.5 m (11.5 ft)
Hip height: 2 m (6.6 ft)
Body mass: 5,500 kg (6.1 t)
Reconstruction: ☐☐☐☐

| 250 | 245 | 240 | 235 | 230 | 225 | 220 | 215 | 210 | 205 | 200 | 195 | 190 | 185 | 180 | 175 | 170 | 165 | 160 | 155 | 150 | 145 | 140 | 135 | 130 | 125 | 120 | 115 | 110 | 105 | 100 | 95 | 90 | 85 | 80 | 75 | 70 | 65 |

TRIASSIC   **JURASSIC**   **CRETACEOUS**

Inawentu oslatus (meaning "broad-mouthed mimic") is so named because, even though it is a titanosaur, it bears a remarkable resemblance to a different group of sauropods, the rebbachisaurids.

Several kinds of rebbachisaurids, such as Nigersaurus, are famous for their unusual jaw morphology, having a very broad, flat snout that seems to have been designed to graze low-level vegetation close to the ground. *Inawentu* seems to have adopted a similar cranial configuration. While there are a few other titanosaurs known to have had "square" jaws, the broad, nearly "duckbill" shaped, upper jaw of *Inawentu* is strikingly unique among titanosaurs.

According to the fossil record as we currently know it, the rebbachisaurids seem to have gone extinct about eight million years before *Inawentu* appeared on the scene. This suggests that the adaptations possessed by *Inawentu* had evolved to fill whichever low-grazing ecological niche the rebbachisaurids had left vacant.

Aside from the morphology of its jaws, *Inawentu* had other adaptations to make grazing at ground level an easier task, including having a proportionately short neck that was made up of only 12 cervical vertebrae—the fewest of any known titanosaur. Additionally, the animal's skull connected to its neck at a right angle, placing the jaws at an optimal angle for low browsing when the neck was held horizontally.

While the majority of titanosaur specimens are known from just a few bones that were collected from jumbled bone beds, the *Inawentu* holotype stands out as being articulated, alone, and preserving a large portion of the animal's body.

The generic name *Inawentu* is an indigenous Mapundungun term for "mimic." The specific name *oslatus* combines the Latin "os" (meaning "mouth") and "latus" (meaning "broad").

CLASSIFICATION

Sauropoda
 Macronaria
  Somphospondyli
   Titanosauria
    Lithostrotia
     Eutitanosauria
      Colossosauria

LOCATION

Argentina

KNOWN REMAINS

Partial skull and skeleton

5 m 10 m 15 m 20 m 25 m

N. gonzalezparejasi

(González Riga et al., 2016)
Length: 28 m (92 ft)
Height: 12 m (39 ft)
Hip height: 4.8 m (16 ft)
Body mass: 40,000 kg (44.1 t)
Reconstruction: ☐☐☐☐

10 m

5 m

| 250 | 245 | 240 | 235 | 230 | 225 | 220 | 215 | 210 | 205 | 200 | 195 | 190 | 185 | 180 | 175 | 170 | 165 | 160 | 155 | 150 | 145 | 140 | 135 | 130 | 125 | 120 | 115 | 110 | 105 | 100 | 95 | 90 | **86** | 80 | 75 | 70 | 65 |

TRIASSIC **JURASSIC** **CRETACEOUS**

When it was described, **Notocolossus gonzalezparejasi** (meaning "González Parejas's southern giant") could boast having the largest intact humerus fossil of any known titanosaur, with a length of 1.76 meters (69 inches). Initial estimates of the animal's size were nearly as record breaking, with some making the animal almost comparable to *Argentinosaurus*. But as is often the case in these situations, subsequent estimates have consistently been more modest.

Two known specimens of *Notocolossus* were described at the same time. One larger specimen consisted of just the humerus, two vertebrae, and a partial pubis. Approximately 400 meters away, the remains of a smaller individual were unearthed, consisting of a series of articulated tail bones and a complete foot. Based on similarities in the tail vertebrae of the two specimens, they were both assigned to the same genus.

The complete fossilized foot is noteworthy, because only a few titanosaur hindfeet are currently known to science. *Notocolossus* was "short footed," having metatarsals that were very compact and stout; each was of roughly the same size. Unlike most sauropods, which possessed three claws on each hindfoot, *Notocolossus* instead only had blunt, stubby protrusions.

Taxonomically, *Notocolossus* is often found to be within **Lognkosauria**—or, if not quite in the clade proper, then being in very close proximity (Navarro et al., 2022; Pérez Moreno et al., 2023).

The generic name *Notocolossus* combines the Greek "notos" (meaning "southern") and the Latin "colossus" (meaning "huge" or "giant"). The specific name *gonzalezparejasi* honors Jorge González Parejas for his work regarding dinosaur studies and site preservation in Mendoza Province.

CLASSIFICATION
Sauropoda
 Macronaria
  Titanosauriformes
  Somphospondyli
   Titanosauria
   Lithostrotia
    Eutitanosauria
    Colossosauria
    Lognkosauria

LOCATION
Argentina

KNOWN REMAINS
Partial skeleton

LITHOSTROTIA

4 m　　　　8 m　　　　12 m　　　　16 m　　　　20 m　　　24

8 m

4 m

A. magnificus

(Bandeira et al., 2016)
Length: 25 m (82 ft)
Height: 9 m (30 ft)
Hip height: 3.8 m (12.5 ft)
Body mass: 40,000 kg (44.1 t)
Reconstruction: ☐☐☐☐

250 245 240 235 230 225 220 215 210 205 200 195 190 185 180 175 170 165 160 155 150 145 140 135 130 125 120 115 110 105 100 95 90 85 80 75 **72** 65

TRIASSIC　　　　**JURASSIC**　　　　**CRETACEOUS**

Austroposeidon magnificus (meaning "Poseidon's southern great one") was, at the time of its description, the largest dinosaur known from Brazil.

The holotype and only known specimen was originally collected in 1953 by pioneering Brazilian paleontologist Llewellyn Ivor Price. The fossil site is currently unworkable, as the area has since been urbanized, so it is unknown whether more substantial remains once existed there. The bones are not in pristine condition, with their shape having been distorted during their time in the ground and then eroded before their collection.

Only a few partial vertebrae make up the specimen. The rearmost neck and foremost back bones are the most intact, with fragments of dorsal (back) vertebrae making up most of the remainder. Fine-grained sandstone encased the fossils, which were likely buried by a low-energy flow on a floodplain. The bones show no signs of immaturity, suggesting the animal was an adult at the time of its death. Computerized tomography scans have shown internal layers that may represent growth marks.

The describers placed *Austroposeidon* as being just stemward of Lognkosauria, in what would now be called Colossosauria. The subsequent majority of studies have excluded *Austroposeidon*, most likely because of the fragmentary nature of its remains. However, one did recover it as being within **Lognkosauria** (Navarro et al., 2022).

The generic name *Austroposeidon* combines the Greek "austro" (meaning "southern") with the name of the ancient Greek deity of earthquakes and the sea, Poseidon. The specific name *magnificus* is Latin for "great" or "elevated," inspired by the animal's size, which the describers estimated to be 25 meters in length.

CLASSIFICATION
Sauropoda
 Macronaria
  Titanosauriformes
   Somphospondyli
    Titanosauria
     Lithostrotia
      Eutitanosauria
       Colossosauria
        Lognkosauria

LOCATION
Brazil

KNOWN REMAINS
Vertebrae

M. neguyelap

(González Riga, 2003)
Length: 20 m (65 ft)
Height: 7.5 m (24.5 ft)
Hip height: 3.5 m (11.5 ft)
Body mass: 16,000 kg (17.6 t)
Reconstruction: ☐☐☐☐

8 m

4 m

| 4 m | 8 m | 12 m | 16 m |

```
250 245 240 235 230 225 220 215 210 205 200 195 190 185 180 175 170 165 160 155 150 145 140 135 130 125 120 115 110 105 100 95 88 85 80 75 70 65
```

TRIASSIC **JURASSIC** **CRETACEOUS**

Mendozasaurus neguyelap (meaning "first beast-lizard from Mendoza") is notable for having uniquely shaped osteoderms. These bony, armor-like projections of the tail are present in multiple titanosaur species, but the "subspherical" shape of the two largest ones recovered from *Mendozasaurus* are considered to be a distinguishing characteristic of the genus. Also peculiar is the wide, bulbous shape of the neck vertebrae's spines (as seen from the front).

The known specimens of *Mendozasaurus* come from a bone bed that was discovered by oil workers; excavations began in 1998. Many of the bones were disarticulated from one another, except for a series of more-or-less connected tail vertebrae, which became the holotype specimen. Only some of these skeletal elements were represented in the original description of the genus; more were revealed in a thorough reanalysis of the remains (González Riga et al., 2018).

Most of the bones in the area are believed to belong to the same individual, although there were at least four separate individuals whose remains have been found, judging by repeated elements and differences in size. The fourth individual is believed to have been a juvenile and is represented by a single metacarpal. Many of the fossils were already partially eroded prior to their initial burial and have also been distorted by geologic forces.

Mendozasaurus is one of the genera that serves as an anchor for the clade **Lognkosauria** (Carballido et al., 2022).

The generic name *Mendozasaurus* refers to Mendoza Province, Argentina, since this was the first nonavian dinosaur to be named from the region. The specific name *neguyelap* combines the indigenous Millcayac terms "neguy" (meaning "first") and "yelap" (meaning "beast").

CLASSIFICATION

Sauropoda
 Macronaria
  Titanosauriformes
   Somphospondyli
    Titanosauria
     Lithostrotia
      Eutitanosauria
       Colossosauria
        Lognkosauria

LOCATION
Argentina

KNOWN REMAINS
Partial skeleton

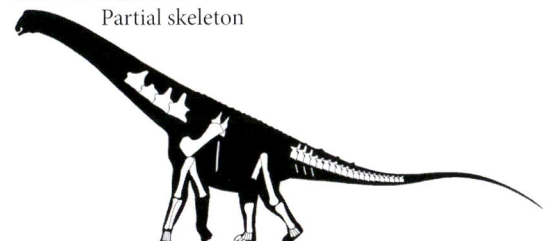

305

LITHOSTROTIA

4 m   8 m   12 m   16 m   20 m   24 m

8 m

4 m

F. dukei

(Calvo et al., 2007b)
Length: 24 m (79 ft)
Height: 11 m (36 ft)
Hip height: 3.8 m (12.5 ft)
Body mass: 38,000 kg (41.9 t)
Reconstruction: ☐☐☐ |

250 245 240 235 230 225 220 215 210 205 200 195 190 185 180 175 170 165 160 155 150 145 140 135 130 125 120 115 110 105 100 95 90 **87** 80 75 70 65

| TRIASSIC | JURASSIC | CRETACEOUS |

Futalognkosaurus dukei (meaning "Duke's giant chief lizard") was, at the time of its discovery, the "most complete giant sauropod ever found." Original estimates of its size were somewhat overblown, placing it in the same size category as *Argentinosaurus*, but *Futalognkosaurus* was nonetheless one of the biggest titanosaurs.

The vertebral and pelvic elements of the specimen were excavated from 2000 through 2005, near Lake Barreales. Additional limb elements, unearthed from the site later, have yet to be properly described (Calvo, 2014; González Riga et al., 2016). The remains were encased in floodplain sediment, originating from what was probably a braided system of rivers or streams.

One of the most notable features of *Futalognkosaurus* has to do with the shape of its cervical (i.e., neck) vertebrae. The bones from the central and rear portions of the neck have unusually tall and long neural spines, which bear a profile akin to a shark's fin (Calvo, 2007).

According to a comprehensive study that proposed updated definitions for the various titanosaurian clades, **Lognkosauria** is literally defined on the basis of *Futalognkosaurus* and *Mendozasaurus*, encompassing those taxa, their common ancestor, and everything in between. After the clade was introduced in the same paper that described *Futalognkosaurus*, several subsequent studies failed to find many other members of the group; however, this has since changed, with numerous analyses now showing Lognkosauria to contain multiple genera (Carballido et al., 2022).

The generic name *Futalognkosaurus* combines the indigenous Mapuche words "futa" (meaning "giant") and "lognko" (meaning "chief"). The specific name *dukei* honors the financial sponsor of the excavation, the Duke Energy Argentina Company.

CLASSIFICATION
Sauropoda
 Macronaria
  Titanosauriformes
   Somphospondyli
    Titanosauria
     Lithostrotia
      Eutitanosauria
       Colossosauria
        Lognkosauria

LOCATION
Argentina

KNOWN REMAINS
Partial skeleton

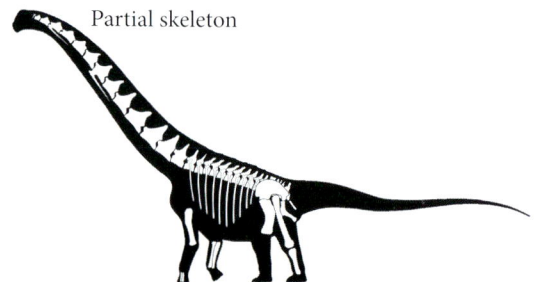

4 m	8 m	12 m	16 m	20 m	24 m	28 m

12 m

8 m

4 m

P. reuili

(Novas et al., 2005)
Length: 30 m (98 ft)
Height: 11.5 m (38 ft)
Hip height: 4.9 m (16 ft)
Body mass: 60,000 kg (66.1 t)
Reconstruction:

| 250 | 245 | 240 | 235 | 230 | 225 | 220 | 215 | 210 | 205 | 200 | 195 | 190 | 185 | 180 | 175 | 170 | 165 | 160 | 155 | 150 | 145 | 140 | 135 | 130 | 125 | 120 | 115 | 110 | 105 | 100 | 95 | 90 | 85 | 80 | 75 | 70 | 65 |

TRIASSIC | **JURASSIC** | **CRETACEOUS**

Puertasaurus reuili (meaning "Puerta and Reuil's lizard") is the current record holder for "widest vertebra." The recovered dorsal vertebra (thought to be the animal's second, in terms of placement), measures 1.68 meters (66 inches) wide—for comparison, the entire skull of a *Tyrannosaurus rex* is only about 1.35 meters in length.

The known dorsal vertebra of *Puertasaurus* is also unique among titanosaurs because the bone's centrum is just as wide as it is long; comparable bones from other species are all longer than they are wide.

The long neck vertebra is only partially complete, but it, too, is remarkably wide. It was also quite insightful at the time of its description, as it was the very first neck vertebra discovered from one of the gargantuan (meaning more than roughly 31,750 kilograms, or 35 tons) lognkosaurs.

Naturally, estimations of an entire animal's size based on so few bones will have a significant margin for error. Initial claims were somewhat exaggerated, according to the consensus drawn by ensuing studies (Paul, 2019). Part of this preliminary inaccuracy can be put down to an insufficient understanding of general titanosaur anatomy at the time. In just the last couple of decades, an explosion of specimens has allowed paleontologists to get a much firmer grasp on the general proportions of titanosaur genera.

Phylogenetically, *Puertasaurus* is usually recovered within **Lognkosauria**—or, if not quite in the clade proper, then being in very close proximity (Navarro et al., 2022; Pérez Moreno et al., 2023).

The generic name *Puertasaurus* honors fossil hunter Pablo Puerta. Likewise, the specific name *reuili* honors fossil hunter Santiago Reuil. These two individuals discovered the fossils in 2001 and also undertook their preparation.

CLASSIFICATION

Sauropoda
 Macronaria
  Titanosauriformes
   Somphospondyli
    Titanosauria
     Lithostrotia
      Eutitanosauria
       Colossosauria
        Lognkosauria

LOCATION

Argentina

KNOWN REMAINS

Vertebrae

6 m	12 m	18 m	24 m	30 m

12 m

6 m

A. huinculensis

(Bonaparte and Coria, 1993)
Length: 35 m (115 ft)
Height: 13 m (43 ft)
Hip height: 5.5 m (18 ft)
Body mass: 75,000 kg (82.7 t)
Reconstruction: ☐ ☐ ☐ ☐

250	245	240	235	230	225	220	215	210	205	200	195	190	185	180	175	170	165	160	155	150	145	140	135	130	125	120	115	110	105	100	95	90	85	80	75	70	65

TRIASSIC	JURASSIC	CRETACEOUS

Argentinosaurus huinculensis (meaning "Argentine lizard from Huincul") is best known for its (possibly) record-breaking size. It is one of the most massive known terrestrial creatures; however, it is impossible to determine its actual size accurately because of the fragmentary nature of its fossils.

The holotype remains were found in 1987 by a local farmer, who mistook the fibula (the only limb bone then recovered) for a petrified log. A larger excavation was launched in 1989, primarily uncovering a number of vertebrae from the exceptionally hard rock; the exact positional sequence for several of the dorsal vertebrae remains unclear (Paul, 2019).

While the recovered bones were undoubtedly colossal, determining the overall size of the animal from these remains was challenging. Matters were helped when a femur was later discovered from the same locality; while most of the bone was technically present, it had been crushed during fossilization (Bonaparte, 1996). The midshaft of a less-distorted femur has also been tentatively assigned to the genus; when extrapolated using the more complete femur as a basis, some estimates place the total length of this bone at having been 2.5 meters—longer than any other known sauropod femur (Mazzetta et al., 2004).

While numerous recent analyses have placed *Argentinosaurus* within **Lognkosauria** (Hechenleitner et al., 2020; Gallina et al., 2022; Santucci and Filippi, 2022), this is not the only valid taxonomic interpretation; it has also been found in more basal locations within Titanosauria (Navarro et al., 2022).

The generic name *Argentinosaurus* honors the South American nation of Argentina. The specific name *huinculensis* refers to the location of the holotype's discovery, Plaza Huincul, of the Rio Limay Formation.

CLASSIFICATION

Sauropoda
 Macronaria
  Titanosauriformes
  Somphospondyli
   Titanosauria
    Lithostrotia
     Eutitanosauria
      Colossosauria
       Lognkosauria

LOCATION

Argentina

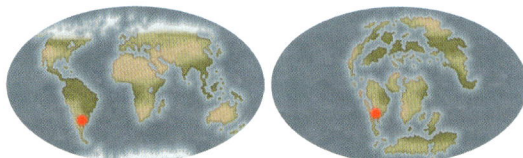

KNOWN REMAINS

Partial skeleton

6 m   12 m   18 m   24 m   30 m

12 m

6 m

P. mayorum

(Carballido et al., 2017)
Length: 31 m (102 ft)
Height: 13 m (42.5 ft)
Hip height: 4.8 m (16 ft)
Body mass: 60,000 kg (66.1 t)
Reconstruction: ☐☐☐☐

250 245 240 235 230 225 220 215 210 205 200 195 190 185 180 175 170 165 160 155 150 145 140 135 130 125 120 115 110 105 101 95 90 85 80 75 70 65

TRIASSIC   **JURASSIC**   **CRETACEOUS**

Patagotitan mayorum (meaning "the Mayo's giant from Patagonia") is among the most massive of titanosaurs, with some initial estimates claiming a size even larger than the famed *Argentinosaurus*; more recent estimates have trimmed those numbers down a bit. However, histological analyses of the various specimens' inner bone tissue indicate that the animals' maximum size might not yet have been reached.

Among the species composing the echelon of largest titanosaurs, *Patagotitan* is definitively known from the most complete set of skeletal material. The remains of at least six individuals have been recovered from the type locality; the lion's share of bones are split between two fairly complete individuals, with isolated remains accounting for the rest. None of the smaller bones from the hands, feet, or skull have been recovered; this is likely due to how the animals were buried in floodplain sediment, as the flow of water was strong enough to wash these small or delicate bones away but not strong enough to move the larger bones.

Aside from its considerable size, *Patagotitan* is also notable for having vertebral neural spines that were anomalously tall, especially at the front portions of the back and tail. The animal's humerus is noted as being slender, serving as a reminder that titanosaur species can have quite different proportions (González Riga et al., 2019).

While numerous recent analyses have placed *Patagotitan* within **Lognkosauria** (Hechenleitner et al., 2020; Gallina et al., 2022; Santucci and Filippi, 2022), this is not the only valid taxonomic interpretation; it has also been found in more basal locations within Titanosauria (Navarro et al., 2022).

The generic name *Patagotitan* refers to the Patagonia region of Argentina. The specific name *mayorum* honors the landowners, the Mayo family, for their cooperation and hospitality.

CLASSIFICATION
Sauropoda
 Macronaria
  Titanosauriformes
   Somphospondyli
    Titanosauria
     Lithostrotia
      Eutitanosauria
       Colossosauria
        Lognkosauria

LOCATION
Argentina

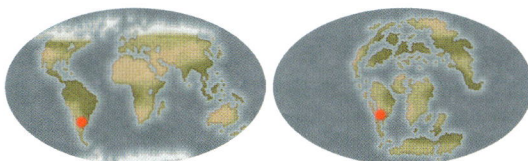

KNOWN REMAINS
Partial skeleton

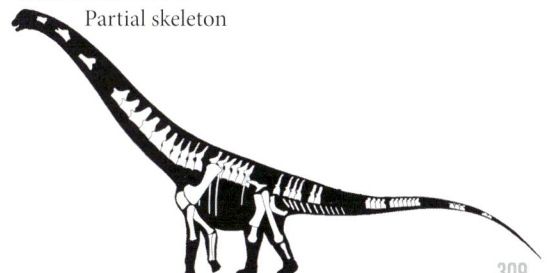

LITHOSTROTIA

J. ganzhouensis

(Mo et al., 2023)
Length: 12.4 m (40.6 ft)
Height: 6.6 m (21.6 ft)
Hip height: 3.2 m (10.5 ft)
Body mass: 3,500 kg (3.9 t)
Reconstruction: ☐ ☐ ☐ ☐

TRIASSIC   JURASSIC   CRETACEOUS

Jiangxititan ganzhouensis (meaning "giant from Ganzhou, Jiangxi") is notable for the unique form of its vertebrae, which have an extremely flattened shape. According to its describers, this is a natural feature of the bones and not the result of any distortional forces that the remains were subjected to after burial.

The describers of *Jiangxititan* conducted a phylogenetic analysis that placed the animal within **Lognkosauria**, which would make it the only non–South American member of that clade. As of the time of this writing, no other studies have had a chance to either corroborate or refute this conclusion. In the past, though, Late Cretaceous sauropod remains that were originally suspected to be titanosaurian have always ended up being reassigned. *Jiangxititan* shares several anatomical traits in common with *Opisthocoelicaudia*.

If *Jiangxititan* does indeed fall within Titanosauriformes, it would be the only Asian member of the clade to possess deeply bifurcated vertebral neural spines. Owing to the compacted shape of the vertebrae in general, portions of the neural arches that would ordinarily face upward in other species instead face sideways in *Jiangxititan*.

The holotype and only known specimen of *Jiangxititan* was discovered during construction work. It consists of seven vertebrae from the region near the base of the animal's neck and their associated partial ribs. The foremost and rearmost vertebrae of this series were damaged during extraction. The bones are in a relatively good preservational state, although the left side of the remains suffered some distortion during fossilization.

The animal's binomial name refers to Ganzhou City in Jiangxi Province, China.

CLASSIFICATION

Sauropoda
 Macronaria
  Titanosauriformes
   Somphospondyli
    Titanosauria
     Lithostrotia
      Eutitanosauria
       Colossosauria
        Lognkosauria

LOCATION

China

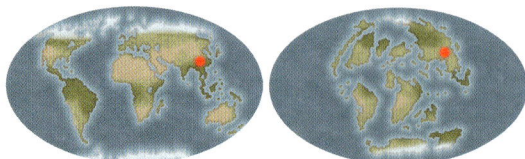

KNOWN REMAINS

Vertebrae

B. mansillai

(Calvo and González Riga, 2019)
Length: 11 m (36 ft)
Height: 4.5 m (15 ft)
Reconstruction: ☐☐☐☐

90—

TRIASSIC · JURASSIC · **CRETACEOUS**

Baalsaurus mansillai is known from a single jaw bone, the right dentary. One of the few "square-jawed" titanosaurs (as opposed to having a U-shaped jaw), most of the animal's teeth were seemingly located in the front "flat" part of the jaw, and the few teeth that extended back were smaller in size. Each peg-like tooth had two or three replacement teeth growing underneath it.

While the describers of Baalsaurus were able to categorize the animal as a **titanosaur**, they were unable to pin down a more precise classification, owing to the lack of available remains.

The generic name Baalsaurus refers to the fossil's location of discovery, the so-called "Baal site" in Argentina, named after the Phoenician and Canaanite deity. The specific name mansillai honors fossil discoverer and museum technician Juan Eduardo Mansilla.

"A." giganteus

(von Huene, 1929)
Length: 30 m (100 ft)
Height: 13 m (42.5 ft)
Reconstruction: ☐☐☐☐

83—

TRIASSIC · JURASSIC · **CRETACEOUS**

"Antarctosaurus" giganteus was named by Friedrich von Huene in the same publication in which he described *Antarctosaurus wichmannianus*. In this manuscript, he himself was hesitant as to whether the fossils should actually be attributed to the same genus as *A. wichmannianus*.

The main claim to fame that "A." giganteus possesses is its sheer size. It has one of the longest confirmed femurs of any known animal, measuring 235 cm in length (Mazzetta et al., 2004), which is surpassed only by fellow titanosaurs *Patagotitan* and *Argentinosaurus*.

Today, some researchers consider this Argentine titanosaur to be an invalid species, lacking any unique, identifiable characteristics. Others do consider it to be valid; among those, some consider the genus assignment as valid, while others call for the erection of a new genus.

LITHOSTROTIA

TENGRISAURUS

T. starkovi

(Averianov and Skutschas, 2017)
Length: 12 m (40 ft)
Height: 5 m (16.5 ft)
Reconstruction: ☐☐☐☐☐

2 m

120

TRIASSIC | JURASSIC | CRETACEOUS

Tengrisaurus starkovi is known from four partial tail vertebrae discovered in Siberia. The first three that were described were found at various times in the 1990s. The fourth was found in 1959 and was attributed in 2021 (Averianov et al., 2021).

The classification of *Tengrisaurus* is quite uncertain. All current analyses place the genus within **Lithostrotia**, but that's where the agreement ends. Various calculations have suggested saltasaurid (Amerianov and Efimov, 2018), rinconsaurian, or basal eutitanosaurian affinities (Averianov et al., 2021).

The generic name *Tengrisaurus* refers to the creator deity Tengri, of the indigenous nomadic belief system. The specific name *starkovi* honors expedition contributor Alexey Starkov.

VAHINY

V. depereti

(Curry Rogers and Wilson, 2014)
Length: 13 m (43 ft)
Height: 5 m (16.5 ft)
Reconstruction: ☐☐☐☐☐

2 m

70

TRIASSIC | JURASSIC | CRETACEOUS

Vahiny depereti is currently only known from two braincases, one of which is from a juvenile. Phylogenetic results place the taxon within **Colossosauria** (Mannion et al., 2019; Wang et al., 2021).

A large number of titanosaur bones have been recovered from the Maevarano geological formation of Madagascar, and the large majority of those have been assigned to *Rapetosaurus*. About 10% of the sauropod bones, though, do not match, and at times have been referred to as "Malagasy Taxon B." From these remains, a distinctive braincase was chosen as the basis of a new genus, *Vahiny*. It is possible that some of the other remaining fossils also pertain to *Vahiny*, but there is currently no way of knowing.

The generic name *Vahiny* (pronounced 'va-heenh') is a Malagasy word for "visitor," reflecting the rarity of the fossils. The specific name *depereti* honors Charles Deperet, a pioneer of Madagascan paleontology.

TRAUKUTITAN

T. eocaudata

(Juárez Valieri and Calvo, 2011)
Length: 24 m (79 ft)
Height: 10 m (33 ft)
Reconstruction: ☐☐☐☐

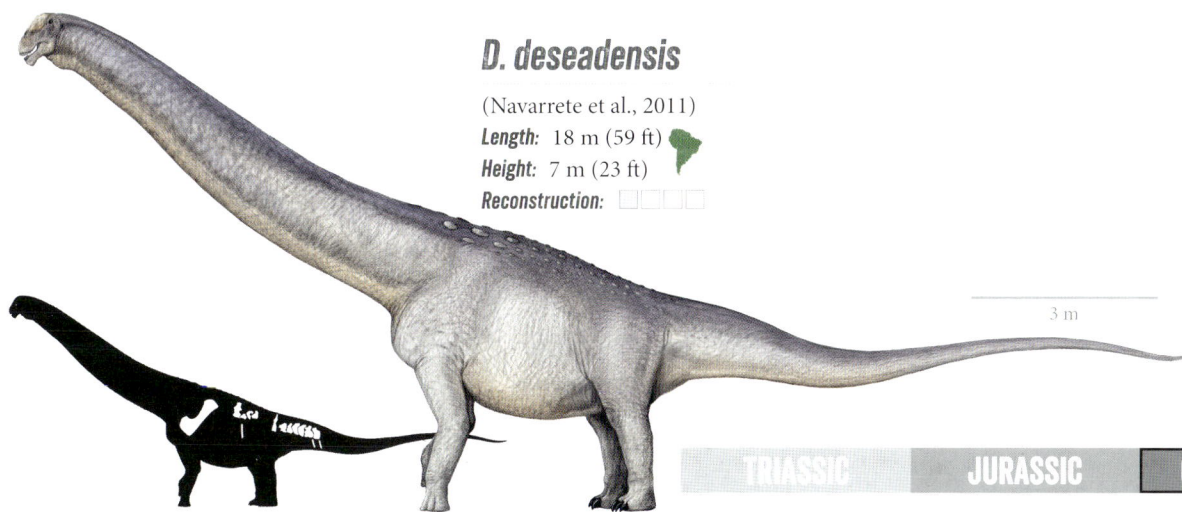

4 m

TRIASSIC **JURASSIC** **CRETACEOUS**

Traukutitan eocaudata (meaning "Trauku's early tailed giant") is known only from some tail vertebrae and thigh bones found in Argentina. These were first described in 1993 but were not assigned to a particular genus (Salgado and Calvo, 1993). (There was also a partial pubis bone present at the time, which has since been lost.) The sparse nature of the remains precludes it from reliable phylogenetic analysis, but it has been suspected of being a **lognkosaur**.

The generic name *Traukutitan* refers to Trauku, the mythological Araucanian (or Mapuche) mountain spirit. The specific name *eocaudata* combines the Greek "eo" (meaning "dawn" or "early") and "caudia" (meaning "tail"), referring to the midtail bone's resemblance to older, more ancestral forms.

DRUSILASAURA

D. deseadensis

(Navarrete et al., 2011)
Length: 18 m (59 ft)
Height: 7 m (23 ft)
Reconstruction: ☐☐☐☐

3 m

TRIASSIC **JURASSIC** **CRETACEOUS**

The holotype remains of *Drusilasaura deseadensis* were first spotted in 2007 by paleontologist Marcelo Tejedor, who was searching for Mesozoic mammals in Argentina. The fossils are in a poor state of preservation; thus, little can be determined about the animal's habits or anatomy.

The describers suspected that *Drusilasaura* belonged within **Lognkosauria**, based on several skeletal characteristics, such as its wide vertebral neural arches, but could not confirm this placement. Several later analyses, though, have done just that (Silva Junior et al., 2022a; Santucci and Filippi, 2022).

The generic name *Drusilasaura* honors excavation volunteer Drusila Ortiz de Zárate, who was also a family member of the landowner. The specific name *deseadensis* refers to where the specimen was found, the valley of the Deseado River.

LITHOSTROTIA

C. diripienda

(Agnolin et al., 2023)
Length: 24 m (80 ft)
Height: 11 m (36 ft)
Reconstruction:

5 m

85

| TRIASSIC | JURASSIC | CRETACEOUS |

Chucarosaurus diripienda (meaning "scrambled indomitable lizard") is only known from a handful of limb and pelvic bones, but these remains are notable for showing "a remarkable morphological variety" that can prove useful in comparative phylogenetic studies. The limb bones are comparatively gracile in comparison with other titanosaurs known from the Huincul geological formation of Argentina, such as the gargantuan *Argentinosaurus*.

Phylogenetically, *Chucarosaurus* has been placed within **Colossosauria**, stemward of Lognkosauria, making it closely related to *Bonitasaura*.

The generic name derives from the indigenous Quechua word "chucaro" (meaning "hard and indomitable animal"). The specific name is Latin for "scrambled".

Q. rusconii

(González Riga and David, 2014)
Length: 15 m (48 ft)
Height: 6 m (19.5 ft)
Reconstruction:

3 m

90

| TRIASSIC | JURASSIC | CRETACEOUS |

Although Argentine titanosaurs are relatively plentiful, ***Quetecsaurus rusconii*** was the first to be found from the Cerro Lisandro geological formation. Although it is represented by only a handful of (mostly broken) bones, it is worth noting that all the fossils were found in close association with one another, meaning that it is reasonably certain to conclude that they came from the same individual. This is advantageous, as many titanosaurian remains come from jumbled bone beds, where it is difficult or impossible to tell which bones came from which animal.

Quetecsaurus is most likely a **lognkosaur** (Hechenleitner et al., 2020; Santucci and Filippi, 2022) or lies just stemward of that clade as a basal colossosaur (Gallina et al., 2022). The generic name derives from the indigenous Millcayac term "quetec" (meaning "fire"). The specific name honors museum director Carlos Rusconi.

Dreadnoughtus

Cretaceous

Late

Early	Late	Paleogene

Cretaceous — Aptian | Albian | Cenomanian | Turon. | Coniac. | Sant. | Campanian | Maastrichtian

Paleogene — Paleocene

125 120 115 110 105 100 95 90 85 80 75 70 65 60 55

Karongasaurus p. 279

Paralititan p. 320

Vahiny p. 312

Jainosaurus p. 318

Antarctosaurus p. 319

Panamericansaurus p. 335

Maxakalisaurus p. 321

Adamantisaurus p. 335

Rinconsaurus p. 322

Muyelensaurus p. 323

Uberabatitan p. 324

Brasilotitan p. 334

Laplatasaurus p. 325

Baurutitan p. 326

Caieiria p. 336

Bravasaurus p. 327

Gondwanatitan p. 328

Overosaurus p. 329

Shingopana p. 330

Punatitan p. 331

Aeolosaurus p. 332

Arrudatitan p. 333

▶ Aeolosaurlnl

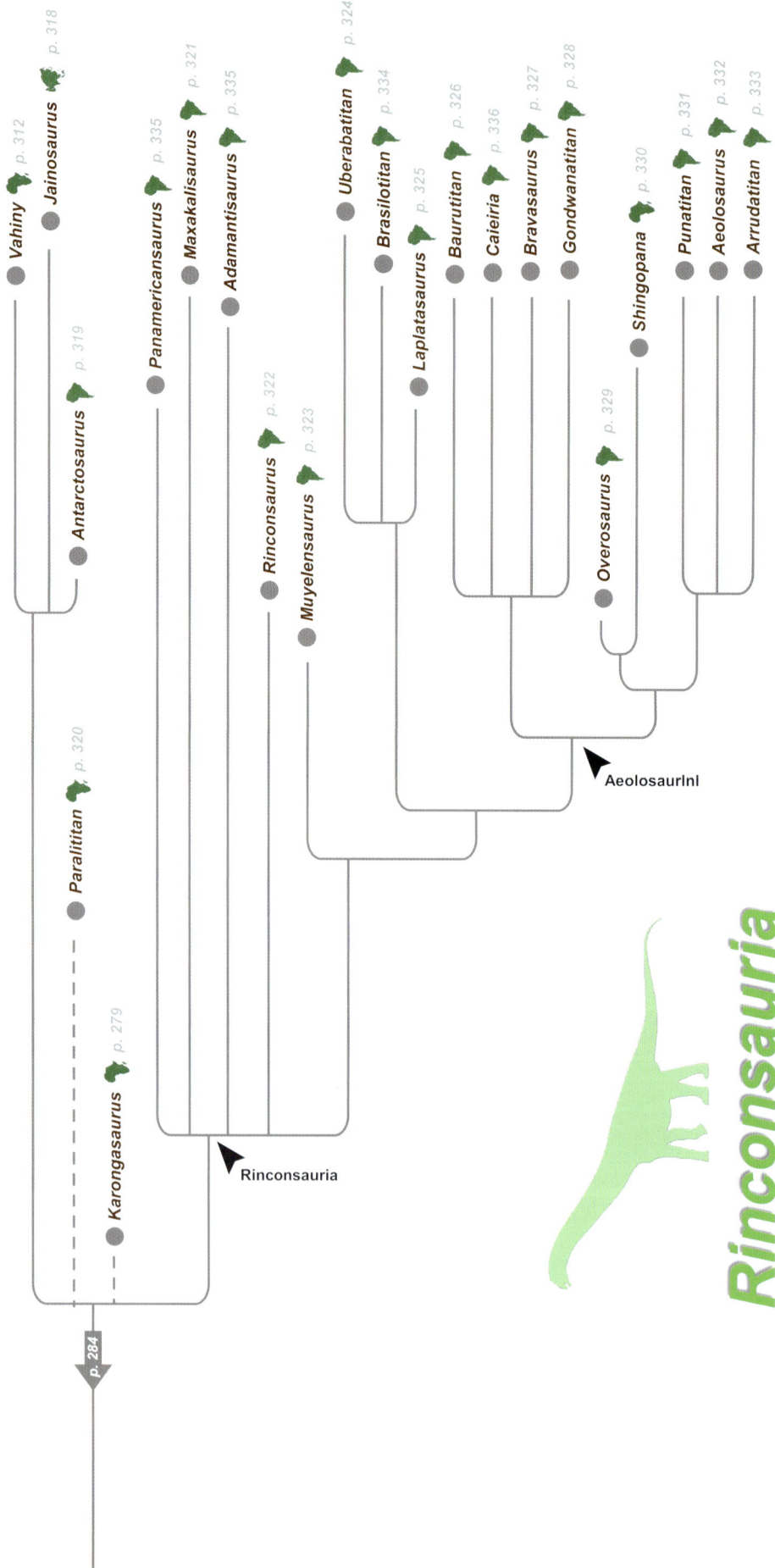

▶ Rinconsauria

p. 284

Rinconsauria

Rinconsauria is currently seen as one of the two main branches within Colossosauria, with the other being Lognkosauria. Unlike the (mostly) gargantuan lognkosaurs, the rinconsaurs tend to be sauropods of a more standard size. The major subfamily Aeolosaurini is now mostly accepted as being a derived clade within Rinconsauria, although historically the position of Aeolosaurini has wandered around the family tree a fair bit.

As a small sister clade to Rinconsauria is a group composed of *Antarctosaurus*, *Vahiny*, and *Jainosaurus*. This geographical hodgepodge has been recovered as a clade by Mannion et al. (2019), Poropat et al. (2021), and Wang et al. (2021).

The position of the fragmentary *Karongasaurus* is very uncertain but might lie somewhere within **Rinconsauria** (Gorscak and O'Connor, 2019). The affinities of *Panamericansaurus* are also less than clear, potentially being a basal member of the clade (Santucci and Filippi, 2022). *Maxakalisaurus* also tends to show a basal placement, as shown by Navarro et al. (2022), which put *Adamantisaurus* in a similar position. *Rinconsaurus* and *Muyelensaurus* form the basis for the clade itself (Carballido et al., 2022). *Paralititan* has a very uncertain placement; see its entry for more details.

A limited cladistic study accompanying the reappraisal of *Laplatasaurus* suggested that it was the sister taxon of *Uberabatitan* (Gallina and Otero, 2015). The often-neglected *Brasilotitan* was recovered in a similar position by Navarro et al. (2022). *Uberabatitan* itself has been found to be just stemward of Aeolosaurini (Silva Junior et al., 2022a; Pérez Moreno et al., 2023) or even within that clade (Hechenleitner et al., 2020).

Where the line gets drawn between Rinconsauria and **Aeolosaurini** varies between studies, based on where the analysis happened to find *Gondwanatitan* and *Aeolosaurus* in relation to one another. For instance, one result from Navarro et al. (2022) has only those two genera as members. Generally, though, *Bravasaurus*, *Punatitan*, and *Arrudatitan* are found within the clade (Hechenleitner et al., 2020; Santucci and Filippi, 2022; Silva Junior et al., 2022a; Pérez Moreno et al., 2023). *Overosaurus* also usually makes the cut, although Santucci and Filippi (2022) found it to be just one step outside the group.

Shingopana, as an African taxon, is often not included in analyses involving the otherwise-exclusive South American rinconsaurs, but its describers consider it to most likely be an aeolosaur, with their most recent work placing it as the sister taxon of *Overosaurus* (Gorscak and O'Connor, 2019). The newly erected *Caieiria* was placed as the sister taxon of *Bravasaurus* by its describers (Silva Junior et al., 2022a). Based on newly incorporated fossil material, *Baurutitan* has also been placed as an aeolosaur (Silva Junior et al., 2022a).

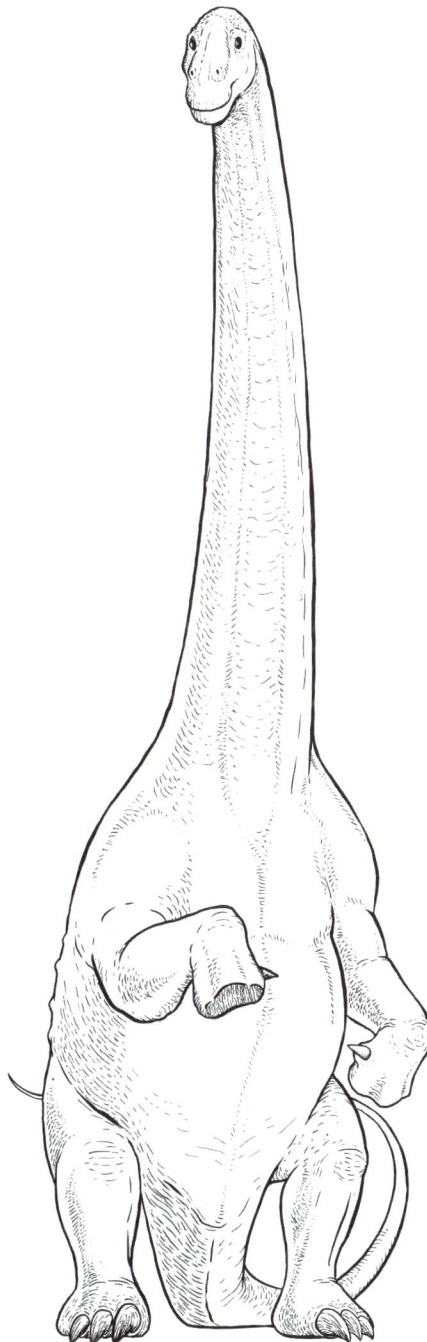

Baurutitan britoi

4 m   8 m   12 m   16 m   20 m

8 m

4 m

J. septentrionalis

(Hunt et al., 1995)
Length: 20 m (66 ft)
Height: 8.5 m (28 ft)
Hip height: 4 m (13 ft)
Body mass: 18,000 kg (19.8 t)
Reconstruction: ☐☐☐☐

250 245 240 235 230 225 220 215 210 205 200 195 190 185 180 175 170 165 160 155 150 145 140 135 130 125 120 115 110 105 100 95 90 85 80 75 70 **68**

| TRIASSIC | JURASSIC | CRETACEOUS |

Jainosaurus septentrionalis (meaning "Jain's northern lizard") was originally named *Antarctosaurus septentrionalis*. The first species of that genus, *A. wichmannianus*, was discovered in Argentina, but despite the great geographical distance between the two locations, Friedrich von Huene named this new species to the same genus in 1933 because the fossils available to him bore no "meaningful" differences from one another.

The question of which bones should be assigned to *Jainosaurus* is a tricky one because much of the sauropod material from the original site came from a large bone bed of disarticulated remains. This is the same location that spawned the type material of the questionable *Titanosaurus indicus*, and designating which bones should be referrable to which animal has inspired numerous debates.

According to a thorough reassessment of *Jainosaurus*, material attributable to the genus includes a brain case, a tail vertebra, numerous rib pieces, shoulder blades, and long bones of the forelimb (Wilson et al., 2009). After the rediscovery of original documentation from the digs in the 1930s, another partial vertebra and the long bones of the hindlimb were added to this collection; these had previously been labeled as "*Titanosaurus* sp." (Wilson et al., 2011).

Some analyses have placed *Jainosaurus* within **Colossosauria**, closely related to *Antarctosaurus* and *Vahiny* (Mannion et al., 2019; Wang et al., 2021).

The generic name *Jainosaurus* honors Indian paleontologist Sohan Lal Jain. The specific name *septentrionalis* means "northern" in Latin, chosen to represent the opposite of "southern" *Antarctosaurus*. ("*Septentrionalis*" combines "septem," or "seven" with "trio" or "plough-ox," referring to the seven stars of Ursa Major in the northern sky).

CLASSIFICATION

Sauropoda
 Macronaria
  Somphospondyli
   Titanosauria
    Lithostrotia
     Eutitanosauria
      Colossosauria

LOCATION

India

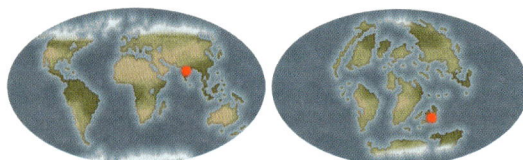

KNOWN REMAINS

Partial skull and skeleton

4 m   8 m   12 m   16 m

A. wichmannianus

(von Huene, 1929)
Length: 17 m (55 ft)
Height: 6.5 m (21 ft)
Hip height: 3.5 m (11.5 ft)
Body mass: 12,000 kg (13.2 t)
Reconstruction:

4 m

TRIASSIC	JURASSIC	CRETACEOUS

250 245 240 235 230 225 220 215 210 205 200 195 190 185 180 175 170 165 160 155 150 145 140 135 130 125 120 115 110 105 100 95 90 **83** 80 75 70 65

Antarctosaurus wichmannianus (meaning "Wichmann's southern lizard") is known from fossilized remains that were discovered in 1912. The bones in question seem to have not been found particularly close to one another, and there are no excavation maps to follow. Consequently, many questions have been raised over the decades as to whether or not all of these bones belonged to the same individual, or even to the same species.

For many years, the squared-off jaw of *Antarctosaurus* was a particular sticking point. It was unlike the jaw of any other known titanosaur, but it did bear a similarity to certain diplodocid mandibles. As a result, it was a popular hypothesis that *Antarctosaurus* was a chimeric specimen that incorporated the remains of both a titanosaur and a diplodocid. However, the 2004 discovery of the square-jawed titanosaur *Bonitasaura* has mostly laid this particular theory to rest. Questions still persist, though, regarding the association of the cranial material with the rest of the skeleton (Upchurch et al., 2004).

This original collection of remains, known as MACN 6904, is now considered the only valid representation of *Antarctosaurus*. Several other species once attributed to the genus are likely not valid, including "*A.*" *brasiliensis*, "*A.*" *jaxarticus*, and "*A.*" *giganteus*. One species, "*A.*" *septentrionalis*, is now known as *Jainosaurus*.

The *Antarctosaurus* holotype material is not often included in phylogenetic studies. However, a few papers have placed it within **Colossosauria**, closely related to *Jainosaurus* and *Vahiny* (Mannion et al., 2019; Wang et al., 2021).

The generic name *Antarctosaurus* is derived from the Greek "antarktos" (meaning "south"), referring to South America. The specific name *wichmannianus* honors fossil discoverer, geologist Ricardo Wichmann.

CLASSIFICATION
Sauropoda
 Macronaria
  Somphospondyli
   Titanosauria
    Lithostrotia
     Eutitanosauria
      Colossosauria

LOCATION
Argentina

KNOWN REMAINS
Partial skull and skeleton

P. stromeri

(Smith et al., 2001)
Length: 27 m (88 ft)
Height: 11 m (36 ft)
Hip height: 4.5 m (14.8 ft)
Body mass: 30,000 kg (33.1 t)
Reconstruction:

Paralititan stromeri (meaning "Stromer's tidal giant") is a large titanosaur known from the same strata as its smaller distant cousin, *Aegyptosaurus*. The proportions of its long humerus hint that it might have had an upward-tilted, *Brachiosaurus*-like posture. The holotype remains were recovered from a fairly small area, along with a partial tooth belonging to a carcharodontid theropod.

The taxonomic position of *Paralititan* is difficult to surmise, partially because of the very incomplete nature of the remains, and partially because the fossils have never received a more detailed description than that which they received in the three-page paper in which they debuted.

The analyses conducted by Vila et al. (2022) and Gorscak et al. (2023) place *Paralititan* within Saltasauridae. This, by itself, is slightly jarring simply because saltasaurids are well known for their diminutive size, and yet *Paralititan* is quite large. It is also worth noting that the "Saltasauridae"

grouping presented in Gorscak et al. (2023) is markedly different than the "classic" version of the family seen in most modern analyses.

Both of these works, though, also place *Paralititan* as being very closely related to *Maxakalisaurus*, which in other studies has been found to be within Rinconsauria (Navarro et al., 2022; Santucci and Filippi, 2022). Additionally, a different version of the analysis of Gorscak et al. (2023) placed *Paralititan* just stemward of Colossosauria, only one step removed from Rinconsauria. Thus, a **colossosaurian** placement for *Paralititan* is plausible, at the very least.

The generic name *Paralititan* derives from the Greek "paralos" (meaning "near the sea"). This was inspired by the specimen having been unearthed from sediments that were deposited in a tidal environment. The specific name *stromeri* honors paleontologist Ernst Stromer von Reichenbach (1871–1952).

CLASSIFICATION
Sauropoda
 Macronaria
  Somphospondyli
   Titanosauria
    Lithostrotia
     Eutitanosauria
      Colossosauria (?)

LOCATION
Egypt

KNOWN REMAINS
Fragments

3 m   6 m   9 m   12 m

6 m

3 m

M. topai

(Kellner et al., 2006)
Length: 13 m (42 ft)
Height: 6.5 m (21 ft)
Hip height: 2.8 m (9 ft)
Body mass: 4,000 kg (4.4 t)
Reconstruction: ☐☐☐☐

250 245 240 235 230 225 220 215 210 205 200 195 190 185 180 175 170 165 160 155 150 145 140 135 130 125 120 115 110 105 100 95 90 85 80 75 **70** 65

TRIASSIC   **JURASSIC**   **CRETACEOUS**

Maxakalisaurus topai (meaning "Topa's lizard of the Maxakali") is one of several titanosaurs known from the Adamantina geological formation. Beginning in 1998, portions of a significant titanosaur skeleton began to be unearthed here. By 2002, the material that would form the holotype for *Maxakalisaurus* was successfully excavated, consisting primarily of fragmentary vertebral elements.

Based on their condition, the bones were most likely exposed for some time before their burial and may even have been trampled. It is very likely that the carcass was scavenged, as evidenced by tooth marks and by shed teeth from both theropods and crocodilians.

The neck vertebrae and some of the tail vertebrae were still articulated when unearthed, while the rest were found within the general area. Most of the remains are thought to have come from a single individual, except for a few shoulder elements that may indicate a second individual being present. Based on a lack of fusion between some vertebral elements, it is likely that the remains belong to specimens that were not yet fully grown.

Later, a partial jaw and some teeth were also found at the site. The jaw is in the more common U-shape, as opposed to the less common squared L-shape. The teeth are peg shaped, which is typical for titanosaurs, but some feature a slight ridge, which is atypical (França et al., 2016).

Phylogenetic results tend to place *Maxakalisaurus* within **Rinconsauria** (Navarro et al., 2022; Santucci and Filippi, 2022), although it has also shown up in Saltasauroidea (Vila et al., 2022).

The generic name *Maxakalisaurus* is meant to honor the Indigenous Maxakali people of Brazil. The specific name *topai* refers to Topa, a deity of the Maxakali.

CLASSIFICATION

Sauropoda
  Macronaria
   Somphospondyli
    Titanosauria
     Lithostrotia
      Eutitanosauria
       Colossosauria
        Rinconsauria

LOCATION

Brazil

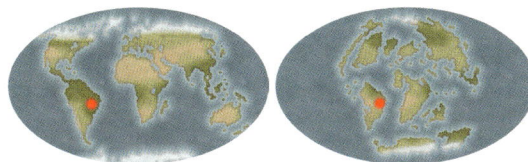

KNOWN REMAINS

Partial cranium and skeleton

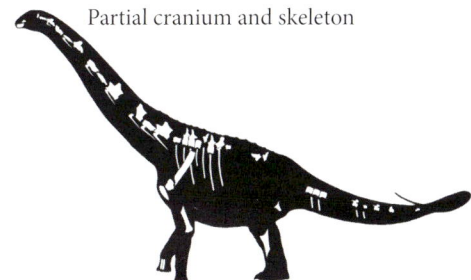

321

| 2 m | 4 m | 6 m | 8 m | 10 m |

R. caudamirus

(Calvo and González Riga, 2003)
Length: 11 m (36 ft)
Height: 4.5 m (15 ft)
Hip height: 2.2 m (7.2 ft)
Body mass: 4,000 kg (4.4 t)
Reconstruction: ☐☐☐☐

| | | | | | | | | | | | | | | | | | | | | | | | | | | | | | | | | | | | | | | |
|250|245|240|235|230|225|220|215|210|205|200|195|190|185|180|175|170|165|160|155|150|145|140|135|130|125|120|115|110|105|100|95|90|85|80|75|70|65|

TRIASSIC | **JURASSIC** | **CRETACEOUS**

Rinconsaurus caudamirus (meaning "amazing tailed lizard from Rincon") is best known for the unusual articulations between its rear tail vertebrae. Bones like this are generally categorized based on the shapes of both the front and rear sites of articulation; each could independently be convex, concave, or flat. Normally, one particular species features little variation in this regard along its vertebral column; the rearmost tail vertebrae of *Rinconsaurus*, however, differ to a highly unusual degree.

Rinconsaurus is known from a mostly jumbled bone bed, with only a few sets of vertebrae remaining in articulation with one another. Multiple analyses have confirmed that only one sauropod species is present at the location. At least four *Rinconsaurus* individuals are present at the site; size differences between fossils could indicate differences in the age of the individual animals (Pérez Moreno et al., 2023).

Recent in-depth analyses have clarified and amended aspects of the original description. Formerly, a few small fragments of skull had been reported, but it has since been determined that these were actually a piece of rib and a bone from a crocodylomorph (Pérez Moreno et al., 2022).

By definition, the clade known as **Rinconsauria** encompasses *Rinconsaurus*, *Muyelensaurus*, and their most recent common ancestor. This clade has come to be widely used in many taxonomic rankings. Many studies have found it to be the sister clade of Lognkosauria, although thorough reviews of the *Rinconsaurus* material suggest that Rinconsauria is rather nested within Lognkosauria (Pérez Moreno et al., 2023).

The generic name *Rinconsaurus* refers to the discovery location, Rincon de los Sauces, in Neuquen Province, Argentina. The specific name *caudamirus* combines the Latin "cauda" (meaning "tail") and "mirus" (meaning "astonishing" or "amazing").

CLASSIFICATION
Sauropoda
 Macronaria
  Somphospondyli
   Titanosauria
    Lithostrotia
     Eutitanosauria
      Colossosauria
       Rinconsauria

LOCATION
Argentina

KNOWN REMAINS
Partial skeleton

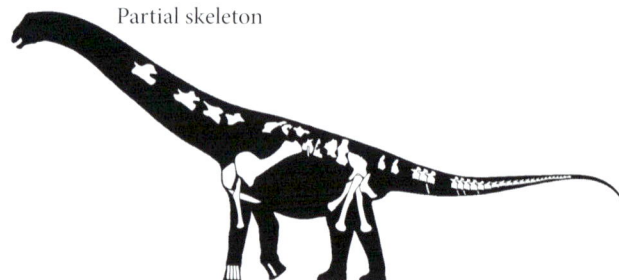

2 m 4 m 6 m 8 m 10 m 12 m

4 m

2 m

M. pecheni

(Calvo et al., 2007a)
Length: 13 m (43 ft)
Height: 5 m (16.5 ft)
Hip height: 2.4 m (8 ft)
Body mass: 5,500 kg (6.1 t)
Reconstruction: ☐☐☐☐

250 245 240 235 230 225 220 215 210 205 200 195 190 185 180 175 170 165 160 155 150 145 140 135 130 125 120 115 110 105 100 95 90 **87** 80 75 70 65

TRIASSIC	JURASSIC	CRETACEOUS

Muyelensaurus pecheni (meaning "Pechen's lizard from Muyelen") is known from a large bone bed that was first discovered in 1997. Approximately 300 titanosaur bones were recovered from the site over the ensuing years, and excavators considered the bones to all have come from a single species—an uncommon occurrence, as different titanosaur species are often notorious for being intermingled throughout such accumulations. (However, one more recent examination has raised the possibility that more than one species was indeed present, and that the type specimen may be chimeric; Carballido et al., 2022).

Based on repeated skeletal elements from the site, at least five individuals are present in the collection, including four adults and one juvenile. *Muyelensaurus* had an overall slender build, with the humerus in particular being quite thin.

The describers of *Muyelensaurus* used it as part of the basis for defining the clade known as **Rinconsauria**, which encompasses *Rinconsaurus*, *Muyelensaurus*, and their most recent common ancestor. This clade has come to be widely used in many taxonomic rankings (which could prove problematic if *Muyelensaurus* is indeed based on a chimeric combination of two separate species). The describers interpreted this group as being a Patagonian clade that was distinct from Aeolosaurini, although subsequent studies have not always found this particular aspect to be the case.

The generic name *Muyelensaurus* derives from the name "Muyelen," which is a name for the Colorado River in the indigenous Mapuche language. The specific name *pecheni* honors Ana Maria Pechen of the National University of Comahue.

CLASSIFICATION
Sauropoda
 Macronaria
  Somphospondyli
   Titanosauria
    Lithostrotia
     Eutitanosauria
      Colossosauria
       Rinconsauria

LOCATION
Argentina

KNOWN REMAINS
Partial skull and skeleton

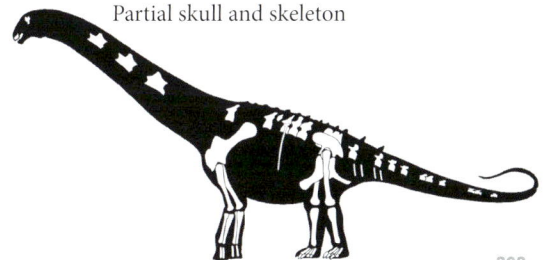

5 m 10 m 15 m 20 m 25 m

10 m

U. ribeiroi

(Salgado and de Souza
Carvalho, 2008)
Length: 25 m (80 ft)
Height: 10 m (33 ft)
Hip height: 4 m (13 ft)
Body mass: 33,000 kg (36.4 t)
Reconstruction: ☐☐☐☐

5 m

250 245 240 235 230 225 220 215 210 205 200 195 190 185 180 175 170 165 160 155 150 145 140 135 130 125 120 115 110 105 100 95 90 85 80 75 70 **67**

| TRIASSIC | JURASSIC | CRETACEOUS |

Uberabatitan ribeiroi (meaning "Ribeiro's giant from Uberaba") is a relatively robust titanosaur and was among the very last of its kind to walk the Earth. Its tibia and fibula had a unique feature, an unusual matching ridge-and-groove structure, the purpose of which is unknown. Another oddity is that the thin strips of bone on the neck vertebrae, which served to separate air sacs, were for some reason segmented, possessing peculiar gaps.

Uberabatitan inhabited a landscape that was predominately hot and arid (Garcia et al., 1998) but that was also known to host temporary streams, presumably caused by brief periods of intense rain (Goldberg and Garcia, 2000). These clues could be key to understanding the origin of the *Uberabatitan* specimens.

Uberabatitan is known from the jumbled remains of several individuals, probably indicating some sort of "mass mortality" event. It is hypothesized that severe drought in

the area led to the deaths of several individuals and that a subsequent rainy episode washed the bones together into the same area.

The bone bed was originally thought to contain the remains of three individuals, but the number has since been revised to five. Three were of intermediate size, one was a juvenile, and one was of much larger stature (Silva Junior et al., 2019).

The taxonomic affinity of *Uberabatitan* has not yet been resolved with any degree of confidence, as several analyses have produced several different results. However, the majority suggest that *Uberabatitan* is not a saltasaurid. Some results have recovered it as a **rinconsaur** (Navarro et al., 2022) or aeolosaur (Silva Junior et al., 2019).

The generic name *Uberabatitan* refers to the city of Uberaba, Brazil. The specific name *ribeiroi* honors researcher Luiz Carlos Borges Ribeiro.

CLASSIFICATION

Sauropoda
 Macronaria
  Somphospondyli
   Titanosauria
    Lithostrotia
     Eutitanosauria
      Colossosauria
       Rinconsauria

LOCATION

Brazil

KNOWN REMAINS

Partial skeleton

2 m   4 m   6 m   8 m

L. araukanicus

(von Huene, 1929)
Length: 10 m (32 ft)
Height: 3.5 m (11.5 ft)
Hip height: 1.8 m (6 ft)
Body mass: 5,000 kg (5.5 t)
Reconstruction: ☐ ☐ ☐ ☐

2 m

250 245 240 235 230 225 220 215 210 205 200 195 190 185 180 175 170 165 160 155 150 145 140 135 130 125 120 115 110 105 100 95 90 85 80 **75** 70 65

TRIASSIC   **JURASSIC**   **CRETACEOUS**

Laplatasaurus araukanicus (meaning "Araucanos' lizard from La Plata") once suffered from being something of a wastebasket taxon, where numerous unrelated fossils were assigned to the genus with little to no justification. As a consequence, our picture of *Laplatasaurus* was artificially inflated and distorted by bones that were not really its own.

The misidentifications began in 1893 when Richard Lydekker included some out-of-place bones in his description of a species of *Titanosaurus*. In 1929, Friedrich von Huene took these individual fossils and placed them in a new genus, *Laplatasaurus*, along with bones from other locations in Argentina and even fossils discovered in Uruguay, Madagascar, and India. Among all these additions, though, he failed to ever designate a holotype specimen. Bonaparte and Gasparini finally rectified this oversight in 1979, giving lectotype status to a tibia and fibula.

A modern review of the unruly mess of supposed *Laplatasaurus* material revealed that most of the material clearly came from different kinds of indeterminate sauropods, and thus the genus was limited, by necessity, to just the lectotype material (Gallina and Otero, 2015).

Based on just the lower leg bones, *Laplatasaurus* was only able to be placed with confidence somewhere within Titanosauria. It was, however, suggested to form a clade with *Uberabatitan*, based on a shared trait of the fibula (Gallina and Otero, 2015). *Uberabatitan*, meanwhile, has been recovered within **Rinconsauria** (Navarro et al., 2022), or more specifically within Aeolosaurini (Hechenleitner et al., 2020).

The generic name *Laplatasaurus* refers to La Plata, Argentina. The specific name *araukanicus* refers to the Indigenous Mapuche people, also sometimes called Araucanos.

CLASSIFICATION
Sauropoda
 Macronaria
  Somphospondyli
   Titanosauria
    Lithostrotia
     Eutitanosauria
      Colossosauria
       Rinconsauria

LOCATION
Argentina

KNOWN REMAINS
Tibia and fibula

8 m   4 m   8 m   12 m   16 m

8 m

4 m

B. britoi

(Kellner et al., 2005)
Length: 18 m (59 ft)
Height: 8 m (26 ft)
Hip height: 3.5 m (11.5 ft)
Body mass: 15,000 kg (16.5)
Reconstruction: ☐ ☐ ☐ ☐

250 245 240 235 230 225 220 215 210 205 200 195 190 185 180 175 170 165 160 155 150 145 140 135 130 125 120 115 110 105 100 95 90 85 80 75 70 65

TRIASSIC   **JURASSIC**   **CRETACEOUS**

Our picture of **Baurutitan britoi** (meaning "Brito's giant from Bauru") has shifted more than once, based on new discoveries.

Baurutitan was originally known only from one set of remains: a set of vertebrae from the region near the base of the tail (known as MCT 1490-R). However, these were far from the only titanosaur fossils unearthed from the Caieira Quarry, discovered by Llewellyn Ivor Price in the 1950s. Two other sets of remains are pertinent, those which originally formed the basis for *Trigonosaurus pricei*: MCT 1488-R (the holotype) and MCT 1719-R (the paratype) (Campos et al., 2005). Overall, there was very little overlap between the bones assigned to each genus, so comparisons between the two were understandably limited.

However, a new (and more complete) set of remains discovered just a few kilometers away showed something surprising: bones that bore a striking resemblance to both MCT 1490-R and MCT 1488-R. The conclusion drawn was

that these two sets of remains, along with the new finds, were all from the same genus. Therefore, the name *Trigonosaurus* was suggested to be considered synonymous with *Baurutitan*. (MCT 1719-R was given its own new genus, *Caieiria*; Silva Junior et al., 2022b.)

Phylogenetic studies have variously placed *Baurutitan* in numerous locations, including Saltasaurine (Navarro et al., 2022) or basal Colossosauria (Hechenleitner et al., 2020), but these placements could be partially attributed to having only incomplete data. The analysis incorporating the newest material placed the animal within **Aeolosaurini** (Silva Junior et al., 2022b).

The generic name *Baurutitan* refers to the Bauru geological unit. The specific name *britoi* honors Brazilian paleontologist Ignácio Aureliano Machado Brito (1938–2001). (The name *Trigonosaurus* refers to the "Triangulo Mineiro" region, and *pricei* honors the fossil's discoverer).

CLASSIFICATION

Sauropoda
  Macronaria
    Somphospondyli
      Titanosauria
        Lithostrotia
          Eutitanosauria
            Colossosauria
              Rinconsauria
                Aeolosaurini

LOCATION

Brazil

KNOWN REMAINS

Partial skeleton

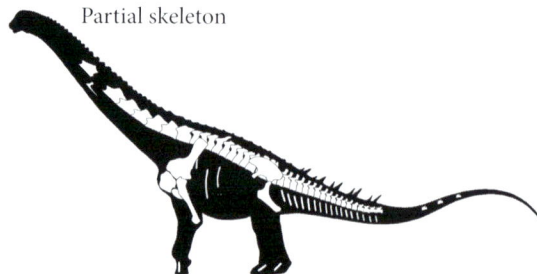

1 m   2 m   3 m   4 m   5 m   6 m   7 m

B. arreirosorum

(Hechenleitner et al., 2020)
Length: 7 m (23 ft)
Height: 2.7 m (9 ft)
Hip height: 1.5 m (5 ft)
Body mass: 3,000 kg (3.3 t)
Reconstruction: ☐☐☐☐☐

2 m

1 m

250 · 245 · 240 · 235 · 230 · 225 · 220 · 215 · 210 · 205 · 200 · 195 · 190 · 185 · 180 · 175 · 170 · 165 · 160 · 155 · 150 · 145 · 140 · 135 · 130 · 125 · 120 · 115 · 110 · 105 · 100 · 95 · 90 · 85 · 80 · 75 · **70** · 65

TRIASSIC	JURASSIC	CRETACEOUS

Bravasaurus arreirosorum (meaning "the arrieros' lizard from Brava") is one of the smallest known species of sauropod. This trait, combined with its stocky limbs and geographic location, could easily lead one to assume that *Bravasaurus* was a saltasaurid, but this is not the case. The describers, along with several subsequent analyses, have found *Bravasaurus* to be placed within **Aeolosaurini** (Santucci and Filippi, 2022; Pérez Moreno et al., 2023).

Bravasaurus is known from two partial sets of remains that were found at the same locality, the site at Quebrada de Santo Domingo. As the vertebral elements displayed complete fusion, it can be determined that the animal was no longer growing and thus was an adult, despite its small stature.

Also present at the locality is an "overwhelming abundance" of fossilized sauropod eggs, making the location "one of the largest nesting sites documented worldwide." This occurrence paints a picture that titanosaurs engaged in philopatry—the practice of regularly returning to the same location to breed or nest. The eggs in question are spherical in shape, a trait typical of titanosaurs, and are relatively small in size.

The same locality also bore fellow aeolosaur, *Punatitan*. Together, they mark the first known occurrence of colossosaurian species in northwest Argentina. Either *Bravasaurus* or *Punatitan* could have been responsible for producing the eggs in the region.

The generic name *Bravasaurus* refers to the lake Laguna Brava, which itself lends its name to the Laguna Brava National Park. The specific name *arreirosorum* refers to the mule-bound "arriero" cargo haulers who once transported cattle across the Andes throughout the nineteenth century.

CLASSIFICATION
Sauropoda
 Macronaria
  Somphospondyli
   Titanosauria
    Lithostrotia
     Eutitanosauria
      Colossosauria
       Rinconsauria
        Aeolosaurini

LOCATION
Argentina

KNOWN REMAINS
Partial skull and skeleton

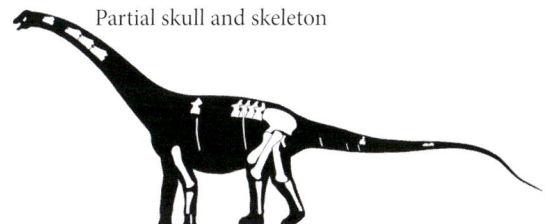

G. faustoi

(Kellner and Azevedo, 1999)

Length: 7 m (23 ft)
Height: 3 m (10 ft)
Hip height: 1.4 m (4.5 ft)
Body mass: 2,500 kg (2.8 t)
Reconstruction: ☐ ☐ ☐

TRIASSIC	JURASSIC	CRETACEOUS

Gondwanatitan faustoi (meaning "Fausto's giant from Gondwana") is known from a single specimen that was discovered on the farm belonging to Yoshitoshi Myzobuchi in 1983. The animal had been buried in fluvial (i.e., river-based) deposits, along with the remains of crocodylomorphs and turtles. Excavations took place until 1986, though it would not be until 1997 when preparatory work began in the lab.

The specimen was housed in the National Museum of Brazil, which suffered a major fire in 2018. The fate of the remains is currently unknown.

Initially after the description of the genus, arguments were presented that *G. faustoi* should actually be considered a species of *Aeolosaurus* (Santucci and Bertini, 2001). However, *Gondwanatitan* is still considered by most to be a distinct and valid designation (Martinelli et al., 2011). One

difference between the two genera has to do with the shape of the middle tail vertebrae, which in *Gondwanatitan* are "heart shaped" when viewed from the rear.

Other aspects of *Gondwanatitan*'s anatomy are a slender and gracile humerus and a tibia that is much straighter than those found in most other titanosaurs.

According to a comprehensive study that proposed updated definitions for the various titanosaurian clades, Aeolosaurini is literally defined on the basis of *Aeolosaurus* and *Gondwanatitan*, encompassing those taxa, their common ancestor, and everything in between (Carballido et al., 2022).

The generic name *Gondwanatitan* refers to the ancient southern continent of Gondwana. The specific name *faustoi* honors Fausto L. de Souza Cunha, who led the specimen's excavation.

CLASSIFICATION

Sauropoda
　Macronaria
　　Somphospondyli
　　　Titanosauria
　　　　Lithostrotia
　　　　　Eutitanosauria
　　　　　　Colossosauria
　　　　　　　Rinconsauria
　　　　　　　　Aeolosaurini

LOCATION

Brazil

KNOWN REMAINS

Partial skeleton

2 m 4 m 6 m 8 m 4 m

2 m

O. paradasorum

(Coria et al., 2013)
Length: 8.5 m (28 ft)
Height: 4 m (13 ft)
Hip height: 1.6 m (5.3 ft)
Body mass: 1,800 kg (2 t)
Reconstruction: ☐☐☐☐

250 245 240 235 230 225 220 215 210 205 200 195 190 185 180 175 170 165 160 155 150 145 140 135 130 125 120 115 110 105 100 95 90 **85** 80 75 70 65

TRIASSIC	JURASSIC	CRETACEOUS

Overosaurus paradasorum (meaning "Parada's lizard from Overo") is known from one of the most complete vertebral series preserved in any titanosaur specimen. From the base of the neck through the middle of the tail, each of the 40 back bones was found in an articulated, lifelike position. Except for the rearmost tail bones, most of the skeletal elements are well preserved and remarkably undamaged.

Based on the arrangement of the remains, the describers hypothesize that *Overosaurus* had a very brachiosaur-like posture, with its back tilting upward toward the front of the animal. Consequently, this would probably indicate that the animal's forelimbs were longer than its hindlimbs, as is the case with many brachiosaurs. This body plan had previously not been observed for any titanosaur; it is possible that distortional forces altered the position of the animal's remains after burial, thus artificially implying a non-lifelike posture.

Although *Overosaurus* was definitely on the small end of the sauropod scale, this was not due to skeletal immaturity. Fusion of the vertebral elements indicates that the specimen was of adult age when it died. Although the specimen was initially stated as being found in the Anacleto Formation, this was later amended to the Bajo de la Carpa Formation.

Taxonomically, studies generally agree that *Overosaurus* falls within **Aeolosaurini** (Hechenleitner et al., 2020; Gallina et al., 2021), although some results place it slightly outside that clade, within Rinconsauria (Navarro et al., 2022; Santucci and Filippi, 2022).

The generic name *Overosaurus* refers to the discovery location, Cerro Overo, in Neuquen Province, Argentina. The specific name *paradasorum* honors the family of fossil discoverer Carlos Parada.

CLASSIFICATION
Sauropoda
 Macronaria
  Somphospondyli
   Titanosauria
    Lithostrotia
     Eutitanosauria
      Colossosauria
       Rinconsauria
        Aeolosaurini

LOCATION
Argentina

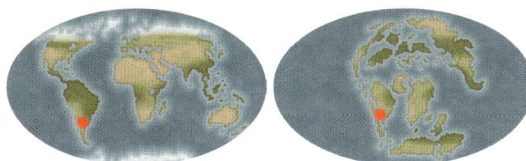

KNOWN REMAINS
Partial skeleton

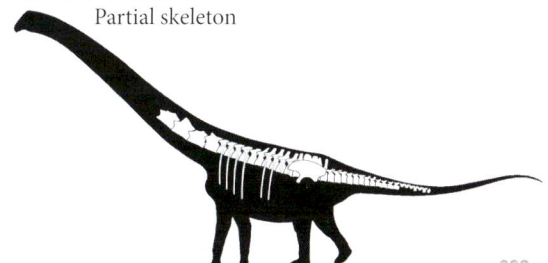

2 m   4 m   6 m   8 m

2 m

S. songwensis

(Gorscak et al., 2017)
Length: 8 m (26 ft)
Height: 3.5 m (11.5 ft)
Hip height: 1.4 m (4.5 ft)
Body mass: 3,000 kg (3.3 t)
Reconstruction: ☐☐☐☐☐

250 245 240 235 230 225 220 215 210 205 200 195 190 185 180 175 170 165 160 155 150 145 140 135 130 125 120 115 110 105 100 95 90 85 80 **73** 70 65

TRIASSIC   **JURASSIC**   **CRETACEOUS**

Shingopana songwensis (meaning "wide neck from Songwe") was discovered in a small concentration of mudstone, among larger sandstone deposits. Erosion along one edge of the area led excavators to assume that more fossils were likely present at one point but were lost to weathering forces.

The well-preserved bones of *Shingopana* were found in a small area, roughly 6 square meters in size, although rib fragments and other shards were spread over a larger region. Excavations took place from 2002 through 2004, and the remains were first mentioned in literature years before their formal description (Santucci and Arruda-Campos, 2011).

Even the bones that were otherwise well preserved contain numerous incursions, believed to have been created by boring insects.

Two other titanosaurs had been known from the Galula Formation of Tanzania prior to the description of

Shingopana, those being *Rukwatitan* and *Mnyamawamtuka*. Based on that information alone, it might be reasonable to assume that the three genera would be closely related within Titanosauria, but this does not seem to be the case. Phylogenetic results place *Shingopana* as being within **Aeolosaurini**, as a closer relative to many South American taxa than to its own geographical neighbors in Africa (Gorscak and O'Connor, 2019). With the separation of the two continents at the time being taken into account, this implies an earlier and broader distribution of aeolosaurs than was previously considered.

The generic name *Shingopana* combines the Kiswahili words "shingo" (meaning "neck") and "pana" (meaning "wide"), inspired by the distinctive "bulbous" neural spine of one of the neck vertebrae. The specific name *songwensis* refers to the local geographic area, Songwe.

CLASSIFICATION
Sauropoda
 Macronaria
  Somphospondyli
   Titanosauria
    Lithostrotia
     Eutitanosauria
      Colossosauria
       Rinconsauria
        Aeolosaurini

LOCATION
Tanzania

KNOWN REMAINS
Mandible and partial skeleton

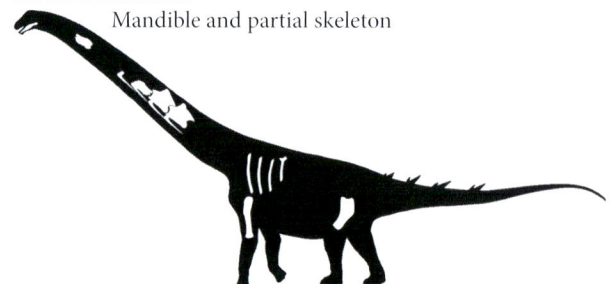

2 m   4 m   6 m   8 m   10 m   12 m

4 m

2 m

P. coughlini

(Hechenleitner et al., 2020)
Length: 14 m (46 ft)
Height: 6 m (20 ft)
Hip height: 2.8 m (9 ft)
Body mass: 6,000 kg (6.6 t)
Reconstruction: ☐☐☐☐

| 250 | 245 | 240 | 235 | 230 | 225 | 220 | 215 | 210 | 205 | 200 | 195 | 190 | 185 | 180 | 175 | 170 | 165 | 160 | 155 | 150 | 145 | 140 | 135 | 130 | 125 | 120 | 115 | 110 | 105 | 100 | 95 | 90 | 85 | 80 | 75 | 70 | 65 |

TRIASSIC   **JURASSIC**   **CRETACEOUS**

Punatitan coughlini (meaning "Coughlin's high-altitude giant") is known from a single set of remains consisting almost entirely of vertebrae and pelvic elements. At the time of description, the sacrum of the specimen was still not yet fully prepared and could possibly reveal more diagnostic characters in the future. The individual was an adult at the time of its death. The remains were likely buried in a flood quickly after death, as they were still semi-articulated and well preserved in fine-grained sandstone.

Also present at the locality is an "overwhelming abundance" of fossilized sauropod eggs, making the location "one of the largest nesting sites documented worldwide." This occurrence paints a picture that titanosaurs engaged in philopatry—the practice of regularly returning to the same location to breed or nest. The eggs in question are spherical in shape, a trait of titanosaurs, and are relatively small in size.

The same locality also bore fellow aeolosaur, *Bravasaurus*. Together, they mark the first known occurrence of colossosaurian species in northwest Argentina. Either *Bravasaurus* or *Punatitan* could have been responsible for producing the eggs in the region. *Punatitan* is significantly larger than *Bravasaurus* and was found at a higher elevation.

The describers, along with several subsequent analyses, have found *Punatitan* to be placed within **Aeolosaurini** (Santucci and Filippi, 2022; Pérez Moreno et al., 2023).

The generic name *Punatitan* derives from "puna," which is "the local name that distinguishes the oxygen-depleted atmosphere typical of the high Andes." The specific name *coughlini* honors fossil discoverer, geologist Tim Coughlin.

CLASSIFICATION
Sauropoda
 Macronaria
  Somphospondyli
   Titanosauria
    Lithostrotia
     Eutitanosauria
      Colossosauria
       Rinconsauria
        Aeolosaurini

LOCATION
Argentina

KNOWN REMAINS
Partial skeleton

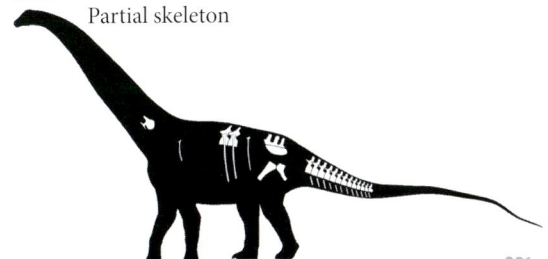

331

8 m   4 m   8 m   12 m   16 m

4 m

A. rionegrinus

(Powell, 1987)
Length: 18 m (59 ft)
Height: 8 m (26 ft)
Hip height: 3 m (10 ft)
Body mass: 15,000 kg (16.5 t)
Reconstruction: ☐☐☐☐

250 245 240 235 230 225 220 215 210 205 200 195 190 185 180 175 170 165 160 155 150 145 140 135 130 125 120 115 110 105 100 95 90 85 80 75 70 65

TRIASSIC   **JURASSIC**   **CRETACEOUS**

Aeolosaurus rionegrinus (meaning "Aeolus's lizard from Rio Negro") and the clade that has formed around it (Aeolosaurini) are very distinctive in regard to the shape of their tail vertebrae.

Simply put, such a vertebra has two main components: the centrum and the spine. Typically, the spine sits more or less directly on top of the centrum. In the tail vertebrae of titanosaurs, though, the spine tends to be shifted anteriorly (i.e., forward) by some degree. In aeolosaurs, this shift is extremely exaggerated, to the point that the rearmost edge of the spine is aligned with the foremost edge of the centrum.

Another quirk of *Aeolosaurus* is that the front portion of its tail seems to curve downward at an awkward angle, making the tail's drooping posture almost seem like something out of an early twentieth-century piece of paleoart. This could possibly have something to do with unique musculature present in the hindlimb (Vidal et al., 2021).

The type species, *A. rionegrinus*, is known from a single specimen, a partial skeleton that includes tail, limb, shoulder, and hip elements. A second species, *A. colhuehuapensis*, named in 2007, is known from a badly eroded set of 21 tail vertebrae (Casal et al., 2007). A third species has since been reassigned to its own genus, *Arrudatitan*. Numerous other remains from the region, mainly bones from the tail, have been tentatively identified as belonging to an indeterminate species of *Aeolosaurus*.

The generic name *Aeolosaurus* refers to the fictional figure of Aeolus, the Keeper of the Winds from Homer's *Odyssey*, inspired by the common windy weather at the dig site during excavation. The specific name *rionegrinus* refers to the Rio Negro Province of Argentina; *colhuehuapensis* refers to Lake Colhué Huapi.

CLASSIFICATION
Sauropoda
 Macronaria
  Somphospondyli
   Titanosauria
    Lithostrotia
     Eutitanosauria
      Colossosauria
       Rinconsauria
        Aeolosaurini

LOCATION
Argentina

KNOWN REMAINS
Partial skeleton

3 m   6 m   9 m   12 m   15 m

6 m

3 m

A. maximus

(Silva Junior et al., 2022a)
Length: 17 m (56 ft)
Height: 8 m (26 ft)
Hip height: 3.2 m (10.5 ft)
Body mass: 12,000 kg (13.2 t)
Reconstruction: ☐☐☐☐

250 245 240 235 230 225 220 215 210 205 200 195 190 185 180 175 170 165 160 155 150 145 140 135 130 125 120 115 110 105 100 95 90 85 80 75 **70** 65

TRIASSIC   **JURASSIC**   **CRETACEOUS**

Arrudatitan maximus (meaning "Arruda's great giant") was originally known as *Aeolosaurus maximus*, the third species named to that genus (Santucci and Arruda-Campos, 2011). Confoundingly, though, multiple phylogenetic computations began to calculate a position for *A. maximus* that was not most closely related to the other two species (Filippi et al., 2013; Bandeira et al., 2016). Consequently, as it no longer made sense for this species to be unified with the other *Aeolosaurus* species, it was given its own new genus.

The holotype specimen was found mostly disarticulated (except for the remnants of the tail and neck) but still associated within a 100 square meter area. The preserved posture of the neck suggests that the animal had enough time after death to arch into a dinosaurian "death pose" before its burial in sand. Although no bite marks are evident

on the bones, teeth from both theropods and crocodyliforms were found near the pelvic region.

As with *Aeolosaurus*, *Arrudatitan*'s tail displays a drooping downward curve, possibly related to the hindlimbs' unique musculature (Vidal et al., 2020).

Phylogenetically, *Arrudatitan* probably fits within **Aeolosaurini** (Pérez Moreno et al., 2023), which makes sense given the similarities with the peculiar tail vertebrae of *Aeolosaurus*. At least one recent work has found the genus to be placed more basally, within the more inclusive Rinconsauria (Navarro et al., 2022).

The generic name *Arrudatitan* honors fossil collector Antonio de Celso Arruda Campos, one of the specimen's original describers. The specific name *maximus* is Latin for "large" or "great."

CLASSIFICATION

Sauropoda
 Macronaria
  Somphospondyli
   Titanosauria
    Lithostrotia
     Eutitanosauria
      Colossosauria
       Rinconsauria
        Aeolosaurini

LOCATION

Brazil

KNOWN REMAINS

Partial skeleton

333

BRASILOTITAN

B. nemophagus

(Machado et al., 2013)

Length: 10 m (33 ft)
Height: 4.5 m (15 ft)
Reconstruction: ☐ ☐ ▮ ☐ ☐

2 m

70 —

TRIASSIC	JURASSIC	CRETACEOUS

Brasilotitan nemophagus is most notable for the shape of its extremely squared-off jaw, which is quite unusual for titanosaurs. Behind the teeth, the jaw forms a "guillotine" shape. The few fragments that are known for *Brasilotitan* were unearthed in 2000 in what was originally considered to be the Adamantina Formation but which has since been revised as the Presidente Prudente Formation (Bandeira et al., 2016). The describers were only able to phylogenetically place the animal somewhere vaguely within Titanosauria. A more recent study has suggested a more detailed placement, within **Rinconsauria** (Navarro et al., 2022).

The generic name *Brasilotitan* refers to the nation of Brazil. The specific name *nemophagus* combines the Greek "némos" (meaning "pasture") and "phagos" (meaning "to eat").

PITEKUNSAURUS

P. macayai

(Filippi and Garrido, 2008)

Length: 12 m (39 ft)
Height: 6.5 m (21 ft)
Reconstruction: ☐ ☐ ☐ ☐

3 m

75 —

TRIASSIC	JURASSIC	CRETACEOUS

Pitekunsaurus macayai (meaning "lizard discovered by Macaya") was discovered in the Anacleto Formation of Argentina in 2004. An unusual feature of the animal is that its rear tail vertebrae feature a number of different shapes of articulation surfaces. It was not a huge sauropod, and the fusion of vertebral elements suggests that the animal was not a juvenile. The associated remains were buried within the deposits of a floodplain.

The describers could get no more taxonomically specific than placing *Pitekunsaurus* within Titanosauria. It has been recovered by subsequent studies as an aeolosaur (Coria et al., 2013), a basal colossosaur (González Riga et al., 2018), and as a **rinconsaur** (Navarro et al., 2022).

The generic name incorporates the indigenous Mapuche term "pitëkun" (meaning "to discover"). The specific name honors fossil discoverer Luis Macaya.

ADAMANTISAURUS

A. mezzalirai

(Santucci and Bertini, 2006)
Length: 18 m (59 ft)
Height: 8 m (26 ft)
Reconstruction:

3 m

TRIASSIC | JURASSIC | CRETACEOUS

Adamantisaurus mezzalirai is known from six articulated caudal (tail) vertebrae that were discovered in 1958. The genus has not been included in many cladistic studies; one large sampling has calculated its position as being within **Rinconsauria** (Navarro et al., 2022).

The generic name refers to the Adamantina geological formation of Brazil, where the fossils were located. The specific name honors fossil discoverer, geologist Sérgio Mezzalira, who first illustrated the remains in 1959, attributing them to "Titanosauridae indet." Mezzalira also considered a nearby femur to belong to the same individual, but based on its size, it has now been excluded from the overall specimen.

PANAMERICANSAURUS

P. schroederi

(Calvo and Porfiri, 2010)
Length: 15 m (50 ft)
Height: 7.5 m (25 ft)
Reconstruction:

3 m

TRIASSIC | JURASSIC | CRETACEOUS

Panamericansaurus schroederi is known from a handful of bones that were found in Argentina in 2003 and were originally attributed to *Aeolosaurus*. Aside from their original (brief) description in a Spanish-language source, they have not been closely studied.

Taxonomic results generally show *Panamericansaurus* to be a member of Aeolosaurini (França et al., 2016; Gorscak and O'Connor, 2019) or, alternatively, some results place it slightly outside that clade, within **Rinconsauria** (Santucci and Filippi, 2022).

The generic name *Panamericansaurus* refers to the Pan American Energy Company for their paleontological support. The specific name *schroederi* honors the landowners, the Schroeder family.

RINCONSAURIA

C. allocaudata

(Silva Junior et al., 2022b)
Length: 18 m (59 ft)
Height: 8 m (26 ft)
Reconstruction:

3 m

70 —

| TRIASSIC | JURASSIC | CRETACEOUS |

The holotype remains of **Caieiria allocaudata** were originally identified as a second specimen of the Brazilian genus *Trigonosaurus*. However, further discoveries revealed that *Trigonosaurus* is likely one and the same as another titanosaur, *Baurutitan*. As such, the 10 tail vertebrae of this specimen (MCT 1719-R) were assigned to their own new genus, *Caieiria*, as they did not match the remains of *Baurutitan*: the neural spines of *Caieiria* point more or less upward, while those of *Baurutitan* point backward. The describers concluded that *Caieiria* is a member of **Aeolosaurini**.

The generic name *Caieiria* refers to "Caieira," the name of the site where the fossils were found. The specific name *allocaudata* combines the Greek "allos" (meaning "strange") and the Latin "cauda" (meaning "tail").

S. saihangaobiensis

(Xu et al., 2006)
Length: 10 m (33 ft)
Height: 4 m (13 ft)
Reconstruction:

2 m

85 —

| TRIASSIC | JURASSIC | CRETACEOUS |

Sonidosaurus saihangaobiensis was discovered in 2001, from the same location as the theropod *Gigantoraptor*. Although it was a small sauropod, it was either fully grown or at least close to it, as evidenced by closed sutures of the vertebrae. The dorsal vertebrae featured unusually tall neural spines compared with other titanosaurs.

The describers concluded that *Sonidosaurus* was a derived titanosaur. A later study speculated that it was "a non- saltasaurid lithostrotian with possible saltasaurid affinities" (Averianov and Sues, 2017), possibly placing it within **Saltasauroidea**.

The generic name refers to the region in Inner Mongolia, Sonid. The specific name refers to the more specific location, Saihangaobi.

1 m   2 m   3 m   4 m   5 m   6 m   7 m   8 m   9 m

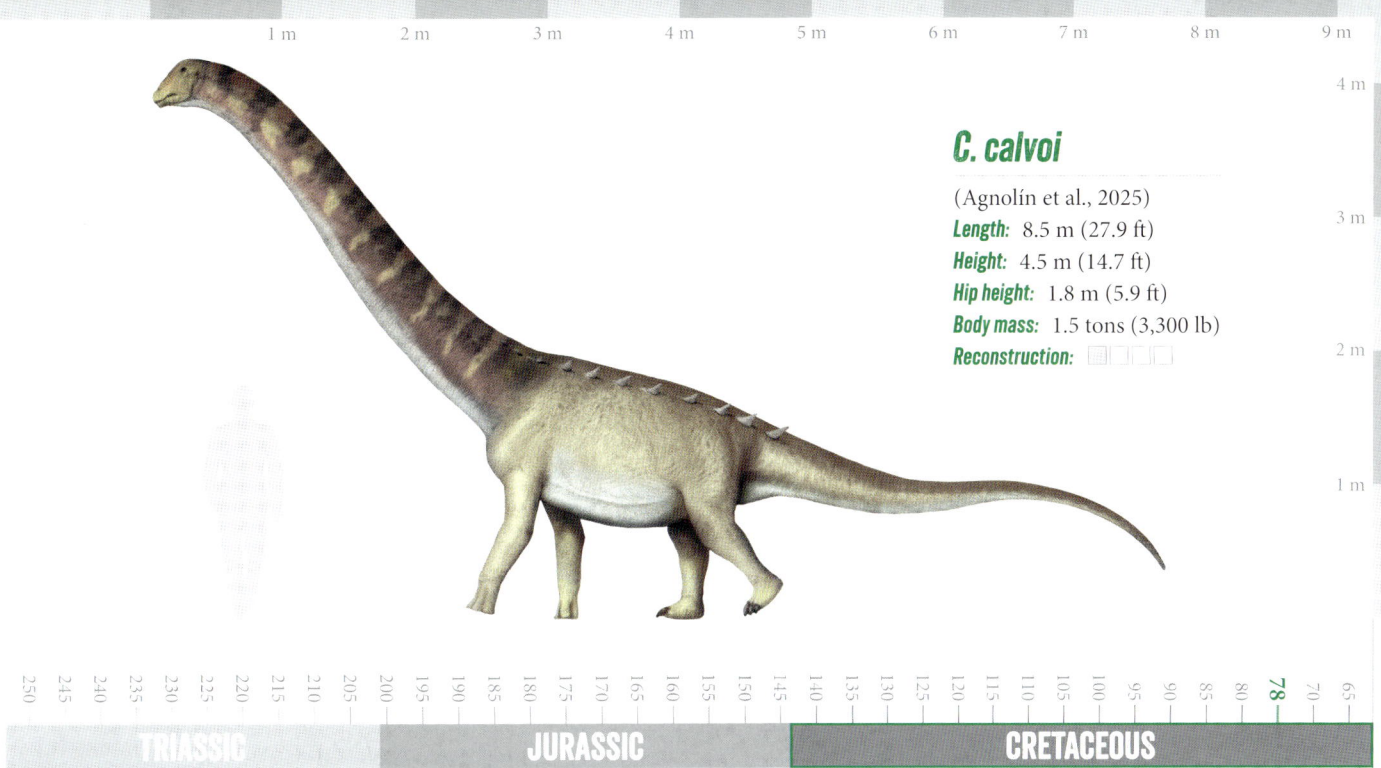

C. calvoi

(Agnolín et al., 2025)
Length: 8.5 m (27.9 ft)
Height: 4.5 m (14.7 ft)
Hip height: 1.8 m (5.9 ft)
Body mass: 1.5 tons (3,300 lb)
Reconstruction: ☐☐☐☐

250 245 240 235 230 225 220 215 210 205 200 195 190 185 180 175 170 165 160 155 150 145 140 135 130 125 120 115 110 105 100 95 90 85 80 **78** 70 65

TRIASSIC	JURASSIC	CRETACEOUS

Chadititan calvoi (meaning "Calvo's salt giant") is a rather small and slim sauropod, with relatively thin and gracile legs. The proportions of its known limb bones (along with those of its closest relatives) hint that rinconsaurs were built differently from other titanosaurs, having a more brachiosaur-like posture.

Chadititan is known from the remains of several different individuals, and each set of remains is ill-preserved and fragmentary. Many of these individuals seem to have been fully grown, so the small size of the *Chadititan* fossils is not simply because they came from juvenile specimens. These bones come from a large assemblage of fossils, which seem to have originated from a lakeside that was situated in an arid environment.

The geological formation that bore *Chadititan*, the Anacleto Formation, has also yielded more than half a dozen other titanosaur genera. This could indicate that the region hosted an astonishing diversity of sauropods that somehow all avoided direct competition with each another. Alternatively, it may turn out that some of these animals, which have been interpreted as different species, may in fact belong to the same species, and that as more complete skeletal remains are uncovered, our picture of this sauropod-heavy ecosystem will become clearer.

The generic name *Chadititan* derives from "chadi," which is the Mapundungum word for "salt," as the fossils were unearthed near salt flats. The specific name *calvoi* honors paleontologist Jorge O. Calvo.

CLASSIFICATION
Sauropoda
  Macronaria
    Somphospondyli
      Titanosauria
        Lithostrotia
          Eutitanosauria
            Colossosauria
              Rinconsauria

LOCATION
Argentina

KNOWN REMAINS
Partial skeleton

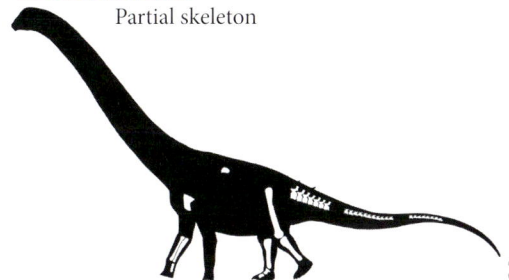

Paleogene

Paleocene

Cretaceous

Late

Early

Maastrichtian

Campanian

Sant.

Coniac.

Turon.

Cenomanian

Albian

Aptian

60

65

70

75

80

85

90

95

100

105

110

115

120

125

Rukwatitan p. 340

Titanomachya p. 359

Isisaurus p. 341

Arackar p. 342

Rapetosaurus p. 343

Udelartitan p. 360

Nullotitan p. 345

Tapuiasaurus p. 344

Bustingorrytitan p. 359

Opisthocoelicaudia p. 346

Nemegtosaurus p. 347

Quaesitosaurus p. 348

Abditosaurus p. 349

Igai p. 350

Alamosaurus p. 351

Pellegrinisaurus p. 352

Ibirania p. 353

Bonatitan p. 354

Rocasaurus p. 358

Neuquensaurus p. 355

Saltasaurus p. 356

Opisthocoelicaudiinae

Saltasaurini

Saltasauridae

Saltasaurinae

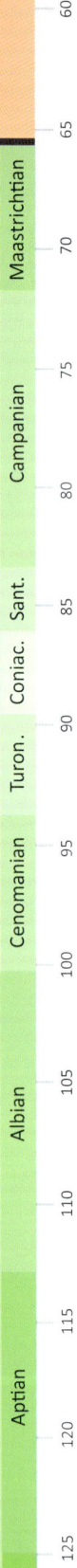

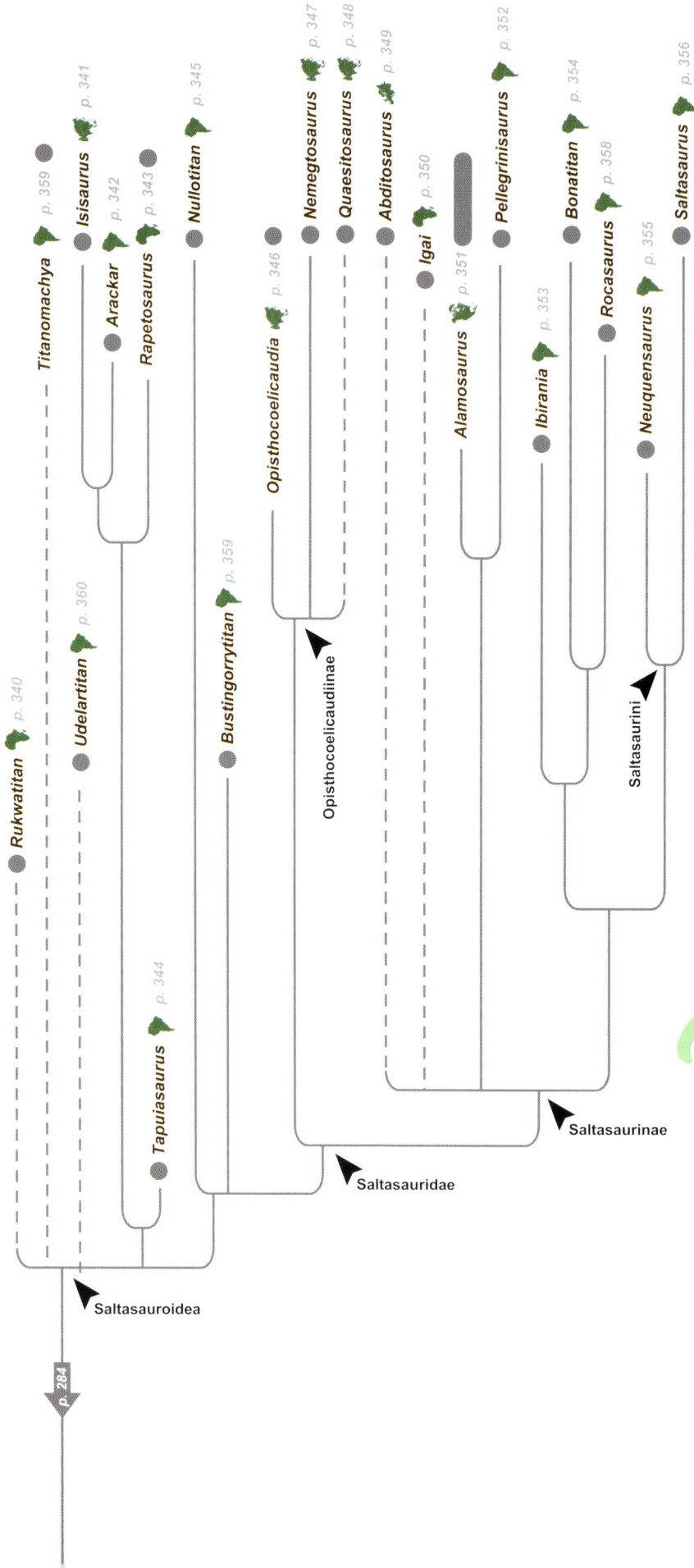

Saltasauroidea

Saltasauroidea

p. 284

Saltasauroids are a diverse group, featuring giants and pipsqueaks, long necks and short necks, and slender and robust species. As with the rest of Titanosauria, the status of the relationships among these species is not yet well understood.

Isisaurus, *Rapetosaurus*, and *Tapuiasaurus* have often been recovered as being very closely related and being near the very base of **Saltasauroidea** (Hechenleitner et al., 2020; Navarro et al., 2022). The describers of *Arackar* considered it to be the sister taxon of *Isisaurus* (Rubilar-Rogers et al., 2021).

Rukwatitan has variously been placed outside Lithostrotia (Gorscak et al., 2014) or within Saltasauroidea (Gorscak and O'Connor, 2019) or Colossosauria (Gorscak et al., 2023). The describers of *Bustingorrytitan* placed it just stemward of Saltasauridae (Simón and Salgado, 2023).

Nemegtosaurus and *Opisthocoelicaudia* are variable in their positioning, although they tend to be placed crownward of the aforementioned taxa (Filippi et al., 2019; Silva Junior et al., 2022), and they tend to be recovered as being very closely related (Cerda et al., 2021), with some analyses going so far as to consider them possibly to be synonymous (Currie et al., 2018). Some alternative topologies place the taxa within a much broader clade, dubbed **Opisthocoelicaudiinae** (Vila et al., 2022). The affinities of *Quaesitosaurus* are most definitely unknown, but it has been suggested that the genus may be synonymous with *Nemegtosaurus* (Currie et al., 2018).

The definitions of **Saltasauridae** and **Saltasaurinae** are dependent on the relative positions of *Saltasaurus* and *Opisthocoelicaudia* (Carballido et al., 2022). Since the relative relationships between these two taxa vary from study to study, the member list for the clades also fluctuates. *Nullotitan* has not been a part of many studies but has been placed as being sandwiched between the *Rapetosaurus* clade and Saltasauridae (Cerda et al., 2021).

Competing theories place *Alamosaurus* as being most closely related to Asian (Averianov and Efimov, 2018; Santucci and Filippi, 2022) or South American (Cerda et al., 2021; Navarro et al., 2022) taxa. Analyses that include both it and *Pellegrinisaurus* have found the two to be closely related (Cerda et al., 2021; Vila et al., 2022).

Rocasaurus, *Neuquensaurus*, and (of course) *Saltasaurus* are near-universally considered to be derived saltasaurids (Díez Díaz et al., 2021; Santucci and Filippi, 2022). *Bonatitan* is much less studied and could be a saltasaurid (Navarro et al., 2022) or could be sister to Saltasauridae (Salgado et al., 2014). The describers of *Ibirania* also place it within this general grouping (Navarro et al., 2022).

The describers of *Igai* place it as a saltasaurid (Gorscak et al., 2023). The describers of *Abditosaurus* also place it as a saltasaurid, although they also place *Paralititan* and *Maxakalisaurus* nearby, which runs contrary to the majority of analyses (Vila et al., 2022). The describers of *Titanomachya* placed the genus within Saltasauroidea but without high resolution (Pérez-Moreno et al., 2024). Various versions of analyses conducted by the describers of *Udelartitan* place it in one of several spots within Saltasauroidea (Soto et al., 2024).

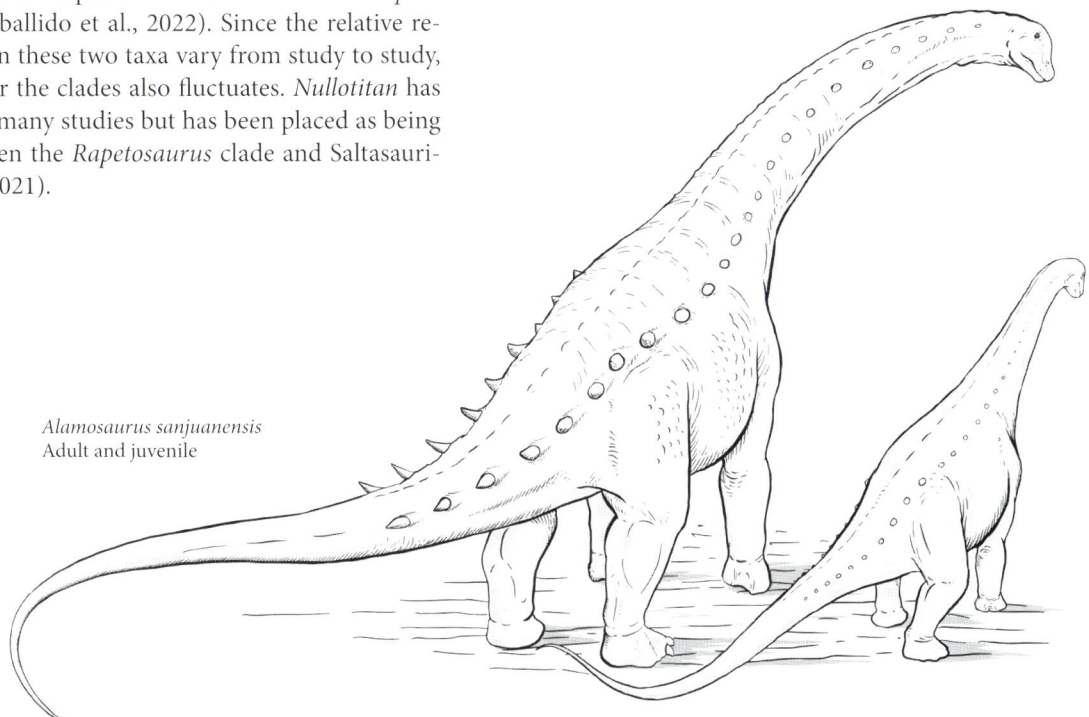

Alamosaurus sanjuanensis
Adult and juvenile

2 m   4 m   6 m   8 m   10 m

4 m

2 m

R. bisepultus

(Gorscak et al., 2014)
Length: 11 m (36 ft)
Height: 4.5 m (15 ft)
Hip height: 2 m (6.5 ft)
Body mass: 6,000 kg (6.6 t)
Reconstruction: ☐☐☐☐

250 245 240 235 230 225 220 215 210 205 200 195 190 185 180 175 170 165 160 155 150 145 140 135 130 125 120 115 110 105 **100** 95 90 85 80 75 70 65

TRIASSIC	JURASSIC	CRETACEOUS

Rukwatitan bisepultus (meaning "twice-buried giant from Rukwa") is one of only a handful of "middle" Cretaceous sauropods from southern Africa. Discovered in 2007, this small and relatively long-limbed titanosaur lived in an area that featured numerous large rivers that would periodically reshape the landscape as their paths meandered. The largest river in the area during this time span is estimated to have been 100 meters wide and 10 meters deep (Roberts et al., 2010). The ecosystem has also preserved mammalian, crocodyliform, and partial theropod remains.

The describers of *Rukwatitan* placed it as a basal titanosaur outside Lithostrotia. A later study by some of the same authors, one which incorporated more taxa, favored the interpretation of *Rukwatitan* being within a clade that is sister to Saltasauridae, thereby placing it within the confines of **Saltasauroidea** (Gorscak and O'Connor, 2019). One of the possible results from a still-later study placed it in Colossosauria (Gorscak et al., 2023).

The generic name *Rukwatitan* refers to Lake Rukwa, within the Rukwa Valley. The specific name *bisepultus* is Latin (meaning "twice buried"); this was inspired by the specimen's unique taphonomy, that is, how it was buried and fossilized.

The remains were found lying atop a layer of mudstone, suggesting that the sauropod's carcass came to rest on a riverbank, lying on its left side. After this period of burial, a brief spate of high-energy river motion unburied most of the animal, washing away smaller skeletal elements while damaging and disassembling the larger ones.

After this period, sand, which eventually became sandstone, covered what was left. Another, much more recent period of erosion, likely by the modern Namba River, had further eroded what still remained by the time paleontologists arrived.

CLASSIFICATION
Sauropoda
 Macronaria
  Titanosauriformes
   Somphospondyli
    Titanosauria
     Lithostrotia
      Eutitanosauria
       Saltasauroidea

LOCATION
Tanzania

KNOWN REMAINS
Partial skeleton

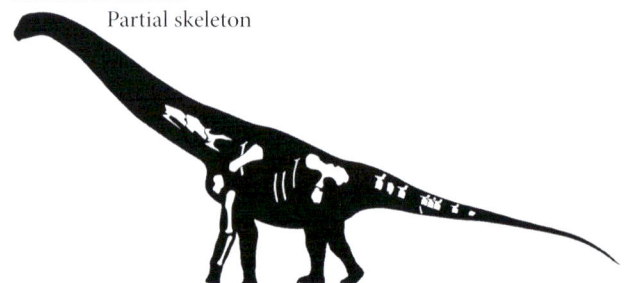

2 m   4 m   6 m   8 m   10 m

I. colberti

4 m

(Wilson and Upchurch, 2003)
Length: 11 m (36 ft)
Height: 5 m (16.5 ft)
Hip height: 2.2 m (7.2 ft)
Body mass: 11,500 kg (12.7 t)
Reconstruction: ▢▢▢▢▢

2 m

250 245 240 235 230 225 220 215 210 205 200 195 190 185 180 175 170 165 160 155 150 145 140 135 130 125 120 115 110 105 100 95 90 85 80 75 **70** 65

TRIASSIC	JURASSIC	CRETACEOUS

Isisaurus colberti (meaning "Colbert's ISI lizard") is an unusually proportioned sauropod. Only a few neck vertebrae are known, but they are unusually short (from front to back) in comparison to their height. The vertebral centra are only twice as long as they are tall, which is a near-record-breaking proportionality. As such, it seems very likely that the overall length of the animal's neck was quite short, for a sauropod. Meanwhile, the vertebral spines of both the neck and back vertebrae are quite tall and exceptionally broad.

Isisaurus has equally odd shoulders and forelimbs. For starters, the humerus is nearly half again as long as the shoulder blade, resulting in an underdeveloped pectoral region and stilt-like front appendages. The forearms were contrastingly stocky, and the ulna is also uniquely shaped, having a triangular cross section. Since these features all undoubtedly served a definite purpose, it seems likely that *Isisaurus* utilized these attributes in order to occupy a distinct ecological niche separate from most other titanosaurs; what exactly that niche was, though, is still a mystery.

The holotype remains from which *Isisaurus* was described were found in 1984, having been buried in freshwater sediment. The skeletal elements were "smeared" quite evenly across a broad swath of area. When the animal was first described in 1997, it was named as yet another species of the prolific *Titanosaurus* genus, as *T. colberti* (Jain and Bandyopadhyay, 1997). However, a later study would go on to conclude that the animal should form a new, distinct genus, *Isisaurus*. Other fragmentary remains from the general region, including some tail vertebrae and a partial ulna, have also been tentatively referred to the genus (Malkani, 2019).

The generic name *Isisaurus* refers to the ISI, the Indian Statistical Institute. The specific name *colberti* honors paleontologist Edwin Harris Colbert.

CLASSIFICATION
Sauropoda
 Macronaria
  Titanosauriformes
   Somphospondyli
    Titanosauria
     Lithostrotia
      Eutitanosauria
       Saltasauroidea

LOCATION
Pakistan

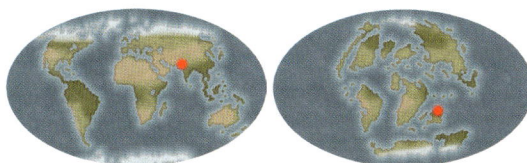

KNOWN REMAINS
Partial skeleton

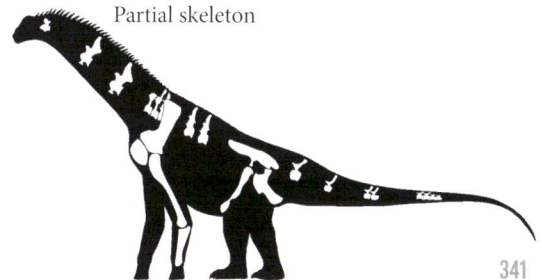

341

1 m   2 m   3 m   4 m   5 m   6 m

2 m

1 m

A. licanantay

(Rubilar-Rogers et al., 2021)
Length: 6 m (21 ft)
Height: 2.5 m (8.2 ft)
Hip height: 1.5 m (5 ft)
Body mass: 1,500 kg (1.7 t)
Reconstruction:

250 245 240 235 230 225 220 215 210 205 200 195 190 185 180 175 170 165 160 155 150 145 140 135 130 125 120 115 110 105 100 95 90 85 80 75 70 65

TRIASSIC	JURASSIC	CRETACEOUS

Arackar licanantay (meaning "skeleton of the Atacamenians") is known from only a handful of bones but nonetheless represents the most complete set of sauropod remains discovered in Chile. The fossils were found together in a small area, covering less than 2 square meters, leading researchers to conclude that they came from the same animal. The specimen was mostly exposed when it was discovered, meaning that other elements of the skeleton may have eroded away before they could be recovered.

The animal was very small by sauropod standards; it was only a subadult at the time it died, as evidenced by a lack of fusion between certain vertebral elements. Consequently, the full size of an adult specimen is unknown.

The specimen was discovered in 1993 and unearthed in 1994. In 2012, the remains were tentatively assigned to

Atacamatitan, the only other named sauropod from Chile (Rubilar-Rogers and Gutstein, 2012).

Notable features of *Arackar* include vertebral neural spines that are positioned at a strong angle pointed toward the rear, as well as limbs that are noticeably less wide splayed than in other derived titanosaurs.

The describers found *Arackar* to be within Lithostrotia, in a small clade most closely related to *Rapetosaurus* and *Isisaurus*. Under later-established clade definitions, this region of the family tree would fall within **Saltasauroidea** (Carballido et al., 2022).

The generic name *Arackar* means "skeleton" or "bones" in the indigenous Kunza language. The specific name *licanantay*, or "lican antay," refers to the Atacamenian people, ancestral inhabitants of the Atacama Desert of Chile.

CLASSIFICATION

Sauropoda
 Macronaria
  Titanosauriformes
   Somphospondyli
    Titanosauria
     Lithostrotia
      Eutitanosauria
       Saltasauroidea

LOCATION

Chile

KNOWN REMAINS

Fragments

4 m   8 m   12 m   16 m

R. krausei

(Curry Rogers and Forster, 2001)
Length: 16 m (54 ft)
Height: 7.5 m (24.5 ft)
Hip height: 3 m (9.8 ft)
Body mass: 10,000 kg (11 t)
Reconstruction: ☐☐☐☐

4 m

250 245 240 235 230 225 220 215 210 205 200 195 190 185 180 175 170 165 160 155 150 145 140 135 130 125 120 115 110 105 100 95 90 85 80 75 70 66

TRIASSIC   **JURASSIC**   **CRETACEOUS**

Rapetosaurus krausei (meaning "Krause's Rapeto lizard") is fairly unremarkable in terms of its anatomy when compared with other eutitanosaurs. Historically, however, the discovery of *Rapetosaurus* was a very crucial event for our understanding of titanosaurs in general.

The first remains attributed to *Rapetosaurus* were originally spotted in 1993. Included among these fossils was a mostly complete skull and jaw—the first ever substantial cranial material that was unequivocally titanosaurian in origin. The skull bears a passing resemblance to a standard diplodocid skull but with several key differences, such as an extremely elongated skull opening in front of the eye socket.

Also retrieved was the mostly complete (though disarticulated) skeleton of a juvenile individual, comprising approximately 75% of the animal's body; this still today ranks among the most complete titanosaurian specimens

ever discovered. Additional cranial and skeletal elements have been attributed to the genus in the ensuing years, including osteoderms (bony armor-like elements) (Curry Rogers et al., 2011) and the partial skeleton of a near infant (Curry Rogers et al., 2016).

Since the most complete of specimens was not fully grown, it is difficult to determine the true upper size of *Rapetosaurus*. One recovered femur, though, is more than twice the size of the juvenile specimen's, suggesting a decently sized sauropod (Curry Rogers and Kulik, 2018).

Rapetosaurus is most often recovered as a basal **saltasauroid**, in the company of *Isisaurus* (Poropat et al., 2021; Navarro et al., 2022).

The generic name *Rapetosaurus* refers to Rapeto, a mythological giant from Malagasy folklore. The specific name *krausei* honors paleontologist David W. Krause.

CLASSIFICATION
Sauropoda
 Macronaria
  Titanosauriformes
  Somphospondyli
   Titanosauria
   Lithostrotia
    Eutitanosauria
    Saltasauroidea

LOCATION
Madagascar

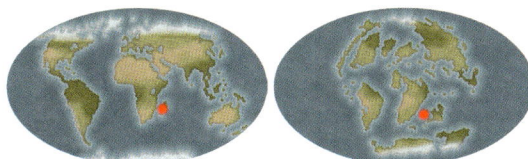

KNOWN REMAINS
Partial skull and skeleton

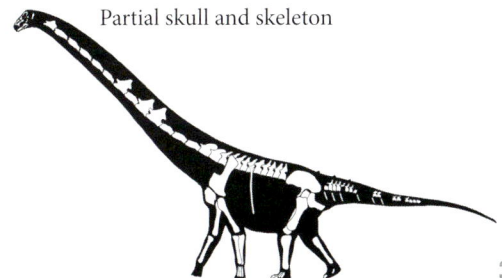

343

2 m   4 m   6 m   8 m   10 m   12 m

4 m

2 m

T. macedoi

(Zaher et al., 2011)
Length: 11 m (35 ft)
Height: 5 m (16.4 ft)
Hip height: 2.4 m (7.9 ft)
Body mass: 3,000 kg (3.3 t)
Reconstruction: ☐ ☐ ☐ ☐

250 245 240 235 230 225 220 215 210 205 200 195 190 185 180 175 170 165 160 155 150 145 140 135 130 125 120 **115** 110 105 100 95 90 85 80 75 70 65

TRIASSIC   **JURASSIC**   **CRETACEOUS**

Tapuiasaurus macedoi (meaning "Macedo's lizard of the tapuia") is the owner of the first titanosaur skull to be discovered in South America. The skull is nearly complete, missing only a region near the "forehead," although it has been slightly deformed by taphonomic processes—the left side (which was facing upward in the ground and thus was less deeply buried) is the most distorted, being compressed and slightly sheared. The skull still contained most of its teeth, as did the attached jaw. Based on unfused elements, the individual was likely not fully grown. The delicate hyoid bones were also preserved.

Other skeletal elements of the torso and limbs were also preserved but have been less thoroughly described, as preparation of the fossils has not yet been completed (Wilson et al., 2016).

Tapuiasaurus is most often recovered as being quite closely related to *Nemegtosaurus* and *Rapetosaurus*, despite living tens of millions of years before either of those genera. This places *Tapuiasaurus* as a very basal member of **Saltasauroidea** (Hechenleitner et al., 2020; Navarro et al., 2022; Santucci and Filippi, 2022). However, an argument has been made that this close grouping may be an artificial by-product of the fact that these are some of the only titanosaurs that preserve significant data regarding the skull; simply by virtue of having these data points, computational analyses might mistakenly group the species tighter than they otherwise might have, given complete data. As such, it is possible that *Tapuiasaurus* has a much more basal position within Titanosauria, possibly even being pre-lithostrotian (Wilson et al., 2016).

The generic name *Tapuiasaurus* derives from the term "tapuia," an indigenous Jês word for the native tribes that lived in the inner part of Brazil. The specific name *macedoi* honors site discoverer Ubirajara Alves Macedo.

CLASSIFICATION

Sauropoda
  Macronaria
   Titanosauriformes
    Somphospondyli
     Titanosauria
      Lithostrotia
       Eutitanosauria
        Saltasauroidea

LOCATION

Brazil

KNOWN REMAINS

Skull and partial skeleton

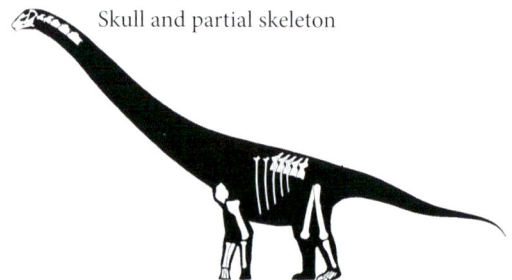

| 4 m | 8 m | 12 m | 16 m | 20 m | 24 m |

N. glaciaris

(Novas et al., 2019)
Length: 25 m (84 ft)
Height: 10.5 m (35.5 ft)
Hip height: 4.5 m (15 ft)
Body mass: 41,000 kg (45.2 t)
Reconstruction: ☐☐☐☐

8 m

4 m

| 250 | 245 | 240 | 235 | 230 | 225 | 220 | 215 | 210 | 205 | 200 | 195 | 190 | 185 | 180 | 175 | 170 | 165 | 160 | 155 | 150 | 145 | 140 | 135 | 130 | 125 | 120 | 115 | 110 | 105 | 100 | 95 | 90 | 85 | 80 | 75 | **70** | 65 |

| TRIASSIC | JURASSIC | CRETACEOUS |

Nullotitan glaciaris (meaning "Nullo's glacier giant") is known from bones that are mostly broken or fragmented. One unique aspect of the animal relates to its forward tail vertebrae, which have unusual pits and depressions; these may have been places for tendons to anchor, meaning that for some reason, *Nullotitan* had a different way of holding or moving its tail compared with any other known sauropod.

In 1980, geologist Francisco Nullo came across some dinosaur fossils in the field and alerted paleontologist José Bonaparte. The following year, Bonaparte recovered a portion of a neck bone from the site and cataloged it as belonging to an indeterminate species of the sauropod *Antarctosaurus*. Later, he interpreted the animal as being related to *Aeolosaurus* (Bonaparte et al., 2002).

In 2019, the original site was revisited, and a host of additional fossils were recovered. Sauropod fossils were found in five separate accumulations, spread over a wide area; many had been eroded out of their sandstone matrix and were collected as surface "float." No repeated skeletal elements were found, and their sizes were consistent with one another, meaning that it is possible that all of the remains came from a single individual. Altogether, these fossils formed the basis for the description of *Nullotitan*.

The describers failed to precisely account for *Nullotitan*'s phylogenetic position but suspected it to be colossosaurian. Somewhat contrastingly, a later study found it more likely to be the sister taxon of Saltasauridae (Cerda et al., 2021), consequently putting the genus within the reaches of **Saltasauroidea**.

The generic name *Nullotitan* honors fossil discoverer Francisco E. Nullo. The specific name *glaciaris* refers to the nearby Perito Moreno Glacier.

CLASSIFICATION
Sauropoda
  Macronaria
    Titanosauriformes
      Somphospondyli
        Titanosauria
          Lithostrotia
            Eutitanosauria
              Saltasauroidea

LOCATION
Argentina

KNOWN REMAINS
Fragments

345

O. skarzynskii

(Borsuk-Białynicka, 1977)
Length: 14 m (46 ft)
Height: 6.5 m (21 ft)
Hip height: 2.5 m (8.2 ft)
Body mass: 8,500 kg (9.4 t)
Reconstruction: ☐☐☐☐☐

| TRIASSIC | JURASSIC | CRETACEOUS |

Opisthocoelicaudia skarzynskii (meaning "Skarzynski's hollow-backed tail") is most often mentioned in connection with *Nemegtosaurus*, as the two sauropods are often speculated to actually be one and the same.

The holotype remains consist of a nearly complete skeleton, from the base of the neck back, making it one of the most complete specimens of any titanosaur. The remains were mostly preserved in articulation, which provides a wealth of anatomical data regarding posture and anatomical positioning. The body was belly-up and held some evidence of being fed upon by carnivores.

The fossils of *Opisthocoelicaudia* were found relatively near those of *Nemegtosaurus*. While *Nemegtosaurus* was described solely by cranial remains, *Opisthocoelicaudia* featured nearly everything except the cranium. With no overlapping material, it was effectively impossible to say for certain whether or not the sets of remains represented the same species.

When additional *Nemegtosaurus* fossils were finally described in 2018, a comparison between the two could finally be better attempted. The researchers concluded that the new femur attributed to *Nemegtosaurus* was effectively identical to that of *Opisthocoelicaudia*, and thus the two genera were synonymous with one another (Currie et al., 2018). However, a significant piece of information has thrown this identification into question, namely that the supposedly original *Opisthocoelicaudia* femur used in the 2018 study was actually not the same bone that researchers thought it was, having been a mislabeled mix-up (Averianov and Lopatin, 2019). Thus, the comparative identification saga has effectively been returned to square one.

The generic name *Opisthocoelicaudia* combines the Greek "opisthe" (meaning "backward" or "behind"), the Greek "koilos" (meaning "hollow"), and the Latin "cauda" (meaning "tail"). The specific name *skarzynskii* honors fossil preparator Wojciech Skarzynski.

CLASSIFICATION
Sauropoda
  Macronaria
    Titanosauriformes
    Somphospondyli
    Titanosauria
      Lithostrotia
      Eutitanosauria
        Saltasauroidea

LOCATION
Mongolia

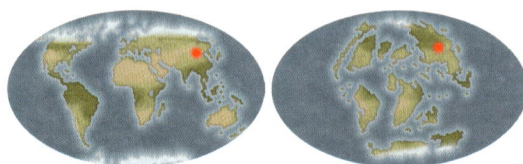

KNOWN REMAINS
Skeleton

2 m   4 m   6 m   8 m   10 m

4 m

N. mongoliensis

(Nowinski, 1971)
Length: 12.5 m (41 ft)
Height: 5 m (16.5 ft)
Hip height: 2.2 m (7.2 ft)
Body mass: 6,000 kg (6.6 t)
Reconstruction: ☐☐☐☐

2 m

250 245 240 235 230 225 220 215 210 205 200 195 190 185 180 175 170 165 160 155 150 145 140 135 130 125 120 115 110 105 100 95 90 85 80 75 **70** 65

| TRIASSIC | JURASSIC | CRETACEOUS |

Nemegtosaurus mongoliensis (meaning "lizard from Nemegt, Mongolia") is one of the few derived titanosaurs that preserves significant skull material and was among the first Late Cretaceous discoveries to do so. As such, with very few other specimens to compare against, *Nemegtosaurus* was initially suspected to be a diplodocid sauropod. As more and more subsequent discoveries were made, though, it eventually became evident that *Nemegtosaurus* was a titanosaur.

The skull and jaw, discovered in 1965, were the animal's only known remains for quite some time. As seems to be the norm among titanosaurs, the rear portion of the upper jaw bows upward toward the cheek rather noticeably. Also standard are the animal's peg-like teeth and its U-shaped jaw, as opposed to the blunt square jaw seen in a few titanosaurs.

Insufficient documentation resulted in the location of the discovery site becoming lost—until 2016, when it was rediscovered after some dedicated sleuthing, which included finding the abandoned remnants of the original quarrying gear and camping supplies. At the site, a femur, ankle bone, claw, and partial tail bone were recovered, and the ends of a tibia and fibula were found protruding from a tough sandstone cliff. It is believed that these bones and the skull came from the same individual (Currie et al., 2018). A set of dorsal vertebrae, which were found nearby in 1949 have now also been tentatively assigned to the genus (Averianov and Lopatin, 2019), along with another (undescribed) skull and jaw (Currie et al., 2018).

Although analyses tend to agree that *Nemegtosaurus* is a non-saltasaurid saltasauroid, pinning down a more precise phylogenetic position has proven challenging (Santucci and Filippi, 2022).

The animal's binomial name refers to the region known as the Nemegt Basin, in Mongolia.

CLASSIFICATION

Sauropoda
 Macronaria
  Titanosauriformes
   Somphospondyli
    Titanosauria
     Lithostrotia
      Eutitanosauria
       Saltasauroidea

LOCATION

Mongolia

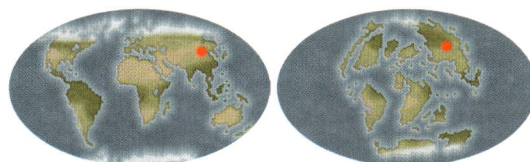

KNOWN REMAINS

Skull and jaw, partial leg, possible vertebrae

SALTASAUROIDEA

2 m   4 m   6 m   8 m

4 m

2 m

Q. orientalis

(Kurzanov and Bannikov, 1983)
Length: 9.5 m (31 ft)
Height: 4 m (13.1 ft)
Hip height: 1.8 m (5.9 ft)
Body mass: 2,500 kg (2.8 t)
Reconstruction: ☐ ☐ ☐ ☐

250 245 240 235 230 225 220 215 210 205 200 195 190 185 180 175 170 165 160 155 150 145 140 135 130 125 120 115 110 105 100 95 90 85 80 75 71 65

TRIASSIC   **JURASSIC**   **CRETACEOUS**

Quaesitosaurus orientalis (meaning "abnormal eastern lizard") is oddly neglected by researchers, given that titanosaur skulls are so rare and that *Quaesitosaurus* is known solely from cranial material.

The fossils in question (consisting of a jaw, the front portion of a skull, and the rear portion of a skull, missing the middle) were unearthed in 1971 during a joint Soviet-Mongolian expedition. This situation is quite similar to fellow titanosaur *Nemegtosaurus*, which is also known primarily from cranial remains and was discovered in the same general region. *Quaesitosaurus* was originally differentiated as being a separate genus from *Nemegtosaurus* based on several traits, one of which being the dimensions of the skull. However, it is now known that the *Nemegtosaurus* skull has been slightly distorted via sideways compression, whereas the *Quaesitosaurus* has been likewise deformed by top-to-bottom compression. These compressions mean that

the perceived differences in the shapes of the skulls might just be an artifact of fossilization and might not reflect their true anatomy (Wilson, 2005).

The two genera were also originally reported as being from slightly different rock formations, which would make them not quite the same age. But recent works have shifted some stratum boundaries, blurring the temporal line that supposedly separated the two sauropods (Currie et al., 2018). As such, it could very well be that *Quaesitosaurus* is synonymous with either *Nemegtosaurus* or *Opisthocoelicaudia*.

The generic name *Quaesitosaurus* is derived from the Latin "quaesitus" (meaning "abnormal" or "uncommon"). This name is based on a perceived abnormality of the skull, namely a large opening in the animal's inner ear; the opening has since been reinterpreted as a result of incomplete fossilization (Wilson, 2005). The specific name *orientalis* is Latin for "eastern."

CLASSIFICATION

Sauropoda
  Macronaria
    Titanosauriformes
    Somphospondyli
    Titanosauria
    Lithostrotia
    Eutitanosauria
    Saltasauroidea

LOCATION

Mongolia

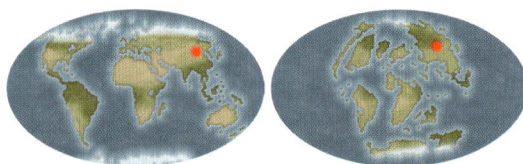

KNOWN REMAINS

Partial skull and jaw

3 m   6 m   9 m   12 m   15 m

6 m

A. kuehnei

(Vila et al., 2022)
Length: 17 m (57 ft)
Height: 7.5 m (24.5 ft)
Hip height: 3.5 m (11.5 ft)
Body mass: 14,000 kg (15.4 t)
Reconstruction: ☐ ☐ ☐ ☐

3 m

250 245 240 235 230 225 220 215 210 205 200 195 190 185 180 175 170 165 160 155 150 145 140 135 130 125 120 115 110 105 100 95 90 85 80 75 **70** 65

TRIASSIC   **JURASSIC**   **CRETACEOUS**

Abditosaurus kuehnei (meaning "Kühne's concealed lizard") is a moderately large titanosaur, which is somewhat unusual in this case since most European titanosaurs from this time were rather pint-sized.

During much of the Mesozoic, large portions of modern-day Europe were composed of small island-like landmasses. As such, with a limited food supply and a smaller habitat, species isolated on these islands tend to shrink, a process known as "insular dwarfism" or "island dwarfism." Numerous titanosaur species from southern Europe have been discovered that are the product of this phenomenon.

However, *Abditosaurus* did not suffer from diminished size, suggesting that the lineage to which it belongs is not native to the European islands. The phylogenetic study conducted by the describers seems to confirm this, as the animal's closest calculated relative is the African genus *Paralititan*, while the other European titanosaurs are not closely related.

The only known specimen of *Abditosaurus* was slowly excavated over several decades. Some remains were unearthed in 1954 and 1955, but subsequent excavations were canceled because of funding issues. Fresh attempts were made in 1984 and 1986, which were cut short partially due to bad weather. Finally, beginning in 2012, the site was fully explored, resulting in most of the animal's articulated neck being transported in a single gargantuan plaster jacket. (By this time, some of the original fossils that had been collected had unfortunately gone missing.) Histological analysis and the presence of ossified cartilaginous elements suggest that the animal was of a very advanced age when it perished.

The generic name *Abditosaurus* derives from the Latin "abditus" (meaning "concealed"), referring to the protracted nature of the specimen's recovery. The specific name *kuehnei* honors fossil discoverer, Walter Georg Kühne (1911–1991).

CLASSIFICATION

Sauropoda
 Macronaria
  Titanosauriformes
   Somphospondyli
    Titanosauria
     Lithostrotia
      Eutitanosauria
       Saltasauroidea
        Saltasauridae

LOCATION

Spain

KNOWN REMAINS

Partial skeleton

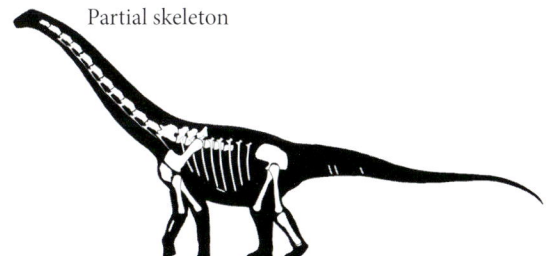

349

2 m   4 m   6 m   8 m   10 m

4 m

2 m

I. semkhu

(Gorscak et al., 2023)
Length: 10 m (33 ft)
Height: 3.5 m (11.5 ft)
Hip height: 2.7 m (8.9 ft)
Body mass: 3,000 kg (3.3 t)
Reconstruction: ☐ ☐ ☐ ☐

250 245 240 235 230 225 220 215 210 205 200 195 190 185 180 175 170 165 160 155 150 145 140 135 130 125 120 115 110 105 100 95 90 85 80 **73** 70 65

TRIASSIC   **JURASSIC**   **CRETACEOUS**

Igai semkhu (meaning "the forgotten lord of the oasis") is known from a poorly preserved set of remains that took nearly half a century to be fully described.

The fragmentary skeleton, consisting mainly of limb and torso elements, was collected in 1977 from Egypt by German paleontologists. From the outset, the remains were subjected to insufficient preservational procedures; over the ensuing decades, they endured further harm from inadequate storage conditions as they were shuffled between various German institutions. During this trek, numerous elements became lost or simply deteriorated to the point of disintegration; when they were finally studied in earnest, several bones were "preserved essentially as loosely compacted sediment." Original quarry maps indicate that at least 30 elements are unaccounted for.

Even in this crumbling state, though, *Igai* serves as an important data point for the understanding of the region's titanosaurs, as only a handful are known in total. Excluding the most basal titanosaurian taxa, less than half a dozen genera of African eutitanosaurs are currently known, so any information that can be gleaned from *Igai* is valuable information.

The describers classified *Igai* as a **saltasaurid**. Specifically, it appears to be more closely related to European titanosaurs than it is to species that hail from the southern regions of Africa. This conforms with a growing idea that there was some sort of equatorial divide in Africa that separated the northern and southern taxa during this time frame.

The generic name *Igai* is named for the enigmatic "lord of the oasis" deity known from some regions of ancient Egypt. The specific name *semkhu* derives from the ancient Egyptian "semekh" (meaning "to forget"), alluding to the extended period of time in which the holotype specimen remained uncared for in storage.

CLASSIFICATION
Sauropoda
 Macronaria
  Titanosauriformes
   Somphospondyli
    Titanosauria
     Lithostrotia
      Eutitanosauria
       Saltasauroidea
        Saltasauridae

LOCATION
Egypt

KNOWN REMAINS
Partial skeleton

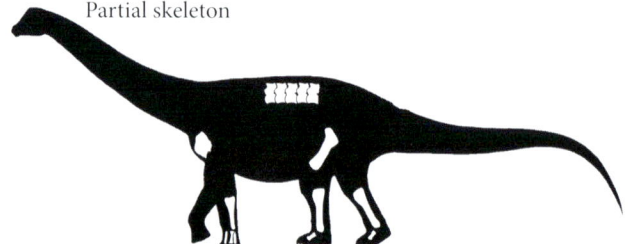

4 m   8 m   12 m   16 m   20 m   24 m   12 m

A. sanjuanensis

(Gilmore, 1922)
Length: 26 m (85 ft)
Height: 11 m (36 ft)
Hip height: 4.5 m (15 ft)
Body mass: 38,000 kg (41.9 t)
Reconstruction: ☐☐☐☐

8 m

4 m

250 245 240 235 230 225 220 215 210 205 200 195 190 185 180 175 170 165 160 155 150 145 140 135 130 125 120 115 110 105 100 95 90 85 80 75 70 **65**

TRIASSIC   **JURASSIC**   **CRETACEOUS**

Alamosaurus sanjuanensis (meaning "Alamo lizard from San Juan") is notable for not only being the sole North American titanosaur currently known but also for being the first sauropod to reappear in North America after the "sauropod hiatus," a roughly 30-million-year period during which sauropods seemingly suffered a regional extinction.

This reappearance begs the question: From where did *Alamosaurus*, or its direct ancestors, immigrate? Numerous phylogenetic studies have found *Alamosaurus* to be most closely related to the Asian taxon *Opisthocoelicaudia* (Poropat et al., 2021; Santucci and Filippi, 2022), suggesting that a crossing utilizing the Bering Land Bridge (e.g., Beringia) would be the most likely origin. However, no northern sauropod fossils from that age have been found, and sauropods in general are thought to have been ill-equipped for high-latitude environments (Chiarenza et al., 2022).

The alternative is that South America is the most likely location for *Alamosaurus*'s ancestral origin, as some studies

have found *Alamosaurus* to be most closely related to taxa from that continent (Navarro et al., 2022), especially those studies that incorporate the South American genus *Pellegrinisaurus* (Cerda et al., 2021).

Alamosaurus is known from numerous fossil specimens. Many of these consist of individual bones, but several feature remains that are much more complete (Jasinski et al., 2011). Bones attributed to the genus have been described from New Mexico, Utah, and Texas; possible specimens have also been found in Wyoming, Montana, and Mexico.

Whereas most titanosaurs are thought to have had a wide-legged stance, traits of the femur of *Alamosaurus* seem to suggest that it had a much narrower stance (Wick and Lehman, 2014).

The generic name *Alamosaurus* refers to the Ojo Alamo geological formation. The specific name *sanjuanensis* refers to San Juan County, New Mexico.

CLASSIFICATION

Sauropoda
  Macronaria
   Titanosauriformes
    Somphospondyli
     Titanosauria
      Lithostrotia
       Eutitanosauria
        Saltasauroidea
         Saltasauridae

LOCATION

North America

KNOWN REMAINS

Nearly complete skeleton

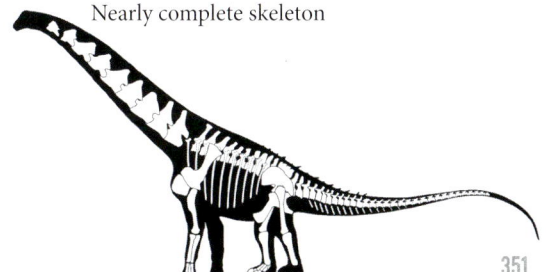

351

SALTASAUROIDEA

3 m   6 m   9 m   12 m   15 m

6 m

3 m

P. powelli

(Salgado, 1996)
Length: 15 m (49 ft)
Height: 6.5 m (21 ft)
Hip height: 3 m (9.8 ft)
Body mass: 8,000 kg (8.8 t)
Reconstruction: ☐ ▯ ▮ ▮

250 245 240 235 230 225 220 215 210 205 200 195 190 185 180 175 170 165 160 155 150 145 140 135 130 125 120 115 110 105 100 95 90 85 80 75 **70** 65

TRIASSIC   **JURASSIC**   **CRETACEOUS**

Pellegrinisaurus powelli (meaning "Powell's lizard from Pellegrini") is a medium-to-large-sized sauropod that possessed some slightly unusual vertebrae. The central portions of its back bones are twice as wide as they are tall; additionally, the neural spines of the tail vertebrae are so low and long that they effectively interlock with the subsequent bones, stiffening the joints between them.

The only known specimen of *Pellegrinisaurus* was discovered in 1975 by Roberto Abel and Jaime Emilio Powell of the Museo Carlos Ameghino of Cipolletti. The sand quarry that yielded the remains was, at the time, considered to be part of the Allen geological formation but is now seen as being a part of the Anacleto Formation, which has also produced several other titanosaur genera.

When Powell briefly described the specimen in 1986, he initially identified it as an indeterminate species of *Epachthosaurus*. The remains consisted of four dorsal (back)

vertebrae, 26 caudal (tail) vertebrae, and part of a femur; this is somewhat mysterious, as original photos from the excavation clearly show more skeletal material being present, such as sacral (hip) vertebrae. The fate of this material is undocumented. Separately, a lone tail vertebra from another locality was once assigned to the genus (Salgado and Bonaparte, 2007) but is no longer considered as such (Cerda et al., 2021).

Phylogenetically, *Pellegrinisaurus* has found itself in several taxonomic positions throughout various analyses. The growing consensus, though, seems to be that the genus is most closely related to *Alamosaurus* and *Baurutitan* (Cerda et al., 2021), which, by some definitions, would make a **saltasaurid** (Navarro et al., 2022).

The generic name *Pellegrinisaurus* refers to Pellegrini Lake. The specific name *powelli* honors paleontologist Jaime E. Powell.

CLASSIFICATION

Sauropoda
 Macronaria
  Titanosauriformes
  Somphospondyli
   Titanosauria
    Lithostrotia
     Eutitanosauria
      Saltasauroidea

LOCATION

Argentina

KNOWN REMAINS

Partial Skelton

1 m 2 m 3 m 4 m 5 m 6 m

2 m

I. parva

(Navarro et al., 2022)
Length: 6 m (19 ft)
Height: 2 m (6.5 ft)
Hip height: 1.5 m (5 ft)
Body mass: 2,000 kg (2.2 t)
Reconstruction:

1 m

250 245 240 235 230 225 220 215 210 205 200 195 190 185 180 175 170 165 160 155 150 145 140 135 130 125 120 115 110 105 100 95 90 85 80 75 70 65

TRIASSIC	JURASSIC	CRETACEOUS

Ibirania parva (meaning "little tree wanderer") is a miniature (or "nanoid") sauropod. Unlike some dwarf sauropod species from Europe, whose diminished sizes were likely due to their restrictively small geographic habitat, *Ibirania* lived on a relatively large landmass, so its small size must have come about for different reasons. The describers surmised that the likely culprit was the aridity of the environment, which resulted in a reduced quantity of vegetation, which in turn necessitated a smaller body that required less food.

Researchers know that the fossils of *Ibirania* are not from immature, still-growing individuals for two main reasons. The first is that the recovered vertebral elements feature complete fusion between their "spine" and "centrum" components, which is a widely accepted indicator of skeletal maturity. Secondly, the researchers conducted a histological

examination, in which the growth patterns in the interior of the limb bones were analyzed, finding that the specimens were mature.

Ibirania is represented by a handful of fragmentary and eroded remains that originate from at least four separate individuals. The specimens were found within a single rock outcrop.

Phylogenetic analysis places *Ibirania* within **Saltasaurinae**, along with several other diminutive South American taxa.

The generic name *Ibirania* derives from the name of the municipality where the fossils were discovered, Ibirá, which itself comes from the Tupi word "ybyrá," which means "tree" or "wood." This is combined with "ania," a modification of the Greek "plania" (meaning "wander"). The specific name *parva* is a feminized version of the Latin "parvus" (meaning "little").

CLASSIFICATION
Sauropoda
 Macronaria
  Titanosauriformes
   Somphospondyli
    Titanosauria
     Lithostrotia
      Eutitanosauria
       Saltasauroidea
        Saltasauridae

LOCATION
Brazil

KNOWN REMAINS
Fragments

2 m

1 m

B. reigi

(Martinelli and Forasiepi, 2004)
Length: 6 m (20 ft)
Height: 2 m (6.5 ft)
Hip height: 1.5 m (5 ft)
Body mass: 2,000 kg (2.2 t)
Reconstruction: ☐☐☐☐

| 250 | 245 | 240 | 235 | 230 | 225 | 220 | 215 | 210 | 205 | 200 | 195 | 190 | 185 | 180 | 175 | 170 | 165 | 160 | 155 | 150 | 145 | 140 | 135 | 130 | 125 | 120 | 115 | 110 | 105 | 100 | 95 | 90 | 85 | 80 | 75 | 70 | 65 |

| TRIASSIC | JURASSIC | CRETACEOUS |

Bonatitan reigi (meaning "Bonaparte's and Reig's giant") is known from a series of disarticulated remains that were collected between 1990 and 1994. Along with these sauropod bones, a smattering of fossils from many different organisms were recovered from this former lagoonal-tidal ecosystem, including reptiles, plesiosaurs, other dinosaurs such as hadrosaurids, and theropods such as dromaeosaurs and carcharodontosaurids, as well as numerous sauropod eggs.

Altogether, the *Bonatitan* remains were originally thought to belong to two separate individuals. However, a more comprehensive analysis of the bones' sizes revealed that at least five individuals were represented by the material. This reevaluation also described several bones that were not mentioned in the original description and corrected a few misidentifications among the elements. The largest individuals were likely still subadults (Salgado et al., 2014).

The remains include two separate braincases, which have been analyzed via CT scan. Results indicate that *Bonatitan* likely had a poor sense of smell and a reduced range of possible head movements (Carabajal, 2012).

Bonatitan often finds itself excluded from taxonomic studies. The describers considered it to be a **saltasaurid**, and at least one subsequent paper has agreed with this assessment (Navarro et al., 2022). This view is not universal, however, as the authors responsible for the species' redescription recovered it in a more basal position, in a clade positioned as the sister of Saltasauridae (Salgado et al., 2014), which would put it within Saltasauroidea.

The generic name *Bonatitan* honors paleontologist Jose F. Bonaparte. The specific name *reigi* honors paleontologist Osvaldo Reig. Both individuals are responsible for paleontological advancements made specifically in South America.

CLASSIFICATION
Sauropoda
 Macronaria
  Titanosauriformes
   Somphospondyli
    Titanosauria
     Lithostrotia
      Eutitanosauria
       Saltasauroidea
        Saltasauridae

LOCATION
Argentina

KNOWN REMAINS
Partial skull and skeleton

1 m	2 m	3 m	4 m	5 m	6 m	7 m	8 m

N. australis

(Powell, 1992)
Length: 8 m (26 ft)
Height: 2.5 m (8.2 ft)
Hip height: 2 m (6.5 ft)
Body mass: 2,300 kg (2.5 t)
Reconstruction: ☐☐☐☐

3 m

2 m

1 m

250–245–240–235–230–225–220–215–210–205–200–195–190–185–180–175–170–165–160–155–150–145–140–135–130–125–120–115–110–105–100–95–90–85–**80**–75–70–65

TRIASSIC | **JURASSIC** | **CRETACEOUS**

The history of **Neuquensaurus australis** (meaning "southern lizard from Neuquen") is inexorably tied up with that of *Titanosaurus*.

The genus *Titanosaurus* is now largely known as a wastebasket taxon—a convenient and inaccurate label that was used in the past to expediently identify numerous fossils that were, it would turn out, not actually from the same kind of animal. Many species within *Titanosaurus* were erected, including *T. australis* (Lydekker, 1893) and *T. robustus* (von Huene, 1929). Predictably, many disparate fossils were labeled as one of these species over the ensuing decades.

Eventually, as steps were taken to straighten out the mess that was *Titanosaurus*, the genus *Neuquensaurus* was proposed for both aforementioned species. Even later still, researchers painstakingly showed how most of the so-called *Neuquensaurus* (i.e., "*T. australis* and *T. robustus*") specimens had been assigned hastily, without rigorous

rationale, and as such, *N. robustus* was rendered a dubious taxon and *N. australis* was whittled down to just a handful of specimens (D'Emic and Wilson, 2011).

The most noteworthy aspect of the animal's anatomy has to do with the set of vertebrae, called the sacrum, that fuse together in order to connect the hip bones. This structure in most titanosaurs is composed of six vertebrae, but *Neuquensaurus* instead has seven, as the bone that would ordinarily be the first vertebra of the tail has instead become incorporated into the sacrum.

Neuquensaurus is often recovered as the sister taxon of *Saltasaurus* within **Saltasauridae** (Navarro et al., 2022), and fossils from both genera have been uncovered from the same localities.

The generic name *Neuquensaurus* refers to the province of Neuquen, Argentina. The specific name *australis* is Latin for "south."

CLASSIFICATION
Sauropoda
 Macronaria
  Titanosauriformes
   Somphospondyli
    Titanosauria
     Lithostrotia
      Eutitanosauria
       Saltasauroidea
        Saltasauridae

LOCATION
Brazil

KNOWN REMAINS
Partial skeleton

S. loricatus

(Bonaparte and Powell, 1980)
Length: 9 m (29 ft)
Height: 3.2 m (10.5 ft)
Hip height: 2 m (6.5 ft)
Body mass: 3,000 kg (3.3 t)
Reconstruction:

TRIASSIC	JURASSIC	CRETACEOUS

Saltasaurus loricatus (meaning "armored lizard from Salta") is arguably most famous for its "armor." Several large-ish "osteoderms" and a few patches of smaller "ossicles" have been described among the known *Saltasaurus* remains, and numerous other titanosaurs are known to have possessed osteoderms of various sizes and structures. So, what purpose did they serve?

Contrary to popular depiction, *Saltasaurus* probably did not rely upon its osteoderms for protection against predators or competitors the way that *Ankylosaurus* likely would have; while these bony knobs would have undoubtedly offered some small level of defense, they were too few in number and likely too far apart to have served as truly effective "armor" (Curry-Rogers et al., 2011).

An alternate hypothesis is that titanosaur osteoderms were used as storage sites for excess minerals. Modern female crocodiles are known to draw upon mineral reserves from their osteoderms when producing eggs (Chinsamy et al., 2016), and titanosaurs are known to have been prodigious egg layers. Since saltasaurids in particular had very pneumatized (i.e., hollow) bones, it could be that drawing minerals from traditional skeletal storage would have been insufficient for nesting females (Cerda et al., 2015). Some titanosaur specimens have even been discovered with hollowed osteoderms, interpreted as having been depleted of their primary reserves (Vidal et al., 2017).

The only known *Saltasaurus* specimens come from a jumbled bone bed, which is composed of at least five adult and subadult individuals (Powell, 2003). It is possible, however, that some of these remains actually belong to *Neuquensaurus* (D'Emic and Wilson, 2011).

The generic name *Saltasaurus* refers to the Salta Province of Argentina. The specific name *loricatus* is a Latin term for "armored."

CLASSIFICATION
Sauropoda
 Macronaria
  Titanosauriformes
   Somphospondyli
    Titanosauria
     Lithostrotia
      Eutitanosauria
       Saltasauroidea
        Saltasauridae

LOCATION
Argentina

KNOWN REMAINS
Partial skeleton

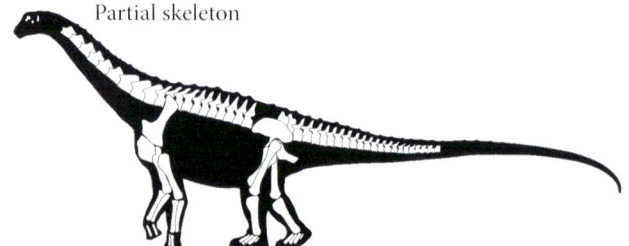

Q. youjiangensis

(Mo et al., 2008)
Length: 12 m (40 ft)
Height: 4.5 m (15 ft)
Reconstruction:

2 m

70

TRIASSIC   JURASSIC   CRETACEOUS

Qingxiusaurus youjiangensis is a Chinese sauropod known from scant remains that were discovered in 1991, including just the sternum, upper arms, and one vertebral fragment (a neural spine that is rather flat and "paddle-like").

The authors thought it probable that *Qingxiusaurus* was a **saltasaurid**. At least one subsequent study has arrived at the same conclusion (Averianov and Sues, 2017).

The generic name *Qingxiusaurus* derives from "Qingxiu" (which is short for "shangqingshuixiu"), which refers to "a picturesque scenery of mountains and water in Guangxi." The specific name *youjiangensis* refers to the Youjiang River.

ZHUCHENGTITAN

Z. zangjiazhuangensis

(Mo et al., 2017)
Length: 15 m (49 ft)
Height: 6 m (20 ft)
Reconstruction:

2 m

73

TRIASSIC   JURASSIC   CRETACEOUS

Zhuchengtitan zangjiazhuangensis is known from a single nearly complete humerus that was unearthed in 2008. A few other sauropod fragments were also recovered, but these were undiagnostic and, as such, cannot be identified with any certainty. Much more common at the locality are the remains of hadrosaurs, particularly *Shantungosaurus*.

The describers categorized *Zhuchengtitan* as belonging to **Saltasauridae**, comparing the shape of the humerus to that of *Saltasaurus*.

The animal's binomial name refers to the city of Zhucheng, China, and a quarry located near the village of Zangjiazhuang.

R. muniozi

(Salgado and Azpilicueta, 2000)
Length: 8 m (26 ft)
Height: 3 m (9.8 ft)
Reconstruction:

1 m

TRIASSIC | JURASSIC | CRETACEOUS

67

Rocasaurus muniozi is a small saltasaurid that shares many characteristics with its more famous cousin, *Saltasaurus*. Indeed, the two genera are universally recovered as being quite closely related (Hechenleitner et al., 2020; Navarro et al., 2022).

Rocasaurus is known from several fragmentary specimens discovered from the same general locality in Argentina. Aside from the femur and bones of the hip, only disarticulated vertebrae are known (Garcia and Salgado,

2012; Rolando et al., 2022). Even so, these bones are informative, as it appears that *Rocasaurus* had the most extensively pneumatized skeleton of any known saltasaurid, a group already well-known for taking this trait to the extreme (Zurriaguz and Cerda 2017).

The generic name *Rocasaurus* refers to the city of General Roca, in Rio Negro Province, Patagonia. The specific name *muniozi* honors museum attendant Juan Carlos Muñoz.

Y. lojaensis

(Apesteguía et al., 2019)
Length: 7 m (23 ft)
Height: 2.5 m (8.2 ft)
Reconstruction:

1 m

TRIASSIC | JURASSIC | CRETACEOUS

67

Yamanasaurus lojaensis (meaning "lizard from Yamana, Loja") is the first nonavian dinosaur to be described and named from Ecuador. The very fragmentary remains were first unearthed by a local farmer.

Based on the animal's size and proportions, the describers considered it likely that *Yamanasaurus* was a part of

Saltasaurinae, which would make it the northernmost representative of that group. Much like other saltasaurids, *Yamanasaurus* was quite small (for a sauropod), with stocky proportions.

The animal's binomial name refers to the region Yamana, located in the Province of Loja, Ecuador.

B. shiva

(Simón and Salgado, 2023)
Length: 30 m (98 ft)
Height: 14.6 m (48 ft)
Reconstruction: ☐☐☐☐

2 m

95

TRIASSIC	JURASSIC	CRETACEOUS

Bustingorrytitan shiva (meaning "Bustingorry's giant of Shiva") is one of the largest titanosaurs, featuring the second-longest humerus of any titanosaur. The bones (consisting of the partial remains of at least four individuals) were discovered prior to 2001. The describers of *Bustingorrytitan* estimated its mass as being 61,054 kilograms (67.3 tons), but this figure was calculated using only the size of the femur, and such measurements have

been known to give overinflated values in other sauropods. Taxonomically, the genus was found to be sister to Saltasauridae.

The generic name *Bustingorrytitan* honors landowner Manuel Bustingorry. The specific name refers to the Hindu deity Shiva, who "destroys and transforms the Universe," alluding to the "faunal turnover that occurred in the middle of the Cretaceous period."

T. gimenezi

(Pérez-Moreno et al., 2024)
Length: 9.6 m (31.5 ft)
Height: 5 m (16.4 ft)
Reconstruction: ☐☐☐

1 m

99

TRIASSIC	JURASSIC	CRETACEOUS

Titanomachya gimenezi is a small yet robust saltasauroid titanosaur. Unlike many titanosaur species, it is described primarily on the basis of limb bones rather than vertebral elements. Although of a similar size as *Saltasaurus*, *Titanomachya* was likely heavier, judging from its known proportions, including a distinctly wide ankle bone.

The animal's generic name refers to the Titanomachy, the battle in Greek mythology fought by the Olympians against the Titans. The specific name *gimenezi* honors paleontologist Olga Giménez.

U. celeste

(Soto et al., 2024)
Length: 15 m (49 ft)
Height: 7 m (23 ft)
Reconstruction: ☐ ☐ ☐ ☐

2 m

95?

TRIASSIC	JURASSIC	CRETACEOUS

Udelartitan celeste (meaning "UdelaR's sky-blue giant") is the first nonavian dinosaur to be named from Uruguay, although titanosaurian remains have been known from the country for quite some time. The mostly disarticulated fossils of *Udelartitan* were first reported upon in 2012 (Soto et al., 2012). Unusually (but not uniquely) its first tail vertebra was biconvex in shape.

The generic name *Udelartitan* derives from the acronym for the Universidad de la República, "UdelaR." The specific name *celeste* is Spanish for "sky blue," a color worn by prominent Uruguayan athletic teams.

Q. pintiquiniestra

(Mocho et al., 2024)
Length: 11.2 m (38 ft)
Height: 6 m (19.7 ft)
Reconstruction: ☐ ☐ ☐ ☐

2 m

72

TRIASSIC	JURASSIC	CRETACEOUS

Qunkasaura pintiquiniestra (meaning "Pintiquiniestra's lizard from Qunca") is one of multiple titanosaur taxa found together in a single bonebed in 2007, showing that more than one saltasauroid lineage was present during this time in what is now Spain. Phylogenetic analysis places it as the sister taxon of *Abditosaurus*.

A *nomen dubium* (plural *nomina dubia*) is a "doubtful name." A taxonomic genus or species may be relegated to this dubious status if it is no longer considered to be a useful or accurate identifier. For example, numerous dinosaur genera were erected in the 1800s and early 1900s based on anatomical traits that, at the time, were considered to be unique to that creature; however, oftentimes it turns out that those traits are actually widespread across numerous different organisms, and consequently, since the original definition is no longer distinguishable or useful, it may be rendered *nomen dubium*. Even in modern times, researchers may disagree on whether or not certain traits warrant the distinction of erecting a new genus or species, and a new taxon might not be recognized by subsequent works.

Some particularly distinctive examples have been included in previous pages, as have genera whose status as *nomen dubium* has not been universally accepted. The remaining dubious sauropodomorph genera of note are detailed below.

Aepisaurus is known from a humerus (MNHN1868–242), a fossil that is now lost. Two other limb bones and a crocodilian tooth were originally referred to the genus but were later removed (Le Loeuff, 1993). *Aepisaurus* could be camarasaurid or titanosaurian in nature (McIntosh, 1990). A common misspelling in literature is "Aepysaurus." *Aepisaurus* derives from the Greek "aipys" (meaning "high").

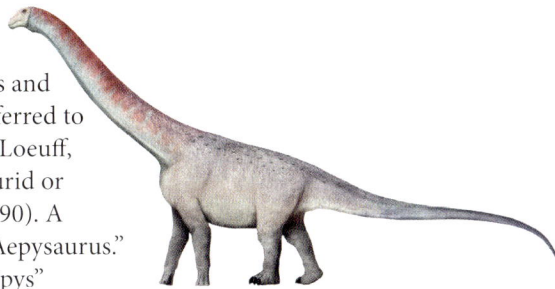

Aepisaurus elephantinus

(Gervais, 1852)
Location: France
Occurrence: Grès verts du Mont Ventoux
Age: 100 MYA (Early Cretaceous)

Asiatosaurus is a "tooth taxon," a genus described on the basis of a single tooth. This practice was fairly common in the 1800s and early 1900s but is now generally frowned upon, as a tooth most likely lacks a sufficient level of defining characteristics. In addition to the holotype (AMNH 6264), other teeth and some vertebrae have also been referred to the genus (Dong, 1973). A second species, A. kwangshiensis, was erected based on teeth, fragmented ribs, and partial vertebrae from the Napai Formation of China (Hou et al., 1975). This material has been interpreted as being euhelopodid in nature (Poropat et al., 2022).

Asiatosaurus mongoliensis

(Osborn, 1924)
Location: Mongolia
Occurrence: Öösh Fm.
Age: 120 MYA (Early Cretaceous)

Balochisaurus was described based on seven caudal vertebrae. Numerous other skeletal elements were subsequently referred to the genus (Malkani, 2021). According to researcher M. Sadiq Malkani, *Balochisaurus* was just one of numerous distinct Pakistani titanosaurs. However, others have regarded the specimens as being dubiously valid, and possibly synonymous with *Jainosaurus* or *Isisaurus* (Wilson et al., 2011). The generic name refers to the Baloch tribes of Pakistan.

Balochisaurus malkani

(Malkani, 2004)
Location: Pakistan
Occurrence: Vitakri Fm.
Age: 70 MYA (Late Cretaceous)

Bothriospondylus was first described based on four dorsal vertebrae. As with many genera from this period in history, the name became a wastebasket taxon, with numerous species being assigned that were based on disparate remains. Most have been reclassified to their own distinct genera or rendered *nomen dubium*, which is the case for the uninformative bones of the type species (Galton and Upchurch, 2004). The generic name combines the Greek "bothrion" (meaning "furrowed"), and "spondylos" (meaning "vertebra"). The specific name is Latin for "undermined."

Bothriospondylus suffossus

(Owen, 1875)
Location: United Kingdom
Occurrence: Kimmeridge Clay
Age: 155 MYA (Late Jurassic)

Brohisaurus was described on the basis of some very fragmentary pieces of ribs, vertebrae, and limb bones. It was proposed as a titanosaur, but later work has suggested indeterminate titanosauriform affinities (Mannion et al., 2013). The name refers to the Indigenous Brohi people and the Kirthar Mountains.

Brohisaurus kirthari

(Malkani, 2003)
Location: Pakistan
Occurrence: Sembar Fm.
Age: 150 MYA (Late Jurassic)

Bruhathkayosaurus is known from several limb and pelvic bones, as well as a tail vertebra. It was originally interpreted as a theropod, which prompted the bombastic claim of it being the largest-ever theropod. The proportions of the absolutely enormous tibia would put *Bruhathkayosaurus* in contention as being the largest-ever land animal … unless the bone was really just a badly damaged femur (Paul, 2019). The fossils were already in poor condition when they were excavated, and they evidently disintegrated before they could be extracted in the laboratory, meaning that there will forever be unanswered questions about the animal (Galton and Ayyasami, 2017). While the fossils were likely genuine (and not a case of misidentified petrified wood) and titanosaurian in nature (Pal and Ayyasami, 2022), there's no guarantee that the remains all came from the same animal, or even genus. The name originates from the Sanskrit word "bruhathkaya" (meaning "huge body") and honors British paleontologist Charles Alfred Matley.

Bruhathkayosaurus matleyi

(Yadagiri and Ayyasami, 1987)
Location: India
Occurrence: Trichinopoly Grp.
Age: 70 MYA (Late Cretaceous)

Campylodoniscus was originally known as *Campylodon*, but it turned out that this name was preoccupied by a kind of fish, prompting a renaming by Haubold and Kuhn (1961). The generic name comes from the Greek terms "kampylos" and "odon" (meaning "bent tooth"). The specific name honors Argentine paleontologist Florentino Ameghino. The genus is known only from a small fragment of the upper jaw and a single tooth. It is potentially a titanosaur (Martínez et al., 2016).

Campylodoniscus ameghinoi

(von Huene, 1929)
Location: Argentina
Occurrence: Bajo Barreal Fm. (?)
Age: 95 MYA (Late Cretaceous) (?)

Cardiodon is based on a single tooth, and a few additional teeth have since been referred to the genus. The genus has occasionally been synonymized with *Cetiosaurus* (Steel, 1970). The inward-curving teeth could belong to some sort of turiasaurian (Royo-Torres et al., 2006). The generic name means "heart tooth," and *rugulosus* is Latin for "having little wrinkles."

Cardiodon rugulosus

(Owen, 1840)
Location: United Kingdom
Occurrence: Forest Marble
Age: 197 MYA (Middle Jurassic)

Chiayusaurus is yet another genus based on a single tooth, with a few additional teeth having since been referred to the genus. The tooth has variously been considered to be near indistinguishable from *Mamenchisaurus* (Russell and Zheng, 1993) or *Euhelopus* (Barrett et al., 2002). Even so, yet another species, C. asianensis, was later erected (Lee et al., 1997). The generic name refers to the city of Jiayuguan, China.

Chiayusaurus lacustris

(Bohlin, 1953)
Location: China
Occurrence: Kalazha Fm.
Age: 120 MYA (Early Cretaceous)

Chondrosteosaurus (meaning "gigantic cartilage-and-bone lizard") is known from two cervical vertebral centra. The name stems from Richard Owen's belief that the bones' pneumatic air chambers had instead once held cartilage. A second species, C. magnus, has been synonymized with *Ornithopsis* (Hulke, 1879).

Chondrosteosaurus gigas

(Owen, 1876)
Location: United Kingdom
Occurrence: Wessex Fm.
Age: 125 MYA (Early Cretaceous)

Clasmodosaurus is known only from three individual teeth. Based on when and where the animal lived, it was most likely a titanosaur (Powell, 2003). The teeth have some features in common with those of *Bonitasaura* (Gallina and Apesteguía, 2011). The generic name comes from the Greek "klasma" ("fragment") and "odon" ("tooth").

Clasmodosaurus spatula

(Ameghino, 1898)
Location: Argentina
Occurrence: Mata Amarilla Fm.
Age: 95 MYA (Late Cretaceous)

Euskelosaurus was described based on vertebral, pelvic, and limb elements (BMNH 1625) that are plateosaurid in nature. However, it eventually became a bit of a wastebasket taxon (Krupandan, 2019), with numerous sets of partial remains being assigned to the genus. Some have since been reclassified, such as the holotype remains of *Antetonitrus*. The binomial name means "good limb lizard" and honors fossil discoverer Alfred Brown.

Fulengia is known from the remains of a very young juvenile (CUP 2037), consisting primarily of the skull. It was first interpreted as *Yunnanosaurus* (Simmons, 1965), then as a lizard (Carroll and Galton, 1977), and then as a sauropodomorph (Evans and Milner, 1989). The genus was rendered *nomen dubium* by Galton and Upchurch (2004).

Gigantosaurus was originally described on the basis of two tail vertebrae, two limb bones, and a foot claw. These remains were found separately, so in truth, they cannot be attributed to the same animal or even the same genus. Nevertheless, *Gigantosaurus* became a popular name, with several different species eventually being erected, most of which have now been more properly reassigned.

Gigantoscelus (meaning "giant shin") is known from the lowermost fragment of a single femur (TrM 65). In 1979, van Heerden considered it to be synonymous with *Euskelosaurus*. The genus was rendered *nomen dubium* by Galton and Upchurch (2004). The specific name honors geologist Gustaaf Adolf Frederik Molengraaff.

Gresslyosaurus (named for paleontologist Amanz Gressly) has had four species attributed to it: *G. ingens*, *G. plieningeri* (von Huene, 1905), *G. robustus* (von Huene, 1905), and *G. torgeri* (Jaekel, 1911). Most often, the genus is now considered to be synonymous with *Plateosaurus* (Galton, 1986), with some species *nomen dubium* (Galton and Upchurch, 2004), although some recent works have again suggested validity (Rauhut et al., 2020).

Gryponyx (meaning "curved claw") has contained three species. *G. transvaalensis* and *G. taylori*, based on scant remains, are well established as *nomen dubium*, likely synonymous with *Massospondylus*. The status of the type species, *G. africanus*, is less certain. It is known from a mostly complete skeleton, lacking the cranium. Some researchers have concluded that it is also a dubious species (Galton and Upchurch, 2004), but others have concluded that it is indeed distinctive (Lü et al., 2010; Yates et al., 2010).

Gspsaurus (referring to the "Geological Survey of Pakistan") is, according to researcher M. Sadiq Malkani, just one of numerous distinct Pakistani titanosaur genera. However, many other researchers have regarded the specimens as being dubiously valid. *Gspsaurus* was initially founded based on skull fragments but has had numerous other elements referred to it (Malkani, 2021).

Hortalotarsus is known from several hindleg elements (AM 455). Reportedly, more of the skeleton was originally present but was destroyed during attempted collection. *Hortalotarsus* has variously been considered *nomen dubium* (Galton and Upchurch, 2004) or synonymous with *Massospondylus* (Barrett et al., 2019).

Euskelosaurus browni

(Huxley, 1866)
Location: South Africa
Occurrence: Lower Elliot Fm.
Age: 210 MYA (Late Triassic)

Fulengia youngi

(Carroll and Galton, 1977)
Location: China
Occurrence: Lufeng Fm.
Age: 190 MYA (Early Jurassic)

Gigantosaurus megalonyx

(Seeley, 1869)
Location: United Kingdom
Occurrence: Kimmeridge Clay
Age: 150 MYA (Late Jurassic)

Gigantoscelus molengraaffi

(van Hoepen, 1916)
Location: South Africa
Occurrence: Bushveld Ss. Fm.
Age: 190 MYA (Early Jurassic)

Gresslyosaurus ingens

(Rütimeyer, 1857)
Location: Switzerland
Occurrence: Trossingen Fm.
Age: 210 MYA (Late Triassic)

Gryponyx africanus

(Broom, 1911)
Location: South Africa
Occurrence: Upper Elliot Fm.
Age: 195 MYA (Early Jurassic)

Gspsaurus pakistani

(Malkani, 2015)
Location: Pakistan
Occurrence: Pab Fm.
Age: 66 MYA (Late Cretaceous)

Hortalotarsus skirtopodus

(Seeley, 1894)
Location: South Africa
Occurrence: Clarens Fm.
Age: 193 MYA (Early Jurassic)

Hypselosaurus (meaning "lofty lizard") is a potential titanosaur that was originally described from two tail vertebrae and three fragmentary limb bones that were first unearthed sometime prior to 1846. These fossils have since gone missing from museum archives (Le Loeuff, 1993). A number of other partial specimens were referred to the genus over time. It is now largely regarded as a dubious genus (Galton and Upchurch, 2004).

"Ischyrosaurus" (meaning "strong lizard") was the name given to a genus of sauropod in 1874, before it became known that this name was already in use. The species *manseli*, based on a single humerus (NHMUK R41626), has variously been attributed to *Ornithopsis* and *Pelorosaurus*. A more recent study found that the bone had potential rebbachisaurid or titanosauriform features (Barrett et al., 2010).

Iuticosaurus (referring to the Germanic Jute tribe), based on two partial tail vertebrae, was originally noted as being similar to *Titanosaurus blanfordi* or *Ornithopsis* (Lydekker, 1887). It was later erected to the new species *Titanosaurus valdensis* (von Huene, 1929), and then to the new genus *Iuticosaurus* (along with a third tail bone), being interpreted as an early titanosaur. However, the bones are now considered to be of an indeterminate titanosauriform (Mannion et al., 2013).

Khetranisaurus (referring to the Khetran people of Pakistan) is, according to researcher M. Sadiq Malkani, just one of numerous distinct Pakistani titanosaur genera. However, many other researchers have regarded the specimens as being dubiously valid. *Khetranisaurus* was initially founded based on a tail vertebra but has had numerous other elements referred to it (Malkani, 2021).

Loricosaurus (meaning "armor lizard") is known primarily from a series of osteoderms. As armored titanosaurs were unknown at the time, they were first thought to have belonged to an ankylosaur. Today, most researchers consider the bones to likely represent *Neuquensaurus* or *Saltasaurus*, or both. The species name derives from the Latin "scutellum" (meaning "little shield").

Macrurosaurus (meaning "large-tailed lizard") was described on the basis of two partial sets of tail vertebrae found in different locations. They are likely titanosauriform in nature but now are otherwise considered to be nondiagnostic. (A second species, *M. platypus*, is part of a nomenclatural mess involving an ankylosaur-sauropod chimera.) The generic name combines the Greek "makros" (meaning "long" or "large") and "oura" (meaning "tail").

Microcoelus (meaning "small cavity") sometimes notably misspelled as "Microsaurus" (Lull, 1910) or "Microsaurops" (Kuhn, 1963) is based on a fragment of a dorsal vertebra (MLP-LY 23), as well as part of a distal caudal and a humerus (which bears a striking resemblance to that of *Neuquensaurus*).

Hypselosaurus priscus

(Matheron, 1869)
Location: France
Occurrence: (?) Near Provence, France
Age: 70 MYA (Late Jurassic)

"Ischyrosaurus" manseli

(Hulke *vide* Lydekker, 1888)
Location: United Kingdom
Occurrence: Kimmeridge Clay
Age: 145 MYA (Late Jurassic)

Iuticosaurus valdensis

(Le Loeuff, 1993)
Location: United Kingdom
Occurrence: Wessex Fm.
Age: 130 MYA (Early Cretaceous)

Khetranisaurus barkhani

(Malkani, 2006)
Location: Pakistan
Occurrence: Pab Fm.
Age: 66 MYA (Late Cretaceous)

Loricosaurus scutatus

(von Huene, 1929)
Location: Argentina
Occurrence: Anacleto Fm.
Age: 71 MYA (Late Cretaceous)

Macrurosaurus semnus

(Seeley, 1869)
Location: United Kingdom
Occurrence: Cambridge Greensand
Age: 100 MYA (Early Cretaceous)

Microcoelus patagonicus

(Lydekker, 1893)
Location: Argentina
Occurrence: Bajo de la Carpa Fm.
Age: 75 MYA (Late Cretaceous)

Morinosaurus (referring to the historic Morini people) is a "tooth taxon" described on the basis of a single fossilized tooth, which is now lost. The tooth has been compared with numerous groups of sauropods, including brachiosaurs and titanosaurs.

Morinosaurus typus

(Sauvage, 1874)
Location: France
Occurrence: (?)
Age: 150 MYA (Late Jurassic)

Morosaurus (meaning "stupid lizard"), first described during the "Bone Wars," featured at least five species at one point in time. Most have now been synonymized with *Camarasaurus*; *M. agilis* has been rebranded as *Smitanosaurus* (Whitlock and Wilson, 2020).

Morosaurus (various)

(Marsh, 1878)
Location: Midwest, USA
Occurrence: Morrison Fm.
Age: 150 MYA (Late Jurassic)

Neosodon (meaning "new tooth") was never given a species name by its describer, who considered the animal to be similar to *Megalosaurus* based on the evaluation of a single tooth. Later, five more teeth were attributed to the genus. It has been suggested that the teeth came from a *Turiasaurus*-like taxon (Royo-Torres et al., 2006).

Neosodon

(Moussaye, 1885)
Location: France
Occurrence: Sables et Grès à Trigonia gibbosa
Age: 146 MYA (Late Jurassic)

Nicksaurus is, according to researcher M. Sadiq Malkani, just one of numerous distinct Pakistani sauropod genera. However, many other researchers have regarded the specimens as being dubiously valid. *Nicksaurus* was initially founded based on various skeletal fragments but has had numerous other elements referred to it.

Nicksaurus razashahi

(Malkani, 2015)
Location: Pakistan
Occurrence: Vitakri Fm.
Age: 66 MYA (Late Cretaceous)

Oplosaurus (probably intended to mean "armed lizard") is known from a single tooth (BMNH R751) that was originally interpreted as belonging to a carnivorous reptile. At one point it was referred to *Pelorosaurus*. A recent review has suggested a potential turiasaur identity for the tooth's owner (Poropat et al., 2022).

Oplosaurus armatus

(Gervais, 1852)
Location: United Kingdom
Occurrence: Wessex Fm.
Age: 130 MYA (Early Cretaceous)

Orosaurus (meaning "mountain lizard") is known from a partial tibia (NHMUK R1626), originally misidentified as a femur. No species name was given until the genus was later renamed *Orinosaurus*, as it was mistakenly believed that the original name was preoccupied (Lydekker, 1889). It was once considered to be synonymous with *Euskelosaurus* (van Heerden, 1979).

Orosaurus capensis

(Huxley, 1867)
Location: South Africa
Occurrence: Lower Elliot Fm.
Age: 210 MYA (Late Triassic)

Pachysuchus (meaning "thick crocodile") was originally interpreted as being a crocodile-like phytosaur, despite the long-standing notion that phytosaurs went extinct at the end of the Triassic. A review of the holotype (IVPP V 40, a portion of the upper jaw) determined that the animal was actually some indeterminate form of early sauropodomorph (Barrett and Xing, 2012).

Pachysuchus imperfectus

(Young, 1951)
Location: China
Occurrence: Lower Lufeng Fm.
Age: 200 MYA (Early Jurassic)

Pakisaurus is, according to researcher M. Sadiq Malkani, just one of numerous distinct Pakistani titanosaur genera. However, many other researchers have regarded the specimens as being dubiously valid (Wilson et al., 2011). *Pakisaurus* was initially founded based on various skeletal fragments but has had numerous other elements referred to it (Malkani, 2017).

Pakisaurus balochistani

(Malkani, 2006)
Location: Pakistan
Occurrence: Vitakri Fm.
Age: 66 MYA (Late Cretaceous)

Qinlingosaurus is known from pelvic elements and three vertebrae. The remains (NWUV 1112) have never been reexamined in subsequent literature after their initial debut. The genus may be valid, but currently all that can be said of them is that they belong to "some sort of sauropod."

Qinlingosaurus luonanensis

(Xue et al., 1996)
Location: China
Occurrence: Hongtuling Fm. (?)
Age: 70 MYA (Late Cretaceous)

Saraikimasoom (meaning "innocent one") is, according to researcher M. Sadiq Malkani, just one of numerous distinct Pakistani titanosaur genera. However, many other researchers have regarded the specimens as being dubiously valid (Wilson et al., 2011). *Saraikimasoom* was initially founded based on various cranial fragments.

Sulaimanisaurus (alternatively "Sulaimansaurus") is, according to researcher M. Sadiq Malkani, just one of numerous distinct Pakistani titanosaur genera. However, many other researchers have regarded the specimens as being dubiously valid (Wilson et al., 2011). *Sulaimanisaurus* was initially described based on several tail vertebrae; additional tail vertebrae were later added (Malkani, 2017).

Ultrasaurus was given its name because it was initially believed to be even larger than *Supersaurus*. This was an error, as a humerus had been mistaken for an ulna, resulting in a sauropod only about 22 meters in length. Further, researchers would go on to conclude that the whole genus is undiagnostic and therefore dubious (Galton and Upchurch, 2004). The holotype (DGBU-1973) also contains some vertebrae.

Ultrasauros was originally intended to be named "Ultrasaurus," and was widely reported as such, but the name got swiped (see above). The genus was mentioned simultaneously during Jensen's description of *Supersaurus*. However, it would later be determined that *Ultrasauros* was merely a chimera of *Brachiosaurus* and *Supersaurus* bones and thus is invalid.

Saraikimasoom vitakri

(Malkani, 2015)
Location: Pakistan
Occurrence: Vitakri Fm.
Age: 66 MYA (Late Cretaceous)

Sulaimanisaurus gingerichi

(Malkani, 2006)
Location: Pakistan
Occurrence: Vitakri Fm.
Age: 66 MYA (Late Cretaceous)

Ultrasaurus tabriensis

(Kim, 1983)
Location: South Korea
Occurrence: Gugyedong Fm.
Age: 105 MYA (Early Cretaceous)

Ultrasauros macintoshi

(Jensen, 1985)
Location: Midwest, USA
Occurrence: Morrison Fm.
Age: 150 MYA (Late Jurassic)

Clasmodosaurus spatula

In order for a new taxon to be erected, it needs to be formally published in a medium that conforms to the guidelines of the International Commission on Zoological Nomenclature (ICZN). Sometimes, proposed dinosaur names can prematurely become known by appearing in the media or in rough draft manuscripts, without ever achieving official status; these *nomina nuda* (singular *nomen nudum*) or "naked names" can nonetheless gain popular usage. Alternatively, some specimens might gain informal names or nicknames that were never actually intended to be used in an official capacity but that have not yet been replaced with a certified alternative.

Some particularly distinctive examples have been included in previous pages. Additional informally named sauropodomorphs of note are detailed below.

"Angloposeidon" is known from a single cervical vertebra (MIWG.7306) discovered in 1993 (Radley, 1997). The specimen has been much studied, and is particularly useful for studying internal pneumatic structures because the bone is broken in half. The informal name was coined by researcher Mike P. Taylor. Although the specimen is thought to be "phylogenetically intermediate between *Brachiosaurus* and *Sauroposeidon*," the specimen's describers did not assign a new genus name to the animal because the fossil's characteristics did not meet the threshold for being "diagnostic" and could potentially belong to one of the other fragmentary sauropod genera known from the same geological formation.

"Baguasaurus" is known from a set of caudal vertebrae, and potentially other undescribed remains. The name was coined by Larramendi and Molina Pérez (2020) based on the city of Bagua, Peru. The animal is possibly saltasaurid in nature and approximately 11 meters in length (Larramendi and Molina Pérez, 2020).

"Barackosaurus," also known as "*Amphicoelias brontodiplodocus*," is the designation given to three impressively complete diplodocid specimens that were uncovered between 2007 and 2010. An informal draft† describing the specimens was made public, although it was not initially clear that the paper was not intended for publication, which caused some degree of confusion. The specimens have yet to be properly described.

"Bashunosaurus kaijiangensis" is known from a partial skeleton (KM 20100) and an additional hip bone (KM 20103). The name originated in the description of *Abrosaurus* (Hui, 1989). The validity of its description publication is disputed, thus the name is *nomen nudum*. It may be synonymous with one of several other genera, such as *Abrosaurus* or *Datousaurus* (Molina-Perez and Larramendi, 2020).

"Biconcavoposeidon" is known from five dorsal vertebrae (AMNH FARB 291) that were unearthed in 1898. It is speculated that they belonged to some sort of brachiosaur. As the name suggests, each vertebra is concave in shape on both the front and back points of connection. Whether this is due to unusual fossil deformation or an unusual anomaly or developmental stage from a previously known taxon, or whether it represents an entirely new taxon, is currently unknown.

"Dachongosaurus yunnanensis" (alternatively "Dachungosaurus") is one of several *nomina nuda* whose names were coined in the 1980s by X. Zhao. Since that time, nothing of the genus has been mentioned. Owing to poor record keeping, it is possible that the partial skeleton of "Dachongosaurus" (purportedly nine dorsal vertebrae, ribs, and a femur; Li et al., 1998) has since been described under a different name.

"Angloposeidon"

(Naish et al., 2004)
Location: England, UK
Occurrence: Wessex Fm.
Age: 125 MYA (Early Cretaceous)

"Baguasaurus"

(Mourier et al., 1988)
Location: Peru
Occurrence: Chota Fm.
Age: 72 MYA (Late Cretaceous)

"Barackosaurus"

(Galiano and Albersdörfer, 2010†)
Location: Wyoming, USA
Occurrence: Morrison Fm.
Age: 150 MYA (Late Jurassic)

"Bashunosaurus"

(Kuang, 2004)
Location: China
Occurrence: Lower Shaximiao Fm.
Age: 165 MYA (Middle Jurassic)

"Biconcavoposeidon"

(Taylor and Wedel, 2017)
Location: Wyoming, USA
Occurrence: Morrison Fm.
Age: 150 MYA (Late Jurassic)

"Dachongosaurus"

(Zhao, 1985)
Location: China
Occurrence: Lower Lufeng Fm.
Age: 200 MYA (Early Jurassic)

"Damalosaurus laticostalis" (alternatively "Damalosaurus" or possibly "D. magnus") is one of several *nomina nuda* whose names were coined in the 1980s by X. Zhao. Since that time, nothing of the genus has been mentioned. Owing to poor record keeping, it is possible that the partial skeleton of "Damalosaurus" (purportedly one rib and possibly unmentioned other material) has since been described under a different name. It is named for Damala Mountain.

"Fendusaurus eldoni" is known from several sets of remains, some of which were originally suspected as belonging to *Ammosaurus* (now synonymized with *Anchisaurus*). At least five or six individuals are represented, possibly from more than one species. These were preliminarily described in the doctoral dissertation of Timothy J. Fedak. The taxon is notable for its elongated cervical vertebrae. "Fendu" is French for "split," while "eldoni" honors "fossil enthusiast" Eldon George.

"Francoposeidon" is, tantalizingly, perhaps the largest known sauropod. Numerous skeletal elements have been under excavation since at least 2012, although the only fossil thus far mentioned in a publication is ANG 10–400, an enormous 220 cm long femur; unconfirmed reports suggest that there are remains from even larger individuals. One source lists the type species as "F. charantensis" (Molina-Pérez and Larramendi, 2020).

"Hisanohamasaurus" was named in a nonscientific publication based on a set of teeth, first thought to be diplodocid in nature, but which were later suggested to be nemegtosaurid (Matsukawa and Obata, 1994).

"Kunmingosaurus wudingensis" (alternatively "K. utingensis") is one of several *nomina nuda* whose names were coined in the 1980s by X. Zhao. Since that time, nothing of the genus has been mentioned. Owing to poor record keeping, it is possible that the partial skeleton of "Kunmingosaurus" (fairly complete vertebral, hindlimb, and pelvic material) has since been described under a different name.

"Lancanjiangosaurus cachuensis" (alternatively "Lanchangjiangosaurus," "Lancangjiangosaurus," or "Lancangosaurus") is one of several *nomina nuda* whose names were coined in the 1980s by X. Zhao. Since that time, nothing of the genus has been mentioned. Owubg to poor record keeping, it is possible that the partial skeleton of "Lancanjiangosaurus" (purportedly skull, jaw, and limb material) (Li et al., 1998) has since been described under a different name.

"Megacervixosaurus tibetensis" (meaning "big-necked lizard") is one of several *nomina nuda* whose names were coined in the 1980s by X. Zhao. Since that time, nothing of the genus has been mentioned. Owing to poor record keeping, it is possible that the partial skeleton of "Megacervixosaurus" (consisting of cervical vertebrae) has since been described under a different name.

"Megapleurocoelus menduckii" is the suggested name for a possible Flagellicaudata sauropod, based on a single dorsal centrum (JP Cr376).

"Damalosaurus"

(Zhao, 1983)
Location: Tibet, China
Occurrence: Middle Daye Grp.
Age: 190 MYA (Early Jurassic)

"Fendusaurus"

(Fedak, 2006)
Location: Canada
Occurrence: McCoy Brook Fm.
Age: 200 MYA (Early Jurassic)

"Francoposeidon"

(Néraudeau et al., 2012)
Location: France
Occurrence: Angeac-Charente bone bed
Age: 130 MYA (Early Cretaceous)

"Hisanohamasaurus"

(Lambert, 1990)
Location: Japan
Occurrence: (?) Near Iwaki, Fukushima
Age: (?)

"Kunmingosaurus"

(Zhao, 1985)
Location: China
Occurrence: Lower Lufeng Fm.
Age: 200 MYA (Early Jurassic)

"Lancanjiangosaurus"

(Zhao, 1985)
Location: Tibet, China
Occurrence: Middle Dapuka Grp.
Age: 170 (?) MYA (Middle Jurassic)

"Megacervixosaurus"

(Zhao, 1983)
Location: Tibet, China
Occurrence: Zonggu Fm.
Age: (?) MYA (Late Cretaceous)

"Megapleurocoelus"

(Singer, 2015)
Location: Morocco
Occurrence: Kem Kem Fm.
Age: 95 MYA (Late Cretaceous)

"Microdontosaurus petersonii" (meaning "small-toothed lizard") is one of several *nomina nuda* whose names were coined in the 1980s by X. Zhao. Since that time, nothing of the genus has been mentioned. Owing to poor record keeping, it is possible that the partial specimen of "Microdontosaurus" (perhaps no more than some teeth) has since been described under a different name. The name is also preoccupied by an ichthyosaur (Gilmore, 1902).

"Moshisaurus" was a name informally used for a poorly preserved partial sauropod humerus. Later works referred it to *Mamenchisaurus* (Dong et al., 1990), while still later works list it as an indeterminate sauropod (Barrett et al., 2002).

"Nurosaurus qaganensis" (alternatively "Nuoerosaurus" (Dong,1991) (meaning "Nur lizard") supposedly boasts fairly substantial skeletal remains, but the animal has never been described in a formal context. This is apparently because, in part, the fossils are a part of a (seemingly unending) traveling exhibit that prevents them from being studied in depth.

"Oharasisaurus" is known from a single "enigmatic" tooth from a juvenile, possibly euhelopodid, sauropod (Larramendi and Molina Pérez, 2020).

"Oshanosaurus youngi" (meaning "lizard from Oshan") is one of several *nomina nuda* whose names were coined in the 1980s by X. Zhao. It is possible that the specimen (a lower jaw) might be the same that was used to establish the similarly named, possible therizinosaur theropod, *Eshanosaurus*.

"Otogosaurus sarulai" (meaning "Sarula's lizard from Otog") is supposedly known from a partial specimen that contains at least a tibia. However, researchers are unsure of the exact source of this information.

"Rutellum impicatum" is a name that predates the 1758 starting date for the ICZN guidelines and thus is not considered a valid modern name. This single tooth (OU 1352) has the historical distinction of being the earliest dinosaur specimen to be given a name. It is now thought to be cetiosaurid in nature (Delair and Sarjeant, 2002).

"Sauropodus" (meaning "reptile foot") is a name announced in a press statement to go along with the partial skeletal remains of a sauropod. The animal has not been properly described, possibly because of a lack of defining characteristics. *Choconsaurus* was discovered in the same locality and described by the same researcher.

"Microdontosaurus"

(Zhao, 1983)
Location: Tibet, China
Occurrence: Middle Dapuka Grp.
Age: 170 (?) MYA (Middle Jurassic)

"Moshisaurus"

(Hisa, 1985)
Location: Japan
Occurrence: Miyako Grp.
Age: (?) MYA (Early Cretaceous)

"Nurosaurus"

(Dong, 1992)
Location: Inner Mongolia
Occurrence: Qagannur Fm.
Age: 129 MYA (Early Cretaceous)

"Oharasisaurus"

(Matsuoka, 2000)
Location: Japan
Occurrence: Kuwajima Fm.
Age: 125 MYA (Early Cretaceous)

"Oshanosaurus"

(Zhao, 1985)
Location: China
Occurrence: Lower Lufeng Fm.
Age: 200 MYA (Early Jurassic)

"Otogosaurus"

(Zhao, 2004?) (Zhao and Tan, 2004?)
Location: China
Occurrence: (?)
Age: (?) (Early Jurassic)

"Rutellum"

(Lhuyd, 1699)
Location: England, UK
Occurrence: Coralline Oolite Fm.
Age: 155 MYA (Late Jurassic)

"Sauropodus"

(Simón, 2001)
Location: Argentina
Occurrence: Huincul Fm.
Age: 95 MYA (Late Cretaceous)

"Sousatitan" (named for the Sousa Basin of Brazil) is the informal name given to a sauropod that is represented by a single fibula (DGEO-CTG-UFPE 7517), the oldest sauropod bone found in Brazil at the time. The bone was found amid a very large dinosaur trackway. Histological examination suggests the individual was a subadult.

"Sugiyamasaurus" is named from several teeth that were unearthed from the same quarry as *Fukuititan*, which was described in 2010; it is possible that the two names apply to the same species. The name was first seen in print in the *Dinosaur Data Book*.

"Tobasaurus" (named for Toba City, Japan) is a possible euhelopodid known primarily from limb bones (MPMF 0014) that were discovered in 1996 (Tomida et al., 2001). The animal is thought to have reached an overall length of 20 meters.

"Xinghesaurus" is the name given to a mounted skeleton on display at a Tokyo museum. The fossils have never appeared in publication; measurements have never been distributed; how much of the skeleton is artificially reconstructed is unknown.

"Yibinosaurus zhoui" was first mentioned in a 2001 guidebook for the Chongqing Natural History Museum. It was then described in the dissertation of Chinese paleontologist Ouyang Hui in 2003; the specimen in question had first been referred to *Gongxianosaurus* (Luo and Wang, 1999).

"Yunxianosaurus hubeinensis" is a potential titanosaur, given an informal name pending further research. The specimen in question is a juvenile individual and includes substantial vertebral, pelvic, and limb remains.

"Sousatitan"

(Ghilardi et al., 2016)
Location: Brazil
Occurrence: Rio Piranhas Fm.
Age: 140 MYA (Early Cretaceous)

"Sugiyamasaurus"

(Lambert, 1990)
Location: Japan
Occurrence: Kitadani Fm.
Age: 125 MYA (Early Cretaceous)

"Tobasaurus"

(Tomida and Tsumura, 2006)
Location: Japan
Occurrence: Matsuo Grp.
Age: 125 MYA (Early Cretaceous)

"Xinghesaurus"

(Hasegawa et al., 2009)
Location: Japan (?)
Occurrence: (?)
Age: (?)

"Yibinosaurus"

(Ouyang, 2003)
Location: China
Occurrence: Ziliujing Fm.
Age: 180 MYA (Early Jurassic)

"Yunxianosaurus"

(Li, 2001)
Location: China
Occurrence: Majiacun Fm.
Age: 85 MYA (Late Cretaceous)

We would like to thank the following researchers for graciously providing access to their invaluable works:

Logan King, Hai-Lu You, Yilun Yu, Agustin G. Martinelli, Alexey V. Lopatin, Frederico Agnolin, Agustín Pérez Moreno, Pedro Mocho, Julian C. G. Silva Junior, and Theo Baptista Ribeiro.

ᕙᕗ

The illustrator is grateful to his DeviantArt watchers for their support and valuable suggestions, which have helped make his drawings more accurate.

Some illustrations are based on the skeletal reconstructions of other paleoartists, including the following:

André de Oliveira Fonseca (https://www.deviantart.com/andreof-gallery): *Paludititan*. Andrew Mcafee: *Katepensaurus*. Ashley Patch (https://www.deviantart.com/plastospleen): *Austrosaurus, Diamantinasaurus*. Asier Larramendi (https://www.deviantart.com/eofauna): *Turiasaurus*. Brian Engh: *Dystrophaeus*. BritishPalaeontology (https://www.deviantart.com/britishpalaeontology): *Cetiosaurus, Sarmientosaurus*. Franoys (https://www.deviantart.com/franoys): *Leinkupal, Galeamopus, Atlantosaurus, Galvesaurus, Soriatitan, Tastavinsaurus, Lohuecotitan*. GetAwayTrike (https://www.deviantart.com/getawaytrike): *Bonitasaura, Jainosaurus, Neuquensaurus*. Gregory S. Paul 2016: *Pantydraco, Plateosaurus, Anchisaurus, Tazoudasaurus, Nebulasaurus*. Gunnar Bivens (https://www.deviantart.com/gunnarbivens): *Janenschia, Jobaria, Ligabuesaurus, Yunmenglong, Normanniasaurus, Baurutitan, Qingxiusaurus*. Henrique Paes (https://www.deviantart.com/randomdinos): *Mussaurus, Vouivria, Lusotitan, Venenosaurus, Triunfosaurus, Austroposeidon, Mendozasaurus, Puertasaurus, Argentinosaurus, Patagotitan, Quetecsaurus, Uberabatitan, Arrudatitan, Tapuiasaurus*. Ijreid (https://www.deviantart.com/ijreid): *Blikanasaurus, Vulcanodon*. Javifel (https://www.deviantart.com/javifel): *Qijianglong, Chuanjiesaurus, Abrosaurus, Erketu, Overosaurus*. Kevin Yan (https://www.deviantart.com/yty2000): *Xinjiangtitan, Daxiatitan, Notocolossus*. Lukas Panzarin: *Brachytrachelopan*. lythronax-argestes (https://www.deviantart.com/lythronax-argestes): *Europatitan, Shingopana, Zhuchengtitan*. Mauricio S. Garcia (https://www.deviantart.com/maurissauro): *Buriolestes, Eoraptor, Pampadromaeus, Panphagia, Nhandumirim, Saturnalia, Chromogisaurus, Bagualosaurus, Thecodontosaurus, Unaysaurus, Macrocollum, Guaibasaurus, Pradhania*. Megalotitan (https://www.deviantart.com/megalotitan): *Sibirotitan, Wintonotitan, Aegyptosaurus*. Nima Sassani (https://www.deviantart.com/paleo-king): *Klamelisaurus, Atlasaurus, Europasaurus, Duriatitan, Ornithopsis, Fusuisaurus, Sonorasaurus, Pelorosaurus, Sauroposeidon, Huanghetitan, Jiutaisaurus, Ruyangosaurus, Euhelopus, Argyrosaurus, Jiangshanosaurus, Dongyangosaurus, Baotianmansaurus, Futalognkosaurus*. Ornithopsis (https://www.deviantart.com/ornithopsis): *Aeolosaurus*. Palaeozoologist (https://www.deviantart.com/palaeozoologist): *Limaysaurus*. PaleoNeolitic: *Atacamatitan*. Pol et al., 2020: *Bagualia*. Rojo-Torres et al., 2017: *Lapparentosaurus*. SassyPaleoNerd (https://www.deviantart.com/sassypaleonerd): *Kaatedocus, Oceanotitan*. Scott Hartman (https://www.skeletaldrawing.com/): *Mbiresaurus, Riojasaurus, Lufengosaurus, Melanorosaurus, Shunosaurus, Amargasaurus, Apatosaurus, Brontosaurus, Supersaurus, Barosaurus, Diplodocus, Nigersaurus, Brachiosaurus, Giraffatitan, Malawisaurus, Dreadnoughtus, Traukutitan, Isisaurus, Rapetosaurus, Opisthocoelicaudia, Nemegtosaurus, Alamosaurus, Saltasaurus*. Slate Weasel: *Omeisaurus, Suuwassea, Dicraeosaurus, Amazonsaurus, Petrobrasaurus, Antarctosaurus giganteus*. SpinoInWonderland (https://www.deviantart.com/spinoinwonderland): *Yunnanosaurus*. Stephen O'Connor (https://www.deviantart.com/steveoc86): *Andesaurus, Huabeisaurus*. TheHellckan (https://www.deviantart.com/thehellckan): *Ibirania*. Tyrannoraptoran (https://www.deviantart.com/tyrannoraptoran): *Maraapunisaurus*. Vidal et al., 2020: *Spinophorosaurus*.

SAUROPODOMORPHA KEY HOLOTYPE, NEOTYPE

Note: Bibliographic information for citations in the tables below can be found in the chapter reference lists, which are available via https://press.princeton.edu/books/ebook/9780691250243/the-princeton-encyclopedia-of-dinosaurs-sauropods-pdf.

TYPE SPECIES	SIZE	GENUS MATERIAL	OCCURRENCE
Arcusaurus pereirabdalorum Yates et al., 2011	**Length:** 2.3 m **Mass:** 42 kg	**BP/1/6235**—partial skull and jaw; BP/1/6XXX Series—fragments (*Yates et al., 2011*)	Upper Elliot Fm., South Africa *Early Jurassic, Sinemurian*
Asylosaurus yalensis Galton, 2007a	**Length:** 2.2 m **Mass:** 14 kg	**YPM 2195**—partial skeleton; YPM 2195a–g—disarticulated bones; BMNH R1542—partial humerus; YPM 56745—manual phalanx; YPM 56725, 56726, 56739—ischia (*Galton, 2007a*)	Uncertain, United Kingdom *Late Triassic, Rhaetian*
Bagualosaurus agudoensis Pretto et al., 2019	**Length:** 2.3 m **Mass:** 15 kg	**UFRGS-PV-1099-T**—partial skull, skeleton (*Pretto et al., 2019*)	Santa Maria Fm., Brazil *Late Triassic, Carnian*
Buriolestes schultzi Cabreira et al., 2016	**Length:** 1.8 m **Mass:** 7.4 kg	**ULBRA-PVT280**—skull and skeleton (*Cabreira et al., 2016*); CAPPA/UFSM 0035—skull and skeleton; ULBRA-PVT056—fragments; CAPPA/UFSM 0179—axis; ULBRA-PVT289—femur (*Müller et al., 2018b*)	Santa Maria Fm., Brazil *Late Triassic, Carnian*
Chromogisaurus novasi Ezcurra, 2010	**Length:** 1.9 m **Mass:** 12 kg	**PVSJ 845**—partial skeleton (*Ezcurra, 2010*)	Ischigualasto Fm., Argentina *Late Triassic, Carnian*
Efraasia minor Galton, 1973	**Length:** 6.5 m **Mass:** 365 kg	**SMNS 11838**—partial skeleton; SMNS 12188–12192—partial skeleton; SMNS 12667—partial skull and skeleton; SMNS 12684—partial skull and skeleton; SMNS 17928—partial skeleton; SMNS 12354—partial skeleton; [potentially SMNS 12216—partial skull; 12668—partial skull; and 14881—skull fragment] (*Langer, 2003*)	Stubensandstein Fm., Germany *Late Triassic, Norian*
Eoraptor lunensis Sereno et al., 1993	**Length:** 1.7 m **Mass:** 10 kg	**PVSJ 512**—skull and skeleton; PVSJ series—partial remains of numerous adults and subadults (*Sereno et al., 2013*)	Ischigualasto Fm., Argentina *Late Triassic, Carnian*
Guaibasaurus candelariensis Bonaparte et al., 1999	**Length:** 3 m **Mass:** 35 kg	**MCN PV2355**—partial skeleton; MCN PV2356—hindlimb (*Bonaparte et al., 1999*); UFRGS PV0725T—skeleton (*Bonaparte et al., 2007*); MCN PV 10112—fragments (*Langer et al., 2010*)	Caturrita Fm. and Santa Maria Fm., Brazil *Late Triassic, Carnian*
Issi saaneq Beccari et al., 2021	**Length:** 4.7 m **Mass:** 260 kg	**NHMD 164741**—skull; NHMD 164758—skull; NHMD 164734, NHMD 164775—partial skeletons that are undescribed and/or unprepared (*Beccari et al., 2021*)	Malmros Klint Fm., Greenland *Late Triassic, Norian*
Jaklapallisaurus asymmetrica Novas et al., 2010	**Length:** 4.5 m **Mass:** 240 kg	**ISI R274**—partial leg and fragments; ISI R279—partial femur (*Novas et al., 2010*)	Upper Maleri Fm., India *Late Triassic, Norian*
Macrocollum itaquii Müller et al., 2018a	**Length:** 3.4 m **Mass:** 95 kg	**CAPPA/UFSM 0001a**—skull and skeleton; CAPPA/UFSM 0001b—skull and skeleton; CAPPA/UFSM 0001c—partial skeleton (*Müller et al., 2018a*)	Candelária Sequence, Brazil *Late Triassic, Norian*
Mbiresaurus raathi Griffin et al., 2022	**Length:** 1.7 m **Mass:** 6 kg	**NHMZ 2222**—skull and skeleton; NHMZ 2547—partial skeleton (*Griffin et al., 2022*)	Pebbly Arkose Fm., Zimbabwe *Late Triassic, Carnian*
Nambalia roychowdhurii Novas et al., 2010	**Length:** 3.7 m **Mass:** 125 kg	**ISI R273 series**—various skeletal elements (*Novas et al., 2010*)	Upper Maleri Fm., India *Late Triassic, Norian/Rhaetian*
Nhandumirim waldsangae Marzola et al., 2018	**Length:** 1 m **Mass:** 1.5 kg	**LPRP/USP 0651**—fragments (*Marzola et al., 2018*)	Santa Maria Fm., Brazil *Late Triassic, Carnian*
Pampadromaeus barberenai Cabreira et al., 2011	**Length:** 1.4 m **Mass:** 3 kg	**ULBRA-PVT016**—skull and skeleton (*Cabreira et al., 2011*); CAPPA/UFSM 0027—femur (*Müller et al., 2016*); CAPPA/UFSM 0028—femur (*Müller et al., 2017*)	Santa Maria Fm., Brazil *Late Triassic, Carnian*
Panphagia protos Martinez and Alcober, 2009	**Length:** 2 m **Mass:** 7.8 kg	**PVSJ 874**—skull and skeleton of subadult (*Martínez and Alcober, 2009*)	Ischigualasto Fm., Argentina *Late Triassic, Carnian*
Pantydraco caducus Galton et al., 2007b	**Length:** 1.5 m **Mass:** 6 kg	**BMNH P24**—partial skull and skeleton; NHMUK P77/1—partial leg; BMNH P19/7—humerus (*Galton et al., 2007b*); NHMUK RU P-series—fragments (*Galton, and Kermack, 2010*); NHMUK R37043—phalanx; BRSUG 28381–1—teeth; R37048a and b—fragments (*Keeble et al., 2018*)	Uncertain, United Kingdom *Late Triassic, Rhaetian*
Plateosaurus trossingensis von Meyer, 1837	**Length:** 9.4 m **Mass:** 2,600 kg	**SMNS 13200**—partial skull and skeleton; many hundreds of fossils (*Fernández and Werneburg, 2022*)	Trossingen Fm., Germany; Marnes Irisees Superieures Fm., France; Klettgau Fm., Switzerland *Late Triassic, Norian*
Ruehleia bedheimensis Galton, 2001	**Length:** 7.6 m **Mass:** 1,200 kg	**MB.R.4718-42**—partial skeleton (*Otero, 2018; Galton, 2001*)	Trossingen Fm., Germany *Late Triassic, Norian*
Saturnalia tupiniquim Langer et al., 1999	**Length:** 1.7 m **Mass:** 6 kg	**MCP 3844-PV**—partial skeleton; MCP 3845-PV—partial skull and skeleton; MCP 3846-PV—partial skeleton (*Langer et al., 1999*)	Santa Maria Fm., Brazil *Late Triassic, Carnian*

TYPE SPECIES	SIZE	GENUS MATERIAL	OCCURRENCE
Thecodontosaurus antiquus Riley and Stutchbury, 1836	**Length:** 3 m **Mass:** 34 kg	**BRSMG Ca7465**—right dentary *(Benton et al., 2000)*; **BRSMG C4529**—left dentary *(Galton, 1985)*; many hundreds of fossils *(Benton et al., 2000; Ballell et al., 2020)*	Magnesian Conglomerate, England *Late Triassic, Rhaetian*
Unaysaurus tolentinoi Leal et al., 2004	**Length:** 2.9 m **Mass:** 65 kg	**UFSM11069**—partial skull and skeleton *(Leal et al., 2004)*	Caturrita Fm., Brazil *Late Triassic, Norian*
Xixiposaurus suni Sekiya, 2010	**Length:** 4.3 m **Mass:** 200 kg	**ZLJ 01018**—skull and partial skeleton *(Sekiya, 2010)*	Lufeng Fm., China *Early Jurassic, Hettangian/Sinemurian*
Yimenosaurus youngi Bai et al., 1990	**Length:** 8.9 m **Mass:** 1,800 kg	**YXV 8701**—skull and partial skeleton; YXV 8702—partial skull and skeleton; possibly elements of eight other skeletons *(Bai et al., 1990)*	Fengjiahe Fm., China *Early Jurassic, Pliensbachian*

MASSOPODA

TYPE SPECIES	SIZE	GENUS MATERIAL	OCCURRENCE
Aardonyx celestae Yates et al., 2010	**Length:** 8.7 m **Mass:** 1,700 kg	**BP/1/6254**—partial maxilla; BP/1/6505—partial maxilla; numerous BP/1-series disarticulated specimens from two immature individuals *(Yates, 2010)*; BP/1/5339—fragments *(Otero and Peyre de Fabrègues, 2022)*	Upper Elliot Fm., South Africa *Early Jurassic, Sinemurian*
Adeopapposaurus mognai Martínez, 2009	**Length:** 2.9 m **Mass:** 60 kg	**PVSJ610**—skull and partial skeleton; PVSJ568—partial skull and skeleton; PVSJ569—partial skeleton; PVSJ570—fragments *(Martínez, 2009)*	Cañon del Colorado Fm., Argentina *Early Jurassic, (?)*
Amygdalodon patagonicus Cabrera, 1947	**Length:** 12 m **Mass:** 3,300 kg	**MLP 46-VIII-21-1/2**—dorsal vertebra; MLP 46-VIII-21-1 series and 36-XI-10-3 series—fragments *(Cabrera, 1947)*	Cerro Carnereo Fm., Argentina *Early Jurassic, Toarcian*
Anchisaurus polyzelus Hitchcock, 1865	**Length:** 3.2 m **Mass:** 55 kg	**ACM 41/109**—partial skeleton; **YPM 1883**—partial skull and skeleton *(Galton, 2012)*; YPM 208—partial rear skeleton *(Marsh, 1889)*; YPM 209—skull and partial skeleton *(Marsh, 1891)*	Portland Fm., United States *Early Jurassic, Hettangian*
Camelotia borealis Galton, 1985	**Length:** 10.2 m **Mass:** 2,700 kg	**NHMUK R2870-R2874, R2876-R2878**—partial skeleton *(Galton, 1985)*	Westbury Fm., United Kingdom *Late Triassic, Rhaetian*
Chuxiongosaurus lufengensis Lü et al., 2010	**Length:** 9.2 m **Mass:** 1,400 kg	**CMY-LT9401**—skull and jaw *(Lü et al., 2010)*	Lufeng Fm., China *Early Jurassic, Hettangian/Sinemurian*
Coloradisaurus brevis Bonaparte, 1978	**Length:** 6.3 m **Mass:** 600 kg	**PVL 3967**—skull and partial skeleton; PVL 5904—partial skeleton *(Apaldetti et al., 2013)*	Los Colorados Fm., Argentina *Late Triassic, Norian*
Eucnemesaurus fortis van Hoepen, 1920	**Length:** 7.8 m **Mass:** 1,200 kg	**TrM 119**—partial skeleton; NMW 1889-XV-39 and 1876-VIIB124—femur; BP/1/6107, 6110–6115, 6220—fragments *(Yates, 2007)*; BP/1/6234—partial skeleton *(McPhee et al., 2015)*	Lower Elliot Fm., South Africa *Late Triassic, Norian/Rhaetian*
Glacialisaurus hammeri Smith and Pol, 2007	**Length:** 6.3 m **Mass:** 600 kg	**FMNH PR 1823**—partial foot; FMNH PR 1822—partial femur (Smith and Pol, 2007)	Hanson Fm., Antarctica *Early Jurassic, Sinemurian/Pliensbachian*
"Gyposaurus" sinensis Young, 1941	**Length:** 2.9 m **Mass:** 58 kg	**IVPP V24**—skull and jaw fragments; IVPP V25—skull fragment; IVPP V26—partial skeleton; IVPP V27—partial skeleton *(Young, 1941)*	Lufeng Fm., China *Early Jurassic, Hettangian/Sinemurian*
Ignavusaurus rachelis Knoll, 2010	**Length:** 4 m **Mass:** 190 kg	**BM HR 20**—partial skeleton *(Knoll, 2010)*; NHMUK PV R37375—several vertebrae *(Bodenham and Barrett, 2020)*	Upper Elliot Fm., Lesotho *Early Jurassic, Hettangian/Sinemurian*
Irisosaurus yimenensis Peyre de Fabrègues et al., 2020	**Length:** 5.7 m **Mass:** 500 kg	**CVEB 21901**—partial skull and skeleton *(Peyre de Fabrègues et al., 2020)*	Fengjiahe Fm., China *Early Jurassic, Hettangian/Sinemurian*
Jingshanosaurus xinwaensis Zhang and Yang, 1995	**Length:** 9.2 m **Mass:** 1,600 kg	**LFGT-ZLJ0113 {aka LV003}**—skull and skeleton *(Zhang and Yang, 1995)*; CXM-LT9401–skull and jaw *(Zhang et al., 2020)*	Lufeng Fm., China *Early Jurassic, Hettangian/Sinemurian*
Kholumolumo ellenbergerorum Peyre de Fabrègues and Allain, 2020	**Length:** 11 m **Mass:** 3,600 kg	**MNHN.F.LES381m**—tibia; MNHN.F.LES series—various fragments *(Peyre de Fabrègues and Allain, 2020)*	Lower Elliot Fm., Lesotho *Late Triassic, Norian/Rhaetian*
Leonerasaurus taquetrensis Pol et al., 2011	**Length:** 3 m **Mass:** 70 kg	**MPEF-PV 1663**—partial skeleton *(Pol et al., 2011)*	Las Leoneras Fm., Argentina *Early Jurassic, (?)*
Leyesaurus marayensis Apaldetti et al., 2011	**Length:** 3.1 m **Mass:** 70 kg	**PVSJ 706**—skull and partial skeleton *(Apaldetti et al., 2011)*	Quebrada del Barro Fm., Argentina *Late Triassic, Rhaetian*
Lishulong wangi (Zhang et al., 2024)	**Length:** 10.9 m **Mass:** 2,900 kg	**LFGT-ZLJ0011**—skull and neck *(Zhang et al., 2024)*	Lufeng Fm., China *Early Jurassic, Sinemurian*
Lufengosaurus huenei Young, 1940	**Length:** 8.9 m **Mass:** 1,750 kg	**IVPP V15**—skull and skeleton; numerous IVPP V-series—fragments *(Young, 1941; 1942; 1948; 1951)*; GM. V15—partial skull *(Young, 1982)*; numerous FMNH (CUP)-series—fragments *(Simmons, 1965; Evans and Milner, 1989)*; YX V0003—partial skeleton *(Xing et al., 2018)*; ZLJ0112—partial skull *(Sekiya and Dong, 2010)*	Lufeng Fm., China *Early Jurassic, Hettangian/Sinemurian*
Massospondylus carinatus Owen, 1854	**Length:** 5.7 m **Mass:** 450 kg	**BP/1/4934**—skull and skeleton *(Yates and Barrett, 2010)*; numerous specimens *(Owen, 1854; Cooper, 1981; Gow, 1990; Gow et al., 1990; Barrett, 2009; Barrett et al., 2019; Chapelle et al., 2020)*	South Africa, Lesotho, Zimbabwe *Early Jurassic, Hettangian/Sinemurian*
Melanorosaurus readi Haughton, 1924	**Length:** 6.5 m **Mass:** 700 kg	**SAM 3449**—partial skeleton; **SAM 3450**—fragments; SAM 3532—fragments; NM QR3314—skull and skeleton; NM QR1551—partial skeletons *(Galton et al., 2005)*	Lower Elliot Fm., South Africa *Late Triassic, Norian/Rhaetian*

Meroktenos thabanensis Peyre de Fabrègues and Allain, 2016	**Length:** 4.8 m **Mass:** 300 kg	MNHN.F.LES16—fragments; MNHN.F.LES351—fragments (*Peyre de Fabrègues and Allain, 2016*)	Lower Elliot Fm., Lesotho *Late Triassic, Norian/Rhaetian*
Musankwa sanyatiensis Barrett et al., 2024	**Length:** 5.7 m **Mass:** 390 kg	NHMZ 2521—partial hindlimb (*Barrett et al., 2024*)	Pebbly Arkose Fm., Zimbabwe *Late Triassic, Norian*
Mussaurus patagonicus Bonaparte and Martin, 1979	**Length:** 8 m **Mass:** 1,350 kg	PVL 4068—skeleton of a post-hatchling; numerous partial remains of hatchlings, juveniles, and adults (*Otero and Pol, 2013*); MPM-PV series—skeletons of numerous individuals of multiple ages (*Pol et al., 2021*); MLP 68-II-27-1—partial skeleton (*Cerda et al., 2022*)	Laguna Colorada Fm., Argentina *Early Jurassic, Pliensbachian*
Ngwevu intloko Chapelle et al., 2019	**Length:** 3.5 m **Mass:** 100 kg	BP/1/4779—skull and partial skeleton (*Chapelle et al., 2019*)	Upper Elliot Fm., South Africa *Early Jurassic, Hettangian/Sinemurian*
Plateosauravus cullingworthi von Huene, 1932	**Length:** 8.3 m **Mass:** 1,400 kg	SAM PK 3341–3344, 3346–3349, 3350, 3356—partial skeleton (*Krupandan, 2019*)	Lower Elliot Fm., South Africa *Late Triassic, Norian/Rhaetian*
Pradhania gracilis Kutty et al. 2007	**Length:** 4.6 m **Mass:** 240 kg	ISI R265—fragmentary skull and skeleton (*Kutty et al., 2007*)	Upper Dharmaram Fm., India *Early Jurassic, Sinemurian*
Qianlong shouhu Han et al., 2024	**Length:** 8.1 m **Mass:** 1,200 kg	GZPM VN001—partial skull and skeleton; GZPM VN002 and 003—partial skeletons; GZPM VN004–008—eggs (*Han et al., 2024*)	Ziliujing Fm., China *Early Jurassic, Sinemurian*
Riojasaurus incertus Bonaparte, 1969	**Length:** 6.8 m **Mass:** 800 kg	PVL 3808—skeleton; MLR 56—skull and skeleton; PVL 3806, 3805, 3815, 3844, 3845, 3846, 3847, 3662, 3663, 3668, 3669, 3526, 3533, 3467, 3472, 3478, 3392, 3393—various (*Pol et al., 2011*)	Los Colorados Fm., Argentina *Late Triassic, Norian*
Sarahsaurus aurifontanalis Rowe et al., 2011	**Length:** 4.3 m **Mass:** 190 kg	TMM 43646-2—skull and skeleton; TMM 43646-3—partial skeleton; MCZ 8893—cranial and skeletal fragments (*Rowe et al., 2011*)	Kayenta Fm., United States *Early Jurassic, Sinemurian/Pliensbachian*
Sefapanosaurus zastronensis Otero et al., 2015	**Length:** 6 m **Mass:** 550 kg	BP/1/386—partial upper foot; BP/1/7409–7455—fragments of at least four individuals (*Otero et al., 2015*)	Lower Elliot Fm., South Africa *Late Triassic, Norian/Rhaetian*
Seitaad ruessi Sertich and Loewen, 2010	**Length:** 3.5 m **Mass:** 100 kg	UMNH VP 18040—partial skeleton (*Sertich and Loewen, 2010*)	Navajo Ss., United States *Early Jurassic, Pliensbachian*
Xingxiulong chengi Wang et al., 2017b	**Length:** 6 m **Mass:** 460 kg	LFGT-D0002—partial skull and skeleton; LFGT-D0001—partial skull and skeleton; LFGT-D0003—partial skull and skeleton (*Wang et al., 2017*)	Lufeng Fm., China *Early Jurassic, Hettangian*
Yizhousaurus sunae Zhang et al., 2018	**Length:** 8.2 m **Mass:** 1,500 kg	BLFGT-ZLJ0033—partial skull and skeleton (*Zhang et al., 2018*)	Lufeng Fm., China *Early Jurassic, Hettangian*
Yunnanosaurus huangi Young, 1942	**Length:** 7 m **Mass:** 600 kg	NGMJ 004546—skull and partial skeleton (*Young, 1942*); numerous specimens (*Young, 1951; Lü et al., 2007; Sekiya et al., 2014*)	Lufeng Fm., China *Early Jurassic, Hettangian*

SAUROPODA

TYPE SPECIES	SIZE	GENUS MATERIAL	OCCURRENCE
Antetonitrus ingenipes Yates and Kitching, 2003	**Length:** 7.8 m **Mass:** 1,250 kg	BP/1/4952—partial skeleton; BP/1/4952b—limb bones and vertebrae (*Yates and Kitching, 2003*); BP/1/4952c—fragments; NM QR1545—partial skeleton (*McPhee et al., 2014*); possibly NM QR1705—partial skeleton (*Krupandan, 2019*)	Upper Elliot Fm., South Africa *Early Jurassic, Hettangian/Sinemurian*
Blikanasaurus cromptoni Galton and van Heerden, 1985	**Length:** 5.4 m **Mass:** 420 kg	SAM K403—lower rear leg and foot (*Galton and van Heerden, 1985*); BP/1/5271—right metatarsal (*Yates, 2008*)	Lower Elliot Fm., South Africa *Late Triassic, Norian/Rhaetian*
Chinshakiangosaurus chunghoensis Dong, 1992	**Length:** 10.1 m **Mass:** 2,750 kg	IVPP V14474—partial skeleton (*Upchurch et al., 2007*)	Fengjiahe Fm., China *Early Jurassic, Hettangian*
Gongxianosaurus shibeiensis He et al., 1998	**Length:** 12.5 m **Mass:** 4,000 kg	**Number unknown**—four specimens of various completeness that preserve the majority of the skeleton, except for skull and hand (*Yaonan and Changsheng, 2000*)	Ziliujing Fm., China *Early Jurassic, Toarcian*
Ingentia prima Apaldetti et al., 2018	**Length:** 6.8 m **Mass:** 850 kg	PVSJ 1086—partial skeleton; PVSJ 1087—partial skeleton (*Apaldetti et al., 2018*)	Quebrada del Barro Fm., Argentina *Late Triassic, Rhaetian*
Isanosaurus attavipachi Buffetaut et al., 2000	**Length:** 13 m **Mass:** 7,000 kg	CH4—skeletal fragments (*Buffetaut et al., 2000*); possibly MH 350—skeletal elements including humerus (*Buffetaut et al., 2002*); possibly CH8–66—partial humerus of young adult (*Jentgen-Ceschino et al., 2020*)	Nam Phong Fm., Thailand *Early Jurassic, (?)*
Kotasaurus yamanpalliensis Yadagiri, 1988	**Length:** 12.4 m **Mass:** 4,500 kg	21/SR/PAL—ilium; S1Y/76 series—partial skeletons of 12 individuals (*Yadagiri, 2001*)	Kota Fm., India *Early Jurassic, Hettangian/Sinemurian*
Lamplughsaura dharmaramensis Kutty et al., 2007	**Length:** 7.8 m **Mass:** 1,200 kg	ISI R257—partial skeleton; ISI R258—partial skull and skeleton; ISI R259—partial skull and skeleton; ISI R261—partial skeleton; ISI R262—partial skeleton (*Kutty et al., 2007*)	Upper Dharmaram Fm., India *Early Jurassic, Sinemurian*
Ledumahadi mafube McPhee et al., 2018	**Length:** 10.8 m **Mass:** 3,400 kg	BP/1/7120—fragmentary skeleton (*McPhee et al., 2018*)	Upper Elliot Fm., South Africa *Early Jurassic, Hettangian/Sinemurian*

Lessemsaurus sauropoides Bonaparte, 1999	**Length:** 10.3 m **Mass:** 2,900 kg	**PVL 4822–1**—neural arches; PVL 4822 series—partial skeleton *(Pol and Powell, 2007)*; CRILAR PV-302—partial pelvis; CRILAR PV-303—partial scapulae; PVL 6580—partial femur *(Apaldetti et al., 2018)*	Los Colorados Fm., Argentina *Late Triassic, Norian*
Ohmdenosaurus liasicus Wild, 1978	**Length:** 7 m **Mass:** 1,100 kg	**Unnumbered**—tibia and ankle *(Wild, 1978)*	Sachrang Fm., Germany *Early Jurassic, Toarcian*
Protognathosaurus oxyodon Zhang, 1988	**Length:** 14 m **Mass:** 6,300 kg	**CV 00732**—partial lower jaw *(Zhang, 1988)*	Xiashaximiao Fm., China *Middle Jurassic, (?)*
Pulanesaura eocollum McPhee et al., 2015	**Length:** 8 m **Mass:** 1,100 kg	**BP/1/6882**—partial vertebra; BP/1 series—various fragments *(McPhee et al., 2015)*	Upper Elliot Fm., South Africa *Early Jurassic, Hettangian/Sinemurian*
Rhoetosaurus brownei Longman, 1926	**Length:** 15 m **Mass:** 9,000 kg	**QM F1659**—partial skeleton *(Nair and Salisbury, 2012)*	Walloon Coal Measures, Australia *Late Jurassic, Oxfordian*
Sanpasaurus yaoi Young, 1944	**Length:** 7.2 m **Mass:** 1,100 kg	**IVPP V156A**—dorsal vertebral series; IVPP V156B—various skeletal elements *(McPhee et al., 2016)*	Ziliujing Fm., China *Early Jurassic, Toarcian*
Schleitheimia schutzi Rauhut et al., 2020	**Length:** 6.5 m **Mass:** 550 kg	**PIMUZ A/III 550**—partial right ilium; PIMUZ A/III series—various fragments *(Rauhut et al., 2020)*	Klettgau Fm., Switzerland *Late Triassic, Norian*
Tazoudasaurus naimi Allain et al., 2004	**Length:** 12.4 m **Mass:** 6,000 kg	**CPSGM To2000-1**—partial skeleton; CPSGM To1 series—skeletal elements of numerous individuals *(Peyer and Allain, 2010)*	Toundoute Continental Series, Morocco *Early Jurassic, Pliensbachian/Toarcian*
Tuebingosaurus maierfritzorum Fernández and Werneburg, 2022	**Length:** 7.8 m **Mass:** 1,000 kg	**GPIT-PV-30787**—partial skeleton *(Fernández and Werneburg, 2022)*	Trossingen Fm., Germany *Late Triassic, Norian*
Volkheimeria chubutensis Bonaparte, 1979	**Length:** 8.7 m **Mass:** 4,000 kg	**PVL 4077**—partial skeleton *(Bonaparte, 1979)*	Cañadón Asfalto Fm., Argentina *Early Jurassic, Toarcian*
Vulcanodon karibaensis Raath, 1972	**Length:** 11 m **Mass:** 3,500 kg	**QG24**—partial skeleton; QG152—partial scapula; QG1406—partial vertebra *(Raath, 1972)*	Upper Forest Sandstone, Zimbabwe *Early Jurassic, Sinemurian*
Zizhongosaurus chuanchengensis Dong et al., 1983	**Length:** 8.4 m **Mass:** 1,800 kg	**IVPP V9067**—fragments *(Dong et al., 1983)*	Ziliujing Fm., China *Middle Jurassic, Aalenian*

EUSAUROPODA

TYPE SPECIES	SIZE	GENUS MATERIAL	OCCURRENCE
Amanzia greppini Schwarz et al., 2020	**Length:** 6.3 m **Mass:** 1,000 kg	**NMB M.H. series**—partial skeleton *(Schwarz et al., 2020)*	Reuchenette Fm., Switzerland *Late Jurassic, Kimmeridgian*
Bagualia alba Pol et al., 2020	**Length:** 9.5 m **Mass:** 2,800 kg	**MPEF-PV 3301**—partial skull and vertebrae; MPEF-PV 3305 through 3348—partial skeletons of three individuals *(Pol et al., 2020)*	Cañadón Asfalto Fm., Argentina *Early Jurassic, Toarcian*
Barapasaurus tagorei Jain et al., 1975	**Length:** 14 m **Mass:** 8,500 kg	**ISIR 50**—sacrum; ISRI series—many assorted skeletal elements *(Bandyopadhyay et al., 2010)*	Kota Fm., India *Early Jurassic, Hettangian/Sinemurian*
Cetiosauriscus stewarti Charig, 1980	**Length:** 16.8 m **Mass:** 11,000 kg	**BMNH R.3078**—partial skeleton *(Charig, 1980)*	Oxford Clay, United Kingdom *Middle Jurassic, Callovian*
Cetiosaurus oxoniensis Phillips, 1871	**Length:** 16 m **Mass:** 11,000 kg	OUMNH J13605–13613, J13615–16, J13619-J13688, J13899, J13614, J13617–8, J13780–1, J13596, LCM G468.1968—various bones *(Upchurch and Martin, 2003)*	Forest Marble, United Kingdom *Middle Jurassic, Bathonian*
Chebsaurus algeriensis Mahammed et al., 2005	**Length:** 8.2 m **Mass:** 1,500 kg	**D001–01 to 78**—partial skull and skeleton of juvenile *(Mahammed et al., 2005)*; D001–79 to D001–118—partial skull and skeleton of juvenile *(Läng and Mahammed, 2009)*	Unnamed, Algeria *Middle Jurassic, (?)*
Ferganasaurus verzilini Alifanov and Averianov, 2003	**Length:** 18 m **Mass:** 15,000 kg	**PIN N 3042/1**—partial skeleton (some material lost) *(Alifanov and Averianov, 2003)*	Balabansai Fm., Kyrgyzstan *Middle Jurassic, Callovian*
Haestasaurus becklesii Mantell, 1852	**Length:** 9 m **Mass:** 3,000 kg	**BMNH R1870**—forelimb, **BMNH R 1868**—skin impression *(Upchurch et al., 2015)*	Hastings Grp., United Kingdom *Early Cretaceous, Berriasian/Valanginian*
Janenschia robusta Wild, 1991	**Length:** 15.3 m **Mass:** 14,000 kg	**SMNS 12144**—hindlimb; MB.R.2090 series—limb and hip bones; MB.R.2095, 2245, 2707—limb bones *(Mannion et al., 2019)*	Tendaguru Fm., Tanzania *Late Jurassic, Tithonian*
Jobaria tiguidensis Sereno et al., 1999	**Length:** 15.2 m **Mass:** 17,000 kg	**MNN TIG3**—partial skeleton; MNN TIG4, TIG5—partial skull and skeletons; MNN TIG6—skeleton; MNN TIG7—braincase *(Sereno et al., 1999)*	Tiourarén Fm., Niger *Middle-Late Jurassic, Bathonian-Oxfordian*
Lapparentosaurus madagascariensis Bonaparte, 1986	**Length:** 12.3 m **Mass:** 10,500 kg	**MAA 91–92**—vertebral fragments; MAA series—skeletal fragments; AND 15–001, 2, and 3—skeletal fragments; possibly MNHN MAJ 289—skeletal fragments *(Raveloson et al., 2019)*	Isalo III Fm., Madagascar *Middle Jurassic, Bathonian*
Losillasaurus giganteus Casanovas et al., 2001	**Length:** 20 m **Mass:** 26,000 kg	**Lo-5**—tail vertebra; Lo-series—various skeletal and cranial elements *(Casanovas et al., 2001)*; MAP-6000 series—various skeletal and cranial elements; SHN 180—tail vertebra *(Royo-Torres et al., 2020)*	Villar del Arzobispo Fm., Spain Praia de Amoreira-Porto Novo Fm, Portugal *Late Jurassic, Kimmeridgian/Tithonian*

Mierasaurus bobyoungi Royo-Torres et al., 2017	**Length:** 11.5 m **Mass:** 4,800 kg	UMNH.VP.26004—partial skull and skeleton; UMNH.VP.26010—juvenile dentary; UMNH.VP.260111—juvenile femur *(Royo-Torres et al., 2017)*	Cedar Mt. Fm., United States *Early Cretaceous, Berriasian/Valanginian*
Moabosaurus utahensis Britt et al., 2017	**Length:** 12 m **Mass:** 6,000 kg	BYU 7510—three dorsal vertebrae; multiple BYU specimens—fragments *(Britt et al., 2017)*	Cedar Mt. Fm., United States *Early Cretaceous, Aptian*
Narindasaurus thevenini Royo-Torres et al., 2020	**Length:** 13.5 m **Mass:** 8,000 kg	MNHN MAJ 400-series—fragments *(Royo-Torres et al., 2020)*	Isalo III Fm., Madagascar *Middle Jurassic, Bathonian*
Nebulasaurus taito Xing et al., 2015a	**Length:** 8 m **Mass:** 1,700 kg	LDRC-v.d.1—braincase *(Xing et al., 2015a)*	Zhanghe Fm., China *Middle Jurassic, Aalenian/Bajocian*
Patagosaurus fariasi Bonaparte, 1979	**Length:** 18.6 m **Mass:** 12,300 kg	PVL 4170—partial skeleton; PVL 4615, PVL 4616, PVL 4171, PVL 4172, PVL 4076, PVL 4075, PVL 4617, MACN-CH 934—various skeletal and cranial remains *(Bonaparte, 1986)*; MPEF-PV 1670–dentary *(Rauhut, 2003)*	Cañadón Asfalto Fm., Argentina *Early–Middle Jurassic, Toarcian/?*
Perijasaurus lapaz Rincón et al., 2022	**Length:** 13 m **Mass:** 6,000 kg	UCMP 37689—partial vertebra *(Rincon et al., 2022)*	La Quinta Fm., Colombia *Early–Middle Jurassic, Toarcian/Aalenian*
Shunosaurus lii Dong et al., 1983	**Length:** 12.5 m **Mass:** 7,000 kg	IVPP V9065—partial skeleton; numerous specimens *(Zhang et al., 1984; Dong et al., 1989; Ma et al., 2022)*	Xiashaximiao Fm., China *Middle Jurassic, Oxfordian*
Spinophorosaurus nigerensis Remes et al., 2009	**Length:** 13.8 m **Mass:** 8,500 kg	GCP-CV-4229 and NMB-1699-R—partial skull and skeleton; NMB-1698-R—partial skeleton *(Remes et al., 2009)*; GCP-CV-BB-15—partial juvenile skeleton *(Páramo and Ortega, 2012)*	Irhazer Grp., Niger *Middle Jurassic, (?)*
Tehuelchesaurus benitezii Rich et al., 1999	**Length:** 16 m **Mass:** 16,000 kg	MPEF-PV 1125—partial skeleton *(Rich et al., 1999)*	Cañadón Asfalto Fm., Argentina *Early–Middle Jurassic, Toarcian/?*
Tendaguria tanzaniensis Bonaparte et al., 2000	**Length:** 16 m **Mass:** 12,000 kg	MB.R.2092.1 (NB4) and MB. R.2092.2 (NB5)—two dorsal vertebrae; MB.R.2091.21 (G45)—cervical vertebra *(Bonaparte et al., 2000)*	Tendaguru Fm., Tanzania *Late Jurassic, Tithonian*
Turiasaurus riodevensis Royo-Torres et al., 2006	**Length:** 24 m **Mass:** 30,000 kg	CPT1195 to CPT1210—partial skeleton; CPT-1211 to 1261—partial skull and skeleton *(Royo-Torres et al., 2006)*; CPT-3300—caudal vertebra *(Royo-Torres et al., 2008)*; CPT-1609 to CPT-1657 and CPT-1661 to 1674—partial skeleton *(Royo-Torres et al., 2009)*	Villar del Arzobispo Fm., Spain *Late Jurassic, Kimmeridgian/Tithonian*
Zby atlanticus Mateus et al., 2014	**Length:** 18 m **Mass:** 19,000 kg	ML 368—forelimb and fragments *(Mateus et al., 2014)*	Lourinhã Fm., Portugal *Late Jurassic, Kimmeridgian*

MAMENCHISAURIDAE

TYPE SPECIES	SIZE	GENUS MATERIAL	OCCURRENCE
Analong chuanjieensis Ren at al., 2018	**Length:** 16.7 m **Mass:** 10,000 kg	LFGT LCD 9701–1—partial skeleton *(Ren et al., 2018)*	Chuanjie Fm., China *Middle Jurassic, (?)*
Anhuilong diboensis Ren et al., 2018	**Length:** 17.5 m **Mass:** 6,300 kg	AGB 5822—forelimb *(Ren et al., 2018)*	Hongqin Fm., China *Middle Jurassic, (?)*
Bellusaurus sui Dong, 1990	**Length:** 12 m **Mass:** 8,000 kg	IVPP V 8299—partial juvenile skull and skeleton; unnumbered, V8300, and IVPP V17768 series—multiple juvenile skeletons *(Moore et al., 2018)*	Shishugou Fm., China *Late Jurassic, Oxfordian*
Chuanjiesaurus anaensis Fang et al., 2000	**Length:** 17 m **Mass:** 11,000 kg	Lfch 1001—partial skeleton *(Fang et al., 2000)*	Chuanjie Fm., China *Middle Jurassic, (?)*
Daanosaurus zhangi Ye et al., 2005	**Length:** 6.9 m **Mass:** 1,200 kg	ZDM0193—partial juvenile skull and skeleton *(Ye et al., 2005)*	Shangshaximiao Fm., China *Late Jurassic, Oxfordian*
Eomamenchisaurus yuanmouensis Lü et al., 2008	**Length:** 15.5 m **Mass:** 4,500 kg	CZMVZA 165—partial skeleton *(Lü et al., 2008)*	Zhanghe Fm., China *Middle Jurassic, Aalenian/Bajocian*
Huangshanlong anhuiensis Huang et al., 2014	**Length:** 16 m **Mass:** 4,900 kg	AGB5818—forelimb *(Huang et al., 2014)*	Hongqin Fm., China *Middle Jurassic, (?)*
Hudiesaurus sinojapanorum Dong, 1997	**Length:** 25 m **Mass:** 18,000 kg	IVPP V. 11120—cervical vertebra *(Dong, 1997)*	Kalazha Fm., China *Late Jurassic, Kimmeridgian/Tithonian*
Jingiella dongxingensis Ren et al., 2024	**Length:** 16 m **Mass:** 6,000 kg	DXJL2021001—skeletal fragments *(Ren et al., 2024)*	Dongxing Fm., China *Late Jurassic, (?)*
Klamelisaurus gobiensis Zhao, 1993	**Length:** 14.5 m **Mass:** 6,000 kg	IVPP V9492—partial skeleton *(Moore et al., 2020)*	Shishugou Fm., China *Middle Jurassic, Callovian*
Mamenchisaurus constructus Young, 1954	**Length:** 20 m **Mass:** 12,000 kg	IVPP V790—partial skeleton *(Young, 1954)*; IVPP V948—partial skeleton *(Young, 1954)*; IVPP V929—partial scapula; IVPP V946—partial skeleton; IVPP V947, V945, V932—fragments *(Young, 1958)*; GCC V 20401—partial skeleton *(Young and Chao, 1972)*; ZDM 0126—partial skull and skeleton *(Ye et al., 2001)*; IVPP V10603—fragments *(Russell and Zheng, 1994)*; ZDM 0083—skull and partial skeleton *(Pi et al., 1996)*; 001–003, AL101–106—vertebrae *(He et al., 1996)*; CV00734—skull and partial skeleton *(Zhang et al., 1998)*; CV00219—partial skeleton *(Zhang et al., 1998)*	Shangshaximiao Fm., China Shishugou Fm., China *Late Jurassic, Oxfordian*

Omeisaurus junghsiensis Young, 1939	**Length:** 16.5 m **Mass:** 6,000 kg	**CV-001**—partial skull; CV series—skull and jaw bones *(Young, 1939)*; numerous specimens see *(Young, 1958; Dong et al., 1983; He et al., 1984, 1988; Dong et al., 1989; Tang et al., 2001; Jiang et al., 2011; Tan et al., 2021)*	Xiashaximiao Fm., China Shangshaximiao Fm., China *Late Jurassic, Oxfordian*
Qijianglong guokr Xing et al., 2015b	**Length:** 13.2 m **Mass:** 5,200 kg	**QJGPM 1001**—partial skull and skeleton *(Xing et al., 2015b)*	Suining Fm., China *Early Cretaceous, Aptian*
Rhomaleopakhus turpanensis Upchurch et al., 2021	**Length:** 20 m **Mass:** 18,000 kg	**IVPP V11121-1**—forelimb *(Upchurch et al., 2021)*	Kalazha Fm., China *Late Jurassic, Kimmeridgian/Tithonian*
Tienshanosaurus chitaiensis Young, 1937	**Length:** 10 m **Mass:** 3,300 kg	**IVPP AS 40002-3**—partial skeleton *(Yang, 1937)*	Shishugou Fm., China *Late Jurassic, Oxfordian*
Tonganosaurus hei Li et al., 2010	**Length:** 11 m **Mass:** 2,000 kg	**MCDUT 14454**—partial skeleton *(Li et al., 2010)*	Yimen Fm., China *Early Jurassic, (?)*
Wamweracaudia keranjei Mannion et al., 2019	**Length:** 16.7 m **Mass:** 7,000 kg	**MB.R.2091.1-30, 3817.1,** and **3817.2 (G1-30)**—tail; MB.R.2094—tail vertebra *(Mannion et al., 2019)*	Tendaguru Fm., Tanzania *Late Jurassic, Tithonian*
Xinjiangtitan shanshanesis Wu et al., 2013	**Length:** 27 m **Mass:** 25,000 kg	**SSV12001**—partial skeleton *(Wu et al., 2013)*	Qiketai Fm., China *Middle Jurassic, Callovian*
Yuanmousaurus jiangyiensis Lü et al., 2006	**Length:** 17 m **Mass:** 11,000 kg	**YMV 601**—partial skeleton *(Lü et al., 2006)*	Zhanghe Fm., China *Middle Jurassic, Aalenian/Bajocian*

DIPLODOCOIDEA

TYPE SPECIES	SIZE	GENUS MATERIAL	OCCURRENCE
Amargasaurus cazaui Salgado and Bonaparte, 1991	**Length:** 13.5 m **Mass:** 3,500 kg	**MACN-N 15**—partial skull and skeleton *(Salgado and Bonaparte, 1991)*; MOZ-Pv 6126—vertebrae *(Windholz et al., 2021)*	La Amarga Fm., Argentina *Early Cretaceous, Barremian/Aptian*
Amargatitanis macni Apesteguía, 2007	**Length:** 12 m **Mass:** 3,100 kg	**MACN PV N53**—fragments *(Gallina, 2016)*	La Amarga Fm., Argentina *Early Cretaceous, Barremian/Aptian*
Amphicoelias altus Cope, 1878	**Length:** 25 m **Mass:** 26,000 kg	**AMNH 5764**—fragments *(Tschopp et al., 2015)*	Morrison Fm., United States *Late Jurassic, Kimmeridgian/Tithonian*
Apatosaurus ajax Marsh, 1877b	**Length:** 23 m **Mass:** 20,000 kg	**YPM 1860**—partial skeleton; CM 3018—partial skeleton; CM 3378—partial skeleton; CM 11162—partial skull; YPM 1861—partial skeleton; numerous specimens of contentious identification *(Tschopp et al., 2015)*	Morrison Fm., United States *Late Jurassic, Kimmeridgian/Tithonian*
Ardetosaurus viator (van der Linden et al., 2024)	**Length:** 19.4 m **Mass:** 6,000 kg	**MAB011899**—partial skeleton *(van der Linden et al., 2024)*	Morrison Fm., Wyoming, USA *Late Jurassic, Kimmeridgian*
Atlantosaurus montanus Marsh, 1877a	**Length:** 23 m **Mass:** 20,000 kg	**YPM 1835**—partial sacrum *(Marsh, 1877a)*	Morrison Fm., United States *Late Jurassic, Kimmeridgian/Tithonian*
Bajadasaurus pronuspinax Gallina et al., 2019	**Length:** 11.5 m **Mass:** 2,200 kg	**MMCh-PV 75**—partial skull and neck vertebrae *(Gallina et al., 2019)*	Bajada Colorada Fm., Argentina *Early Cretaceous, Berriasian/Valanginian*
Barosaurus lentus Marsh, 1890	**Length:** 27 m **Mass:** 10,000 kg	**YPM 429**—partial skeleton; AMNH 6341—partial skeleton; possibly YPM 419—partial rear foot; possibly AMNH 7530—partial skull and skeleton; AMNH 7535—partial skeleton; possibly CM 11984—partial skeleton *(Tschopp et al., 2015)*; ROM 3670 *(Ahmed, 2021)*	Morrison Fm., United States *Late Jurassic, Kimmeridgian/Tithonian*
Brachytrachelopan mesai Rauhut et al., 2005	**Length:** 8.5 m **Mass:** 3,000 kg	**MPEF-PV 1716**—partial skeleton *(Rauhut et al., 2005)*	Cañadón Cálcareo Fm., Argentina *Late Jurassic, Oxfordian/?*
Brontosaurus excelsus Marsh, 1879	**Length:** 21 m **Mass:** 16,000 kg	**YPM 1980**—partial skeleton; YPM 1981—partial skeleton; CM 566—fragments; UW 15556 (previously CM 563)—partial skeleton; BYU 1252-18531 (provisionally)—partial skeleton; Tate-001—partial skeleton *(Tschopp et al., 2015)*	Morrison Fm., United States *Late Jurassic, Kimmeridgian/Tithonian*
Dicraeosaurus hansemanni Janensch, 1914	**Length:** 15.5 m **Mass:** 6,000 kg	**"Skeleton m"**—partial skeleton; numerous partial specimens *(Janensch, 1914, 1921, 1929, 1961; Schwarz-Wings and Böhm, 2012)*	Tendaguru Fm., Tanzania *Late Jurassic, Tithonian*
Dinheirosaurus lourinhanensis Bonaparte and Mateus, 1999	**Length:** 21 m **Mass:** 8,800 kg	**ML 414**—partial skeleton *(Bonaparte and Mateus, 1999)*	Lourinhã Fm., Portugal *Late Jurassic, Kimmeridgian*
Diplodocus longus Marsh, 1878	**Length:** 26 m **Mass:** 13,500 kg	**YPM 1920**—partial tail *(Marsh, 1878)*; CM 84—partial skeleton; CM 94—partial skeleton; NMMMNH 3690—partial skeleton; AMNH 233—partial skeleton; DMNS 1494—partial skeleton; USNM 10865—partial skeleton *(Tschopp et al., 2015)*; CM 11255, CM 3452, CM 11161, MOR 7029, SMM P84.15.3, CMC VP14128—skulls *(Woodruff et al., 2018)*; numerous others	Morrison Fm., United States *Late Jurassic, Kimmeridgian/Tithonian*
Dyslocosaurus polyonychius McIntosh et al., 1992	**Length:** 16 m **Mass:** 7,500 kg	**AC 663**—partial legs *(McIntosh et al., 1992)*	Morrison Fm., United States *Late Jurassic, Kimmeridgian/Tithonian*
Galeamopus hayi Tschopp et al., 2015	**Length:** 23 m **Mass:** 10,000 kg	**HMNS 175** (formerly CM 662)—partial skull and skeleton; AMNH 969—skull; USNM 2673—partial skull; SMA 0011—partial skull and skeleton *(Tschopp and Mateus, 2017)*; HMNS 17512,27 *(Woodruff et al., 2018)*	Morrison Fm., United States *Late Jurassic, Kimmeridgian/Tithonian*

Haplocanthosaurus priscus Hatcher, 1903	**Length:** 16 m **Mass:** 13,000 kg	CM 572—partial skeleton; CM 879—partial skeleton; CMNH 10380—partial skeleton; MWC 8028—partial skeleton; potentially several other specimens *(Forster and Wedel, 2014; Curtice et al., 2023)*	Morrison Fm., United States *Late Jurassic, Kimmeridgian/Tithonian*
Kaatedocus siberi Tschopp and Mateus, 2013	**Length:** 13 m **Mass:** 2,000 kg	SMA 0004—partial skull and neck *(Tschopp and Mateus, 2012)*; SMA D16–3—partial skull; possibly AMNH 7530—partial skeleton *(Tschopp et al., 2015)*	Morrison Fm., United States *Late Jurassic, Kimmeridgian/Tithonian*
Leinkupal laticauda Gallina et al., 2014	**Length:** 11 m **Mass:** 1,700 kg	MMCH-Pv 63–1—tail vertebra; MMCH-Pv 63–2 through 8—various vertebrae *(Gallina et al., 2014)*; MMCh-Pv series—fragments *(Gallina et al., 2019)*; MMCh-Pv- 232—partial braincase *(Garderes et al., 2022)*	Bajada Colorada Fm., Argentina *Early Cretaceous, Berriasian/Valanginian*
Lingwulong shenqi Xu et al., 2018	**Length:** 17 m **Mass:** 8,000 kg	LM V001a—partial skull; LM V001b—partial skeleton; IVPP V23704—teeth; LGP V002 through V006—partial skeletons *(Xu et al., 2018)*	Yanan Fm., China *Middle Jurassic, Aalenian*
Pilmatueia faundezi Coria et al., 2019	**Length:** 14.5 m **Mass:** 4,200 kg	MLL-Pv-005—dorsal vertebra; MLL-Pv series—vertebrae and femur *(Coria et al., 2019)*; MLL-Pv-010—various elements *(Windholz et al., 2022)*	Mulichinco Fm., Argentina *Early Cretaceous, Valanginian*
Smitanosaurus agilis Whitlock and Mantilla, 2020	**Length:** 22 m **Mass:** 17,000 kg	USNM 5384—partial skull and vertebrae *(Whitlock and Mantilla, 2020)*	Morrison Fm., United States *Late Jurassic, Kimmeridgian/Tithonian*
Supersaurus vivianae Jensen, 1985	**Length:** 39 m **Mass:** 27,000 kg	BYU 12962—partial skeleton (same individual as other BYU specimens from the locality); WDC DMJ-021—partial skeleton *(Tschopp et al., 2015)*	Morrison Fm., United States *Late Jurassic, Kimmeridgian/Tithonian*
Suuwassea emilieae Harris and Dodson, 2004	**Length:** 16.2 m **Mass:** 7,000 kg	ANS 21122—partial skull and skeleton *(Harris and Dodson, 2004)*	Morrison Fm., United States *Late Jurassic, Kimmeridgian/Tithonian*
Tharosaurus indicus Bajpai et al., 2023	**Length:** 12 m **Mass:** 3,000 kg	RWR-241A–K—fragments *(Bajpai et al., 2023)*	Jaisalmer Fm., India *Middle Jurassic, Bathonian*
Tornieria africana Fraas, 1908	**Length:** 24 m **Mass:** 13,500 kg	SMNS 12141a and other elements from "Skeleton A"—fragments; "Skeleton k"—fragments; MB.R series—fragments *(Remes, 2006)*	Tendaguru Fm., Tanzania *Late Jurassic, Tithonian*

REBBACHISAURIDAE

TYPE SPECIES	SIZE	GENUS MATERIAL	OCCURRENCE
Agustinia ligabuei Bonaparte, 1999	**Length:** 16.7 m **Mass:** 9,800 kg	MCF-PVPH-110—fragmentary skeleton *(Bellardini et al., 2022)*	Lohan Cura Fm., Argentina *Early Cretaceous, Aptian/Albian*
Amazonsaurus maranhensis Carvalho et al., 2003	**Length:** 10.5 m **Mass:** 2,500 kg	Collection of differently numbered fragments *(Carvalho et al., 2003)*	Itapecuru Fm., Brazil *Early Cretaceous, Aptian/Albian*
Cathartesaura anaerobica Gallina and Apesteguía, 2005	**Length:** 14.8 m **Mass:** 7,000 kg	MPCA-232—fragments *(Gallina and Apesteguía, 2005)*	Huincul Fm., Argentina *Late Cretaceous, Cenomanian*
Comahuesaurus windhauseni Carballido et al. 2012	**Length:** 14 m **Mass:** 5,800 kg	MOZ-PV series—fragments *(Carballido et al. 2012)*	Lohan Cura Fm., Argentina *Early Cretaceous, Aptian/Albian*
Demandasaurus darwini Fernández-Baldor et al., 2011	**Length:** 10.8 m **Mass:** 2,400 kg	MDS–RVII series—partial skeleton *(Fernández-Baldor et al., 2011)*	Castrillo de la Reina Fm., Spain *Early Cretaceous, Barremian/Aptian*
Dzharatitanis kingi Averianov and Sues, 2021	**Length:** 15 m **Mass:** 9,000 kg	USNM 538127—caudal vertebra *(Averianov and Sues, 2021)*	Bissekty Fm., Uzbekistan *Late Cretaceous, Turonian*
Histriasaurus boscarollii Dalla Vecchia, 1998	**Length:** 11.5 m **Mass:** 3,300 kg	WN-V6—partial vertebra *(Dalla Vecchia, 1998)*	unnamed, Croatia *Early Cretaceous, Hauterivian/Barremian*
Itapeuasaurus cajapioensis Lindoso et al., 2019	**Length:** 7.5 m **Mass:** 800 kg	UFMA.1.10 series—fragments *(Lindoso et al., 2019)*	Alcântara Fm., Brazil *Late Cretaceous, Cenomanian*
Katepensaurus goicoecheai Ibiricu et al., 2013	**Length:** 13 m **Mass:** 4,700 kg	UNPSJB-PV 1007—vertebrae and fragments *(Ibiricu et al., 2015)*	Bajo Barreal Fm., Argentina *Late Cretaceous, Cenomanian*
Lavocatisaurus agrioensis Canudo et al., 2018	**Length:** 10.2 m **Mass:** 2,300 kg	MOZ-Pv1232—partial skull and skeleton; MOZ-Pv series—fragments (juvenile) *(Canudo et al., 2018)*	Rayoso Fm., Argentina *Early Cretaceous, Aptian/Albian*
Limaysaurus tessonei Salgado et al., 2004	**Length:** 15 m **Mass:** 7,500 kg	MUCPv-205—partial skeleton; MUCPv-206—partial skeleton; MUCPv-153—partial skeleton *(Calvo and Salgado, 1995; Salgado et al., 2004)*	Candeleros Fm., Argentina Huincul Fm., Argentina *Late Cretaceous, Cenomanian*
Maraapunisaurus fragillimus Cope, 1878	**Length:** 35 m **Mass:** 70,000 kg	AMNH 5777—partial vertebra *(Carpenter, 2018)*	Morrison Fm., United States *Late Jurassic, Kimmeridgian/Tithonian*
Nigersaurus taqueti Sereno et al., 1999	**Length:** 9 m **Mass:** 1,900 kg	MNN CDF512—skull and skeleton *(Sereno et al., 1999)*; MNN GAD513, MNN GAD515 through 518—partial skulls and skeletons *(Sereno et al., 2007)*	Elrhaz Fm., Niger *Early Cretaceous, Aptian/Albian*
Nopcsaspondylus alarconensis Apesteguía, 2007	**Length:** 11.5 m **Mass:** 3,000 kg	Unnumbered—partial vertebra *(Apesteguía, 2007)*	Candeleros Fm., Argentina *Late Cretaceous, Cenomanian*
Rayososaurus agrioensis Bonaparte, 1996	**Length:** 13.1 m **Mass:** 4,900 kg	MACN-N 41—scapula, partial leg; UFMA 1.10 series—vertebral fragments *(Carballido et al., 2010)*	Candeleros Fm., Argentina *Late Cretaceous, Cenomanian*

Rebbachisaurus garasbae Lavocat, 1954	**Length:** 26 m **Mass:** 12,000 kg	**MNHN-MRS (various)**—skeletal fragments; NMC 50844—vertebral fragments *(Wilson and Allain, 2015)*	Kem Kem Beds, Morocco *Late Cretaceous, Cenomanian*
Sidersaura marae Lerzo et al., 2024	**Length:** 21 m **Mass:** 20,000 kg	**MMCh-PV 70**—partial skull and skeleton; several partial specimens *(Lerzo et al., 2024)*	Huincul Fm., Argentina *Late Cretaceous, Cenomanian*
Tataouinea hannibalis Fanti et al., 2013	**Length:** 15 m **Mass:** 6,400 kg	**ONM DT 1–36**—partial hip and tail *(Fanti et al., 2013)*	Aïn el Guettar Fm., Tunisia *Early Cretaceous, Albian*
Xenoposeidon proneneukos Taylor and Naish, 2007	**Length:** 14.8 m **Mass:** 7,000 kg	**BMNH R2095**—partial vertebra *(Taylor and Naish, 2007)*	Ashdown Beds Fm., United Kingdom *Early Cretaceous, Berriasian/Valanginian*
Zapalasaurus bonapartei Salgado et al., 2006	**Length:** 10 m **Mass:** 2,300 kg	**Pv-6127-MOZ**—partial skeleton *(Salgado et al., 2006)*	La Amarga Fm., Argentina *Early Cretaceous, Barremian/Aptian*

MACRONARIA

TYPE SPECIES	SIZE	GENUS MATERIAL	OCCURRENCE
Abrosaurus dongpoi Ouyang, 1989	**Length:** 9.5 m **Mass:** 4,100 kg	**ZDM5038**—skull; ZDM5033—partial skull *(Ouyang, 1989)*	Tendaguru Fm., Tanzania *Late Jurassic, Tithonian*
Abydosaurus mcintoshi Chure et al., 2010	**Length:** 20 m **Mass:** 26,000 kg	**DINO 16488**—skull; partial remains of several individuals *(Chure et al., 2010)*	Cedar Mt. Fm., United States *Early Cretaceous, Albian*
Aragosaurus ischiaticus Sanz et al., 1987	**Length:** 18 m **Mass:** 7,900 kg	**ZH series**—partial skeleton *(Sanz et al., 1987)*	Villar del Arzobispo Fm., Spain *Late Jurassic, Kimmeridgian/Tithonian*
Atlasaurus imelakei Monbaron et al., 1999	**Length:** 15.5 m **Mass:** 21,000 kg	**No number**—skeleton and partial skull *(Monbaron et al., 1999)*; private collection—tail *(Reuters, 2018)*	Guettioua Ss., Morocco *Middle Jurassic, Bathonian*
Brachiosaurus altithorax Riggs, 1903	**Length:** 26 m **Mass:** 50,000 kg	**FMNH P 25107**—partial skeleton; USNM 5730—partial skull; BYU 4744—partial skeleton; USNM 21903—humerus; BYU 12866—vertebra; BYU 12867—vertebra; OMNH 01138—metacarpal; BYU 9462—scapula *(Taylor, 2009)*; USNM PAL337859 and PAL337923—partial vertebrae *(Smithsonian Institution Archives)*; KUVP 129724, 142200, 133862, 144767—partial foot *(Maltese et al., 2018)*; FHPR 17108—humerus *(Katz, 2020)*; BYU 9754—forelimb and fragments *(D'Emic and Carrano, 2019)*	Morrison Fm., United States *Late Jurassic, Kimmeridgian/Tithonian*
Camarasaurus supremus Cope, 1877b	**Length:** 20 m **Mass:** 30,000 kg	**AMNH 5760**—partial skeleton; more than fifty sets of remains *(Ikejiri, 2005; Woodruff and Foster, 2017; Woodruff et al., 2021)*	Morrison Fm., United States *Late Jurassic, Kimmeridgian/Tithonian*
Cedarosaurus weiskopfae Tidwell et al., 1999	**Length:** 15 m **Mass:** 10,500 kg	**DMNH 39045**—partial skeleton; FMNH PR 977—hindlimb *(D'Emic, 2013)*	Cedar Mt. Fm., United States *Early Cretaceous, Albian*
Dashanpusaurus dongi Peng et al., 2005	**Length:** 13.8 m **Mass:** 5,200 kg	**ZDM 5028**—partial skeleton; ZDM 5027—partial skeleton *(Ren et al., 2023)*	Xiashaximiao Fm., China *Late Jurassic, Oxfordian*
Datousaurus bashanensis Dong and Tang, 1984	**Length:** 14 m **Mass:** 10,000 kg	**IVPPV 7262, 7263**—partial skull and skeleton; CV 00740—partial skull; ZDM 5021—skeleton *(Dong and Tang, 1984)*	Xiashaximiao Fm., China *Late Jurassic, Oxfordian*
Dinodocus mackesoni Owen, 1884	**Length:** 15 m **Mass:** 10,300 kg	**NHMUK 14695**—partial forelimb *(Owen, 1884)*	Atherfield Clay Fm., United Kingdom *Early Cretaceous, Aptian*
Duriatitan humerocristatus Barrett et al., 2010	**Length:** 15 m **Mass:** 10,000 kg	**BMNH 44635**—humerus *(Barrett et al., 2010)*; MG 4976—humerus *(Mocho et al., 2016)*; MDS-VPCR 214—humerus *(Fernández-Baldor et al., 2020)*	Kimmeridge Clay, United Kingdom *Late Jurassic, Kimmeridgian*
Dystrophaeus viaemalae Cope, 1877a	**Length:** 18.7 m **Mass:** 10,500 kg	**USNM 2364**—fragments *(Cope, 1877a)*	Morrison Fm., United States *Late Jurassic, Kimmeridgian/Tithonian*
Eucamerotus foxi Blows, 1995	**Length:** 14 m **Mass:** 8,500 kg	**BMNH R2522**—vertebral arch; BMNH R89, R90, R2524—dorsal vertebrae *(Blows, 1995)*; R88, R91, R2523, R406, R708—dorsal vertebrae *(Upchurch et al., 2011)*	Wessex Fm., United Kingdom *Early Cretaceous, Barremian*
Europasaurus holgeri Sander et al., 2006	**Length:** 6 m **Mass:** 800 kg	**DFMMh/FV 291 series**—partial skull and skeleton *(Sander et al. 2006)*; material from at least 21 additional individuals *(Carballido et al., 2020)*	Süntel Fm., Germany *Late Jurassic, Kimmeridgian*
Fushanosaurus qitaiensis Wang et al., 2019	**Length:** 20 m **Mass:** 24,000 kg	**FH000101**—right femur *(Wang et al., 2019)*	Shishugou Fm., China *Late Jurassic, Oxfordian*
Fusuisaurus zhaoi Mo et al., 2006	**Length:** 25 m **Mass:** 35,000 kg	**NHMG 6729**—fragments *(Mo et al., 2006)*; LCL 63—humerus *(Mo et al., 2020)*	Napai Fm., China *Early Cretaceous, Aptian/Albian*
Galvesaurus herreroi Barco et al., 2005	**Length:** 18 m **Mass:** 14,000 kg	**CLH-16**—dorsal vertebra; series of additional fragments *(Pérez-Pueyo et al., 2019)*	Villar del Arzobispo Fm., Spain *Late Jurassic, Kimmeridgian/Tithonian*
Giraffatitan brancai Paul, 1988	**Length:** 25 m **Mass:** 48,000 kg	**MB.R.2180** {aka HMN SI}—partial skull and skeleton; MB.R.2181 {aka HMN SII}—partial skull and skeleton; MB.R.5000—tail; MB.R.3736—partial tail; MB.R.2921 {aka HMN Aa}–partial tail; HMN t—atlas; HMN be 1—axis; HMN t1—juvenile skull; HMN S116—skull; HMN S66—partial juvenile skull; HMN WJ 4170—juvenile upper jaw; HMN Sa9—scapula; HMN XV2—fibula *(Taylor, 2009; Wedel and Taylor, 2013; Kosch et al., 2014)*	Tendaguru Fm., Tanzania *Late Jurassic, Tithonian*

Type Species	Size	Genus Material	Occurrence
Lourinhasaurus alenquerensis *Dantas et al., 1998*	**Length:** 17.5 m **Mass:** 19,000 kg	MG LNEG—partial skeleton *(Mocho et al., 2014)*	Lourinhã Fm., Portugal *Late Jurassic, Kimmeridgian*
Lusotitan atalaiensis *Antunes and Mateus, 2003*	**Length:** 24 m **Mass:** 34,000 kg	MIGM series—partial skeleton *(Antunes and Mateus, 2003)*	Lourinhã Fm., Portugal *Late Jurassic, Kimmeridgian*
Ornithopsis hulkei *Seeley, 1870*	**Length:** 17.4 m **Mass:** 15,500 kg	**NHMUK 28632**—partial vertebra *(Upchurch et al., 2011)*	Wessex Fm., United Kingdom *Early Cretaceous, Barremian*
Pelorosaurus brevis *Mantell, 1850*	**Length:** 14 m **Mass:** 9,000 kg	NHMUK 28626—humerus; NHMUK R2544–2547—four anterior caudal vertebrae; NHMUK R2548–2550—three chevrons *(Upchurch et al., 2011)*	Grinstead Clay Fm., United Kingdom *Early Cretaceous, Valanginian*
Rugocaudia cooneyi *Woodruff, 2012*	**Length:** 14 m **Mass:** 6,300 kg	**MOR 334**—partial tail and fragments *(Woodruff, 2012)*	Cloverly Fm., United States *Early Cretaceous, Aptian/Albian*
Sonorasaurus thompsoni *Ratkevich, 1998*	**Length:** 18 m **Mass:** 18,000 kg	**ASDM 500**—partial skeleton *(Ratkevich, 1998)*	Turney Ranch Fm., United States *Late Cretaceous, Cenomanian*
Soriatitan golmayensis *Royo-Torres et al., 2017*	**Length:** 11 m **Mass:** 4,000 kg	**MNS 2001/122**—fragments *(Royo-Torres et al., 2017)*	Golmayo Fm., Spain *Early Cretaceous, Hauterivia/Barremian*
Venenosaurus dicrocei *Tidwell et al., 2001*	**Length:** 13.5 m **Mass:** 8,000 kg	DMNH 40932—fragments *(Tidwell et al., 2001)*; DMNH 40930—fragments (juvenile) *(Tidwell and Wilhite, 2005)*; possible specimens *(Britt et al., 2009)*	Cedar Mt. Fm., United States *Early Cretaceous, Aptian*
Vouivria damparisensis *Mannion et al., 2017*	**Length:** 15 m **Mass:** 9,000 kg	**MNHN.F.1934.6 DAM**—partial skeleton *(Mannion et al., 2017)*	Calcaires de Clerval Fm., France *Late Jurassic, Oxfordian*
Yuzhoulong qurenensis *Dai et al., 2022*	**Length:** 9.5 m **Mass:** 3,000 kg	**CLGRP V0013**—partial skull and skeleton *(Dai et al., 2022)*	Xiashaximiao Fm., China *Late Jurassic, Oxfordian*

SOMPHOSPONDYLI

TYPE SPECIES	SIZE	GENUS MATERIAL	OCCURRENCE
Algoasaurus bauri *Broom, 1904*	**Length:** 7 m **Mass:** 1,000 kg	**Unnumbered** and **AMNH 5631**—fragments; SAM-PK-K1500—tail vertebra *(McPhee et al., 2016)*	Kirkwood Fm., South Africa *Early Cretaceous, Berriasian/Valanginian*
Angolatitan adamastor *Mateus et al., 2011*	**Length:** 13 m **Mass:** 6,000 kg	**MGUAN-PA3**—forelimb *(Mateus et al., 2011)*	Itombe Fm., Angola *Late Cretaceous, Coniacian*
Arkharavia heterocoelica *Alifanov and Bolotsky, 2010*	**Length:** 6.3 m **Mass:** 1,000 kg	**AEIM 2/418**—anterior caudal vertebra *(Alifanov and Bolotsky, 2010)*	Udurchurkan Fm., Russia *Late Cretaceous, Maastrichtian*
Astrophocaudia slaughteri *D'Emic, 2013*	**Length:** 11 m **Mass:** 3,000 kg	**SMU 61732**—partial skeleton; 203/73655—tooth *(D'Emic, 2013)*	Paluxy Fm., United States *Early Cretaceous, Albian*
Australodocus bohetii *Remes, 2007*	**Length:** 20 m **Mass:** 18,000 kg	**MB.R.2455 [G 70]**—neck vertebra; MB.R.2454 [G 69]—neck vertebra *(Whitlock, 2011)*	Udurchurkan Fm., Russia *Late Cretaceous, Maastrichtian*
Austrosaurus mckillopi *Longman, 1933*	**Length:** 18 m **Mass:** 13,000 kg	QM F2316—fragments; KK F1020—fragments *(Poropat et al., 2017)*	Tendaguru Fm., Tanzania *Late Jurassic, Tithonian*
Brontomerus mcintoshi *Taylor et al., 2011*	**Length:** 13.2 m **Mass:** 7,000 kg	OMNH 66430—partial ilium; OMNH series—fragments *(Taylor et al., 2011)*	Cedar Mt. Fm., United States *Early Cretaceous, Aptian*
Chubutisaurus insignis *del Corro, 1975*	**Length:** 18.5 m **Mass:** 14,000 kg	MACN 18222—partial skeleton *(Carballido et al., 2011)*	Cerro Barcino Fm., Argentina *Late Cretaceous, Cenomanian*
Dongbeititan dongi *Wang et al., 2007*	**Length:** 14 m **Mass:** 6,000 kg	**DNHM D2867**—partial skeleton *(Wang et al., 2007)*	Yixian Fm., China *Early Cretaceous, Barremian*
Europatitan eastwoodi *Fernández-Baldor et al., 2017*	**Length:** 21.5 m **Mass:** 22,000 kg	**MDS-OTII,1** through **MDS-OTII,32**—partial skeleton *(Fernández-Baldor et al., 2017)*	Castrillo de la Reina Fm., Spain *Early Cretaceous, Barremian/Aptian*
Fukuititan nipponensis *Azuma and Shibata, 2010*	**Length:** 16 m **Mass:** 6,000 kg	**FPDM-V8468**—partial skeleton *(Azuma and Shibata, 2010)*	Kitadani Fm., Japan *Early Cretaceous, Aptian*
Garumbatitan morellensis *Mocho et al., 2024*	**Length:** 18 m **Mass:** 14,000 kg	**SAV series**—partial skeleton; several others *(Mocho et al., 2023)*	Arcillas de Morella Fm., Spain *Early Cretaceous, Barremian*
Huanghetitan liujiaxiaensis *You et al., 2006*	**Length:** 18 m **Mass:** 12,000 kg	**GSLTZP02–001**—fragments *(You et al., 2006)*	Hekou Grp., China *Early Cretaceous, (?)*
Jiutaisaurus xidiensis *Wu et al., 2006*	**Length:** 16.3 m **Mass:** 5,900 kg	**CAD-02**—partial tail *(Wu et al., 2006)*	Quantou Fm., China *Early Cretaceous, Albian*
Liaoningotitan sinensis *Zhou et al., 2018*	**Length:** 13.3 m **Mass:** 4,500 kg	**PMOL-AD00112**—partial skull and skeleton *(Zhou et al., 2018)*	Yixian Fm., China *Early Cretaceous, Barremian*
Ligabuesaurus leanzai *Bonaparte et al., 2006*	**Length:** 23 m **Mass:** 27,000 kg	MCF-PHV-233—partial skull and skeleton *(Bonaparte et al., 2006)*; MCF-PVPH-228, 261, 744, 908—fragments *(Bellardini et al., 2022)*	Lohan Cura Fm., Argentina *Early Cretaceous, Aptian/Albian*
Liubangosaurus hei *Mo et al., 2010*	**Length:** 21 m **Mass:** 26,000 kg	**NHMG8152**—dorsal vertebrae *(Mo et al., 2010)*	Napai Fm., China *Early Cretaceous, Aptian/Albian*

Malarguesaurus florenciae González Riga et al., 2009	**Length:** 20 m **Mass:** 16,000 kg	**IANIGLA-PV 110**—partial skeleton; **IANIGLA-PV 111**—fragments *(González Riga et al., 2009)*	Portezuelo Fm., Argentina *Late Cretaceous, Turonian/Coniacian*
Oceanotitan dantasi Mocho et al., 2019	**Length:** 16 m **Mass:** 10,000 kg	**SHN 181**—fragments *(Mocho et al., 2019)*	Praia da Amoreira-Porto Novo Fm., Portugal *Late Jurassic, Kimmeridgian/Tithonian*
Padillasaurus leivaensis Carballido et al., 2015	**Length:** 16 m **Mass:** 1?,000 kg	**JACVM 0001**—vertebrae *(Carballido et al., 2015)*	Paja Fm., Colombia *Early Cretaceous, Barremian*
Paluxysaurus jonesi Rose, 2007	**Length:** 20 m **Mass:** 12,500 kg	**FWMSH 93B-10 series**—partial skull and skeleton *(Rose, 2007)*	Cloverly Fm., United States *Early Cretaceous, Albian*
Pukyongosaurus millenniumi Dong et al., 2001	**Length:** 14 m **Mass:** 8,500 kg	**PKNUG-G.102-109**—fragments *(Dong et al., 2001)*; PKNU GS08–05—caudal vertebra *(Paik et al., 2011)*	Hasandong Fm., South Korea *Early Cretaceous, Aptian*
Sauroposeidon proteles Wedel et al., 2000	**Length:** 29 m **Mass:** 40,000 kg	**OMNH 53062**—neck vertebrae *(Wedel et al., 2000)*; YPM series—fragments; UM 20800—shoulder *(D'Emic and Foreman, 2012)*	Antlers Fm., Cloverly Fm., Twin Mts. Fm., United States *Early Cretaceous, Aptian/Albian*
Sibirotitan astrosacralis Averianov et al., 2018	**Length:** 21.5 m **Mass:** 21,500 kg	**PM TGU series**—fragments *(Averianov et al., 2017)*; PIN no. 929/14—axis *(Averianov and Lopatin, 2022)*	Ilek Fm., Russia *Early Cretaceous, Barremian*
Tastavinsaurus sanzi Canudo et al., 2008	**Length:** 15 m **Mass:** 7,500 kg	**MPZ 99/9**—partial skeleton *(Canudo et al., 2008)*; CPT series—partial skeleton *(Royo-Torres et al., 2012)*	Xert Fm., Forcall Fm., Spain *Early Cretaceous, Aptian*
Triunfosaurus leonardii Carvalho et al., 2017	**Length:** 21 m **Mass:** 17,000 kg	**UFRJ-DG-498 series**—fragments *(Carvalho et al., 2017)*	Rio Piranhas Fm., Brazil *Early Cretaceous, Berriasian/Valanginian*
Wintonotitan wattsi Hocknull et al., 2009	**Length:** 15 m **Mass:** 7,000 kg	**QM F7292**—partial skeleton; QM F10916—caudal vertebrae *(Poropat et al., 2015)*	Winton Fm., Australia *Late Cretaceous, Cenomanian/Turonian*

EUHELOPODIDAE

TYPE SPECIES	SIZE	GENUS MATERIAL	OCCURRENCE
Erketu ellisoni Ksepka and Norell, 2006	**Length:** 16 m **Mass:** 5,500 kg	**IGM 100/1803**—neck and lower rear leg *(Ksepka and Norell, 2010)*	Bayan Shireh Fm. (?), Mongolia *Late Cretaceous, (?)*
Euhelopus zdanskyi Romer, 1956	**Length:** 13 m **Mass:** 6,000 kg	**PMU.R233** {aka PMU 24705}—skull and partial skeleton; PMU 24706—partial skeleton *(Poropat, 2013)*	Mengyin Fm., China *Early Cretaceous, Berriasian/Valanginian*
Gobititan shenzhouensis You et al., 2003	**Length:** 15 m **Mass:** 8,000 kg	**IVPP 12579**—tail and hindlimb *(You et al., 2003)*	Zhonggou Fm., China *Early Cretaceous, Albian*
Phuwiangosaurus sirindhornae Martin et al., 1994	**Length:** 19 m **Mass:** 11,000 kg	**SM PW1-1 through 22**—partial skeleton *(Martin et al., 1994)*; SM PW1–23 and 24—vertebrae *(Suteethorn et al., 2010)*; SM K11—partial skeleton *(Suteethorn et al., 2009)*; various other remains *(Klein et al., 2009)*	Sao Khua Fm., Thailand *Early Cretaceous, Valanginia/Hauterivian*
Qiaowanlong kangxii You and Li, 2009	**Length:** 19 m **Mass:** 13,000 kg	**FRDC GJ 07-14**—partial skeleton *(You and Li, 2009)*	Xiagou Fm., China *Early Cretaceous, Aptian*
Ruyangosaurus giganteus Lü et al., 2009	**Length:** 27 m **Mass:** 46,000 kg	**41HIII-0002**—fragments *(Lü et al., 2009)*; various referred specimens—partial skeleton *(Lü et al., 2014)*	Haoling Fm., China *Early Cretaceous, Aptian/Albian*
Silutitan sinensis Wang et al., 2021	**Length:** 20 m **Mass:** 22,000 kg	**IVPP V27874**—partial neck *(Wang et al., 2021)*	Shengjinkou Fm., China *Early Cretaceous, (?)*
Tangvayosaurus hoffeti Allain et al., 1999	**Length:** 19 m **Mass:** 17,000 kg	**TV4–1 through TV4–36**—partial skeleton; TV2–40—left fibula; TV2–39—left tibia; TV2–6—caudal vertebra *(Allain et al., 1999)*	Grès supérieurs Fm., Laos *Early Cretaceous, Aptian/Albian*
Yongjinglong datangi Li et al., 2014	**Length:** 15 m **Mass:** 7,500 kg	**GSGM ZH (08)-04**—partial skeleton *(Li et al., 2014)*	Hekou Grp., China *Early Cretaceous, (?)*
Yunmenglong ruyangensis Lü et al., 2013	**Length:** 27 m **Mass:** 29,000 kg	**41HIII-0006**—vertebrae and femur *(Lü et al., 2013)*	Haoling Fm., China *Early Cretaceous, Aptian/Albian*

TITANOSAURIA

TYPE SPECIES	SIZE	GENUS MATERIAL	OCCURRENCE
Abdarainurus barsboldi Averianov and Lopatin, 2020	**Length:** 15 m **Mass:** 10,000 kg	**PIN 5669/1**—tail vertebrae *(Averianov and Lopatin, 2020)*	Alagteeg Fm., Mongolia *Late Cretaceous, (?)*
Aegyptosaurus baharijensis Stromer, 1932	**Length:** 16 m **Mass:** 7,500 kg	**1912VIII61**—fragments (destroyed) *(Stromer, 1932)*	Baharija Fm., Egypt *Late Cretaceous, Cenomanian*
Andesaurus delgadoi Calvo and Bonaparte, 1991	**Length:** 18 m **Mass:** 20,000 kg	**MUCPv 132**—partial skeleton *(Calvo and Bonaparte, 1991)*	Candeleros Fm., Argentina *Late Cretaceous, Cenomanian*

Argyrosaurus superbus Lydekker, 1893	**Length:** 21 m **Mass:** 26,000 kg	**MLP 77-V-29-1**—forelimb *(Mannion and Otero, 2012)*	Lago Colhué Huapi Fm., Argentina *Late Cretaceous, Campanian/ Maastrichtian*
Atacamatitan chilensis Kellner et al., 2011	**Length:** 13 m **Mass:** 4,400 kg	**SGO-PV-961**—fragments *(Kellner et al., 2011)*	Tolar Fm., Chile *Late Cretaceous, (?)*
Australotitan cooperensis Hocknull et al., 2021	**Length:** 24 m **Mass:** 35,000 kg	**EMF102**—partial skeleton; **EMF164**—fragments; **EMF105**—femur; **EMF165**—humerus *(Hocknull et al., 2021)*	Winton Fm., Australia *Late Cretaceous, Cenomanian/Turonian*
Baotianmansaurus henanensis Zhang et al., 2009	**Length:** 22 m **Mass:** 21,000 kg	**41HIII-0200**—fragments *(Zhang et al., 2009)*	Xiaguan Fm., China *Late Cretaceous, (?)*
Barrosasaurus casamiquelai Salgado and Coria, 2009	**Length:** 18 m **Mass:** 13,500 kg	**MCF-PVPH-447**—fragments *(Salgado and Coria, 2009)*	Anacleto Fm., Argentina *Late Cretaceous, Campanian*
Borealosaurus wimani You et al., 2004	**Length:** 15 m **Mass:** 5,500 kg	**LPM0167** through **LPM0170**—fragments *(You et al., 2004)*	Sunjiawan Fm., China *Early Cretaceous, Albian*
Choconsaurus baileywillisi Simón et al., 2017	**Length:** 21 m **Mass:** 18,000 kg	**MMCh-PV series**—partial skeleton *(Simón et al., 2017)*	Huincul Fm., Argentina *Late Cretaceous, Cenomanian*
Daxiatitan binglingi You et al., 2008	**Length:** 25 m **Mass:** 23,000 kg	**GSLTZP03–001**—partial skeleton *(You et al., 2008)*	Hekou Grp., China *Early Cretaceous, (?)*
Diamantinasaurus matildae Hocknull et al., 2009	**Length:** 16 m **Mass:** 20,000 kg	**AODF 0603**—partial skeleton; **AODF 0836**—partial skull and skeleton *(Poropat et al., 2016, 2021)*; **AODF 0663**—partial skeleton *(Rigby et al., 2022)*; **AODF 0906**—partial skull and skeleton *(Poropat et al., 2023)*	Winton Fm., Australia *Late Cretaceous, Cenomanian/Turonian*
Dongyangosaurus sinensis Lü et al., 2008	**Length:** 15 m **Mass:** 7,000 kg	**DYM 04888**—partial skeleton *(Lü et al., 2008)*	Jinhua Fm., China *Early Cretaceous, Albian*
Gandititan cavocaudatus Han et al., 2024	**Length:** 14 m **Mass:** 7,500 kg	**JXGM-F-V1**—partial skeleton *(Han et al., 2024)*	Zhoutian Fm., China *Late Cretaceous, Cenomanian/Turonian*
Gannansaurus sinensis Lü et al., 2013	**Length:** 26 m **Mass:** 25,000 kg	**GMNH F10001**—two vertebrae *(Lü et al., 2013)*	Nanxiong Fm., China *Late Cretaceous, Campanian/ Maastrichtian*
Hamititan xinjiangensis Wang et al., 2021	**Length:** 17 m **Mass:** 22,000 kg	**HM V22**—partial tail *(Wang et al., 2021)*	Shengjinkou Fm., China *Early Cretaceous, (?)*
Huabeisaurus allocotus Pang and Cheng, 2000	**Length:** 20 m **Mass:** 15,500 kg	**HBV-20001**—partial skeleton; **HBV-20002**—humerus *(Pang and Cheng, 2000)*	Huiquanpu Fm., China *Late Cretaceous, (?)*
Jiangshanosaurus lixianensis Tang et al., 2001	**Length:** 18 m **Mass:** 12,500 kg	**M1322**—fragments *(Tang et al., 2001)*	Jinhua Fm., China *Early Cretaceous, Albian*
Kaijutitan maui Filippi et al., 2019	**Length:** 19 m **Mass:** 15,000 kg	**MAU-Pv-CM-522**—partial skull and skeleton *(Filippi et al., 2019)*	Sierra Barrosa Fm., Argentina *Late Cretaceous, Coniacian*
Karongasaurus gittelmani Gomani, 2005	**Length:** 9.5 m **Mass:** 2,000 kg	**MAL-175**—mandible; **MAL series**—teeth *(Gomani, 2005)*	Dinosaur Beds Fm., Malawi *Early Cretaceous, Aptian*
Mnyamawamtuka moyowamkia Gorscak and O'Connor, 2019	**Length:** 9 m **Mass:** 1,700 kg	**RRBP 05834**—partial skeleton *(Gorscak and O'Connor, 2019)*	Galula Fm., Tanzania *Early Cretaceous, Aptian/Albian*
Mongolosaurus haplodont Gilmore, 1933	**Length:** 16 m **Mass:** 6,900 kg	**AMNH 6710**—fragments *(Gilmore, 1933)*	On Gong Fm., China *Early Cretaceous, Barremian/Aptian*
Ninjatitan zapatai Gallina et al., 2021	**Length:** 15 m **Mass:** 6,000 kg	**MMCh-Pv228**—fragments *(Gallina et al., 2021)*	Bajada Colorada Fm., Argentina *Early Cretaceous, Berriasian/Valanginian*
Normanniasaurus genceyi Le Loeuff et al., 2013	**Length:** 12 m **Mass:** 4,400 kg	**MHNH-2013**—fragments *(Le Loeuff et al., 2013)*	Poudingue Ferrigineux Fm., France *Early Cretaceous, Albian*
Petrustitan hungaricus (Díez Díaz et al., 2025)	**Length:** 11.3 m **Mass:** 3,300 kg	**NHMUK R.3853**—tibia and fibula *(Díez Díaz et al., 2025)*	Sânpetru Fm., Romania *Late Cretaceous, Maastrichtian*
Petrobrasaurus puestohernandezi Filippi et al., 2011	**Length:** 18 m **Mass:** 13,000 kg	**HAU-Pv-PH-449**—partial skeleton *(Filippi et al., 2011)*	Plottier Fm., Argentina *Late Cretaceous, Coniacian/Santonian*
Ruixinia zhangi Mo et al., 2023	**Length:** 12 m **Mass:** 4,000 kg	**ELDM EL-J009**—partial skeleton *(Mo et al., 2022)*	Yixian Fm., China *Early Cretaceous, Barremian*
Sarmientosaurus musacchioi Martínez et al., 2016	**Length:** 12 m **Mass:** 10,000 kg	**MDT-PV 2**—skull and fragments *(Martínez et al., 2016)*	Bajo Barreal Fm., Argentina *Late Cretaceous, Cenomanian*
Savannasaurus elliottorum Poropat et al., 2016	**Length:** 15 m **Mass:** 20,000 kg	**AODF 660**—partial skeleton *(Poropat et al., 2016)*	Winton Fm., Australia *Late Cretaceous, Cenomanian/Turonian*
Tambatitanis amicitiae Saegusa and Ikeda, 2014	**Length:** 18 m **Mass:** 8,000 kg	**MNHAH D-1029280**—partial skull and skeleton *(Saegusa and Ikeda, 2014)*	Sasayama Grp., Japan *Early Cretaceous, Albian*
Tiamat valdecii Pereira et al., 2024	**Length:** 9 m **Mass:** 8,000 kg	**FRJ-DG [636, 638, 606, 635, 591, 527, 704]-R**—caudal vertebrae *(Pereira et al., 2024)*	Açu Fm., Brazil *Early/Late Cretaceous, Albian/ Cenomanian*

| Titanosaurus indicus
Lydekker, 1877 | **Length:** 15 m
Mass: 8,000 kg | **Unnumbered**—caudal vertebra *(Lydekker, 1877)* | Lameta Fm., India
Late Cretaceous, Maastrichtian |
| Xianshanosaurus shijiagouensis
Lü et al., 2009 | **Length:** 15 m
Mass: 7,000 kg | **KLR-07-62 series**—fragments *(Lü et al., 2009)* | Haoling Fm., China
Early Cretaceous, Aptian/Albian |

LITHOSTROTIA

TYPE SPECIES	SIZE	GENUS MATERIAL	OCCURRENCE
Ampelosaurus atacis Le Loeuff, 1995	**Length:** 16 m **Mass:** 15,000 kg	**MDE-C3-247**—dorsal vertebrae; numerous individual bones *(Le Loeuff, 1995, 2005)*	Grès à Reptiles Fm., Grès de Saint-Chinian Fm., Marnes Rouges Inférieures Fm., France *Late Cretaceous, Campanian*
"Antarctosaurus" giganteus von Huene, 1929	**Length:** 30 m **Mass:** 45,000 kg	**MLP 26–316**—fragments *(von Huene, 1929)*	Plottier Fm., Argentina *Late Cretaceous, Coniacian/Santonian*
Argentinosaurus huinculensis Bonaparte and Coria, 1993	**Length:** 35 m **Mass:** 75,000 kg	**MCF-PVPH-1**—partial skeleton *(Bonaparte and Coria, 1993)*; MCF-PVPH unnumbered—femur *(Bonaparte, 1996)*; MLP-DP 46-VIII-21-3—femur *(Mazzetta et al., 2004)*	Huincul Fm., Argentina *Late Cretaceous, Cenomanian*
Atsinganosaurus velauciensis Garcia et al., 2010	**Length:** 8 m **Mass:** 2,000 kg	**VBN.93.01**—dorsal vertebrae; numerous VBN series bones *(Garcia et al., 2010; Díez Díaz et al., 2018, 2021)*	Grès à Reptiles Fm., France *Late Cretaceous, Campanian*
Austroposeidon magnificus Bandeira et al., 2016	**Length:** 25 m **Mass:** 40,000 kg	**MCT 1628-R**—vertebral fragments *(Bandeira et al., 2016)*	Presidente Prudente Fm., Brazil *Late Cretaceous, Campanian/Maastrichtian*
Baalsaurus mansillai Calvo and González Riga, 2019	**Length:** 11 m **Mass:** 6,000 kg	**MUCPv-1460**—dentary *(Calvo and González Riga, 2019)*	Portezuelo Fm., Argentina *Late Cretaceous, Coniacian*
Bonitasaura salgadoi Apesteguía, 2004	**Length:** 10 m **Mass:** 5,500 kg	**MPCA 460**—partial skull and skeleton; MPCA 467–partial skeleton; 468—tibia *(Gallina and Apesteguía, 2015)*	Bajo de la Carpa Fm., Argentina *Late Cretaceous, Santonian*
Chucarosaurus diripienda Agnolin et al., 2023	**Length:** 24 m **Mass:** 28,000 kg	**MPCA PV 820**—fragments; MPCA PV 821—fragments *(Agnolin et al., 2023)*	Huincul Fm., Argentina *Late Cretaceous, Cenomanian*
Dreadnoughtus schrani Lacovara et al., 2014	**Length:** 28 m **Mass:** 48,000 kg	**MPM-PV 1156**—partial skeleton; MPM-PV 3546—partial skeleton *(Lacovara et al., 2014)*; MPM-PV-1156-49—soft tissue *(Schroeter et al., 2022)*	Mata Amarilla Fm., Argentina *Late Cretaceous, Cenomanian*
Drusilasaura deseadensis Navarrete et al., 2011	**Length:** 18 m **Mass:** 13,000 kg	**MPM-PV 2097**—partial skeleton *(Navarrete et al., 2011)*	Bajo Barreal Fm., Argentina *Late Cretaceous, Cenomanian*
Elaltitan lilloi Mannion and Otero, 2012	**Length:** 20 m **Mass:** 23,000 kg	**PVL 4628**—partial skeleton *(Mannion and Otero, 2012)*	Lago Colhué Huapi Fm., Argentina *Late Cretaceous, Coniacian/Santonian*
Epachthosaurus sciuttoi Powell, 1990	**Length:** 13 m **Mass:** 5,000 kg	**MACN-CH 1317**—dorsal vertebra; MACN-CH 18689—cast of pelvic region; UNPSJB-PV 920—partial skeleton *(Martínez et al., 2004)*; UNPSJB-PV 1006—dorsal vertebra; UNPSJB-PV 956—pelvic fragments *(Casal and Ibiricu, 2010)*	Bajo Barreal Fm., Argentina *Late Cretaceous, Cenomanian*
Futalognkosaurus dukei Calvo et al., 2007b	**Length:** 24 m **Mass:** 38,000 kg	**MUCPv-323**—partial skeleton *(Calvo et al., 2007; Calvo, 2014)*	Portezuelo Fm., Argentina *Late Cretaceous, Coniacian*
Garrigatitan meridionalis Díez Díaz et al., 2021	**Length:** 14 m **Mass:** 8,000 kg	**MMS/VBN.09.170**—sacrum; MMS/VBN series—fragments *(Díez Díaz et al., 2021)*	Grès à Reptiles Fm., France *Late Cretaceous, Campanian*
Inawentu oslatus Filippi et al., 2024	**Length:** 10 m **Mass:** 5,500 kg	**MAU-Pv-LI-595**—skull and partial skeleton *(Filippi et al., 2024)*	Bajo de la Carpa Fm., Argentina *Late Cretaceous, Santonian*
Jiangxititan ganzhouensis Mo et al., 2023	**Length:** 12.4 m **Mass:** 3,500 kg	**NHMG 034062**—vertebrae *(Mo et al., 2023)*	Nanxiong Fm., China *Late Cretaceous, Maastrichtian*
Lirainosaurus astibiae Sanz et al., 1999	**Length:** 6 m **Mass:** 4,000 kg	**MCNA 7458**—caudal vertebra *(Sanz et al., 1999)*; numerous skeletal elements *(Sanz et al., 1999; Díez Díaz et al., 2013a, b, 2015)*	Marnes Rouges Inférieures Fm., Spain *Late Cretaceous, Campanian/Maastrichtian*
Lohuecotitan pandafilandi Díez Díaz et al., 2016	**Length:** 11 m **Mass:** 3,000 kg	**HUE-EC-01**—partial skeleton *(Díez Díaz et al., 2016)*; possibly MPCM-HUE-8741—braincase *(Knoll et al., 2019)*	Villalba de la Sierra Fm., Spain *Late Cretaceous, Campanian/Maastrichtian*
Magyarosaurus dacus von Huene, 1932	**Length:** 6 m **Mass:** 520 kg	**NHM-R.3861**—dorsal vertebrae; potentially numerous other bones *(Grigorescu, 2010; Stein et al., 2010)*	Sânpetru Fm., Romania *Late Cretaceous, Maastrichtian*
Malawisaurus dixeyi Jacobs et al., 1993	**Length:** 11 m **Mass:** 3,000 kg	**SAM 7405**—caudal vertebra; numerous individual fossil specimens *(Gomani, 2005)*	Dinosaur Beds Fm., Malawai *Early Cretaceous, Aptian*
Mansourasaurus shahinae Sallam et al., 2018	**Length:** 8.5 m **Mass:** 1,300 kg	**MUVP 200**—partial skeleton *(Sallam et al., 2018)*	Quseir Fm., Egypt *Late Cretaceous, Campanian*

Mendozasaurus neguyelap González Riga, 2003	**Length:** 20 m **Mass:** 16,000 kg	**IANIGLA-PV 065**—partial skeleton; IANIGLA-PV series—various bones *(González Riga et al., 2018)*	Sierra Barrosa Fm., Argentina *Late Cretaceous, Coniacian*
Menucocelsior arriagadai Rolando et al., 2022	**Length:** 9 m **Mass:** 6,000 kg	**MPCN-PV-798**—partial skeleton *(Rolando et al., 2022)*	Allen Fm., Argentina *Late Cretaceous, Campanian/ Maastrichtian*
Narambuenatitan palomoi Filippi et al., 2011	**Length:** 13 m **Mass:** 4,000 kg	**MAU-Pv-N-425**—partial skeleton *(Filippi et al., 2011)*	Anacleto Fm., Argentina *Late Cretaceous, Campanian*
Notocolossus gonzalezparejasi González Riga et al., 2016	**Length:** 28 m **Mass:** 40,000 kg	**UNCUYO-LD 301**—fragments; UNCUYO-LD 302—partial skeleton *(González Riga et al., 2016)*	Plottier Fm., Argentina *Late Cretaceous, Coniacian/Santonian*
Paludititan nalatzensis Csiki et al., 2010	**Length:** 8.5 m **Mass:** 1,300 kg	**UBB NVM1**—partial skeleton *(Cski et al., 2010)*; UBB SPM-4, UBB PGR1–6, -14, -15—vertebrae *(Mocho et al., 2022)*	Sânpetru Fm., Romania *Late Cretaceous, Maastrichtian*
Patagotitan mayorum Carballido et al., 2017	**Length:** 31 m **Mass:** 60,000 kg	**MPEF-PV 3400**—partial skeleton; MPEF-PV 3399—partial skeleton; MPEF-PV series—various bones *(Carballido et al., 2017)*	Cerro Barcino Fm., Argentina *Early Cretaceous, Albian*
Puertasaurus reuili Novas et al., 2005	**Length:** 30 m **Mass:** 60,000 kg	**MPM 10002**—vertebrae *(Novas et al., 2005)*	Mata Amarilla Fm., Argentina *Late Cretaceous, Cenomanian*
Quetecsaurus rusconii González Riga and David, 2014	**Length:** 15m **Mass:** 7,000 kg	**UNCUYO-LD-300**—fragments *(González Riga and David, 2014)*	Cerro Lisandro Fm., Argentina *Late Cretaceous, Turonian*
Tengrisaurus starkovi Averianov and Skutschas, 2017	**Length:** 12 m **Mass:** 4,200 kg	**ZIN PH 7/13**—caudal vertebra *(Averianov and Skutschas, 2017)*; ZIN PH 14/13, ZIN PH 8/13, BM 38/7120—caudal vertebrae *(Averianov et al., 2021)*	Murtoi Fm., Russia *Early Cretaceous, Barremian/Aptian*
Traukutitan eocaudata Juárez Valieri and Calvo, 2011	**Length:** 24 m **Mass:** 26,000 kg	**MUCPv 204**—partial skeleton *(Juárez Valieri and Calvo, 2011)*	Bajo de la Carpa Fm., Argentina *Late Cretaceous, Santonian*
Vahiny depereti Curry Rogers and Wilson, 2014	**Length:** 13 m **Mass:** 5,400 kg	**UA 9940**—braincase; FMNH PR 3046—juvenile partial braincase *(Curry Rogers and Wilson, 2014)*	Maevarano Fm., Madagascar *Late Cretaceous, Maastrichtian*
Volgatitan simbirskiensis Averianov and Efimov, 2018	**Length:** 20 m **Mass:** 15,000 kg	**UPM 976**—tail vertebrae *(Averianov and Efimov, 2018)*	(?) Fm., Russia *Early Cretaceous, Hauterivian*

RINCONSAURIA

TYPE SPECIES	SIZE	GENUS MATERIAL	OCCURRENCE
Adamantisaurus mezzalirai Santucci and Bertini, 2006	**Length:** 18 m **Mass:** 14,400 kg	**MUGEO 1282**—caudal vertebrae; MUGEO 1289, 1295—chevrons *(Santucci and Bertini, 2006)*	Adamantina Fm., Brazil *Late Cretaceous, Campanian/ Maastrichtian*
Aeolosaurus rionegrinus Powell, 1987	**Length:** 18 m **Mass:** 15,000 kg	**MJG-R 1**—partial skeleton *(Powell, 1987)*; UNPSJB-PV 959 series—tail vertebrae *(Casal et al., 2007)*	Lago Colhué Huapi Fm., Argentina Angostura Colorada Fm., Argentina *Late Cretaceous, Campanian/ Maastrichtian*
Antarctosaurus wichmannianus von Huene, 1929	**Length:** 17 m **Mass:** 12,000 kg	**MACN 6904**—partial skull and skeleton *(von Huene, 1929)*	Anacleto Fm., Argentina *Late Cretaceous, Campanian*
Arrudatitan maximus Silva Junior et al., 2022a	**Length:** 17 m **Mass:** 12,000 kg	**MPMA 12-0001-97**—partial skeleton *(Silva et al., 2022a)*	Adamantina Fm., Brazil *Late Cretaceous, Campanian/ Maastrichtian*
Baurutitan britoi Kellner et al., 2005	**Length:** 18 m **Mass:** 15,000 kg	**MCT 1490-R**—anterior tail *(Kellner et al., 2005)*; MCT 1488-R—partial skeleton; CPPLIP series—partial skeleton *(Silva Junior et al., 2022b)*	Serra da Galga Fm., Brazil *Late Cretaceous, Maastrichtian*
Brasilotitan nemophagus Machado et al., 2013	**Length:** 10 m **Mass:** 5,300 kg	**UFRG MN 7371-V**—fragments; MPM 126R—tooth *(Machado et al., 2013)*	Presidente Prudente Fm., Brazil *Late Cretaceous, Campanian/ Maastrichtian*
Bravasaurus arreirosorum Hechenleitner et al., 2020	**Length:** 7 m **Mass:** 3,000 kg	**CRILAR-Pv 612**—partial skull and skeleton; CRILAR-Pv 613—fragments *(Hechenleitner et al., 2020)*	Ciénaga del Río Huaco Fm., Argentina *Late Cretaceous, Campanian/ Maastrichtian*
Caieiria allocaudata Silva Junior et al., 2022b	**Length:** 18 m **Mass:** 15,000 kg	**MCT 1719-R**—caudal vertebrae *(Silva Junior et al., 2022)*	Serra da Galga Fm., Brazil *Late Cretaceous, Maastrichtian*
Chaditipan calvoi (Agnolín et al., 2025)	**Length:** 8.5 m **Mass:** 1,500 kg	**MPCN-Pv 1034**—partial skeleton; MPCN-Pv 1034 through 1041—various fragments *(Agnolín et al., 2025)*	Anacleto Fm., Argentina *Late Cretaceous, Campanian*
Gondwanatitan faustoi Kellner and Azevedo, 1999	**Length:** 7 m **Mass:** 2,500 kg	**MN 4111-V**—partial skeleton *(Kellner and de Azevedo, 1999)*	Presidente Prudente Fm., Brazil *Late Cretaceous, Campanian/ Maastrichtian*
Jainosaurus septentrionalis Hunt et al., 1995	**Length:** 20 m **Mass:** 18,000 kg	**GSI K27/497**—partial skull; collection of other bones and fragments *(Wilson et al., 2009, 2011)*	Lameta Fm., India *Late Cretaceous, Maastrichtian*
Laplatasaurus araukanicus von Huene, 1929	**Length:** 10 m **Mass:** 5,000 kg	**MLP-CS 1128**—right tibia, and **MLP-CS 1127**—right fibula *(Gallina and Otero, 2015)*	Anacleto Fm., Argentina *Late Cretaceous, Campanian*

Maxakalisaurus topai *Kellner et al., 2006*	**Length:** 13 m **Mass:** 4,000 kg	**MN 5013-V**—partial skull and skeleton *(Kellner et al., 2006)*; MN 7048-V, MN 7049-V, MN 7050-V—shoulder elements; MBC-38-PV, MBC-42-PV—jaw and teeth *(França et al., 2016)*	Adamantina Fm., Brazil *Late Cretaceous, Campanian/ Maastrichtian*
Muyelensaurus pecheni *Calvo et al., 2007a*	**Length:** 13 m **Mass:** 5,500 kg	**MRS-PV 207**—braincase; MRS-PV series—partial skull and skeleton *(Calvo et al., 2007a)*	Plottier Fm., Argentina *Late Cretaceous, Coniacia/Santonian*
Overosaurus paradasorum *Coria et al., 2013*	**Length:** 8.5 m **Mass:** 1,800 kg	**MAU-Pv-CO-439**—partial skeleton *(Coria et al., 2013)*	Bajo de la Carpa Fm., Argentina *Late Cretaceous, Santonian*
Panamericansaurus schroederi *Calvo and Porfiri, 2010*	**Length:** 15 m **Mass:** 12,000 kg	**MUCPv-417**—fragments *(Calvo and Porfiri, 2010)*	Allen Fm., Argentina *Late Cretaceous, Campanian/ Maastrichtian*
Paralititan stromeri *Smith et al., 2001*	**Length:** 27 m **Mass:** 30,000 kg	**CGM 81119**—partial skeleton *(Smith et al., 2001)*	Baharija Fm., Egypt *Late Cretaceous, Cenomanian*
Pitekunsaurus macayai *Filippi and Garrido, 2008*	**Length:** 12 m **Mass:** 4,000 kg	**MAU-Pv-AG-446**—partial skeleton *(Filippi and Garrido, 2008)*	Anacleto Fm., Argentina *Late Cretaceous, Campanian*
Punatitan coughlini *Hechenleitner et al., 2020*	**Length:** 14 m **Mass:** 6,000 kg	**CRILAR-Pv 614**—partial skeleton *(Hechenleitner et al., 2020)*	Ciénaga del Río Huaco Fm., Argentina *Late Cretaceous, Campanian/ Maastrichtian*
Rinconsaurus caudamirus *Calvo and González Riga, 2003*	**Length:** 11 m **Mass:** 4,000 kg	**MRS-Pv 26**—partial skeleton; MRS-PV series—partial skeleton *(Pérez Moreno et al., 2022, 2023)*	Bajo de la Carpa Fm., Argentina *Late Cretaceous, Santonian*
Shingopana songwensis *Gorscak et al., 2017*	**Length:** 8 m **Mass:** 3,000 kg	**RRBP 02100**—mandible and partial skeleton *(Gorscak et al., 2017)*	Galula Fm., Tanzania *Late Cretaceous, (?)*
Sonidosaurus saihangaobiensis *Xu et al., 2006*	**Length:** 10 m **Mass:** 3,300 kg	**LH V0010**—partial skeleton *(Xu et al., 2006)*	Iren Dabasu Fm., China *Late Cretaceous, Campanian/ Maastrichtian*
Uberabatitan ribeiroi *Salgado and de Souza Carvalho, 2008*	**Length:** 25 m **Mass:** 33,000 kg	**CPP-UrHo**—partial skeleton; CPP-UrB—partial skeleton; CPP-UrC—fragments *(Salgado and de Souza Carvalho, 2008)*	Serra da Galga Fm., Brazil *Late Cretaceous, Maastrichtian*

SALTASAUROIDEA

TYPE SPECIES	SIZE	GENUS MATERIAL	OCCURRENCE
Abditosaurus kuehnei *Vila et al., 2022*	**Length:** 17 m **Mass:** 14,000 kg	**Various MNCN and MCD specimens**—partial skeleton *(Vila et al., 2022)*	Conques Fm., Spain *Late Cretaceous, Maastrichtian*
Alamosaurus sanjuanensis *Gilmore, 1922*	**Length:** 26 m **Mass:** 38,000 kg	**USNM 10486**—scapula; USNM 10487—ischium *(Gilmore, 1922)*; numerous mostly isolated fragments *(Lucas and Sullivan, 2000; Jasinski et al., 2011; Poropat, 2013)*; USNM 15560—partial skeleton *(Gilmore, 1946)*; TMM 41541–1—partial skeleton *(Wick and Corrick, 2015)*; BIBE 45854—neck *(Tykoski and Fiorillo, 2017)*; TMM 43621–1—partial juvenile skeleton *(Lehman and Coulson, 2002)*	Ojo Alamo Fm., Black Peaks Fm., Javelina Fm., North Horn Fm., United States *Late Cretaceous, Maastrichtian*
Arackar licanantay *Rubilar-Rogers et al., 2021*	**Length:** 6 m **Mass:** 1,500 kg	**SNGM-1**—fragments *(Rubilar-Rogers et al., 2021)*	Hornitos Fm., Chile *Late Cretaceous, Campanian/ Maastrichtian*
Bonatitan reigi *Martinelli and Forasiepi, 2004*	**Length:** 6 m **Mass:** 2,000 kg	**MACN-PV RN 821**—partial skull and skeleton; MACN-PV RN 1061 *(Martinelli and Forasiepi, 2004)*	Allen Fm., Argentina *Late Cretaceous, Campanian/ Maastrichtian*
Bustingorrytitan shiva *Simón and Salgado, 2023*	**Length:** 30 m **Mass:** 67,000 kg	**MMCH-Pv 59/1-40**—partial skeleton; MMCH-Pv 60/1–6—partial skeleton; MMCH-Pv 61—femur; MMCH-Pv 62—partial hindlimb *(Simón and Salgado, 2023)*	Huincul Fm., Argentina *Late Cretaceous, Cenomanian*
Ibirania parva *Navarro et al., 2022*	**Length:** 6 m **Mass:** 2,000 kg	**LPP-PV-0200 through 0207**—fragments; additional other fragments *(Navarro et al., 2022)*	São José do Rio Preto Fm., Brazil *Late Cretaceous, Santonian/Campanian*
Igai semkhu *Gorscak et al., 2023*	**Length:** 10 m **Mass:** 3,000 kg	**MB.R.Vb-621–640**—partial skeleton *(Gorscak et al., 2023)*	Quseir Fm., Egypt *Late Cretaceous, Campanian*
Isisaurus colberti *Wilson and Upchurch, 2003*	**Length:** 11 m **Mass:** 11,500 kg	**ISI R335/1-65**—partial skeleton *(Wilson and Upchurch, 2003)*; possibly MSM-17-4 through MSM-22-4—tail vertebrae *(Malkani, 2019)*	Lameta Fm., India *Late Cretaceous, Maastrichtian*
Nemegtosaurus mongoliensis *Nowinski, 1971*	**Length:** 12.5 m **Mass:** 6,000 kg	**Z. PAL MgD-I/9**—skull and jaw *(Nowinski, 1971)*; MPCD100/413—partial skeleton; possibly MPC-D100/402—skull and jaw *(Currie et al., 2018)*; possibly PIN 3837/P821—dorsal vertebrae *(Averianov and Lopatin, 2019)*	Nemegt Fm., Mongolia *Late Cretaceous, Maastrichtian*
Neuquensaurus australis *Powell, 1992*	**Length:** 8 m **Mass:** 2,300 kg	**MLP Ly 1–7**—vertebrae; MCS 5—partial skeleton; MLP CS 1400, 1402, and 1407—vertebrae; PVL 4017–18—partial sacrum *(Salgado et al., 2005; D'Emic and Wilson, 2011)*	Anacleto Fm., Lecho Fm., Argentina *Late Cretaceous, Campanian/ Maastrichtian*
Nullotitan glaciaris *Novas et al., 2019*	**Length:** 25 m **Mass:** 41,000 kg	**MPM 21542**—partial skeleton; **MACN-PV 18644**—neck vertebra; MPM 21546—tail vertebra; MPM 21545—fragments; MPM 21548—tibia; MPM 21549—tail vertebra; MPM 21547—tail vertebrae *(Novas et al., 2019)*	Chorrillo Fm., Argentina *Late Cretaceous, Campanian/ Maastrichtian*

Opisthocoelicaudia skarzynskii Borsuk-Białynicka, 1977	**Length:** 14 m **Mass:** 8,500 kg	**MPC-D100/404** {aka ZPAL MgD-I/48}—skeleton; ZPAL MgD-I/25c—juvenile shoulder *(Borsuk-Białynicka, 1977)*	Nemegt Fm., Mongolia *Late Cretaceous, Maastrichtian*
Pellegrinisaurus powelli Salgado, 1996	**Length:** 15 m **Mass:** 8,000 kg	**MPCA 1500**—partial skeleton *(Salgado, 1996)*	Anacleto Fm., Argentina *Late Cretaceous, Campanian/* *Maastrichtian*
Qingxiusaurus youjiangensis Mo et al., 2008	**Length:** 12 m **Mass:** 6,300 kg	**NHMG 8499**—fragments *(Mo et al., 2008)*	unnamed, China *Late Cretaceous, (?)*
Quaesitosaurus orientalis Kurzanov and Bannikov, 1983	**Length:** 9.5 m **Mass:** 2,500 kg	**PIN 3906/2**—partial skull and jaw *(Kurzanov and Bannikov, 1983)*	Nemegt Fm., Mongolia *Late Cretaceous, Maastrichtian*
Qunkasaura pintiquiniestra (Mocho et al., 2024)	**Length:** 11.2 m **Mass:** 4,400 kg	**HUE-EC-04**—partial skeleton *(Mocho et al., 2024)*	Villalba de la Sierra Fm, Spain *Late Cretaceous, Campanian*
Rapetosaurus krausei Curry Rogers and Forster, 2001	**Length:** 16 m **Mass:** 10,000 kg	**UA 8698**—partial skull and jaw; **FMNH PR series**—partial juvenile skeleton *(Curry Rogers and Forster, 2001)*; UA 9998—partial infant *(Curry Rogers et al., 2016)*; various specimens *(Curry Rogers and Forster, 2004; Curry Rogers et al., 2011; Curry Rogers and Kulik, 2018)*	Maevarano Fm., Madagascar *Late Cretaceous, Maastrichtian*
Rocasaurus muniozi Salgado and Azpilicueta, 2000	**Length:** 8 m **Mass:** 2,000 kg	**MPCA-Pv 46**—partial skeleton; MPCA–Pv series—vertebrae *(Garcia and Salgado, 2012)*; MPCN-PV-800—vertebrae *(Rolando et al., 2022)*	Allen Fm., Argentina *Late Cretaceous, Campanian/* *Maastrichtian*
Rukwatitan bisepultus Gorscak et al., 2014	**Length:** 11 m **Mass:** 6,000 kg	**RRBP 07409**—partial skeleton; RRBP 03151—right humerus *(Gorscak et al., 2014)*	Galula Fm., Tanzania *Late Cretaceous, (?)*
Saltasaurus loricatus Bonaparte and Powell, 1980	**Length:** 9 m **Mass:** 3,000 kg	**PVL 4017–92**—partial pelvis; PVL. 4017 series—partial skeletons *(Bonaparte and Powell, 1980)*	Lecho Fm., Argentina *Late Cretaceous, Maastrichtian*
Tapuiasaurus macedoi Zaher et al., 2011	**Length:** 11 m **Mass:** 3,000 kg	**MZSP-PV 807**—skull and partial skeleton *(Zaher et al., 2011)*	Quiricó Fm., Brazil *Early Cretaceous, Aptian*
Titanomachya gimenezi Pérez-Moreno et al., 2024	**Length:** 9.6 m **Mass:** 5,400 kg	**MPEF Pv 11547**—partial skeleton *(Pérez-Moreno et al., 2024)*	La Colonia Fm., Argentina *Late Cretaceous, Campanian/* *Maastrichtian*
Udelartitan celeste Soto et al., 2024	**Length:** 15 m **Mass:** 8,000 kg	**FC-DPV 3595**—tail vertebrae; FC-DPV 1900—partial skeleton *(Soto et al., 2024)*	Guichón Fm., Uruguay *Late Cretaceous, (?)*
Yamanasaurus lojaensis Apesteguía et al., 2019	**Length:** 7 m **Mass:** 1,500 kg	**YM-UTPL_002**—fragments *(Apesteguía et al., 2019)*	Río Playas Fm., Equador *Late Cretaceous, Campanian/* *Maastrichtian*
Zhuchengtitan zangjiazhuangensis Mo et al., 2017	**Length:** 15 m **Mass:** 11,000 kg	**ZJZ-57**—humerus *(Mo et al., 2017)*	Wangshi Grp., China *Late Cretaceous, Campanian/* *Maastrichtian*

ORIGINAL GENUS	SYNONYMIZED
Aetonyx palustris Broom, 1911	*Massospondylus carinatus* Barrett et al., 2019
Aliwalia rex Galton, 1985	*Eucnemesaurus fortis* Yates, 2006
Ammosaurus major Marsh, 1889	*Anchisaurus polyzelus* Sereno, 1999
Amphisaurus polyzelus Hitchcock, 1865	*Anchisaurus polyzelus* Marsh, 1885
Aristosaurus erectus Hoepen, 1920	*Massospondylus carinatus* Barrett et al., 2019
Cathetosaurus lewisi Jensen, 1988	*Massospondylus carinatus* McIntosh et al., 1996
Dimodosaurus poligniensis Pidancet and Chopard, 1862	*Plateosaurus longiceps* Galton, 1998
Dromicosaurus gracilis Van Hoepen, 1920	*Massospondylus carinatus* Barrett et al., 2019
Dromicosaurus gracilis Hoepen, 1920	*Massospondylus carinatus* Barrett et al., 2019
Dystylosaurus edwini Jensen, 1985	*Supersaurus vivianae* Curtice and Stadtman, 2001
Elosaurus parvus Peterson and Gilmore, 1902	*Brontosaurus parvus* Tschopp et al., 2015
Eobrontosaurus yahnahpin Bakker, 1998	*Brontosaurus yahnahpin* Tschopp et al., 2015
Gripposaurus sinensis Barrett et al., 2007	*Lufengosaurus huenei* Wang et al., 2017
Leptospondylus capensis Owen, 1854	*Massospondylus carinatus* Barrett et al., 2019

ORIGINAL GENUS	SYNONYMIZED
Maojandino alami Malkani, 2015	*Gspsaurus pakistani* Malkani, 2020
Marisaurus jeffi Malkani, 2015	*Gspsaurus pakistani* Malkani, 2020
Megadactylus polyzelus Hitchcock, 1865	*Anchisaurus polyzelus* Marsh, 1882
Morosaurus Marsh, 1878	*Camarasaurus* Galton and Upchurch, 2004
Pachysaurus von Huene, 1907	*Plateosaurus engelhardti* Moser, 2003
Pachyspondylus orpenii Owen, 1854	*Massospondylus carinatus* Barrett et al., 2019
Seismosaurus halli Gillette, 1991	*Diplodocus hallorum* Lucas et al., 2006
Sellosaurus von Huene, 1907	*Plateosaurus engelhardti* Moser, 2003
Strenusaurus procerus Bonaparte, 1969	*Riojasaurus incertus* Galton, 1985
Tawasaurus minor Young, 1982	*Lufengosaurus* Sereno, 1991
Teratosaurus minor von Huene, 1908	*Efraasia minor* Galton, 1973
Yaleosaurus colurus von Huene, 1932	*Anchisaurus polyzelus* Yates, 2010
Zigongosaurus fuxiensis Hau et al., 1976	*Mamenchisaurus fuxiensis* Hou et al., 1976 **or** *Omeisaurus fuxiensis* McIntosh, 1990

SELECTED BIBLIOGRAPHY

A full list of references can found via https://press.princeton.edu/books/ebook/9780691250243/the-princeton-encyclopedia -of-dinosaurs-sauropods-pdf

An, X., Xu, X., Han, F., et al. (2023). A new juvenile sauropod specimen from the Middle Jurassic Dongdaqiao Formation of East Tibet. *PeerJ*, *11*, e14982.

Cabreira, S. F., Kellner, A. W. A., Dias-da-Silva, S., et al. (2016). A unique Late Triassic dinosauromorph assemblage reveals dinosaur ancestral anatomy and diet. *Current Biology*, *26*(22), 3090–3095.

Carballido, J. L., Otero, A., Mannion, P. D., Salgado, L., and Moreno, A. P. (2022). Titanosauria: A critical reappraisal of its systematics and the relevance of the South American record. In *South American Sauropodomorph Dinosaurs: Record, Diversity and Evolution*, edited by Otero, A., Carballido, J. L., and Pol, D., 269–298. Cham: Springer International.

Ezcurra, M. D. (2010). A new early dinosaur (Saurischia: Sauropodomorpha) from the Late Triassic of Argentina: A reassessment of dinosaur origin and phylogeny. *Journal of Systematic Palaeontology*, *8*(3), 371–425.

Hechenleitner, E. M., Leuzinger, L., Martinelli, A. G., et al. (2020). Two Late Cretaceous sauropods reveal titanosaurian dispersal across South America. *Communications Biology*, *3*(1), 622.

Mannion, P. D., Upchurch, P., Jin, X., and Zheng, W. (2019). New information on the Cretaceous sauropod dinosaurs of Zhejiang Province, China: Impact on Laurasian titanosauriform phylogeny and biogeography. *Royal Society Open Science*, *6*(8), 191057.

Mannion, P. D., Upchurch, P., Schwarz, D., and Wings, O. (2019). Taxonomic affinities of the putative titanosaurs from the Late Jurassic Tendaguru Formation of Tanzania: Phylogenetic and biogeographic implications for eusauropod dinosaur evolution. *Zoological Journal of the Linnean Society*, *185*(3), 784–909.

Moore, A. J., Upchurch, P., Barrett, P. M., Clark, J. M., and Xing, X. (2020). Osteology of *Klamelisaurus gobiensis* (Dinosauria, Eusauropoda) and the evolutionary history of Middle–Late Jurassic Chinese sauropods. *Journal of Systematic Palaeontology*, *18*(16), 1299–1393.

Navarro, B. A., Ghilardi, A. M., Aureliano, T., et al. (2022). A new nanoid titanosaur (Dinosauria: Sauropoda) from the Upper Cretaceous of Brazil. *Ameghiniana*, *59*(5), 317–354.

Peyre de Fabrègues, C., Bi, S., Li, H., Li, G., Yang, L., and Xu, X. (2020). A new species of early-diverging Sauropodiformes from the Lower Jurassic Fengjiahe Formation of Yunnan Province, China. *Scientific Reports*, *10*(1), 10961.

Pol, D., Ramezani, J., Gomez, K., et al. (2020). Extinction of herbivorous dinosaurs linked to Early Jurassic global warming event. *Proceedings of the Royal Society B: Biological Sciences*, *287*(1939), 20202310.

Rauhut, O. W., Holwerda, F. M., and Furrer, H. (2020). A derived sauropodiform dinosaur and other sauropodomorph material from the Late Triassic of Canton Schaffhausen, Switzerland. *Swiss Journal of Geosciences*, *113*, 1–54.

Ren, X. X., Jiang, S., Wang, X. R., et al. (2023). Re-examination of *Dashanpusaurus dongi* (Sauropoda: Macronaria) supports an early Middle Jurassic global distribution of neosauropod dinosaurs. *Palaeogeography, Palaeoclimatology, Palaeoecology*, *610*, 111318.

Santucci, R. M., and Filippi, L. S. (2022). Last titans: Titanosaurs from the Campanian–Maastrichtian age. In *South American Sauropodomorph Dinosaurs: Record, Diversity and Evolution*, edited by Otero, A., Carballido, J. L., and Pol, D., 341–391. Cham: Springer International.

Silva Junior, J. C., Martinelli, A. G., Iori, F. V., Marinho, T. S., Hechenleitner, E. M., and Langer, M. C. (2021). Reassessment of *Aeolosaurus maximus*, a titanosaur dinosaur from the Late Cretaceous of Southeastern Brazil. *Historical Biology*, *34*(3), 403–411.

Weishampel, D. B., Dodson, P., and Osmólska, H., eds. (2004). *The Dinosauria*. 2nd ed. Oakland: University of California Press.

Aardonyx 42, 43, 65, 373

Abdarainurus 254, 255, 258, 381

Abditosaurus 338, 339, 349, 385

Abrosaurus 211, 267, 379

Abydosaurus 186, 187, 200, 240, 379

Adamantisaurus 316, 317, 335, 384

Adeopapposaurus 42, 43, 49, 373

Aegyptosaurus 254, 255, 260, 381

Aeolosaurini 316, 317

Aeolosaurus 316, 317, 332, 384

Aepisaurus 361

Aetonyx 387

Agustinia 166, 167, 170, 378

Alamosaurus 338, 339, 351, 385

Algoasaurus 239, 380

Aliwalia 44, 387

Amanzia 96, 97, 115, 375

Amargasaurus 140, 141, 152, 377

Amargatitanis 140, 141, 149, 377

Amazonsaurus 166, 167, 183, 378

Ammosaurus 60, 387

Ampelosaurus 284, 285, 296, 382

Amphicoelias 140, 141, 155, 377

Amphisaurus 60, 387

Amygdalodon 71, 373

Analong 120, 129, 376

Anchisauria 42, 43

Anchisaurus 42, 43, 60, 373, 387

Andesauridae 257

Andesauroidea 219

Andesaurus 254, 255, 257, 381

"Angloposeidon" 367

Angolatitan 214, 215, 225, 380

Anhuilong 120, 121, 136, 376

Antarctosaurus 316, 317, 318, 319, 384

"*Antarctosaurus*" giganteus 311, 382

Antetonitrus 74, 75, 78, 374

Apatosaurinae 140, 141

Apatosaurus 140, 141, 153, 377

Arackar 338, 339, 342, 385

Aragosaurus 186, 187, 211, 379

Arcusaurus 40, 372

Ardetosaurus 165, 377

Argentinosaurus 284, 285, 308, 383

Argyrosaurus 254, 255, 261, 381

Aristosaurus 51, 387

Arkharavia 239, 380

Arrudatitan 316, 317, 333, 384

Asiatosaurus 361

Astrodon 208

Astrophocaudia 214, 215, 222, 380

Asylosaurus 18, 19, 31, 372

Atacamatitan 281, 381

Atlantosaurus 164, 377

Atlasaurus 186, 187, 192, 379

Atsinganosaurus 284, 285, 297, 383

Australodocus 214, 215, 217, 380

Australotitan 254, 255, 269, 381

Austroposeidon 284, 285, 304, 383

Austrosaurus 214, 215, 220, 380

Baalsaurus 311, 383

Bagualia 96, 97, 108, 375

Bagualosauria 18

Bagualosaurus 18, 19, 28, 372

"Baguasaurus" 367

Bajadasaurus 139, 140, 141, 146, 377

Balochisaurus 361

Baotianmansaurus 254, 255, 280, 381

"Barackosaurus" 367

Barapasaurus 96, 97, 99, 375

Barosaurus 140, 141, 162, 377

Barrosasaurus 281, 381

"Bashunosaurus" 210, 367

Baurutitan 316, 317, 326, 384

Bellusaurus 120, 126, 376

"Biconcavoposeidon" 367

Blikanasaurus 74, 75, 93, 374

Bonatitan 338, 339, 354, 385

Bonitasaura 185, 284, 285, 301, 383

Borealosaurus 254, 255, 262, 381

Bothriospondylus 361

Brachiosauridae 186, 187

Brachiosaurus 186, 187, 201, 213, 379

Brachytrachelopan 140, 141, 150, 377

Brasilotitan 316, 317, 334, 384

Bravasaurus 316, 317, 327, 384

Brohisaurus 361

Brontomerus 214, 215, 223, 380

Brontosaurus 140, 141, 154, 377

Bruhathkayosaurus 362

Buriolestes 18, 19, 20, 372

Bustingorrytitan 338, 339, 359, 385

Caieiria 316, 317, 336, 384

Camarasauridae 186

Camarasauromorpha 186

Camarasaurus 3, 9, 186, 187, 190, 379

Camelotia 42, 43, 71, 373

Campylodoniscus 362

Cardiodon 362

Cathartesaura 166, 167, 176, 378

Cathetosaurus 387

Cedarosaurus 186, 187, 203, 379

Cetiosauridae 96, 97

Cetiosauriscus 96, 97, 102, 375

Cetiosaurus 96, 97, 103, 375

Chadititan 337, 384

Chebsaurus 96, 97, 104, 375

Chiayusaurus 362

Chinshakiangosaurus 74, 75, 83, 374

Choconsaurus 254, 255, 263, 382

Chondrosteosaurus 362

Chromogisaurus 18, 19, 27, 372

Chuanjiesaurus 120, 128, 376

Chubutisaurus 214, 215, 224, 380

Chucarosaurus 314, 383

Chuxiongosaurus 69, 373

Clasmodosaurus 362

Coloradisaurus 42, 43, 53, 373

Colossosauria 284, 285

Comahuesaurus 166, 167, 183, 378

Daanosaurus 120, 135, 376

"Dachongosaurus" 367

"Damalosaurus" 368

Dashanpusaurus 186, 187, 189, 379

Datousaurus 210, 379

Daxiatitan 254, 255, 273, 382

Demandasaurus 166, 167, 181, 378

Diamantinasauria 254, 255

Diamantinasaurus 254, 255, 270, 382

Dicraeosauridae 140, 141

Dicraeosaurus 140, 141, 151, 377

Dimodosaurus 387

Dinheirosaurus 140, 141, 161, 377

Dinodocus 209, 379

Diplodocidae 140, 141

Diplodocimorpha 140

Diplodocinae 140, 141

Diplodocoidea 140, 141

Diplodocus 140, 141, 163

Dongbeititan 214, 215, 226, 380

Dongyangosaurus 254, 255, 267, 382

Dreadnoughtus 284, 285, 292, 315, 383

Dromicosaurus 387

Drusilasaura 284, 313, 383

Duriatitan 186, 187, 195, 379
Dyslocosaurus 164, 377
Dystrophaeus 210, 379
Dystylosaurus 387
Dzharatitanis 166, 167, 173, 378

Efraasia 18, 19, 32, 372
Elaltitan 284, 285, 288, 383
Elosaurus 387
Eobrontosaurus 387
Eomamenchisaurus 137, 376
Eoraptor 18, 19, 21, 372
Epachthosaurus 284, 285, 291, 383
Erketu 242, 243, 252, 381
Eucamerotus 186, 187, 196, 379
Eucnemesaurus 42, 43, 44, 373
Euhelopodidae 242, 243, 381
Euhelopus 242, 243, 250, 381
Europasaurus 186, 187, 193, 240, 379
Europatitan 214, 215, 219, 380
Eusauropoda 96, 97, 375
Euskelosaurus 363
Eutitanosauria 284, 285

"*Fendusaurus*" 368
Ferganasaurus 96, 97, 101, 375
Flagellicaudata 140, 141
"*Francoposeidon*" 368
Fukuititan 241, 380
Fulengia 363
Fushanosaurus 212, 379
Fusuisaurus 186, 187, 198, 379
Futalognkosaurus 284, 306, 383

Galeamopus 140, 141, 159, 377
Galvesaurus 186, 187, 194, 379
Gandititan 254, 255, 282, 382
Gannansaurus 279, 382
Garrigatitan 284, 285, 298, 383
Garumbatitan 214, 228, 380
Gigantosaurus 363
Gigantoscelus 363
Giraffatitan 186, 187, 202, 240, 379
Glacialisaurus 42, 43, 52, 373
Gobititan 242, 243, 251, 381
Gondwanatitan 316, 317, 328, 384
Gongxianosaurus 74, 75, 82, 374
Gravisauria 74, 75
Gresslyosaurus 363
Gripposaurus 387
Gryponyx 363

Gspsaurus 363
Guaibasauridae 27
Guaibasaurus 18, 19, 36, 372
"*Gyposaurus*" 70, 373

Haestasaurus 118, 375
Hamititan 254, 255, 265, 382
Haplocanthosaurus 140, 141, 142, 377
"*Hisanohamasaurus*" 368
Histriasaurus 166, 167, 184, 378
Hortalotarsus 363
Huabeisaurus 254, 255, 259, 382
Huanghetitan 214, 215, 236, 380
Huangshanlong 120, 121, 136, 376
Hudiesaurus 120, 134, 376
Hypselosaurus 364

Ibirania 338, 339, 353, 385
Igai 338, 339, 350, 385
Ignavusaurus 42, 43, 48, 373
Inawentu 284, 285, 302, 383
Ingentia 74, 75, 79, 374
Irisosaurus 42, 43, 63, 373
Isanosaurus 74, 75, 86, 374
"*Ischyrosaurus*" 364
Isisaurus 338, 339, 341, 385
Issi 18, 19, 37, 372
Itapeuasaurus 166, 167, 182, 378
Iuticosaurus 364

Jainosaurus 316, 317, 318, 384
Jaklapallisaurus 18, 19, 33, 372
Janenschia 96, 97, 119, 375
Jiangshanosaurus 254, 255, 266, 382
Jiangxititan 284, 310, 383
Jingiella 120, 121, 138, 376
Jingshanosaurus 42, 43, 58, 69, 373
Jiutaisaurus 240, 380
Jobaria 10, 96, 97, 117, 375

Kaatedocus 140, 141, 156, 377
Kaijutitan 254, 255, 264, 382
Karongasaurus 279, 316, 317, 382
Katepensaurus 166, 167, 177, 378
Khebbashia 166, 167
Khetranisaurus 364
Kholumolumo 42, 43, 46, 373
"*Kholumolumosaurus*" 46
Klamelisaurus 120, 132, 376
Kotasaurus 74, 75, 87, 374
"*Kunmingosaurus*" 368

Lamplughsaura 74, 75, 76, 374
"*Lancanjiangosaurus*" 368
Laplatasaurus 316, 317, 325, 384
Lapparentosaurus 96, 97, 106, 375
Laurasiformes 231
Lavocatisaurus 166, 167, 172, 378
Ledumahadi 74, 75, 80, 374
Leinkupal 140, 141, 158, 377
Leonerasaurus 42, 43, 61, 373
Leptospondylus 387
Lessemsauridae 74, 75
Lessemsaurus 74, 75, 77, 374
Leyesaurus 42, 43, 50, 373
Liaoningotitan 214, 215, 229, 380
Ligabuesaurus 214, 215, 227, 380
Limaysaurinae 166, 167
Limaysaurus 166, 167, 175, 378
Lingwulong 140, 141, 145, 377
Lirainosaurinae 284, 285
Lirainosaurus 284, 285, 295, 383
Lishulong 73, 373
Lithostrotia 284, 285
Liubangosaurus 214, 215, 230, 380
Lognkosauria 284, 285
Lohuecotitan 284, 285, 299, 383
Loricosaurus 364
Losillasaurus 96, 97, 113, 375
Lourinhasaurus 186, 187, 191, 379
Lufengosaurus 42, 43, 54, 373
Lusotitan 186, 187, 204, 379

Macrocollum 18, 19, 35, 372
Macronaria 186, 187
Macrurosaurus 364
Magyarosaurus 284, 285, 289, 383
Malarguesaurus 214, 215, 238, 380
Malawisaurus 4, 284, 285, 287, 383
Mamenchisauridae 120, 121
Mamenchisaurus 17, 120, 121, 131, 138, 376
Mansourasaurus 284, 285, 294, 383
Maojandino 387
Maraapunisaurus 166, 167, 168, 378
Marisaurus 387
Massopoda 42, 43
Massospondylidae 42, 43
Massospondylus 42, 43, 51, 373
Maxakalisaurus 316, 317, 321, 384
Mbiresaurus 18, 19, 22, 372
"*Megacervixosaurus*" 368
Megadactylus 387
"*Megapleurocoelus*" 368

Melanorosauridae 71, 77
Melanorosaurus 42, 43, 68, 373
Mendozasaurus 284, 305, 383
Menucocelsior 284, 285, 300, 383
Meroktenos 42, 43, 67, 373
Microcoelus 364
"*Microdontosaurus*" 369
Mierasaurus 96, 97, 110, 375
Mnyamawamtuka 254, 255, 276, 382
Moabosaurus 96, 97, 111, 375
Mongolosaurus 253, 382
Morinosaurus 365
Morosaurus 365
"*Moshisaurus*" 369
Musankwa 42, 43, 72, 373
Mussaurus 42, 43, 62, 374
Muyelensaurus 316, 317, 323, 384

Nambalia 18, 19, 40, 372
Narambuenatitan 284, 285, 293, 383
Narindasaurus 96, 97, 116, 375
Nebulasaurus 96, 97, 119, 375
Nemegtosaurus 185, 338, 339, 347,
    385
Neosauropoda 96
Neosodon 365
Neuquensaurus 338, 339, 355, 385
Ngwevu 42, 43, 55, 374
Nhandumirim 18, 19, 25, 372
Nicksaurus 365
Nigersaurinae 167
Nigersaurus 166, 167, 178, 378
Ninjatitan 254, 255, 268, 382
Nopcsaspondylus 166, 167, 184, 378
Normanniasaurus 280, 382
Notocolossus 284, 285, 303, 383
Nullotitan 338, 339, 345, 385
"*Nurosaurus*" 269

Oceanotitan 214, 215, 216, 380
"*Oharasisaurus*" 369
Ohmdenosaurus 94, 374
Omeisaurus 120, 121, 123, 138, 376
Opisthocoelicaudia 338, 339, 346, 385
Opisthocoelicaudiinae 338, 339
Oplosaurus 365
Orinosaurus 365
Ornithopsis 186, 187, 197, 379
Orosaurus 365
"*Oshanosaurus*" 369
"*Otogosaurus*" 369
Overosaurus 316, 317, 329, 384

Pachysaurus 387
Pachyspondylus 387
Pachysuchus 365
Padillasaurus 214, 215, 235, 380
Pakisaurus 365
Paludititan 284, 285, 290, 383
Paluxysaurus 214, 215, 233, 380
Pampadromaeus 18, 19, 23, 372
Panamericansaurus 316, 317, 335, 384
Panphagia 18, 19, 24, 372
Pantydraco 18, 19, 29, 372
Paralititan 316, 317, 320, 384
Patagosaurus 96, 97, 105, 376
Patagotitan 284, 309, 383
Pellegrinisaurus 338, 339, 352, 385
Pelorosaurus 209, 379
Perijasaurus 96, 97, 100, 376
Petrobrasaurus 254, 255, 277, 382
Petrustitan 282, 382
Phuwiangosaurus 242, 243, 245, 381
Pilmatueia 140, 141, 148, 378
Pitekunsaurus 334, 384
Plateosauravus 42, 43, 70, 374
Plateosauria 18, 19
Plateosauridae 18, 19
Plateosaurus 18, 19, 38, 372
Pleurocoelus 208
Pradhania 69, 374
Prosauropoda 43
Protognathosaurus 95, 374
Protognathus 95
Puertasaurus 284, 307, 383
Pukyongosaurus 214, 215, 221, 380
Pulanesaura 74, 75, 81, 375
Punatitan 316, 317, 331, 384

Qianlong 42, 43, 72, 374
Qiaowanlong 242, 243, 246, 381
Qijianglong 120, 125, 376
Qingxiusaurus 357
Qinlingosaurus 365
Quaesitosaurus 338, 339, 348, 385
Quetecsaurus 284, 314, 384
Qunkasaura 360, 385

Rapetosaurus 185, 338, 339, 343, 385
Rayososaurus 166, 167, 174, 378
Rebbachisauridae 166, 167
Rebbachisaurinae 166, 167
Rebbachisaurus 166, 167, 179, 378
Rhoetosaurus 74, 75, 92, 375
Rhomaleopakhus 120, 127, 376

Rinconsauria 316, 317, 384
Rinconsaurus 316, 317, 322
Riojasauridae 42, 43
Riojasaurus 42, 43, 45, 374
Rocasaurus 338, 339, 358, 385
Ruehleia 41, 372
Rugocaudia 212, 379
Ruixinia 254, 255, 275, 382
Rukwatitan 338, 339, 340, 386
"*Rutellum*" 369
Ruyangosaurus 242, 243, 248, 381

Saltasauridae 338, 339
Saltasaurinae 338, 339
Saltasaurini 338, 339
Saltasauroidea 338, 339, 385
Saltasaurus 338, 339, 356
Sanpasaurus 74, 75, 88, 375
Sarahsaurus 42, 43, 47, 374
Saraikimasoom 366
Sarmientosaurus 240, 254, 255, 271,
    382
Saturnalia 18, 19, 26, 372
Saturnaliidae 18, 19
Sauropoda 74, 75, 374
Sauropodiformes 42, 43
Sauropodomorpha 18, 19, 372
"*Sauropodus*" 369
Sauroposeidon 214, 215, 232, 381
Savannasaurus 254, 255, 272, 382
Schleitheimia 74, 75, 85, 375
Sefapanosaurus 42, 43, 66, 374
Seismosaurus 387
Seitaad 42, 43, 59, 374
Sellosaurus 32, 387
Shingopana 316, 317, 330, 385
Shunosaurus 12, 96, 97, 98, 376
Sibirotitan 214, 215, 234, 381
Sidersaura 166, 167, 185, 378
Silutitan 242, 243, 249, 381
Smitanosaurus 140, 141, 143, 378
Somphospondyli 214, 215, 380
Sonidosaurus 336, 385
Sonorasaurus 186, 187, 206, 379
Soriatitan 186, 187, 205, 380
"*Sousatitan*" 370
Spinophorosaurus 96, 97, 107, 376
Strenusaurus 387
"*Sugiyamasaurus*" 370
Sulaimanisaurus 366
Supersaurus 140, 141, 160, 378
Suuwassea 140, 141, 144, 378

Tambatitanis 254, 255, 256, 382
Tangvayosaurus 242, 243, 244, 381
Tapuiasaurus 185, 338, 339, 344, 386
Tastavinsaurus 214, 215, 231, 381
Tataouinea 166, 167, 180, 378
Tawasaurus 387
Tazoudasaurus 74, 75, 89, 375
Tehuelchesaurus 96, 97, 109, 376
Tendaguria 96, 97, 114, 376
Tengrisaurus 312, 384
Teratosaurus 32, 388
Tharosaurus 140, 141, 147, 378
Thecodontosaurus 18, 19, 30, 372
"*Thotobolosaurus*" 46
Tiamat 254, 255, 283, 382
Tienshanosaurus 120, 137, 376
Titanomachya 338, 339, 359, 386
Titanosauria 254, 255, 381
Titanosauriformes 186
Titanosaurus 278, 382
"*Tobasaurus*" 370
Tonganosaurus 120, 121, 122, 377
Tornieria 140, 141, 157, 378
Traukutitan 313, 384

Trigonosaurus 326, 336
Triunfosaurus 214, 215, 218, 381
Tuebingosaurus 74, 75, 84, 375
Turiasauria 96, 97
Turiasaurus 12, 96, 97, 112, 376

Uberabatitan 316, 317, 324, 385
Udelartitan 338, 339, 360, 386
Ultrasauros 160, 366
Ultrasaurus 366
Unaysauridae 18, 19
Unaysaurus 18, 19, 34

Vahiny 312, 316, 317, 384
Venenosaurus 186, 187, 207, 380
Volgatitan 284, 285, 286, 384
Volkheimeria 74, 91, 375
Vouivria 186, 187, 199, 380
Vulcanodon 74, 75, 90, 375
Vulcanodontidae 74, 75

Wamweracaudia 120, 124, 377
Wintonotitan 214, 215, 237, 381

Xenoposeidon 166, 167, 169, 378
Xianshanosaurus 254, 255, 274, 382
"*Xinghesaurus*" 370
Xingxiulong 42, 43, 56, 374
Xinjiangtitan 120, 133, 377
Xixiposaurus 41, 373

Yaleosaurus 60, 388
Yamanasaurus 358, 386
"*Yibinosaurus*" 370
Yimenosaurus 18, 19, 39, 373
Yizhousaurus 42, 43, 64, 374
Yongjinglong 242, 243, 247, 381
Yuanmousaurus 120, 121, 130, 377
Yunmenglong 242, 243, 253, 381
Yunnanosaurus 42, 43, 57, 374
"*Yunxianosaurus*" 370
Yuzhoulong 186, 187, 188, 380

Zapalasaurus 166, 167, 171, 378
Zby 96, 97, 118, 376
Zhuchengtitan 357, 386
Zigongosaurus 388
Zizhongosaurus 95, 375